D1074868

THE MATHEMATICS OF

PHYSICS AND CHEMISTRY

BY

HENRY MARGENAU

Eugene Higgins Professor of Physics and Natural Philosophy
Yale University

and

GEORGE MOSELEY MURPHY

Chairman, Department of Chemistry
Washington Square College
New York University

SECOND EDITION

ROBERT E. KRIEGER PUBLISHING COMPANY
HUNTINGTON, NEW YORK
1976

First Published, May 1943
Second Edition, January 1956
Tenth Reprint 1976

Printed and Published by
ROBERT E. KRIEGER PUBLISHING CO., INC.
645 NEW YORK AVENUE
HUNTINGTON, NEW YORK 11743

Library of Congress Cataloging in Publication Data

Margenau, Henry, 1901-
    The mathematics of physics and chemistry.

    Reprint of the ed. published by Van Nostrand,
Princeton, N. J.
    Includes bibliographies and index.
    1. Mathematics—1961-   2. Mathematical physics.
3. Chemistry, Physical and theoretical—Mathematics.
I. Murphy, George Moseley, 1905-1968, joint author.
II. Title.
QA37.2.M356 1977   510'.2'453   76-18724
ISBN 0-88275-423-8

PREFACE TO 1977 REPRINT

It is difficult to write a preface authorizing the reprinting of a mathematics book 33 years after its original publication. A publisher's assessment of its continued usefulness and its likely sale may flatter an author's vanity, but this hardly justifies his decision to save it from a natural death. There must be reasons why a book of this age still appears to be useful, and I find them in the nature of the subject it treats.

In every field of experimental science the accomplishments of one generation are likely to be destroyed or greatly modified by the next in consequence of new empirical discoveries. Mathematics and logic, the pure sciences, never suffer destruction by new evidence because what is internally consistent has permanent stability; their fate is not so much radical change as obsolescence. Thus, the contents of the book are still true, and the question to be answered is whether they continue to be useful.

Some of modern mathematics has moved away from the analytical approaches featured in this book; emphasis has shifted to modern algebras, basic issues of set and number theory, topology and computer science. But it is perhaps a curious fact that the physicist and the chemist have made limited use of these new developments, and that practically the entire contents of this volume still form a large part of the basis on which their work is done. It is from this realization that I draw comfort in permitting the book to be reprinted. Errors, which abounded in the first edition, have been eliminated during the decades of its use.

I record with sadness that my coauthor, George M. Murphy, can not add his voice to this brief preface. He died prematurely in 1963.

HENRY MARGENAU

New Haven, Conn.
April, 1976

iii

PREFACE TO THE SECOND EDITION

In the second edition the main plan of the book has been left unchanged. Small amounts of material have been added in a great number of places, and improvements have been attempted at many points. It was felt that a lack in the first edition was its omission of the theory of Laplace and Fourier transforms. This has been remedied in Chapter 8. The discussion of numerical calculations, integral equations and group theory has likewise been augmented by removal of unnecessary items and some replacements.

Ambiguities, errors and pedagogical faults have been sought in an endeavor to eliminate them. If we have partly succeeded in this task, we owe it to a host of readers and our students who have given us the benefit of their advice and criticism. In this respect we are particularly grateful to many scientists at the Navy Electronics Laboratory who prepared a detailed list of errors soon after the first edition of the book appeared. A similar and very useful list of errata was sent by Professor Pentti Salomaa of the University of Turku, Finland, to whom we express our indebtedness. Dr. M. H. Greenblatt of R.C.A. suggested an improvement in Chapter 12 of which we have made use. Finally, we acknowledge stimulus and aid coming from the careful work of Professors Tsugihiko Sato and Makoto Kuminune of Japan who, in translating the first edition, discovered a number of inaccuracies which have now been corrected.

H. M.
G. M. M.

New Haven, Conn.
November, 1955

PREFACE

The authors' aim has been to present, between the covers of a single book, those parts of mathematics which form the tools of the modern worker in theoretical physics and chemistry. They have endeavored to do this by steering a middle course between the mere recording of facts and formulas which is typical of handbook treatments, and the ponderous development which characterizes treatises in special fields. Therefore, as far as space permitted, all results have been embedded in the logical texture of proofs. Occasionally, when full demonstrations are lengthy or not particularly illuminating with respect to the subject at hand, they have been omitted in favor of references to the literature. Except for the first chapter, which is primarily a survey, proofs have always been given where omission would destroy the continuity of treatment.

Arbitrary selection of topics has been necessary for lack of space. This was based partly on the authors' opinions as to the relevance of various subjects, partly on the results of consultations with colleagues. The degree of difficulty of the treatment is such that a Senior majoring in physics or chemistry would be able to read most parts of the book with understanding.

While inclusion of large collections of routine problems did not seem conformable to the purpose of the book, the authors have felt that its usefulness might be augmented by two minor pedagogical devices: the insertion here and there of fully worked examples illustrative of the theory under discussion, and the dispersal, throughout the book, of special problems confirming, and in some cases supplementing, the ideas of the text. Answers to the problems are usually given.

The degree of rigor to which we have aspired is that customary in careful scientific demonstrations, not the lofty heights accessible to the pure mathematician. For this we make no apology; if the history of the exact sciences teaches anything it is that emphasis on extreme rigor often engenders sterility, and that the successful pioneer depends more on brilliant hunches than on the results of existence theorems. We trust, of course, that our effort to avoid rigor mortis has not brought us dangerously close to the opposite extreme of sloppy reasoning.

A careful attempt has been made to insure continuity of presentation within each chapter, and as far as possible throughout the book. The diversity of the subjects has made it necessary to refer occasionally to chapters ahead. Whenever this occurs it is done reluctantly and in order to avoid repetition.

As to form, considerations of literacy have often been given secondary rank in favor of conciseness and brevity, and no great attempt has been made to disguise individual authorship by artificially uniformising the style.

The authors have used the material of several of the chapters in a number of special courses and have found its collection into a single volume convenient. To venture a few specific suggestions, the book, if it were judged favorably by mathematicians, would serve as a foundation for courses in applied mathematics on the senior and first year graduate level. A thorough introductory course in quantum mechanics could be based on chapter 2, parts of 3, 8 and 10, and chapter 11.   Chapters 1, 10 and parts of 11 may be used in a short course which reviews thermodynamics and then treats statistical mechanics.   Reading of chapters 4, 9, and 15 would prepare for an understanding of special treatments dealing with polyatomic molecules, and the liquid and solid state.   Since ability to handle numerical computations is very important in all branches of physics and chemistry, a chapter designed to familiarize the reader with all tools likely to be needed in such work has been included.

The index has been made sufficiently complete so that the book can serve as a ready reference to definitions, theorems and proofs.   Graduate students and scientists whose memory of specific mathematical details is dimmed may find it useful in review.   Last, but not least, the authors have had in mind the adventurous student of physics and chemistry who wishes to improve his mathematical knowledge through self-study.

HENRY MARGENAU
GEORGE M. MURPHY

New Haven, Conn.
March, 1943

CONTENTS

CHAPTER 1

THE MATHEMATICS OF THERMODYNAMICS

Most of the chapters of this book endeavor to treat some single mathematical method in a systematic manner. The subject of thermodynamics, being highly empirical and synoptic in its contents, does not contain a very uniform method of analysis. Nevertheless, it involves mathematical elements of considerable interest, chiefly centered about partial differentiation. Rather than omit these entirely from consideration, it seemed well to devote the present chapter to them. Of necessity, the treatment is perhaps less systematic than elsewhere. It is placed at the beginning because most readers are likely to have some familiarity with the subject and because the mathematical methods are simple. (A reading of the first chapter is not essential for an understanding of the remainder of the book.)

1.1. Introduction.—The science of thermodynamics is concerned with the laws that govern the transformations of energy of one kind into another during physical or chemical changes. These changes are assumed to occur within a *thermodynamic system* which is completely isolated from its surroundings. Such a system is described by means of *thermodynamic variables* which are of two kinds. *Extensive variables* are proportional to the amount of matter which is being considered; typical examples are the volume or the total energy of the system. Variables which are independent of the amount of matter present, such as pressure or temperature, are called *intensive variables.*

It is found experimentally that it is not possible to change all of these variables independently, for if certain ones of them are held constant, the remaining ones are automatically fixed in value. Mathematically, such a situation is treated by the method of *partial differentiation.* Furthermore, a certain type of differential, called the *exact differential* and an integral, known as the *line integral* are of great importance in the study of thermodynamics. We propose to describe these matters in a general way and to apply them to a few specific problems. We assume that the reader is familiar with the general ideas of thermodynamics and refer him to other sources[1] for a more complete treatment of the physical details.

[1] A representative set of references on thermodynamics will be found at the end of this chapter. Although not easy to read, serious students of the subject should be familiar with the work of J. Willard Gibbs, Transactions of the Conn. Acad., 1875–1878; " Collected Works," Vol. I, Longmans, Green and Co., New York, 1928; " A Commentary on the Scientific Writings of J. W. Gibbs," 2 vols., Yale University Press, New Haven, 1937.

1.2. Differentiation of Functions of Several Independent Variables.—If z is a single-valued function of two real, independent variables, x and y,

$$z = f(x,y)$$

z is said to be an *explicit function* of x and y. The relation between the three variables may be represented by plotting x, y and z along the axes of a Cartesian coordinate system, the result being a surface. If we wish to study the motion of some point (x,y) over the surface, there are three possible cases: (a) x varies and y remains constant; (b) y varies, x remaining constant; (c) both x and y vary simultaneously.

In the first and second cases, the path of the point will be along the curves produced when planes, parallel to the XZ- or YZ-coordinate planes, intersect the original surface. If x is increased by the small quantity Δx and y remains constant, z changes from $f(x,y)$ to $f(x + \Delta x,y)$, and the *partial derivative* of z with respect to x at the point (x,y) is defined by

$$f_x(x,y) = \lim_{\Delta x \to 0} \frac{f(x + \Delta x,y) - f(x,y)}{\Delta x}$$

The following alternative notations are often used

$$f_x(x,y) = z_x(x,y) = \left(\frac{\partial f}{\partial x}\right)_y = \left(\frac{\partial z}{\partial x}\right)_y \tag{1-1}$$

where the constancy of y is indicated by the subscript. Since both x and y are completely independent, the partial derivative is evaluated by the usual method for the differentiation of a function of a single variable, y being treated as a constant.

Defining the partial derivative of z with respect to y (x remaining constant) in a similar way, we may write

$$f_y(x,y) = z_y(x,y) = \left(\frac{\partial f}{\partial y}\right)_x = \left(\frac{\partial z}{\partial y}\right)_x \tag{1-2}$$

If z is a function of more than two variables

$$z = f(x_1, x_2, \cdots, x_n)$$

the simple geometric interpretation is lacking, but such a symbol as:

$$\left(\frac{\partial f}{\partial x_1}\right)_{x_2, x_3, \cdots, x_n}$$

still means that the function is to be differentiated with respect to x_1 by the usual rules, all other variables being considered as constants.

Since the partial derivatives are themselves functions of the independent variables, they may be differentiated again to give second and higher derivatives

$$f_{xx} = \frac{\partial}{\partial x}\left(\frac{\partial z}{\partial x}\right) = \frac{\partial^2 z}{\partial x^2}$$

$$f_{xy} = \frac{\partial}{\partial x}\left(\frac{\partial z}{\partial y}\right) = \frac{\partial^2 z}{\partial x \partial y}$$

$$f_{yx} = \frac{\partial}{\partial y}\left(\frac{\partial z}{\partial x}\right) = \frac{\partial^2 z}{\partial y \partial x} \tag{1-3}$$

$$f_{yy} = \frac{\partial}{\partial y}\left(\frac{\partial z}{\partial y}\right) = \frac{\partial^2 z}{\partial y^2} \quad \text{etc.}$$

It is not always true that $f_{xy} = f_{yx}$; but the order of differentiation *is* immaterial if the function and its derivatives are continuous. Since this is usually the case in physical applications, quantities such as f_{xy}, f_{yx} or f_{xxy}, f_{xyx}, f_{yxx} will be considered identical in the present treatment.

1.3. Total Differentials.—In the third case of sec. 1.2, both x and y vary simultaneously or, in geometric language, the point moves along a curve determined by the intersection with $z = f(x,y)$ of a surface which is neither parallel with the XZ- nor YZ- coordinate plane. Since x and y are independent, both Δx and Δy approach zero as Δz approaches zero. In that case the change in z caused by increments Δx and Δy, called the *total differential* of z, is given by

$$dz = \left(\frac{\partial z}{\partial x}\right)_y dx + \left(\frac{\partial z}{\partial y}\right)_x dy \tag{1-4}$$

If it happens that x and y depend on a single independent variable u (it might be the arc length of the curve along which the point moves, or the time),

$$z = f(x,y); \quad x = F_1(u); \quad y = F_2(u)$$

then, from (4)

$$\frac{dz}{du} = \left(\frac{\partial z}{\partial x}\right)_y \frac{dx}{du} + \left(\frac{\partial z}{\partial y}\right)_x \frac{dy}{du} \tag{1-5}$$

For the special case,

$$z = f(x,y); \quad x = F(y); \quad y \text{ independent}$$

$$\frac{dz}{dy} = \left(\frac{\partial z}{\partial x}\right)_y \frac{dx}{dy} + \left(\frac{\partial z}{\partial y}\right)_x \tag{1-6}$$

An important generalization of these results arises when $x, y, \cdots$ are not independent variables but are each functions of a finite number of independent variables, $u, v, \cdots$

$$f = f(x,y,z, \cdots)$$
$$x = F_1(u,v,w, \cdots)$$
$$y = F_2(u,v,w, \cdots)$$
$$\cdots\cdots\cdots\cdots\cdots$$

Then, from (4)

$$df = \left(\frac{\partial f}{\partial u}\right)_{v,\,w,\,\cdots} du + \left(\frac{\partial f}{\partial v}\right)_{u,\,w,\,\cdots} dv + \cdots \qquad (1\text{-}7)$$

and from (5)

$$\left(\frac{\partial f}{\partial u}\right)_{v,\,w,\,\cdots} = \left(\frac{\partial f}{\partial x}\right)_{y,\,z,\,\cdots}\left(\frac{\partial x}{\partial u}\right)_{v,\,w,\,\cdots}$$
$$+ \left(\frac{\partial f}{\partial y}\right)_{x,\,z,\,\cdots}\left(\frac{\partial y}{\partial u}\right)_{v,\,w,\,\cdots} + \cdots \qquad (1\text{-}8)$$

with similar expressions for $(\partial f/\partial v)$, $(\partial f/\partial w)$, $\cdots$. When these are put into (7) we obtain

$$df = \left[\frac{\partial f}{\partial x}\frac{\partial x}{\partial u} + \frac{\partial f}{\partial y}\frac{\partial y}{\partial u} + \cdots\right] du + \left[\frac{\partial f}{\partial x}\frac{\partial x}{\partial v} + \frac{\partial f}{\partial y}\frac{\partial y}{\partial v} + \cdots\right] dv + \cdots$$
$$= \left[\frac{\partial x}{\partial u} du + \frac{\partial x}{\partial v} dv + \cdots\right]\frac{\partial f}{\partial x} + \left[\frac{\partial y}{\partial u} du + \frac{\partial y}{\partial v} dv + \cdots\right]\frac{\partial f}{\partial y} + \cdots \quad (1\text{-}9)$$

Since $u, v, \cdots$ are independent variables, we may write

$$dx = \frac{\partial x}{\partial u} du + \frac{\partial x}{\partial v} dv + \cdots$$

$$dy = \frac{\partial y}{\partial u} du + \frac{\partial y}{\partial v} dv + \cdots$$

$$(1\text{-}10)$$

Comparing coefficients in (9) and (10), we finally obtain

$$df = \frac{\partial f}{\partial x} dx + \frac{\partial f}{\partial y} dy + \cdots \qquad (1\text{-}11)$$

The difference between (7) and (11) should be noted: in the former equation the partial derivatives are taken with respect to the independent variables, while in the latter, with respect to the dependent variables. The important conclusion may thus be drawn that the total differential may be written either in the form (7) or (11); that is, df may be composed additively of terms $\frac{\partial f}{\partial x} dx, \cdots$, regardless of whether x is a dependent or an independent variable.

1.4. Higher Order Differentials.—*Differentials* of the second, third and higher orders are defined by

$$d^2f = d(df); \quad d^3f = d(d^2f); \quad \cdots; \quad d^nf = d(d^{n-1}f)$$

If there are two variables x and y, we obtain from (4)

$$d^2f = d(df) = d\left(\frac{\partial f}{\partial x}\right) dx + \left(\frac{\partial f}{\partial x}\right) d(dx) + d\left(\frac{\partial f}{\partial y}\right) dy + \left(\frac{\partial f}{\partial y}\right) d(dy)$$

However,

$$d\left(\frac{\partial f}{\partial x}\right) = \frac{\partial}{\partial x}\left(\frac{\partial f}{\partial x}\right) dx + \frac{\partial}{\partial y}\left(\frac{\partial f}{\partial x}\right) dy = \frac{\partial^2 f}{\partial x^2} dx + \frac{\partial^2 f}{\partial x \partial y} dy$$

with a similar expression for $d\left(\frac{\partial f}{\partial y}\right)$, hence

$$d^2f = \frac{\partial^2 f}{\partial x^2}(dx)^2 + \frac{2\partial^2 f}{\partial x \partial y} dxdy + \frac{\partial^2 f}{\partial y^2}(dy)^2 + \frac{\partial f}{\partial x} d^2x + \frac{\partial f}{\partial y} d^2y$$

If x and y are independent variables, $d^2x = d^3x = \cdots d^nx = \cdots d^ny = 0$, and the n-th order differential becomes

$$d^nf = \frac{\partial^n f}{\partial x^n} dx^n + \binom{n}{1}\frac{\partial^n f}{\partial x^{n-1}\partial y} dx^{n-1}dy + \cdots + \binom{n}{k}\frac{\partial^n f}{\partial x^{n-k}\partial y^k} dx^{n-k}dy^k$$

$$+ \cdots + n\frac{\partial^n f}{\partial x \partial y^{n-1}} dxdy^{n-1} + \frac{\partial^n f}{\partial y^n} dy^n \qquad (1\text{--}12)$$

where the $\binom{n}{k}$ are the binomial coefficients, $\binom{n}{k} = \binom{n}{n-k} = n!/k!(n-k)!$

(Cf. sec. 12.2.)

Example. Calculate dp and d^2p for a gas obeying van der Waals' equation:

$$p = \frac{RT}{V-\beta} - \frac{\alpha}{V^2}$$

$$\left(\frac{\partial p}{\partial T}\right)_V = \frac{R}{V-\beta}; \quad \left(\frac{\partial p}{\partial V}\right)_T = -\frac{RT}{(V-\beta)^2} + \frac{2\alpha}{V^3}$$

$$\left(\frac{\partial^2 p}{\partial T^2}\right)_V = 0; \quad \left(\frac{\partial^2 p}{\partial V^2}\right)_T = \frac{2RT}{(V-\beta)^3} - \frac{6\alpha}{V^4}$$

$$\frac{\partial}{\partial V}\left(\frac{\partial p}{\partial T}\right) = -\frac{R}{(V-\beta)^2} = \frac{\partial}{\partial T}\left(\frac{\partial p}{\partial V}\right)$$

$$dp = \frac{R}{(V - \beta)} dT + \left[\frac{2\alpha}{V^3} - \frac{RT}{(V - \beta)^2} \right] dV$$

$$d^2p = \left[\frac{2RT}{(V - \beta)^3} - \frac{6\alpha}{V^4} \right] (dV)^2 - \frac{2R}{(V - \beta)^2} dV dT$$

1.5. Implicit Functions.—In the preceding discussion, the dependence of one variable on another has been given in explicit form, as $x = f(y)$. Let us assume the relation between the variables to be given in *implicit* form such as $f(x,y) = 0$. If it is now desired to compute dy/dx, one could solve $f(x,y) = 0$ for y and then differentiate. This procedure, which is often needlessly complicated, may however be avoided, for, according to (4),

$$df = \left(\frac{\partial f}{\partial x} \right)_y dx + \left(\frac{\partial f}{\partial y} \right)_x dy = 0 \qquad (1\text{--}13)$$

and

$$\frac{dy}{dx} = - \frac{\left(\frac{\partial f}{\partial x} \right)_y}{\left(\frac{\partial f}{\partial y} \right)_x}$$

If the equations for a circle, $x^2 + y^2 - a^2 = 0$, or an ellipse, $x^2/a^2 + y^2/b^2 - 1 = 0$ are taken for $f(x,y) = 0$, the advantage of using this method to obtain derivatives is at once evident.

If an implicit relation is given between three variables, $F(x,y,z) = 0$, any one may be considered to depend on the other two, for there are three possible relations

$$x = f(y,z); \quad y = g(x,z); \quad z = h(x,y)$$

If x be taken as the dependent variable, then

$$dF = F_x dx + F_y dy + F_z dz = 0$$

At constant y, $dy = 0$, so that

$$\left(\frac{\partial x}{\partial z} \right)_y = - \frac{F_z}{F_x} \qquad (1\text{--}14)$$

at constant z, $dz = 0$, hence

$$\left(\frac{\partial x}{\partial y} \right)_z = - \frac{F_y}{F_x} \qquad (1\text{--}15)$$

A third possibility arises if two relations are given between **three** variables

$$f(x,y,z) = 0$$

$$g(x,y,z) = 0$$

Then

$$df = f_x dx + f_y dy + f_z dz = 0$$

$$dg = g_x dx + g_y dy + g_z dz = 0$$

Solving these two equations, we obtain (see sec. 10.9)

$$dx : dy : dz = \begin{vmatrix} f_y\, f_z \\ g_y\, g_z \end{vmatrix} : \begin{vmatrix} f_z\, f_x \\ g_z\, g_x \end{vmatrix} : \begin{vmatrix} f_x\, f_y \\ g_x\, g_y \end{vmatrix}$$

Further examples of the properties of implicit functions and **their derivatives** will be found in the discussion of thermodynamic quantities.

1.6. Implicit Functions in Thermodynamics.—The simplest thermodynamic systems are homogeneous fluids or solids, subjected to no external stresses except a constant hydrostatic pressure. Investigation shows that for all such systems, there is an *equation of state* or *characteristic equation* of the form

$$f(p,V,T) = 0 \tag{1--16}$$

where p is the pressure exerted by the system, V is its volume and T, its temperature on some suitable scale.   From (16), an equation of the form of (13) may then be obtained.

$$df = (\partial f/\partial p)_{V,T}\, dp + (\partial f/\partial V)_{p,T}\, dV + (\partial f/\partial T)_{p,V}\, dT = 0$$

Setting dp, dV, dT equal to zero, successively, there results a set of equations similar to (14) and (15)

$$\left(\frac{\partial V}{\partial T}\right)_p = -\frac{(\partial f/\partial T)_{p,V}}{(\partial f/\partial V)_{p,T}} = \frac{1}{(\partial T/\partial V)_p}$$

$$\left(\frac{\partial T}{\partial p}\right)_V = -\frac{(\partial f/\partial p)_{T,V}}{(\partial f/\partial T)_{p,V}} = \frac{1}{(\partial p/\partial T)_V} \tag{1--17}$$

$$\left(\frac{\partial p}{\partial V}\right)_T = -\frac{(\partial f/\partial V)_{p,T}}{(\partial f/\partial p)_{T,V}} = \frac{1}{(\partial V/\partial p)_T}$$

Three possible products may be found by multiplying **any pair** of these equations and removing the common terms.   A typical **one** is

$$\left(\frac{\partial p}{\partial V}\right)_T \left(\frac{\partial V}{\partial T}\right)_p = -\left(\frac{\partial p}{\partial T}\right)_V \tag{1--18}$$

The product of all three derivatives is

$$\left(\frac{\partial p}{\partial V}\right)_T \left(\frac{\partial V}{\partial T}\right)_p \left(\frac{\partial T}{\partial p}\right)_V = -1 \qquad (1\text{--}19)$$

These results are of considerable importance since they are verified by experiment, the derivatives being proportional to such physical quantities as the coefficients of compressibility, thermal expansion and temperature increase with pressure.

1.7. Exact Differentials and Line Integrals.—It is often required, in thermodynamic problems, to find values of a function $u(x,y)$ at two points (x_1,y_1) and (x_2,y_2) by integration of an equation

$$du(x,y) = M(x,y)dx + N(x,y)dy \qquad (1\text{--}20)$$

between the limits u_1 and u_2.

The attempted integration results in such a symbol as $\displaystyle\int_{x_1}^{x_2} M(x,y)dx$,

which is meaningless unless y can be eliminated by a relation, $y = f(x)$. This is equivalent to specifying the path in the XY-plane along which the integration is performed, hence integrals of (20) are known as *line integrals*. There are many of these paths, the value of the definite integral differing in general, for each. The situation is particularly simple when du is a *total differential*, or, as it is often called, a *complete* or *exact differential*. Comparison of (4) with (20) shows that in this case

$$M(x,y) = \partial u/\partial x; \quad N(x,y) = \partial u/\partial y \qquad (1\text{--}21)$$

Moreover, since the order of differentiation is of no importance, it follows that

$$\partial M/\partial y = \partial^2 u/\partial x \partial y = \partial N/\partial x \qquad (1\text{--}22)$$

Inspection of (21) shows that u may be found by integration even when a functional relation between x and y is unknown. In other words, the line integral is independent of the path; it depends only on the values of x and y at the upper and lower limits. The function u is then said to be a *point function*.

In thermodynamics, it frequently happens that the upper and lower limits are the same, that is, the integration is performed around a complete *cycle*. If the differential du is exact, then the value of the line integral is zero; if du is inexact, integration around a closed cycle gives a result not equal to zero.

1.8. Exact and Inexact Differentials in Thermodynamics.—Examples of exact and inexact differentials are readily found in thermodynamics. Consider a mole of an ideal gas, whose equation of state is $pV = RT$. Let the initial conditions be V_1, p_1 and T_1 and the final conditions be V_2, p_2 and T_2. Calculate the change in volume and the work done in going

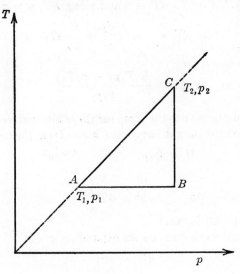

FIG. 1–1

from the initial to the final state, the integration being along two different paths in each case. Since $V = f(p,T)$,

$$dV = \left(\frac{\partial V}{\partial T}\right)_p dT + \left(\frac{\partial V}{\partial p}\right)_T dp$$

$$= \frac{R}{p} dT - \frac{RT}{p^2} dp \qquad (1\text{--}23)$$

Let the first equation of path (AC in Fig. 1) be

$$T - T_1 = \left(\frac{T_2 - T_1}{p_2 - p_1}\right)(p - p_1) = \frac{\Delta T}{\Delta p}(p - p_1)$$

Then $dT = \dfrac{\Delta T}{\Delta p} dp$ and (23) becomes

$$dV = R\left[\frac{\Delta T}{\Delta p}\frac{dp}{p} - \left(T_1 - \frac{\Delta T}{\Delta p}p_1\right)\frac{dp}{p^2} - \frac{\Delta T}{\Delta p}\frac{dp}{p}\right]$$

or, on integration,

$$V_2 - V_1 = \Delta V = \frac{R(T_2 p_1 - p_2 T_1)}{p_1 p_2}$$

The second path will be considered as consisting of two parts: AB and BC (cf. Fig. 1).

Along path AB, $T = T_1$, $dT = 0$ and along BC, $p = p_2$, $dp = 0$, hence

$$dV = -RT_1 \frac{dp}{p^2} + \frac{R}{p_2} dT,$$

or

$$\Delta V = \frac{R(T_2 p_1 - p_2 T_1)}{p_1 p_2}$$

The change in volume is thus the same for these alternative paths.

A similar conclusion might have been drawn from the test for exactness:

$$M = R/p; \quad N = -RT/p^2$$

$$\frac{\partial M}{\partial p} = -\frac{R}{p^2} = \frac{\partial N}{\partial T}$$

which shows that (23) is exact.

The mechanical work done by an expanding gas is

$$dW = pdV \tag{1-24}$$

regardless of the shape of the container and provided that the expansion is performed reversibly[2] in the thermodynamic sense. Combining (24) with (23) we obtain

$$dW = p \left(\frac{\partial V}{\partial T} \right)_p dT + p \left(\frac{\partial V}{\partial p} \right)_T dp$$

$$= RdT - \frac{RT}{p} dp \tag{1-25}$$

It is clear that dW is inexact since

$$M = R; \quad N = -\frac{RT}{p}; \quad \frac{\partial M}{\partial p} = 0 \neq \frac{\partial N}{\partial T} = -\frac{R}{p}$$

By path AC,

$$dW = R \left[dT - \left(T_1 - \frac{\Delta T}{\Delta p} p_1 \right) \frac{dp}{p} - \frac{\Delta T}{\Delta p} dp \right]$$

and, on integration,

$$W_2 - W_1 = \Delta W_1 = R \left(\frac{\Delta T}{\Delta p} p_1 - T_1 \right) \ln \frac{p_2}{p_1}$$

[2] Here and elsewhere in this chapter, we assume that all processes are performed reversibly when such requirement is needed for the argument. For discussions of reversibility, texts on thermodynamics should be consulted.

Along paths AB and BC,

$$dW = R\left[-T_1\frac{dp}{p} + dT\right]$$

or

$$\Delta W_2 = R\left[-T_1\ln\frac{p_2}{p_1} + \Delta T\right]$$

Comparison of ΔW_1 and ΔW_2 shows that the work is different along the two paths.

Heat absorbed or evolved in a process, dQ, also depends on the path. The expression for the inexact differential with p and T as independent variables is

$$dQ = \left(\frac{\partial Q}{\partial T}\right)_p dT + \left(\frac{\partial Q}{\partial p}\right)_T dp$$

$$= C_p dT + \Lambda_p dp \tag{1-26}$$

where C_p and Λ_p are the continuous functions of T and p, known as the heat capacity at constant pressure and the latent heat of change of pressure, respectively.

Problem. Connect the points p_1, V_1 and p_2, V_2 of Fig. 1 with a circular arc. Integrate (23) along this path.

1.9. The Laws of Thermodynamics.—There are obvious advantages in expressing the laws of thermodynamics in terms of quantities which are independent of the path.[3] As we have seen, both dQ and dW are inexact, but the difference between them, a function known as the *internal energy*

$$dU = dQ - dW \tag{1-27}$$

is an exact differential. This equation[4] often serves as a statement of the first law of thermodynamics. By combining (25) and (26) we may also write

$$dU = \left[C_p - p\frac{\partial V}{\partial T}\right]dT + \left[\Lambda_p - p\frac{\partial V}{\partial p}\right]dp \tag{1-28}$$

with the additional requirement of exactness from (22)

$$\frac{\partial}{\partial p}\left[C_p - p\frac{\partial V}{\partial T}\right] = \frac{\partial}{\partial T}\left[\Lambda_p - p\frac{\partial V}{\partial p}\right] \tag{1-29}$$

[3] This fact was recognized by Clausius, " The Mechanical Theory of Heat," translated by W. R. Browne, Macmillan & Co., London, 1879, who discusses the laws of thermodynamics from this standpoint.

[4] Note that $+dQ$ means heat *absorbed* and $+dW$ work *done by* the system. Minus signs indicate heat *evolved* or work *done on* the system.

These two equations are a more satisfactory definition of the first law than (27) since they show the essential fact that the internal energy, dU, is an exact differential. The inexactness of dQ and dW is sometimes indicated[5] by stating the first law in the form of (27) with symbols such as $đQ$, DQ, or δQ on the right.

The second law of thermodynamics is based upon an attempt to find a function of dQ which is an exact differential. From (27) and (24),

$$dQ = dU + dW = dU + pdV \tag{1-27a}$$

but $U = f(V,T)$, hence

$$dU = \left(\frac{\partial U}{\partial V}\right) dV + \left(\frac{\partial U}{\partial T}\right) dT$$

and

$$dQ = \left(\frac{\partial U}{\partial T}\right) dT + \left(p + \frac{\partial U}{\partial V}\right) dV \tag{1-30}$$

In passing from an initial state, V_1, T_1, to a final state, V_2, T_2, the integral on the right of (30) cannot be evaluated without further information, since the second term contains both p and V. In the special case of an ideal gas where $pV = RT$ and $(\partial U)/(\partial V)_T = 0$, (30) becomes

$$dQ = \left(\frac{\partial U}{\partial T}\right)_V dT + \frac{RTdV}{V} \tag{1-31}$$

The first term on the right of this expression is the *heat capacity* at constant volume and depends on the temperature alone. If therefore we make the further restriction of constant temperature, that is, assume the process to be *isothermal*, the integral may be obtained. The form of (31) suggests that if we divide by T, the resulting equation

$$\frac{dQ}{T} = \frac{1}{T}\left(\frac{\partial U}{\partial T}\right)_V dT + \frac{RdV}{V}$$

may also be integrated when T changes. The more general inexact differential (26) when divided by T is also exact, the quantity S so defined being the *entropy*

$$dS = \frac{dQ}{T} = \frac{C_p}{T} dT + \frac{\Lambda_p}{T} dp \tag{1-32}$$

The condition for exactness

$$\frac{\partial}{\partial p}\left(\frac{C_p}{T}\right) = \frac{\partial}{\partial T}\left(\frac{\Lambda_p}{T}\right) \tag{1-33}$$

---

[5] The question of a suitable notation for use in thermodynamics has been discussed by Tunell, G., *J. Phys. Chem.* **36**, 1744 (1932); *J. Chem. Phys.* **9**, 191 (1941); see also, Menger, K., *Am. J. Phys.* **18**, 89 (1950).

together with (32) serve as basis for a statement of the *second law*. Our arguments concerning the first and second laws are intended only to show their property of exactness. The most satisfactory formulation of these laws is probably that of Carathéodory. We consider this subject in sec. 1.15.

The functions dU and dS may be combined by using (24), (27) and (32), to give

$$dU = TdS - pdV \qquad (1\text{--}34)$$

Since $\qquad\qquad U = f(S,V) \qquad\qquad (1\text{--}35)$

and dU is exact, we may also write

$$dU = \left(\frac{\partial U}{\partial S}\right)_V dS + \left(\frac{\partial U}{\partial V}\right)_S dV \qquad (1\text{--}36)$$

Comparison of (34) with (36) shows that

$$T = \left(\frac{\partial U}{\partial S}\right)_V ; \quad p = -\left(\frac{\partial U}{\partial V}\right)_S$$

The importance of (35) arises from the fact that if U is known as a function of two independent variables, S and V, it is possible to calculate numerical values of p, T and U for any thermodynamic state when S and V are given. A quantity like U thus furnishes more information than the equation of state, for the latter will only give p, V and T; in order to obtain U and S, the heat capacity as a function of temperature must also be given. It is not necessary to choose S and V as the independent variables in (35) or (36), in fact any pair of the set: p, V, T, S (or of the functions to be defined immediately) may be taken, but the resulting exact differential is simpler when S and V are selected.

When the conditions of a specific problem suggest another pair of independent variables, it is more convenient to define additional thermodynamic functions. These are given in the following relations, where the symbol as used by Gibbs precedes the one now customary.[6]

The *heat content* or *enthalpy*, $\chi = H = U + pV$

$$dH = dU + pdV + Vdp = TdS + Vdp \qquad (1\text{--}37)$$

The *work content* or *Helmholtz free energy*, $\psi = A = U - TS$

$$dA = dU - TdS - SdT = -SdT - pdV \qquad (1\text{--}38)$$

---

[6] Gibbs preferred S and V as independent variables for reasons given in loc. cit., footnote on page 34.

The *free energy* or *Gibbs thermodynamic potential,*

$$\zeta = F = U - TS + pV$$
$$dF = dU - TdS - SdT + pdV + Vdp$$
$$= -SdT + Vdp \tag{1-39}$$

As in the case of dU, any pair of the set: p, V, T, S, U, H, A, F may be chosen as independent variable, but the exact differential is simpler when expressed in terms of the functions shown in the last equation of (37), (38) or (39). Since most experimental work is done at constant pressure rather that at constant volume, it is obvious that H and F (where the pressure is one of the independent variables) are more generally useful than U and A. The whole of the thermodynamics of systems of constant composition may be developed, however, using any one of the following sets of variables: (1) U, S, V; (2) H, S, p; (3) A, T, V; (4) F, T, p.

It is frequently necessary to have some means of predicting the direction in which a system spontaneously approaches a state of thermodynamic equilibrium. Let us consider two bodies, one at a temperature T_1 and the other at a lower temperature T_2. Then if the whole system is surrounded by adiabatic walls so that no heat enters it, we may write

$$dS_1 = -\frac{dQ}{T_1} \; ; \quad dS_2 = \frac{dQ}{T_2}$$

where dQ is the heat absorbed by the colder body. The total entropy of the system thus increases, for

$$dS = dS_1 + dS_2 = dQ \frac{(T_1 - T_2)}{T_1 T_2} > 0$$

Clearly $dS = 0$ when thermal equilibrium is reached. From (39), we also see that at constant temperature and pressure, $dF = 0$ when equilibrium is established. Since the entropy reaches a maximum, the free energy simultaneously reaches a minimum. In Table 1, we collect the criteria

TABLE 1. DEPENDENT VARIABLE BECOMES A MINIMUM

Independent Variables Fixed	Dependent Variable
T, p	F
T, V	A
S, p	H
S, V	U
S, U	V
A, T	V
A, V or F, p	T

DEPENDENT VARIABLE BECOMES A MAXIMUM

U, V or H, p	S
F, T or H, S	p

for spontaneous approach to equilibrium when various pairs of the independent variables are held constant.

Problem a. Find expressions for S, H, V, A, U in terms of set (4).

Ans.
$$S = -\partial F/\partial T; \quad H = F - T\partial F/\partial T;$$
$$V = \partial F/\partial p; \quad A = F - p\partial F/\partial p;$$
$$U = F - T\frac{\partial F}{\partial T} - p\frac{\partial F}{\partial p}$$

Problem b. Verify the following equations which are known as *Maxwell's relations:*

$$\left(\frac{\partial T}{\partial V}\right)_S = -\left(\frac{\partial p}{\partial S}\right)_V \; ; \; \left(\frac{\partial S}{\partial V}\right)_T = \left(\frac{\partial p}{\partial T}\right)_V$$

$$\left(\frac{\partial T}{\partial p}\right)_S = \left(\frac{\partial V}{\partial S}\right)_p \; ; \; \left(\frac{\partial S}{\partial p}\right)_T = -\left(\frac{\partial V}{\partial T}\right)_p$$

1.10. Systematic Derivation of Partial Thermodynamic Derivatives.—
With the addition of Q and W, we have ten important thermodynamic quantities. The heat capacities are not included in the list, since by their definitions: $C_p = (\partial Q/\partial T)_p$, $C_V = (\partial Q/\partial T)_V$, they may be readily determined from the set of ten just mentioned. We now wish to describe methods of obtaining all first order partial derivatives of the form $(\partial x/\partial y)_z$ where x, y and z are any members of the set. It is immediately apparent that there are a large number of them for there are ten ways of choosing x, leaving nine and eight ways, respectively, of choosing y and z, a total of 720 first derivatives. When all possible relations between the first derivatives are included, the total number of equations is increased enormously for, in general, a selected derivative may be written in terms of three other derivatives which are independent of each other as the following considerations show. Suppose $x = f(y,w)$, then

$$dx = \left(\frac{\partial x}{\partial y}\right)_w dy + \left(\frac{\partial x}{\partial w}\right)_y dw$$

and

$$\left(\frac{\partial x}{\partial y}\right)_z = \left(\frac{\partial x}{\partial y}\right)_w + \left(\frac{\partial x}{\partial w}\right)_y \left(\frac{\partial w}{\partial y}\right)_z$$

There are, of course, many cases where there are relations between fewer than four derivatives but neglecting these, the total number of equations obtainable is the number of combinations of 720 derivatives taken four at a time, 720!/4! 716! or approximately 10^{10}. Although many of the relations are of little use, it is convenient to devise a systematic method for obtaining any of them.

The best known of these methods is that of Bridgman[7] which is simple

---

[7] Bridgman, P. W., "Condensed Collection of Thermodynamic Formulas," Harvard University Press, Cambridge, Mass., 1926.

and often used.   It will be described only briefly since it is a special case of a more general procedure which we give in sec. 1.13.   It is unnecessary to compute the 10^{10} relations because any one of them could be obtained if the 720 first derivatives were tabulated in terms of the same set of three independent derivatives.   The particular choice of the three is arbitrary Bridgman having taken

$$\left(\frac{\partial V}{\partial T}\right)_p, \quad \left(\frac{\partial V}{\partial p}\right)_T, \quad \left(\frac{\partial Q}{\partial T}\right)_p$$

because these are directly obtainable by experiment.   One could then pick any four derivatives, write them in terms of the chosen three and eliminate the three derivatives from the four equations.   The result would be a single equation containing the four derivatives.

The 720 derivatives could then be classified into ten groups by holding one quantity constant and varying the other nine.   Within the group containing derivatives at constant z,

$$\left(\frac{\partial x}{\partial y}\right)_z = \frac{\left(\dfrac{\partial x}{\partial w}\right)_z}{\left(\dfrac{\partial y}{\partial w}\right)_z} \tag{1-40}$$

which follows by writing according to (11)

$$dx = \left(\frac{\partial x}{\partial w}\right)_z dw + \left(\frac{\partial x}{\partial z}\right)_w dz$$

$$dy = \left(\frac{\partial y}{\partial w}\right)_z dw + \left(\frac{\partial y}{\partial z}\right)_w dz \tag{1-41}$$

setting $dz = 0$ and dividing one equation by the other.   It should be remembered that even if x and y are not functions of w and z it is still possible to have inexact differentials of the form of (41), hence the present arguments apply to dQ and dW as well as to the remaining eight thermodynamic functions.   Upon adopting the abbreviations

$$\left(\frac{\partial x}{\partial w}\right)_z = (\partial x)_z$$

$$\left(\frac{\partial y}{\partial w}\right)_z = (\partial y)_z$$

any derivative at constant z may be written in purely formal fashion by taking the ratio of the proper pair, or

$$\left(\frac{\partial x}{\partial y}\right)_z = \frac{(\partial x)_z}{(\partial y)_z}$$

The task of computing the 72 derivatives in this group is thus reduced to calculation of the nine quantities $(\partial x)_z$, $(\partial y)_z$, $\cdots$. The latter are easily found when several of the derivatives $(\partial x/\partial y)_z$ are known in terms of the fundamental three for it proves possible to split the former into numerator and denominator by inspection.

If each of the remaining groups were treated in a similar way, 90 expressions of the form $(\partial x)_z$, $(\partial y)_z$, $(\partial x)_y$, $\cdots$ would be obtained but in every case $(\partial x)_y = -(\partial y)_x$ so that the final list need contain only 45 relations; they are given by Bridgman (loc. cit.) in convenient tables.[8] The following examples show their use. Let it be required to calculate $(\partial T/\partial p)_H$. From the tables, $(\partial T)_H = V - T(\partial V/\partial T)_p$, $(\partial p)_H = -C_p$, thus

$$\left(\frac{\partial T}{\partial p}\right)_H = \frac{1}{C_p}\left[-V + T\left(\frac{\partial V}{\partial T}\right)_p\right]$$

Many alternative forms are easily found, for example,

$$(\partial T/\partial S)_p = T/C_p; \quad (\partial T/\partial p)_S = \frac{T}{C_p}\left(\frac{\partial V}{\partial T}\right)_p; \quad (\partial S/\partial p)_H = -V/T$$

hence,

$$\left(\frac{\partial T}{\partial p}\right)_H = \left(\frac{\partial S}{\partial p}\right)_H\left(\frac{\partial T}{\partial S}\right)_p + \left(\frac{\partial T}{\partial p}\right)_S$$

Additional examples, tables for a few of the second derivatives, and extension of the method to include mechanical variables other than pressure have also been given by Bridgman.

A further amplification of the method has been presented by Goranson[9] whose tables include the following cases: (1) one-component unit mass systems (constant total mass); (2) one-component variable mass systems or two-component unit mass systems; (3) two-component variable mass systems or three-component unit mass systems; (4) three-component variable mass systems or four-component unit mass systems. Simplified methods for constructing such tables have been proposed by several authors.[10]

1.11. Thermodynamic Derivatives by Method of Jacobians.—A more general method which is based on the properties of functional determinants

[8] For abbreviated tables, see, for example, Slater, "Introduction to Chemical Physics," McGraw-Hill Book Co., New York, 1939; or Glasstone, loc. cit.

[9] Goranson, Roy W., "Thermodynamic Relations in Multi-component Systems," Carnegie Institution of Washington, Washington, D. C., 1930.

[10] Lerman, F., *J. Chem. Phys.* **5**, 792 (1937); Tobolsky, A., ibid., **10**, 644 (1942); Bent, H. A., ibid., **21**, 1408 (1953); Carroll, B. and Lehrman, A., *J. Chem. Ed.* **24**, 389 (1947).

or Jacobians has been described by Shaw.[11] The mathematical basis on which it is founded will be discussed in detail in order to explain the construction of the required table and its application to specific examples.

1.12. Properties of the Jacobian.—The *Jacobian*[12] of x and y with respect to two independent variables, u and v, is defined by

$$J(x,y/u,v) \;=\; \partial(x,y)/\partial(u,v) \;=\;$$

$$\begin{vmatrix} \left(\dfrac{\partial x}{\partial u}\right)_v & \left(\dfrac{\partial x}{\partial v}\right)_u \\[2ex] \left(\dfrac{\partial y}{\partial u}\right)_v & \left(\dfrac{\partial y}{\partial v}\right)_u \end{vmatrix} = \left(\dfrac{\partial x}{\partial u}\right)_v \left(\dfrac{\partial y}{\partial v}\right)_u - \left(\dfrac{\partial x}{\partial v}\right)_u \left(\dfrac{\partial y}{\partial u}\right)_v \qquad (1\text{–}42)$$

When the independent variables are discernible from the context, the Jacobian may be abbreviated as $J(x,y)$, the second form of (42) being reserved for cases where it is necessary to give the independent variables explicitly. The following properties are obtained directly from the definition of the Jacobian:

$$J(u,v) \;=\; -J(v,u) \;=\; 1;$$

$$J(x,x) = 0; \quad J(k,x) = 0; \quad k, \text{ any constant} \qquad (1\text{–}43)$$

$$J(x,y) = J(y,-x) = J(-y,x) = -J(y,x)$$

A further important property of the Jacobian arises if x and y are explicit functions of z and w, which in turn are explicit functions of u and v. Writing $\partial(x,y)/\partial(z,w)$ and $\partial(z,w)/\partial(u,v)$ in determinant form, using the rule for the multiplication of determinants, the abbreviations $(\partial x/\partial z)_w = x_z$ and so on, we have

$$\begin{vmatrix} x_z & x_w \\ y_z & y_w \end{vmatrix} \times \begin{vmatrix} z_u & z_v \\ w_u & w_v \end{vmatrix} = \begin{vmatrix} x_z z_u + x_w w_u & x_z z_v + x_w w_v \\ y_z z_u + y_w w_u & y_z z_v + y_w w_v \end{vmatrix}$$

A typical element of the product

$$x_z z_u + x_w w_u = \left(\dfrac{\partial x}{\partial z}\right)_w \left(\dfrac{\partial z}{\partial u}\right)_v + \left(\dfrac{\partial x}{\partial w}\right)_z \left(\dfrac{\partial w}{\partial u}\right)_v = \left(\dfrac{\partial x}{\partial u}\right)_v$$

[11] Shaw, A. N., *Phil. Trans. Roy. Soc.* (London) **A234**, 299–328 (1935).

[12] The properties of determinants, which are used here, are discussed in Chapter 10.

the last form resulting from (8), hence

$$\frac{\partial(x,y)}{\partial(z,w)} \times \frac{\partial(z,w)}{\partial(u,v)} = \begin{vmatrix} x_u & x_v \\ y_u & y_v \end{vmatrix} = \frac{\partial(x,y)}{\partial(u,v)} \qquad (1\text{--}44)$$

In the important special case, $y = v$,

$$\frac{\partial(x,y)}{\partial(u,y)} = \begin{vmatrix} x_u & x_y \\ y_u & y_y \end{vmatrix} = x_u y_y = \left(\frac{\partial x}{\partial u}\right)_y \qquad (1\text{--}45)$$

for

$$y_y = \left(\frac{\partial y}{\partial y}\right)_u = 1 \quad \text{and} \quad y_u = \left(\frac{\partial y}{\partial u}\right)_y = 0$$

Since many thermodynamic functions are of the form $f(x,y,z) = 0$, where any one variable is determined by the other two, we may write from (4),

$$dz = \left(\frac{\partial z}{\partial x}\right)_y dx + \left(\frac{\partial z}{\partial y}\right)_x dy$$

or using (45)

$$dz = \frac{\partial(z,y)}{\partial(x,y)} dx + \frac{\partial(z,x)}{\partial(y,x)} dy$$

Expressing each of these variables in terms of two new independent variables, r and s, and using the abbreviations $J(z,y) = \partial(z,y)/\partial(r,s)$, etc., (44) enables us to write

$$dz = \frac{J(z,y)}{J(x,y)} dx + \frac{J(z,x)}{J(y,x)} dy$$

If we multiply by $J(x,y)$,

$$J(z,y)dx + J(x,z)dy + J(y,x)dz = 0 \qquad (1\text{--}46)$$

since $J(x,y) = -J(y,x)$, etc., from (43). If two more variables, u and v, are related to r and s in the same way, (46) may be divided by du at constant v, giving

$$J(z,y)\left(\frac{\partial x}{\partial u}\right)_v + J(x,z)\left(\frac{\partial y}{\partial u}\right)_v + J(y,x)\left(\frac{\partial z}{\partial u}\right)_v = 0$$

So that finally, again because of (45)

$$J(z,y)J(x,v) + J(x,z)J(y,v) + J(y,x)J(z,v) = 0 \qquad (1\text{--}47)$$

Problem. If r, s are functions of x, y, z and the latter in turn are functions of the independent variables u, v show that

$$J(r,s/u,v) = J(r,s/x,y)J(x,y/u,v) + J(r,s/y,z)J(y,z/u,v) + J(r,s/z,x)J(z,x/u,v).$$

1.13. Application to Thermodynamics.—This last equation is the important one which determines all of the thermodynamic partial derivatives, for if two independent variables, r and s, are chosen which completely determine the others, x, y, z, v, then any one Jacobian, for example $J(x,y)$, is given in terms of five others.   But if r and s are taken from the set x, y, z, v, then $J(x,y)$ is given in terms of only four others, since by (47) $J(r,s) = \partial(r,s)/\partial(r,s) = 1$.

Let us choose p, V, T and S for x, y, z and v, respectively, so that

$$J(T,V)J(p,S) + J(p,T)J(V,S) + J(V,p)J(T,S) = 0 \qquad (1\text{--}48)$$

One more reduction is possible since from (34),

$$(\partial U/\partial V)_S = -p; \quad (\partial U/\partial S)_V = T$$

and

$$(\partial^2 U/\partial S\partial V) = (\partial T/\partial V)_S = -(\partial p/\partial S)_V$$

In Jacobian notation,

$$J(T,S)/J(V,S) = -J(p,V)/J(S,V)$$

Finally since $J(V,S) = -J(S,V)$ from (43), we obtain

$$J(T,S) = J(p,V)$$

When the following abbreviations

$$a = J(V,T)$$
$$b = J(p,V) = J(T,S)$$
$$c = J(p,S) \qquad\qquad (1\text{--}49)$$
$$l = J(p,T)$$
$$n = J(V,S)$$

are substituted into (48) and (43) is used to change the signs, we have

$$b^2 + ac - nl = 0 \qquad (1\text{--}50)$$

It is convenient to list the various Jacobians in rows and columns, $J(x,y)$ occurring at the intersection of row x with column y.   The upper left-hand block of such a table is immediately filled by using the definitions (49), the rule for the change of signs, and the fact that $J(x,x) = 0$ from

(43). The entries for the lower left-hand corner of the table are obtained by writing the definitions of dU, dH, etc., in Jacobian form. For example, since

$$dU = TdS - pdV$$

$$J(U,z) = TJ(S,z) - pJ(V,z)$$

where z is any required variable. Hence, if z is taken as p and then as V

$$J(U,p) = TJ(S,p) - pJ(V,p) = -Tc + pb$$

$$J(U,V) \doteq TJ(S,V) - pJ(V,V) = -Tn$$

the last forms following from the part of the table which is already filled or from the definitions in (49). The upper right-hand corner may be filled at the same time, without further calculation, by changing all signs. The table is completed by using relations already found, as for example

$$J(A,H) = -J(H,A) = -SJ(T,H) - pJ(V,H)$$

$$= -S(Tb - Vl) - p(Tn - Vb)$$

$$= -T(Sb + pn) + V(Sl + pb)$$

The final result is shown in Table 2. The use of it is typified by the following examples.

Example 1. *Evaluate* $(\partial F/\partial T)_V$ *in terms of other partial derivatives with* T *and* V *as independent variables.* In Jacobian notation and from Table 2

$$(\partial F/\partial T)_V = J(F,V)/J(T,V) = -\frac{Sa + Vb}{a} = -S - Vb/a$$

But

$$b/a = J(p,V)/J(V,T) = -J(p,V)/J(T,V) = -(\partial p/\partial T)_V$$

hence,

$$(\partial F/\partial T)_V = -S + V(\partial p/\partial T)_V$$

Example 2. *Transform the result of the preceding example into derivatives with* p *and* S *as independent variables.* If the previous result is used, the term a causes trouble, since with p and S as independent variables, we obtain $a = J(V,T) = \partial(V,T)/\partial(p,S)$, a relation which cannot be reduced to a single derivative. In general, as we have shown, any partial derivative may be expressed in terms of not more than three other derivatives of thermodynamic functions. We therefore use (50), which gives $a = (nl - b^2)/c$, or,

$$(\partial F/\partial T)_V = -S - Vbc/(nl - b^2)$$

TABLE 2

	p	V	T	S	U	H	A	F	Q	W
p	0	b	l	c	$Tc-pb$	Tc	$-Sl-pb$	$-Sl$	Tc	pb
V	$-b$	0	a	n	Tn	$Tn-Vb$	$-Sa$	$-Sa-Vb$	Tn	0
T	$-l$	$-a$	0	b	$Tb+pa$	$Tb-Vl$	pa	$-Vl$	Tb	$-pa$
S	$-c$	$-n$	$-b$	0	pm	$-Vc$	$Sb+pm$	$Sb-Vc$	0	$-pm$
U	$-Tc+pb$	$-Tn$	$-Tb-pa$	$-pm$	0	$-TVc$ $-p(Tn-Vb)$	$T(Sb+pm)$ $+pSa$	$T(Sb-Vc)$ $+p(Sa+Vb)$	$-pTn$	$-pTn$
H	$-Tc$	$-Tn+Vb$	$-Tb+Vl$	Vc	TVc $+p(Tn-Vb)$	0	$T(Sb+pm)$ $-V(Sl+pb)$	$T(Sb-Vc)$ $-VSl$	TVc	$p(Vb-Tn)$
A	$Sl+pb$	Sa	$-pa$	$-Sb-pm$	$-T(Sb+pm)$ $-pSa$	$-T(Sb+pm)$ $+V(Sl+pb)$	0	$-SVl$ $-p(Sa+Vb)$	$-T(Sb+pm)$	pSa
F	Sl	$Sa+Vb$	Vl	$-Sb+Vc$	$-T(Sb-Vc)$ $-p(Sa+Vb)$	$-T(Sb-Vc)$ $+VSl$	SVl $+p(Sa+Vb)$	0	$T(Vc-Sb)$	$p(Sa+Vb)$
Q	$-Tc$	$-Tn$	$-Tb$	0	pTn	$-TVc$	$T(Sb+pm)$	$T(Sb-Vc)$	0	$-pTn$
W	$-pb$	0	pa	pm	pTn	$p(Tn-Vb)$	$-pSa$	$-p(Sa+Vb)$	pTn	0

$$J(x,y) = \left(\frac{\partial x}{\partial r}\right)_s \left(\frac{\partial y}{\partial s}\right)_r - \left(\frac{\partial y}{\partial r}\right)_s \left(\frac{\partial x}{\partial s}\right)_r ; \quad b^2 + ac - nl = 0$$

But

$$b = J(p,V) \doteq \partial(p,V)/\partial(S,p) = -(\partial V/\partial S)_p$$

$$c = J(p,S) = \partial(p,S)/\partial(S,p) = -1$$

$$l = J(p,T) = \partial(p,T)/\partial(S,p) = -(\partial T/\partial S)_p$$

$$n = J(V,S) = \partial(V,S)/\partial(S,p) = -(\partial V/\partial p)_S$$

hence,

$$(\partial F/\partial T)_V = -S - V\left[\frac{(\partial V/\partial S)_p}{(\partial T/\partial S)_p(\partial V/\partial p)_S - (\partial V/\partial S)_p^2}\right]$$

This procedure may be repeated using other quantities, such as T and S, V and p, and so on, as independent variables. The difficulty in choosing the proper form of the original relation may usually be removed in the following way. Referring to the definitions of a, b, c, l and n, it is seen that each can be reduced to unity by a proper choice of the independent variables. For example, if the latter are chosen as V and T, $a = 1$, since $a = J(V,T)$. In the previous case, $c = -1$, and it was found advisable to use some quantity other than a. The situation may be summed up in the following directions. In case one of the letters in the top line of the set $\begin{bmatrix} a & c & l & n \\ c & a & n & l \end{bmatrix}$ equals unity, do not use the one directly beneath it but transform to another by means of (50). In this way, the resulting expression will usually contain only three different partial derivatives. The omission of b from the above list arises from the fact that even if $b = 1$, only single derivatives will occur.

Example 3. *Solve for $(\partial p/\partial T)_V$ in terms of C_V, C_p and $\mu = (\partial T/\partial p)_H$, the Joule-Thomson coefficient.* Problems of this sort frequently arise where it is desired to express a partial thermodynamic derivative in terms of other quantities, which are measured directly. The usual process of obtaining the relationship is tedious and complex. From the table, it is found that

$$C_V = (\partial Q/\partial T)_V = Tn/a$$

$$C_p = (\partial Q/\partial T)_p = Tc/l$$

$$\mu = (\partial T/\partial p)_H = (Tb - Vl)/Tc$$

$$(\partial p/\partial T)_V = -b/a$$

Since there are three relations given and only two letters in the last derivative, it is convenient to write this in the form

$$(\partial p/\partial T)_V = -b^2/ab$$

and to solve for a, b and b^2 in terms of C_V, C_p and μ. Using (50) to obtain a relation between C_p and b^2, we have

$$C_p = T(nl - b^2)/al$$

$$a = Tn/C_V; \quad b^2 = la(C_V - C_p)/T; \quad b = l(\mu C_p + V)/T$$

and finally

$$(\partial p/\partial T)_V = (C_p - C_V)/(C_p\mu + V)$$

Example 4. *Determine $(\partial U/\partial V)_T$ for a gas obeying (i) the ideal gas law, $pV = RT$; (ii) van der Waals' equation, $(p + \alpha/V^2)(V - \beta) = RT$.* In problems of this sort, the resulting formulas usually contain no more than one partial derivative instead of three as in the earlier cases. From Table 2,

$$\left(\frac{\partial U}{\partial V}\right)_T = -\frac{Tb}{a} - p$$

If p and V are taken as independent variables,

$$b = 1; \quad a = J(V,T) = \frac{\partial(V,T)}{\partial(p,V)} = -\left(\frac{\partial T}{\partial p}\right)_V$$

(i)
$$a = -\frac{V}{R}; \quad \left(\frac{\partial U}{\partial V}\right)_T = 0$$

(ii)
$$a = -\frac{(V - \beta)}{R}; \quad \left(\frac{\partial U}{\partial V}\right)_T = \frac{RT}{(V - \beta)} - p = \frac{\alpha}{V^2}$$

In Shaw's paper (loc. cit.), auxiliary tables are given to simplify the calculations for the following cases: the ideal and van der Waals' gas, the saturated vapor, black-body radiation.

The Jacobian method has been extended by Shaw to include second derivatives and to apply to systems of variable composition. For these applications, as well as more detail on the use of the tables, the original paper should be consulted.[13]

Problem. Prove the following relations:

(a) $\mu = \left(\dfrac{\partial T}{\partial p}\right)_H = \dfrac{1}{C_p}\left[T\left(\dfrac{\partial V}{\partial T}\right)_p - V\right]$

(b) $C_V - C_p = T\left(\dfrac{\partial V}{\partial T}\right)_p^2 \Big/ \left(\dfrac{\partial V}{\partial p}\right)_T$

1.14. Thermodynamic Systems of Variable Mass.—The development of thermodynamics up to the time of Gibbs may be briefly summarized by the equation of Clausius (34) which combined the two laws. The subject

---

[13] The Jacobian method has been described and illustrated with numerous examples by Sherwood, T. K., and Reed, C. E., "Applied Mathematics in Chemical Engineering," McGraw-Hill Book Co., New York, 1939; see also, Crawford, F. H., *Am. J. Phys.* **17**, 1 (1949).

was thus confined to systems of *constant total mass*. Gibbs showed how this equation could be extended to include systems of variable mass.[14] If we consider a system composed of several substances whose masses are m_1, m_2, $\cdots$ we may change the internal energy not only by varying the entropy and the volume but also by varying the relative masses. Thus in place of (35) we have

$$U = U(S,V,m_1,m_2,\cdots,m_n)$$

and in place of (36)

$$dU = \left(\frac{\partial U}{\partial S}\right)_{V,m_1,m_2,\cdots} dS + \left(\frac{\partial U}{\partial V}\right)_{S,m_1,m_2,\cdots} dV$$

$$+ \left(\frac{\partial U}{\partial m_1}\right)_{S,V,m_2\cdots} dm_1 + \left(\frac{\partial U}{\partial m_2}\right)_{S,V,m_1,\cdots} dm_2 + \cdots \qquad (1\text{-}51)$$

If we write

$$\left(\frac{\partial U}{\partial m_i}\right)_{V,S,m_1,\cdots} = \mu_i \qquad (1\text{-}52)$$

we have

$$dU = TdS - pdV + \mu_1 dm_1 + \mu_2 dm_2 + \cdots \qquad (1\text{-}53)$$

If dU is eliminated from (53) by using in turn equations (37), (38) and (39) we obtain

$$\mu_i = \left(\frac{\partial H}{\partial m_i}\right)_{S,p,m_1,m_2,\cdots} = \left(\frac{\partial A}{\partial m_i}\right)_{V,T,m_1,m_2,\cdots} = \left(\frac{\partial F}{\partial m_i}\right)_{p,T,m_1,m_2,\cdots} \qquad (1\text{-}54)$$

The partial derivatives defined by any of these equivalent expressions were called by Gibbs the *chemical potentials*. We may also convert (53) into the equation

$$dF = -SdT + Vdp + \mu_1 dm_1 + \mu_2 dm_2 + \cdots \qquad (1\text{-}55)$$

At constant temperature and pressure and for a reversible process, as we have shown, $dF = 0$; hence according to (55) the condition for equilibrium reads

$$dF = \mu_1 dm_1 + \mu_2 dm_2 + \cdots = 0 \qquad (1\text{-}56)$$

From this equation we may derive the celebrated *phase rule* of Gibbs. Let us understand by *phase* a homogeneous part of a system separated from the rest of the system by recognizable boundaries. Thus a mixture of ice, liquid water, and steam is a system of three phases. The number of

---

[14] His results also included other variables such as electric, magnetic, and gravitational fields as well as surface phenomena.

components is the least number of independently variable constituents required to express the composition of each phase.   In our previous example there is only one component.   In a system composed of an aqueous solution of sugar there are two components for it is necessary to specify the amounts of both water and sugar present.   Finally we need a definition of *degree of freedom.*   It is the number of variables (such as temperature, pressure, composition of the components) which is required to describe completely the system at equilibrium.   For example, liquid water in the presence of water vapor is a system of one degree of freedom, for we may vary either the temperature or the pressure but we cannot change both simultaneously for then either the liquid or the vapor disappears.

Suppose a system contains C components and P phases, then an equation of the form of (55) will hold for each phase.   Since F like S and V is an extensive variable, it follows from (55) that the chemical potentials must be independent of the masses, so that we may integrate (56) term by term obtaining

$$F = \mu_1 m_1 + \mu_2 m_2 + \cdots + \mu_C m_C \qquad (1\text{-}57)$$

Differentiation of this equation results in

$$dF = \mu_1 dm_1 + \mu_2 dm_2 + \cdots + \mu_C dm_C$$
$$+ \, m_1 d\mu_1 + m_2 d\mu_2 + \cdots + m_C d\mu_C$$

When it is subtracted from (56) we get

$$m_1 d\mu_1 + m_2 d\mu_2 + \cdots + m_C d\mu_C = 0 \qquad (1\text{-}58)$$

Equilibrium can be established only when an equation of this form holds for each of the P phases.   But there are $C + 2$ variables T, p, μ_1, μ_2, $\cdots$, μ_C, hence the number of degrees of freedom f is

$$f = C + 2 - P \qquad (1\text{-}59)$$

This simple equation has been of inestimable value in the study and interpretation of heterogeneous equilibrium by the chemist, physicist and metallurgist.[15]

1.15. The Principle of Carathéodory.—In most textbooks of thermodynamics, the order of presentation parallels the historical development of the subject.   For this reason, considerable attention is paid to several kinds of ideal or imaginary machines.   The customary procedure is to cite, first of all, the impossibility of constructing perpetual motion machines of various types;   when this is granted it is possible to state the conditions

---

[15] Such applications, where graphical methods are normally used, are discussed by Ricci, J. E., " The Phase Rule and Heterogeneous Equilibrium," D. Van Nostrand Co., Inc., New York, 1951.   Some mathematical methods for treating multicomponent systems have been given by Dahl, L. A., *J. Phys. & Colloid Chem.* **52**, 698 (1948); **54**, 547 (1950).

under which real machines may operate and to derive the whole body of positive assertions which are incorporated into the science of thermodynamics. The critical student may feel the need of a more logical and formal approach, and this will now be given.

We have attempted to emphasize in sec. 1.9 one important mathematical consequence of the laws of thermodynamics, namely, that functions such as dU and dS are exact differentials. We now wish to discuss a more fundamental mathematical property of these laws which was discovered by Carathéodory. His arguments[16] are derived from the geometric behavior of a certain differential equation and its solution. As a result, he is able to obtain in a purely formal way the laws of thermodynamics without recourse to fictitious machines or such objectionable concepts as the flow of heat. We cannot reproduce here the complete theory[17] but shall only give the mathematical details of his treatment of the second law.

Let us assume that a thermodynamic system is composed of n separate parts, each one of which is characterized by its pressure and volume. Further, suppose that the whole system is surrounded by adiabatic walls or thermal insulators while the individual parts of the system are separated from each other by walls that are perfect conductors of heat. As a result of experiment, it is found that there is no observable change in the system (i.e., equilibrium has been reached) when the following conditions are met:

$$f_1(p_1, V_1) = f_2(p_2. V_2) = \cdots = f_n(p_n, V_n) = F(\vartheta) \qquad (1\text{-}60)$$

The relation $f_i(p_i, V_i) = F(\vartheta)$ for the i-th part of the system is, of course, an equation of state, and ϑ is the temperature of the whole system on some suitable empirical scale. According to the first law (see eq. 27a)

$$dQ = dU + pdV = 0 \qquad (1\text{-}61)$$

the whole system being adiabatic. Moreover, a similar equation holds for each part of the system:

$$dQ_i = dU_i + p_i dV_i \qquad (1\text{-}62)$$

and

$$dU = \sum_{i=1}^{n} dU_i; \quad dQ = \sum_{i=1}^{n} dQ_i \qquad (1\text{-}63)$$

As we have shown, dQ_i is not an exact differential. However, it depends on only two variables, and under these conditions an infinite number

[16] Carathéodory, C., *Math. Ann.* **67,** 355 (1909).

[17] Carathéodory's theory has been reviewed by Born, M., *Physik. Z.* **22,** 218, 249, 282 (1922) and by Landé, A., " Handbuch der Physik," Vol. IX, Chapter 4, J. Springer, Berlin, 1926. See also, Buchdahl, H. A., *Am. J. Phys.* **17,** 41, 44, 212 (1949).

of integrating denominators exist.[18]   Hence eq. (62) may be converted into an exact differential.   Let an integrating denominator be t_i, so that

$$d\phi_i = dQ_i/t_i \qquad (1\text{-}64)$$

is exact.   Clearly ϕ_i is then a function of the state of the system, hence we may change (61) in such a manner that the independent variables are ϑ and ϕ_i instead of U and V.   The result of this transformation is

$$dQ = \sum_{i=1}^{n}\left[\left(\frac{\partial U_i}{\partial \phi_i} + p_i \frac{\partial V_i}{\partial \phi_i}\right)d\phi_i + \left(\frac{\partial U_i}{\partial \vartheta} + p_i \frac{\partial V_i}{\partial \vartheta}\right)d\vartheta\right] = 0 \qquad (1\text{-}65)$$

The quantity dQ is not exact, nor is it to be taken for granted that it can be made exact by the use of an integrating denominator if dQ contains more than two variables.   As a matter of fact, the procedure is possible only when the differential equation $dQ = 0$ (known as a *Pfaff equation*) possesses a solution, as we shall show in sec. 2.18.   In that case (and we shall here be interested in no other), there is an integrating denominator t such that

$$d\phi = dQ/t \qquad (1\text{-}66)$$

is exact, even when there are n variables.   More important for our present needs is the conclusion drawn from simple geometric considerations that if there is an integrating denominator, then there are in the neighborhood of any point P many other points which are not accessible from P along the path $dQ = 0$.   This formal mathematical consequence of the properties of the Pfaff equation is known as the principle of Carathéodory.   It is exactly what we need for thermodynamics.   Consider, for example, a gas at a given pressure, p_1 and volume, V_1.   We may expand or compress this gas adiabatically (i.e., along the path $dQ = 0$), but the final state of the system will be characterized by variables p_2, V_2 which we cannot choose at will.   There are many values of p and V which we are not able to realize adiabatically.

We refer the reader again to sec. 2.18 for the conditions under which equations like (65) have a solution, hence an integrating denominator. We proceed here with the physical results which may be obtained when we know that the integrating denominator exists.   In order to simplify the situation let us assume that the thermodynamic system is composed of only two parts.   This restriction does not mean that there is any loss in generality of the final results since all our arguments could easily be extended to cover a system of any number of parts.   With $n = 2$, it follows

---

[18] The proof of this fact as well as other mathematical conclusions reached here are given in sec. 2.18.   Except for the proofs, the present section is complete in itself.

from (63), (64) and (66) that

$$td\phi = t_1 d\phi_1 + t_2 d\phi_2 \tag{1-67}$$

If we take as in (65), ϕ_1, ϕ_2 and ϑ as independent variables we see that

$$\frac{\partial \phi}{\partial \phi_1} = \frac{t_1}{t} \; ; \quad \frac{\partial \phi}{\partial \phi_2} = \frac{t_2}{t} \; ; \quad \frac{\partial \phi}{\partial \vartheta} = 0 \tag{1-68}$$

The last equation of (68) shows that ϕ depends on ϕ_1 and ϕ_2 but not on ϑ, so that according to the other two equations of (68), the ratios t_1/t and t_2/t are also independent of ϑ:

$$\frac{\partial}{\partial \vartheta}\left(\frac{t_1}{t}\right) = 0; \quad \frac{\partial}{\partial \vartheta}\left(\frac{t_2}{t}\right) = 0$$

This result may be written:

$$\frac{1}{t_1}\frac{\partial t_1}{\partial \vartheta} = \frac{1}{t_2}\frac{\partial t_2}{\partial \vartheta} = \frac{1}{t}\frac{\partial t}{\partial \vartheta} \tag{1-69}$$

Now t_1 is a function of the state of the first member of the system and therefore could depend only on ϕ_1 and ϑ, while t_2 could depend only on ϕ_2 and ϑ. However, (69) indicates that t_1 and t_2 must actually satisfy the following equation

$$\frac{d \ln t_1}{d\vartheta} = \frac{d \ln t_2}{d\vartheta} = \frac{d \ln t}{d\vartheta} = g(\vartheta) \tag{1-70}$$

where $g(\vartheta)$ is a function which is common to all systems in thermal contact, not dependent on any special properties of the substances which compose the system. Integrating (70), we obtain

$$\ln t = \int g(\vartheta)d\vartheta + \ln A(\phi) \tag{1-71}$$

where the integration constant $\ln A$ depends only on the quantity ϕ. Note that we have dropped the subscripts from t and ϕ so that eq. (71) refers to any thermodynamic system and t is the appropriate integrating denominator for the particular system under consideration. We see from (71) the important fact that this denominator can be separated into two parts, one depending only on the empirical temperature ϑ and the other only on variables of the state of the system such as ϕ whose differential is exact.

Let us rewrite (71) in the form

$$t = A e^{\int_a^\vartheta g(\vartheta)d\vartheta} \tag{1-72}$$

and define the *absolute temperature* T by the relation

$$T(\vartheta) = Ce^{\int g(\vartheta)d\vartheta} \qquad (1\text{--}73)$$

The constant C relating ϑ and T may be determined by requiring that between two fixed points, say the boiling point and freezing point of water, T shall increase by 100 units. It should be noticed that there is no additive constant in (73), so that if C is positive, the smallest value of T is zero, and there is no upper limit for T.

If our thermodynamic system contains only one part, we may use (72), (73) and (66) to write

$$dQ = td\phi = \frac{TAd\phi}{C} \qquad (1\text{--}74)$$

Also, if we put

$$S = \frac{1}{C} \int A(\phi)d\phi + \text{const.} \qquad (1\text{--}75)$$

we obtain the well-known expression for the second law of thermodynamics which defines a change in entropy, dS:

$$dQ = TdS \qquad (1\text{--}76)$$

The entropy is immediately seen to be a function of the state of the system, constant along an adiabatic path ($dQ = 0$). It is determined except for an additive constant. We also note from (76) that the absolute temperature is an integrating denominator of the inexact differential dQ.

When the system is made up of two parts which are in thermal contact, eqs. (67) and (74) may be combined to give

$$Ad\phi = A_1 d\phi_1 + A_2 d\phi_2 \qquad (1\text{--}77)$$

We know that A_1 is a function of ϕ_1 and that A_2 is a function of ϕ_2. We want to prove that A is a function of ϕ which in turn depends on ϕ_1 and ϕ_2. Let us *assume* that $A = A(\phi)$. Then

$$\frac{\partial A}{\partial \phi_1} = \frac{\partial A}{\partial \phi}\frac{\partial \phi}{\partial \phi_1} \;;\; \frac{\partial A}{\partial \phi_2} = \frac{\partial A}{\partial \phi}\frac{\partial \phi}{\partial \phi_2}$$

If we eliminate $\partial A/\partial \phi$ from these two equations we obtain

$$\frac{\partial A}{\partial \phi_1}\frac{\partial \phi}{\partial \phi_2} - \frac{\partial A}{\partial \phi_2}\frac{\partial \phi}{\partial \phi_1} = 0 \qquad (1\text{--}78)$$

This result is often written in the Jacobian notation of sec. 1.12

$$J(A,\phi/\phi_1,\phi_2) = 0$$

It tells us[19] that if A is a function of ϕ, $J(A,\phi) = 0$ and conversely if $J(A,\phi) = 0$, then A is a function of ϕ. We can easily prove in our case that the Jacobian does vanish. Differentiation of (77) results in

$$A \frac{\partial \phi}{\partial \phi_1} = A_1, \quad A \frac{\partial \phi}{\partial \phi_2} = A_2$$

$$\frac{\partial A}{\partial \phi_1} \frac{\partial \phi}{\partial \phi_2} + A \frac{\partial^2 \phi}{\partial \phi_1 \partial \phi_2} = 0; \quad \frac{\partial A}{\partial \phi_2} \frac{\partial \phi}{\partial \phi_1} + A \frac{\partial^2 \phi}{\partial \phi_2 \partial \phi_1} = 0$$

hence by subtraction we obtain (78). Thus A is a function of ϕ. Under these conditions we have an equation similar to (76) for each part of the thermodynamic system, and since $dQ = \sum dQ_i$, we finally conclude from (75) and (77) that $dS = \sum dS_i$.

[19] This result which may be applied in the case of n variables is often useful. If the n functions $y_1, y_2, \cdots, y_n$ are not independent of each other the Jacobian vanishes; if $J = 0$, then the n functions are related by some equation $f(y_1, y_2, \cdots, y_n) = 0$; see sec. 3.13.

REFERENCES

An excellent summary, including an interesting account of the history of thermodynamic development, has been given by Partington, J. R., " An Advanced Treatise on Physical Chemistry," Vol. I, Longmans, Green and Co., New York, 1949. A large number of literature references, especially to the older source material, is included.

The following list of texts, although far from complete, contains both elementary and advanced treatments of thermodynamics.

Glasstone, S., " Thermodynamics for Chemists," D. Van Nostrand Co., Inc., New York, 1947.

Guggenheim, E. A., " Thermodynamics—An Advanced Treatise for Chemists and Physicists," Interscience Publishers, Inc., New York, 1949.

Klotz, I. M., " Chemical Thermodynamics," Prentice-Hall, Inc., New York, 1950.

Paul, M. A., " Principles of Chemical Thermodynamics," McGraw-Hill Book Co., Inc., New York, 1951.

Prigogine, I. and DeFay, T., translated by D. H. Everett, " Treatise on Thermodynamics. Chemical Thermodynamics," Vol. I, Longmans, Green and Co., Inc., New York, 1954. " Surface Tension and Adsorption," Vol. II and " Irreversible Phenomena," Vol. III, in preparation.

Rossini, F. D., " Chemical Thermodynamics," John Wiley and Sons, Inc., New York, 1950.

Steiner, L. E., " Introduction to Chemical Thermodynamics," Second Edition, McGraw-Hill Book Co., Inc., New York, 1948

CHAPTER 2

ORDINARY DIFFERENTIAL EQUATIONS

2.1. Preliminaries.—The customary classification distinguishes two main types: *ordinary* and *partial* differential equations. The former contain only one independent variable and, as a consequence, total derivatives. They represent a relation between the primitive of the dependent variable (y), its various derivatives, and functions of the independent variable (x). Partial differential equations whose study will be reserved for Chapter 7, contain several independent variables and hence partial derivatives. Concerning terminology, the following is to be noted in connection with ordinary differential equations.

The *order* of a differential equation is the order of its highest derivative; its *degree* is the degree (or power) of the derivative of highest order after the equation has been rationalized, i.e., after fractional powers of all derivatives have been removed. Thus the equation

$$\frac{d^2y}{dx^2} + \left(\frac{dy}{dx}\right)^2 + xy = 0$$

is of the second order and the first degree, while

$$\frac{d^2y}{dx^2} + \sqrt{\frac{dy}{dx}} + xy = 0$$

is of the second order and the second degree. If the dependent variable and all its derivatives occur in the first degree and not multiplying each other, the equation is said to be *linear*. The solution of an equation of n-th order involves, in principle, the carrying out of n quadratures or integrations. Since each of them introduces one arbitrary constant, the final expression for the dependent variable will contain n arbitrary constants. However, a solution in which one or more of these constants are given specific values, for instance the value zero, will also satisfy the differential equation. In view of this consideration two types of solutions of an ordinary differential equation of n-th order may be distinguished: (1) the *complete* or *general* solution which contains its full complement of n independent[1] arbitrary constants; (2) *particular* solutions, obtainable from the general one by fixing one or more of the constants. In addition to these, differential equations of degree higher than the first frequently possess solutions, known as *singular* ones, which cannot be formed from the general solution in this manner. An example of these will be discussed briefly in sec. 2.6; they are rarely of interest in physical or chemical applications.

FIRST ORDER EQUATIONS

An equation of the first order can always be solved although the solution may sometimes not be expressible in terms of familiar or named functions. Methods of solution applicable in the most frequently occurring cases will now be given, and the discussion of each method will be followed by a list of problems, arising in physics and chemistry, which lead to differential equations solvable by the scheme in question.

2.2. The Variables are Separable.—This is true when the equation, which may originally appear in the form $f_1(x,y) \dfrac{dy}{dx} + f_2(x,y) = 0$, is reducible to

$$f(x)dx + g(y)dy = 0$$

Such an equation can be integrated at once and leads to a relation between y and x.

Examples.

a. *Organic growth; radioactive decay.*

Bacterial cultures in an unlimited nutritive medium grow at a time rate proportional to the number of bacteria present at any moment. Hence if the time t is regarded as independent variable and N, the number of bacteria present at time t as dependent variable,

$$\frac{dN}{dt} = \alpha N$$

α being the rate of growth per bacterium. This may be written

$$\frac{dN}{N} = \alpha dt$$

---

[1] Arbitrary constants are said to be independent if two or more of them cannot be replaced by an equivalent single one. Thus the constants c_1 and c_2 in the functions: $ax + c_1 + c_2$ and $c_1 e^{x+c_2}$ are not independent because these functions may be written $ax + c$ and ce^x, respectively.

This distinction is elementary. A more adequate analysis would focus attention upon independent *solutions* of the differential equation rather than independent constants. Solutions are independent when the so-called *Wronskian* determinant fails to vanish. This matter is treated in sec. 3.13.

which, on integration, yields $\ln N = \alpha t + c$, or $N = Ce^{\alpha t}$. If the original number of bacteria at $t = 0$ is N_0, the constant C must have the value N_0 to conform to this physical condition.

Radioactive atoms decay at a rate proportional to the number of atoms, N, present at any moment, t. Hence $dN/dt = -\lambda N$, which has the solution $N = N_0 e^{-\lambda t}$. The disintegration constant λ measures the time rate of decay per atom. It is a fundamental quantity characteristic of each radioactive substance.

b. *Flow of water from an orifice.*

A vertical tank of uniform cross-section A is filled with water to an initial height h_0. Water flows out through a hole of area a. It is desired to find the height of the water, h, in the tank as a function of the time, t. The volume flowing out in time dt is $avdt$, where v is the velocity of the water at the orifice at time t. The loss of height in the tank is dh, hence the loss of volume $A\,dh$. Therefore

$$avdt = -A\,dh$$

But the velocity is related to the height by Torricelli's formula: $v = c\sqrt{2gh}$. The empirical constant c would be unity if there were no obstruction and no " vena contracta " near the orifice; for ordinary small holes with sharp edges it is 0.6. Thus

$$ac\sqrt{2gh}\,dt = -A\,dh$$

or

$$\frac{dh}{\sqrt{h}} = -c\,\frac{a}{A}\sqrt{2g}\,dt$$

On integrating this we have

$$\sqrt{h} = \sqrt{h_0} - \frac{c}{2}\frac{a}{A}\sqrt{2g}\,t$$

where the constant of integration has been so adjusted that $h = h_0$ at $t = 0$.

c. *Heat flow.*

When heat flows through a body the temperature, T, is in general a complicated function of the coordinates within the body. In simple cases, however, it may depend only on a single coordinate, x (distance from a heated plane, or distance from a point source of heat). In that case, the rate at which heat crosses an area A perpendicular to x is given by

$$R = -kA\frac{dT}{dx} \tag{2-1}$$

and R is constant because of the continuity of flow. The quantity k is known as the thermal conductivity.

(α) If the body is a slab with plane parallel faces, one of which is maintained at a temperature T_1, integration of (1) leads to

$$T_1 - T = \frac{Rx}{kA}$$

x being the distance from the heated face. From this one obtains the elementary relation

$$R = kA \frac{T_1 - T_2}{d} \tag{2-2}$$

for the heat transfer across a plate of thickness d.

(β) If a heat source is placed at the center of a sphere, the temperature is a function of r alone. Here $A = 4\pi r^2$, and (1) reads $-4\pi k r^2 (dT/dr) = R$, which gives

$$T = \frac{1}{4\pi k} \frac{R}{r} + C$$

In this case, the temperature is not a linear function of the distance from the source as it was in (α).

(γ) At constant external temperature the thickness of ice on quiescent water increases as the square root of the time. To show this we write (2) in the form

$$R \equiv \frac{dH}{dt} = kA \frac{\Delta T}{x}$$

where x now represents the thickness of ice and dH the quantity of heat transported away from the lower surface of the ice in time dt. This, however, is proportional to the thickness dx which is added on to the already existing layer in time dt. Hence $dx/dt = C/x$, C representing a constant. From this it follows by integration that

$$x^2 \sim t$$

d. *Salt dissolving in water.*
When x_0 grams of salt are placed in M grams of water at time $t = 0$, how many grams will remain undissolved at time t? The rate of solution, dx/dt, is proportional, (a) to the number of grams, x, undissolved at time t, (b) to the difference between the saturation concentration, X/M, and the actual concentration, $(x_0 - x)/M$. (X is the number of grams of salt that would produce saturation.) Thus

$$-\frac{dx}{dt} = kx \cdot \left(\frac{X}{M} - \frac{x_0 - x}{M} \right) = \frac{k}{M} [(X - x_0)x + x^2] \tag{2-3}$$

To solve, we write

$$-\frac{dx}{(X - x_0)x + x^2} \equiv -\frac{1}{X - x_0}\left(\frac{dx}{x} - \frac{dx}{X - x_0 + x}\right) = \frac{k}{M}\,dt$$

Integration then leads to: $\ln \dfrac{X - x_0 + x}{x} + c = \dfrac{X - x_0}{M}\,kt.$ When the constant c is adjusted so that $x = x_0$ at $t = 0$, the result is

$$\ln \frac{(X - x_0 + x)x_0}{xX} = \frac{X - x_0}{M}\,kt$$

If $x_0 = X$, then the solution is $\dfrac{1}{x} - \dfrac{1}{x_0} = (k/M)t$, as one may easily verify by going back to equation (3).

e. *Atmospheric pressure at any height.*
The increment of pressure between two points in the atmosphere differing in height by dh is $dP = -\rho g dh$, if ρ is the density at height h. But ρ is related to P by the expression $P\rho^{-\gamma} = P_0\rho_0^{-\gamma}$, which is valid for adiabatic expansion of air if γ is taken to be 1.4.[2] The quantities P_0 and ρ_0 are the sea level values of P and ρ. Therefore

$$dP = -\left(\frac{P}{P_0}\right)^{1/\gamma} \rho_0 g dh$$

and this, on integration, gives $\left(\dfrac{P}{P_0}\right)^{\frac{\gamma-1}{\gamma}} = 1 - \dfrac{\gamma - 1}{\gamma}\dfrac{\rho_0 g h}{P_0}$, the constant of integration being adjusted so that $P = P_0$ at $h = 0$.

f. *Homogeneous gas reactions.*
Chemical reactions involving but a single phase are said to be homogeneous. Among these there may be distinguished unimolecular, bimolecular, termolecular reactions and so on. In the unimolecular case, the number of molecules undergoing a chemical change is at any instant proportional to the number of molecules present. The decomposition of nitrogen pentoxide into oxygen and nitrogen tetroxide ($2N_2O_5 \rightarrow O_2 + N_2O_4$) is an example of this kind, the differential equation being similar to that describing radioactive decay (Example a).

In a bimolecular reaction, of which there are numerous examples, substances A and B form molecules of type C. If a and b are the original concentrations of A and B respectively, and x is the concentration of C at a given instant, then

$$\frac{dx}{dt} = k(a - x)(b - x)$$

---

[2] γ is the ratio of the specific heat at constant pressure to that at constant volume.

To integrate this equation, the expression $\dfrac{1}{(a - x)(b - x)}$ is resolved into

the partial fractions $\dfrac{1}{a - b}\left[\dfrac{1}{b - x} - \dfrac{1}{a - x}\right]$. We then have

$$\frac{1}{a - b} \int \left(\frac{dx}{b - x} - \frac{dx}{a - x}\right) = \int kdt$$

whence

$$\frac{1}{a - b} \ln \frac{a - x}{b - x} = kt + c$$

Since $x = 0$ at $t = 0$, $c = \dfrac{1}{a - b} \ln \dfrac{a}{b}$, so that

$$\frac{b(a - x)}{a(b - x)} = e^{(a-b)kt}$$

From this, the *reaction rate* is seen to be

$$k = \frac{1}{t(a - b)} \ln \frac{b(a - x)}{a(b - x)}$$

The concentration of substance C is

$$x = \frac{a(1 - e^{(a-b)kt})}{\left(1 - \dfrac{a}{b} e^{(a-b)kt}\right)}$$

When the original concentrations a and b are equal, the expression for k becomes indeterminate, but on putting $b = a + \epsilon$ and letting ϵ approach zero, an expansion of the logarithm yields

$$k = \frac{1}{at} \frac{x}{a - x}$$

which is also seen to be a solution of the differential equation

$$\frac{dx}{dt} = k(a - x)^2$$

Other types of reactions will be dealt with in the problems on p. 40. As to terminology, we note that a rate law for multimolecular reactions of the form

$$\frac{dx}{dt} = k(a_1 - x)^{n_1}(a_2 - x)^{n_2} \cdots (a_s - x)^{n_s}$$

is often said to describe a reaction of the n-th order, where

$$n = \sum_1^s n_i$$

g. *Clapeyron's equation.*

Any phase change of a substance which takes place at constant pressure and temperature conforms to Clapeyron's equation:

$$\frac{dP}{dT} = \frac{l}{T(V_f - V_i)}$$

Here l represents the latent heat of the process, V_f and V_i the volume per mole of the final and the initial phase respectively, and P the pressure. This equation may be applied to the process of sublimation, yielding an approximate expression for the vapor pressure as a function of the temperature. In that case l, the latent heat of sublimation of the solid, is nearly constant over a range of temperatures, and V_i, the volume of the solid, may be neglected in comparison with that of the vapor, V_f. The vapor, though not a perfect gas, will be taken to satisfy $V_f = RT/P$. Clapeyron's equation then becomes

$$\frac{dP}{dT} = \frac{lP}{RT^2}$$

which on integration gives

$$P = ce^{-l/RT}$$

an equation often called the Clausius-Clapeyron equation. This result is found to be valid over small ranges of temperature, for the vapor pressure of both solids and liquids. A more refined result may be obtained by introducing for l a more adequate approximation.

h. *Centrifuge problem.*

When a cylinder of height h, filled with fluid, is rotating about its axis, the pressure within the fluid will not be constant but will depend on r. Consider a cylindrical shell of fluid of thickness dr, the surfaces of which are coaxial with the rotating vessel. The net force pushing inward on this shell is $2\pi r h \, dP$. This must equal the centripetal force due to the angular speed ω, namely $m\omega^2 r$, where m, the mass of the fluid, is given by $2\pi r h \, dr \cdot \rho$. Hence

$$2\pi r h \, dP = 2\pi r h \rho \, dr \cdot \omega^2 r$$

(α) If the fluid is a liquid, the density, ρ, is constant and the solution is

$$P = \tfrac{1}{2}\rho\omega^2 r^2 + P_0$$

(β) If the fluid is a gas, $P = c\rho$ (since $PV = \text{const.}$), the solution is then

$$P = P_0 e^{\omega^2 r^2/2c}$$

i. *Soap film.*

If a soap film is stretched between two circular wires, both having their planes perpendicular to the line joining their centers, it will form a figure of revolution about that line. At every point such as P (cf. Fig. 1) the horizontal force acting around a vertical section of the film is the same. Hence

$$2\pi y T \cos \theta = \text{const.}$$

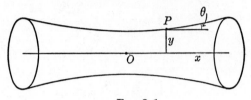

FIG. 2–1

where T is the surface tension of the film. But

$$\cos \theta = \left[1 + \left(\frac{dy}{dx} \right)^2 \right]^{-1/2}$$

so that

$$y \left[1 + \left(\frac{dy}{dx} \right)^2 \right]^{-1/2} = c$$

T being a constant. Solving for the derivative,

$$\frac{dy}{dx} = \left(\frac{y^2}{c^2} - 1 \right)^{1/2}$$

which leads to

$$y = c \cosh \frac{x + c_1}{c}$$

The constants c and c_1 may be expressed in terms of the distance between the wires and their radius. The longitudinal section of the film is seen to be a catenary.

The examples above seem sufficient to illustrate the method under discussion. The problems leading to separable first order equations are very numerous.

Problems.

　　a. *Helmholtz' equation.*

If a circuit has resistance R and inductance L, the current I in it obeys the differential equation

$$L \frac{dI}{dt} + RI = E$$

where E is the impressed or external electromotive force. Show that the *growth* of a current $(E = \text{const.}, I = 0 \text{ at } t = 0)$ is described by

$$I = \frac{E}{R}(1 - e^{-(R/L)t})$$

and the *decay* $(E = 0, I = I_0 \text{ at } t = 0)$ by

$$I = I_0 e^{-(R/L)t}$$

b. Solve the equation for *termolecular reactions:*

$$\frac{dx}{dt} = k(a - x)(b - x)(c - x).$$

Ans. $\left(1 - \frac{x}{a}\right)^{c-b}\left(1 - \frac{x}{b}\right)^{a-c}\left(1 - \frac{x}{c}\right)^{b-a} = e^{(c-b)(a-c)(b-a)kt}$

c. Solve the equation for *opposing unimolecular and bimolecular reactions:*

$$\frac{dx}{dt} = k_1(a - x) - k_2 x^2$$

under the condition $x = 0$ at $t = 0$.

Ans. $\frac{a}{x} = \frac{k_2}{k_1} A \coth A k_2 t + \frac{1}{2}$ where $A^2 = \frac{k_1}{k_2}\left(a + \frac{1}{4}\frac{k_1}{k_2}\right)$

Show that, when equilibrium is established $(t = \infty)$,

$$\frac{x^2}{a - x} = \frac{k_1}{k_2}$$

d. Solve the equation for *consecutive unimolecular reactions* of the type

$$\underset{A \to B \to C}{\overset{k_1 \quad k_2}{}}$$

that is,

$$\frac{dn_1}{dt} = -k_1 n_1, \quad \frac{dn_2}{dt} = k_1 n_1 - k_2 n_2$$

Ans. $n_3 = (n_1 + n_2 + n_3)\left\{1 - \frac{k_2}{k_2 - k_1}e^{-k_1 t} + \frac{k_1}{k_2 - k_1}e^{-k_2 t}\right\}$

where n_3 = amount of C present at t.

e. A projectile is fired vertically into the air with initial velocity V. (1) Find its speed at any height; (2) find the time at which it will have traversed a distance r. Note: the differential equation to be solved is

$$\frac{dv}{dt} \equiv v\frac{dv}{dr} = -\frac{gR^2}{r^2}$$

where g = acceleration due to gravity, R = radius of the earth.

Ans. (1) $\qquad v = \left[V^2 - 2gR\left(1 - \frac{R}{r}\right)\right]^{1/2}$

(2) If $V^2 > 2gR$

$$t = (V^2 - 2gR)^{-1} \left\{ \left[\left(V^2 - 2gR + \frac{2gR^2}{r} \right)^{1/2} - V \right] r \right.$$

$$\left. - \frac{2gR^2}{(V^2 - 2gR)^{1/2}} \left[\ln \frac{\left(V^2 - 2gR + \dfrac{2gR^2}{r} \right)^{1/2} + (V^2 - 2gR)^{1/2}}{V + (V^2 - 2gR)^{1/2}} + \tfrac{1}{2} \ln \frac{r}{R} \right] \right\}$$

2.3. The Differential Equation is, or Can be Made, Exact. Linear Equations.—A differential equation, written in the form

$$A\,dx + B\,dy = 0 \tag{2-4}$$

where A and B are functions of x and y, is said to be exact if the left-hand side is an exact differential. The necessary and sufficient condition for this to be true was shown in sec. 1.7 to be equivalent to the *Cauchy relations*

$$\frac{\partial A}{\partial y} = \frac{\partial B}{\partial x}$$

The equations considered in the foregoing section, where A was a function of x alone and B a function of y alone, are exact in the trivial sense that $\partial A/\partial y = \partial B/\partial x = 0$.

Differential equations occurring in practice are rarely exact, but every equation of the form (4) can be made exact and then integrated. The device for doing this is to multiply it by a suitable factor known as the *integrating factor*. For instance, the equation

$$\frac{dy}{y} + \left(\frac{1}{x} - \frac{x}{y} \right) dx = 0$$

is not exact. It becomes exact on multiplication by xy. For it then takes the form

$$d\left(xy - \frac{x^3}{3} \right) = 0$$

which has the solution:

$$xy - \frac{x^3}{3} = \text{const.}$$

While an integrating factor exists for every equation of the form (4), it is not always easy to find. If the equation is *linear*, however, that is if it can be written

$$\frac{dy}{dx} + f(x)y = g(x) \tag{2-5}$$

an integrating factor is always available. It is $e^{\int f\,dx}$. On application of this factor eq. (5) becomes

$$\frac{d}{dx}(ye^F) = g(x)e^F$$

where the abbreviation $F(x) = \displaystyle\int^x f(\xi)d\xi$ has been used. The solution is, clearly,

$$y = e^{-F}\left[\int e^F g\, dx + c\right] \tag{2-6}$$

This result is most useful, for the occurrence of linear equations is very frequent.

Examples.

a. *Circuit containing inductance and resistance* (Helmholtz' equation).

This problem has already been discussed, but it may be instructive to solve the differential equation also by the method of eq. (6). We have

$$\frac{dI}{dt} + \frac{RI}{L} = \frac{E}{L} \tag{2-7}$$

Thus

$$f = \frac{R}{L} \quad\text{and}\quad F = \frac{R}{L}t; \quad g = \frac{E}{L}$$

so that

$$I = e^{-(R/L)t}\left[\int \frac{E}{L} e^{(R/L)t} dt + c\right] = \frac{E}{R} + ce^{-(R/L)t}$$

and this agrees with our previous result (Problem a).

b. *Circuit with inductance and resistance; variable electromotive force.*

The present method involves the solution of eq. (7) when E is a function of the time, in which case the equation can no longer be separated. Let us assume that

$$E = E_0 \sin \omega t$$

We then have

$$f = \frac{R}{L}; \quad F = \frac{R}{L}t; \quad g = \frac{E_0}{L}\sin \omega t$$

Hence[3]

$$I = \frac{e^{-(R/L)t}}{L} E_0 \int e^{(R/L)t} \sin \omega t \, dt + c e^{-(R/L)t}$$

$$= \frac{E_0}{L} \frac{1}{\omega'^2 + \omega^2} (\omega' \sin \omega t - \omega \cos \omega t) + c e^{-(R/L)t}$$

where ω' has been written for R/L, a quantity having the dimensions of a frequency. To fix the constant we assume that $I(0) = 0$, in which case

$$I = \frac{E_0}{L} \frac{1}{\omega'^2 + \omega^2} (\omega' \sin \omega t - \omega \cos \omega t + \omega e^{-\omega't})$$

The last term represents transient currents which disappear as soon as $t \gg \dfrac{1}{\omega'}$.

c. *Radioactive decay of mother and daughter substances.*

Let A be the number of atoms of the mother substance (e.g., UI) and B the number of atoms of the daughter substance (e.g., UX_1) at time t, A_0 being the original value of A at $t = 0$. Let λ_A and λ_B be the decay constants as defined in sec. 2.2a. The two substances satisfy the two differ-

---

[3] Here and elsewhere, there occurs the integral $\displaystyle\int e^{\omega't} \sin \omega t \, dt$. This is easily evaluated if the sine is written as an exponential:

$$\sin x = \frac{1}{2i} (e^{ix} - e^{-ix}).$$

Thus

$$\int e^{\omega't} \sin \omega t \, dt = \frac{1}{2i} \int [e^{(\omega'+i\omega)t} - e^{(\omega'-i\omega)t}] dt$$

$$= \frac{1}{2i} \left\{ \frac{e^{(\omega'+i\omega)t}}{\omega'+i\omega} - \frac{e^{(\omega'-i\omega)t}}{\omega'-i\omega} \right\} = \frac{e^{\omega't}}{2i} \left\{ \frac{(\omega'-i\omega)e^{i\omega t} - (\omega'+i\omega)e^{-i\omega t}}{\omega'^2 + \omega^2} \right\}$$

$$= \frac{e^{\omega't}}{\omega'^2 + \omega^2} (\omega' \sin \omega t - \omega \cos \omega t) = -\frac{e^{\omega't}}{(\omega'^2 + \omega^2)^{1/2}} \cos (\omega t + \beta)$$

$$\beta \equiv \tan^{-1} \frac{\omega'}{\omega}.$$

Similarly:

$$\int e^{\omega't} \cos \omega t \, dt = \frac{e^{\omega't}}{\omega'^2 + \omega^2} (\omega' \cos \omega t + \omega \sin \omega t)$$

$$= \frac{e^{\omega't}}{(\omega'^2 + \omega^2)^{1/2}} \sin (\omega t + \beta)$$

ential equations

$$\frac{dA}{dt} = -\lambda_A A; \quad \frac{dB}{dt} = -\lambda_B B + \lambda_A A$$

When the solution of the first, $A = A_0 e^{-\lambda_A t}$, is substituted in the second there results

$$\frac{dB}{dt} + \lambda_B B = \lambda_A A_0 e^{-\lambda_A t}$$

an equation which is linear in B and can be solved by formula (6). The solution is:

$$B = e^{-\lambda_B t} \left\{ \int \lambda_A A_0 e^{(\lambda_B - \lambda_A)t} dt + c \right\}$$

$$= \frac{\lambda_A}{\lambda_B - \lambda_A} A_0 (e^{-\lambda_A t} - e^{-\lambda_B t})$$

if we assume that $B(0) = 0$. Note that B will reach a maximum at time $t = \dfrac{\ln \lambda_A - \ln \lambda_B}{\lambda_A - \lambda_B}$.

Problem. A circuit contains capacitance C, resistance R, and is subject to an electromotive force E. Calculate the instantaneous value of the electric charge q on the condenser, noting that it satisfies the differential equation

$$R\frac{dq}{dt} + \frac{q}{C} = E$$

Ans. For $E = E_0 \sin \omega t$,

$$q = \frac{E_0}{R} \frac{1}{\omega'^2 + \omega^2} (\omega' \sin \omega t - \omega \cos \omega t + \omega e^{-\omega' t}), \quad \omega' = \frac{1}{RC}$$

2.4. Equations Reducible to Linear Form.—Of some mathematical interest is an equation of the form

$$\frac{dy}{dx} + f(x)y = g(x)y^n \tag{2-8}$$

because it can be made linear by the substitution $y = u^{1/1-n}$. This converts (8) into

$$\frac{du}{dx} + (1 - n)fu = (1 - n)g$$

which can be solved by the method of the preceding section. Eq. (8) is often called Bernoulli's equation.

2.5. Homogeneous Differential Equations.—A first order equation is said to be homogeneous[4] if, the equation being written in the form

$$A\,dx + B\,dy = 0$$

A and B are homogeneous functions of the same degree, i.e.,

$$A(tx,ty) = t^{\alpha}A(x,y); \quad B(tx,ty) = t^{\alpha}B(x,y)$$

If this is true we can substitute $y = vx$, obtaining

$$A(x,y) = A(x,vx) = x^{\alpha}A(1,v); \quad B(x,y) = x^{\alpha}B(1,v)$$

The original equation,

$$\frac{dy}{dx} = -\frac{A}{B}$$

is converted into

$$v + x\frac{dv}{dx} = -\frac{A(1,v)}{B(1,v)} \equiv f(v)$$

by this substitution, and this equation is separable, yielding

$$\frac{dv}{f(v) - v} = \frac{dx}{x}$$

Example. *Lines of force.*

An equation closely related to the homogeneous type, and tractable by the

---

[4] A remark on the use of the word " homogeneous " in mathematics seems in order, for the term is used with several different meanings in different contexts. The following definitions correspond to the chief usages.

1. Homogeneous function: $f(x_1,x_2,\cdots x_n)$ is said to be homogeneous in all its variables if, for any parameter, t, $f(tx_1,tx_2,\cdots tx_n) = t^{\alpha}f(x_1,x_2,\cdots x_n)$. α is the " degree " of the homogeneous function.

2. Homogeneous equations: A set of simultaneous linear algebraic equations of the form

$$\sum_{i=1}^{n} a_{ji}x_i = c_j; \quad j = 1, 2, \cdots, n$$

in which the a's are constants is said to be homogeneous if all c's are zero.

3. Homogeneous differential equations: (Two usages of the term!)

a. A first order equation of the form $A\,dx + B\,dy = 0$ is said to be homogeneous if $A(x,y)$ and $B(x,y)$ are homogeneous functions of the same degree.

b. In general, $F(x,y,y',y'',\cdots) = 0$ is said to be homogeneous if F is a homogeneous function of y and all its derivatives, *not necessarily of x.* Thus

$$f_n(x)\frac{d^n y}{dx^n} + f_{n-1}(x)\frac{d^{n-1}y}{dx^{n-1}} + \cdots f_1(x) \cdot y = 0$$

is homogeneous and linear. If the right-hand side of this equation were not zero but equal to a function of x, the equation would still be linear but no longer homogeneous.

substitution here described, is the differential equation for lines of force. A line of force is defined as that curve which is tangent, at every point through which it passes, to the force at that point. The present analysis is applicable to attracting mass points, attracting or repelling electric

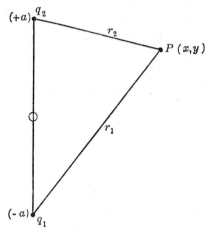

FIG. 2-2

charges, and magnetic poles. Let it be desired, for example, to find the lines of force due to two charges, q_1 and q_2, a distance $2a$ apart. (Cf. Fig. 2.) If we restrict our consideration to the plane containing the charges and the point P, then, for every point in this plane, the definition of a line of force requires that

$$\frac{dy}{dx} = \frac{F_y}{F_x} = \frac{\dfrac{q_1}{r_1^3}(y+a) + \dfrac{q_2}{r_2^3}(y-a)}{\dfrac{q_1}{r_1^3}x + \dfrac{q_2}{r_2^3}x} \tag{2-9}$$

If a were zero, this would reduce to $dy/dx = y/x$, an equation which has for its solution all straight lines through the origin. These, as is well known, represent the lines of force due to a point charge. In general, however, eq. (9) reads

$$\frac{q_1}{r_1^3}[xdy - (y+a)dx] + \frac{q_2}{r_2^3}[xdy - (y-a)dx] = 0 \tag{2-9a}$$

This equation misses being homogeneous by the presence of the quantity a. But a simple artifice will help. If we introduce two new dependent variables, $y_1 = y + a$ and $y_2 = y - a$, so that $dy_1 = dy_2 = dy$; $r_1 = (x^2 + y_1^2)^{1/2}$, $r_2 = (x^2 + y_2^2)^{1/2}$, eq. (9a) takes the form

$$q_1 \frac{xdy_1 - y_1dx}{(x^2 + y_1^2)^{3/2}} + q_2 \frac{xdy_2 - y_2dx}{(x^2 + y_2^2)^{3/2}} = 0$$

each part of which is homogeneous. Now put $y_1 = v_1x$, $y_2 = v_2x$ so that

$$x^2dv = xdy - ydx$$

The result is then simply

$$q_1 \frac{dv_1}{(1 + v_1^2)^{3/2}} + q_2 \frac{dv_2}{(1 + v_2^2)^{3/2}} = 0$$

When this is integrated, we immediately obtain the equation of the lines of force due to the two charges:

$$q_1 \frac{v_1}{(1 + v_1^2)^{1/2}} + q_2 \frac{v_2}{(1 + v_2^2)^{1/2}} = \frac{q_1y_1}{r_1} + \frac{q_2y_2}{r_2} = \text{const.}$$

2.6. Note on Singular Solutions. Clairaut's Equation.—A first order equation of degree higher than the first may have a special kind of solution which is not obtainable by specifying the constants in its general solution. Thus consider

$$y = x\frac{dy}{dx} + \left(\frac{dy}{dx}\right)^2 \tag{2-10}$$

This equation may be solved by the following artifice. Differentiate once more, thus converting it into a second order equation, which, however, can easily be handled by the methods already discussed. The result is

$$\frac{dy}{dx} = \frac{dy}{dx} + x\frac{d^2y}{dx^2} + 2\frac{dy}{dx}\frac{d^2y}{dx^2}$$

or

$$\left(x + 2\frac{dy}{dx}\right)\frac{d^2y}{dx^2} = 0 \tag{2-11}$$

If now the first factor be cancelled, the equation is

$$\frac{d^2y}{dx^2} = 0$$

and has the solution $y = c_1x + c_2$. This, however, is too general a result since it contains two constants of integration, a circumstance brought about by the arbitrary procedure of converting the original first order into a second order equation before solving. To satisfy eq. (10), it is necessary to substitute this solution and adjust c_2 in conformity with its demands. It is then seen that $c_2 = c_1^2$, and

$$y = cx + c^2$$

is the general solution of eq. (10).

But eq. (11) can also be satisfied by equating the first factor on the left to zero. This leads to

$$x + 2\frac{dy}{dx} = 0, \quad \text{or} \quad y = -\frac{x^2}{4} + c$$

This will satisfy eq. (10) if $c = 0$. Thus

$$y = -\frac{x^2}{4}$$

is another solution of the original differential equation, but one which is not derivable from its complete solution. It is called a *singular* solution. Inspection will show that it represents the *envelope* of all the straight lines which correspond to the complete solution. This is generally the meaning of singular solutions.

An equation of the form

$$y = x\frac{dy}{dx} + f\left(\frac{dy}{dx}\right)$$

is known to mathematicians as *Clairaut's equation*. Eq. (10) is a specimen of this type. Clairaut's equation can always be handled by the method here used and has the general solution

$$y = cx + f(c)$$

EQUATIONS OF HIGHER ORDER

A general method for solving certain differential equations of higher order will be presented in secs. 2.10–12. It seems appropriate, however, to discuss first a few special types of differential equations which can be solved by elementary means. While the theory given in this section is applicable to equations of any order, emphasis will be placed solely on second order equations because of their prominence in mathematical physics.

2.7. Linear Equations with Constant Coefficients; Right-Hand Member Zero.—In discussing this type of equation it becomes convenient to introduce a new notation; we write $D = d/dx$. A symbol such as D, which is meaningless unless applied to a function of x, and which is therefore not a mathematical quantity in the usual sense, bears the name " operator." In the present connection D may be regarded as nothing more than an abbreviation. Later, however, when the mathematics of quantum mechanics is to be studied, it will be found that operators such as D are entities of considerable significance which give rise to an operator algebra quite different in many respects from ordinary algebra. For the present we merely observe that a differential equation of the type under discussion in its most general form may be written:

$$D^n y + a_1 D^{n-1} y + a_2 D^{n-2} y + \cdots a_n y = 0 \qquad (2\text{-}12)$$

The a's are constants; the order of the equation is n. Consider now the differential equation

$$(D - r_1)(D - r_2) \cdots (D - r_n)y = 0 \qquad (2\text{-}13)$$

which must be understood to mean that the successive application of $d/dx - r_n$, $d/dx - r_{n-1}$, etc., upon y is to yield zero, the r's being constants. It is clear that (12) and (13) become identical when the r's are chosen to be the roots of the algebraic equation

$$r^n + a_1 r^{n-1} + a_2 r^{n-2} + \cdots + a_n = 0 \qquad (2\text{-}14)$$

Let us then attempt to solve (13). A particular solution of that equation is easily found, for if y satisfies

$$(D - r_n)y = 0$$

it will also satisfy (13), since further differentiations and multiplications by r will leave the right-hand side unchanged. But $(D - r_n)y = 0$ has the solution $y = c_n e^{r_n x}$, hence this is a particular solution of (13).

Furthermore, we observe that the order of the " factors " $(D - r_i)$ appearing in (13) is insignificant. Hence any factor may be written last, and this means that $c_{n-1} e^{r_{n-1} x}$ is also a particular solution, and so on. On adding all particular solutions, i.e., on putting

$$y = \sum_i c_i e^{r_i x} \qquad (2\text{-}15)$$

there results a solution with n independent arbitrary constants, and this must therefore be the complete solution. To summarize: in order to solve (12), first determine the roots of (13), which is known as the *auxiliary equation*. If these roots are denoted by r_i, the general solution is (15).

One point is to be noted. If the coefficients a appearing in (12) are functions of x, the decomposition into factors leading to (13) cannot be made by solving the auxiliary equation. The reason is that then the r's will also be functions of x, and

$$(D - r_1)(D - r_2)y \neq (D - r_2)(D - r_1)y$$

as the reader may easily verify. This state of affairs is expressed succinctly by saying that the operators $(D - r_1)$ and $(D - r_2)$ are *commutative* only if the r's are constants. For variable r's the order of the factors in (13) is also essential, so that the whole method of solution here discussed must fail.

Returning to the case of constant coefficients, one minor difficulty must be considered. Suppose that two roots of the auxiliary equation are equal. If they are called r_1 the supposedly general solution will contain the part $(c_1 + c_2)e^{r_1 x}$ which is equivalent to $ce^{r_1 x}$. One arbitrary constant has been lost and the solution obtained is no longer complete. To remove this fault we consider the two factors of (13) which gave rise to it and study the equation

$$(D - r_1)^2 y = 0 \tag{2-16}$$

One solution is certainly $y = e^{r_1 x}$. Let us look for a general solution of the form $y = f(x)e^{r_1 x}$. On substitution of this into (16) there results the following differential equation for $f(x)$:

$$\frac{d^2 f}{dx^2} = 0$$

Hence $f = c_1 x + c_2$, and the complete solution of (16) reads

$$y = (c_1 x + c_2)e^{r_1 x}$$

This shows that, when two roots of the auxiliary equation are equal and have the value r_1, the part of the solution $(c_1 + c_2)e^{r_1 x}$ occurring in (15) must be replaced by $(c_1 x + c_2)e^{r_1 x}$. An extension of this argument leads to the general result: If r_i is a g-fold root of the auxiliary equation, the complete solution of (12) is

$$y = c_1 e^{r_1 x} + c_2 e^{r_2 x} + \cdots c_i(1 + a_1 x + a_2 x^2 + \cdots + a_{g-1}x^{g-1})e^{r_i x} + \cdots$$

Examples.

a. *Simple harmonic motion.*

When the force on a particle of mass m moving along the y-axis is equal to $-ky$, Newton's second law of motion reads:

$$m\frac{d^2 y}{dt^2} = -ky$$

Here k, the force per unit of displacement of the particle, is known as the *stiffness* of the oscillator. If we denote the positive constant k/m by ω^2, the equation becomes $d^2 y/dt^2 + \omega^2 y = 0$. The roots of the auxiliary equation $r^2 + \omega^2 = 0$ are $r_1 = i\omega$, $r_2 = -i\omega$. Hence by (15)

$$y = c_1 e^{i\omega t} + c_2 e^{-i\omega t}$$

The constants c_1 and c_2 may of course be complex. This result may be written in two other, but equivalent, forms. On expanding the exponentials in sines and cosines we obtain

$$y = (c_1 + c_2) \cos \omega t + (c_1 - c_2)i \sin \omega t = C_1 \cos \omega t + C_2 \sin \omega t$$

This last result may also be stated as follows:

$$y = A \sin (\omega t + \delta) = A' \cos (\omega t + \delta')$$

where the new constants A, δ, and A', δ' are related to C_1 and C_2 by $A \sin \delta = C_1$, $A \cos \delta = C_2$; $A' \cos \delta' = C_1$, $-A' \sin \delta' = C_2$, or conversely $A^2 = A'^2 = C_1^2 + C_2^2$, $\delta = \tan^{-1} C_1/C_2$, $\delta' = \tan^{-1} C_2/C_1$.

b. *Chain sliding over a smooth peg.*
The chain (cf. Fig. 3) is sliding over the peg, the right end moving downward. Let the displacement of this end from 0, the point it would occupy in equilibrium, be y. If the linear density of the chain is λ, and its total length l, the mass to be accelerated is $l\lambda$. The resultant force is $2\lambda yg$. Hence, from Newton's second law,

$$l\lambda \frac{d^2y}{dt^2} = 2\lambda gy, \quad \text{or} \quad \frac{d^2y}{dt^2} - \frac{2g}{l} y = 0$$

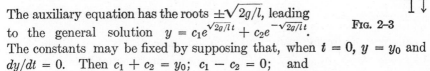

The auxiliary equation has the roots $\pm\sqrt{2g/l}$, leading to the general solution $y = c_1 e^{\sqrt{2g/l}\,t} + c_2 e^{-\sqrt{2g/l}\,t}$.

FIG. 2–3

The constants may be fixed by supposing that, when $t = 0$, $y = y_0$ and $dy/dt = 0$. Then $c_1 + c_2 = y_0$; $c_1 - c_2 = 0$; and

$$y = \frac{y_0}{2} (e^{\sqrt{2g/l}\,t} + e^{-\sqrt{2g/l}\,t}) = y_0 \cosh \sqrt{\frac{2g}{l}}\, t$$

c. *Damped simple harmonic motion.*
When the motion of the oscillator considered in example (a) is damped, there is present, besides the restoring force $-ky$, a damping force proportional (at small velocities) to $-l(dy/dt)$, the negative sign indicating that the force *retards* the motion; l is known as the damping constant. The differential equation describing the motion is

$$\frac{d^2y}{dt^2} + 2b \frac{dy}{dt} + \omega^2 y = 0 \qquad (2\text{–}17)$$

if b is written for the constant quantity $l/2m$. The auxiliary equation has the roots $-b \pm \sqrt{b^2 - \omega^2}$ so that the general solution becomes

$$y = c_1 e^{(-b+\sqrt{b^2-\omega^2})t} + c_2 e^{(-b-\sqrt{b^2-\omega^2})t}$$

To adjust the constants in conformity with physical conditions we suppose that, at $t = 0$, $y = y_0$ and $dy/dt = 0$. Then with the use of the abbreviation $R = \sqrt{b^2 - \omega^2}$

$$y = \frac{y_0}{2} e^{-bt}\left[\left(1 + \frac{b}{R}\right)e^{Rt} + \left(1 - \frac{b}{R}\right)e^{-Rt}\right] \qquad (2\text{-}18)$$

Several special cases are of interest in this connection.

(α) $b > \omega$. R is then real, but smaller than b. Hence both terms of (18) represent an exponential decrease. The motion is not oscillatory.

(β) $b = \omega$. Then $R = 0$, and $y = y_0 e^{-bt}(1 + bt)$. The motion is not oscillatory; it is said to be critically damped.

(γ) $b < \omega$. Then R is imaginary and may be written $R = i\omega'$, $\omega'^2 = \omega^2 - b^2$. Eq. (18) now reads

$$y = y_0 e^{-bt}\left(\cos \omega' t + \frac{b}{\omega'} \sin \omega' t\right)$$

or, in equivalent form,

$$y = \frac{\omega}{\omega'} y_0 e^{-bt} \sin (\omega' t + \delta)$$

where $\delta = \tan^{-1} \omega'/b$. This represents a damped sinusoidal motion of period $T = 2\pi/\sqrt{\omega^2 - b^2}$; the amplitude decreases exponentially as e^{-bt}.

d. *Natural oscillations in an electrical circuit.*

In a circuit containing R, L, and C, the sum of the " partial " electromotive forces due to inductance, resistance and capacitance equals the external e.m.f. If the latter is zero (natural oscillations) we have

$$L \frac{dI}{dt} + RI + \frac{q}{C} = 0$$

or, remembering that $I = dq/dt$,

$$\frac{d^2q}{dt^2} + \frac{R}{L}\frac{dq}{dt} + \frac{1}{LC} q = 0$$

This equation is of the form (17); the constants are $b = R/2L$, $\omega = (LC)^{-1/2}$. The solutions are already given in the foregoing example. In particular, if oscillations are to take place, $\omega > b$, i.e., $2\sqrt{L/C} > R$. In that case

$$q = \left(1 - \frac{R^2 C}{4L}\right)^{-1/2} q_0 e^{-(R/2L)t} \sin\left(\sqrt{\frac{1}{LC} - \frac{R^2}{4L^2}}\, t + \delta\right)$$

and

$$\delta = \tan^{-1} \sqrt{\frac{4L}{CR^2} - 1}$$

The initial conditions here are that at $t = 0$, the condenser has a charge q_0 and there is no current.

2.8. Linear Equations with Constant Coefficients; Right-Hand Member a Function of x.—We now restrict our considerations to differential equations of the second order. In terms of the notation of the foregoing section, the problem is to solve

$$(D^2 + a_1 D + a_2)y = f(x) \tag{2-19}$$

If the roots of the auxiliary equation are r_1 and r_2, this equation takes the form

$$(D - r_1)(D - r_2)y = f(x) \tag{2-20}$$

Put $(D - r_2)y \equiv u$, so that $(D - r_1)u = f(x)$. This is a linear first order equation which can be solved by the method of sec. 3. It gives

$$u = e^{r_1 x} \int e^{-r_1 x} f(x)dx + c_1 e^{r_1 x} = e^{r_1 x}(\varphi(x) + c_1)$$

if we define $\displaystyle\int_0^{x^1} e^{-r_1 \xi} f(\xi)d\xi = \varphi(x)$. If this is substituted back into the definition of u, the result is $(D - r_2)y = e^{r_1 x}(\varphi(x) + c_1)$, an equation which may again be treated in accordance with formula (6). Hence

$$y = e^{r_2 x} \int e^{(r_1 - r_2)x} [\varphi(x) + c_1]dx + c_2 e^{r_2 x}$$

$$= e^{r_2 x} \int e^{(r_1 - r_2)x} \varphi(x)dx + \frac{c_1}{r_1 - r_2} e^{r_1 x} + c_2 e^{r_2 x}$$

On changing the meaning of the constant c_1, we write the solution of (19)

$$y = e^{r_2 x} \int e^{(r_1 - r_2)x} \varphi(x)dx + c_1 e^{r_1 x} + c_2 e^{r_2 x} \tag{2-21}$$

The form of this solution is interesting. The last two terms are identical with the solution of the homogeneous equation. They are called the *complementary function*, while the remainder, $e^{r_2 x} \int e^{(r_1 - r_2)x} \varphi(x)dx$, is known as the *particular integral*. Thus the "inhomogeneity" of the equation, $f(x)$, makes its appearance in the particular integral only. It is sometimes possible to find the particular integral of an equation like (19) by inspection, that is, by selecting any function which will satisfy the equation. When this is available one can make use of the fact just noted and form the complete solution by adding to this function the general solution of the homogeneous equation. Usually, however, the straightforward calculation of the particular integral is hardly more difficult.

The particular integral can be written in a form which is often more convenient in practice. On performing a partial integration we find

$$\int e^{(r_1-r_2)x}\varphi(x)dx = \frac{e^{(r_1-r_2)x}\varphi(x)}{r_1-r_2} - \int \frac{e^{(r_1-r_2)x}}{r_1-r_2}\frac{d\varphi}{dx}dx$$

$$= \frac{e^{(r_1-r_2)x}}{r_1-r_2}\int e^{-r_1x}f(x)dx - \int \frac{e^{-r_2x}}{r_1-r_2}f(x)dx$$

because $d\varphi/dx = e^{-r_1x}f(x)$. The particular integral then becomes

$$\frac{1}{r_1-r_2}\left\{e^{r_1x}\int e^{-r_1x}f(x)dx - e^{r_2x}\int e^{-r_2x}f(x)dx\right\}$$

and finally

$$y = \frac{1}{r_1-r_2}\left\{e^{r_1x}\int e^{-r_1x}f(x)dx - e^{r_2x}\int e^{-r_2x}f(x)dx\right\} + c_1e^{r_1x} + c_2e^{r_2x}$$

$$(2-22)$$

Examples.

a. *Forced oscillations of a mechanical or electrical system.*

The equation to be considered is (17) but with a function of t instead of zero on the right. In most applications this function, which represents the impressed force divided by the mass of the oscillating system in the mechanical case, is a sinusoidal function of the time. Hence we are dealing with the differential equation

$$\frac{d^2y}{dt^2} + 2b\frac{dy}{dt} + \omega^2y = f_0\sin \alpha t \qquad (2-23)$$

As in sec. 7, example (c), the auxiliary equation has the roots

$$r_1 = -b + \sqrt{b^2-\omega^2}, \quad r_2 = -b-\sqrt{b^2-\omega^2}$$

If again we denote $\sqrt{b^2-\omega^2}$ by R, the particular integral is

$$\text{P.I.} = \frac{e^{(-b+R)t}}{2R}\int e^{(b-R)t}f_0\sin \alpha t dt - \frac{e^{-(b+R)t}}{2R}\int e^{(b+R)t}f_0\sin \alpha t dt$$

The integrals here may be evaluated by means of the formulas on p. 43.

When this is done and the terms are suitably collected,

$$\text{P.I.} = \frac{f_0}{2R} \left\{ \left[\frac{b - R}{(b - R)^2 + \alpha^2} - \frac{b + R}{(b + R)^2 + \alpha^2} \right] \sin \alpha t - \right.$$
$$\left. \left[\frac{\alpha}{(b - R)^2 + \alpha^2} - \frac{\alpha}{(b + R)^2 + \alpha^2} \right] \cos \alpha t \right\}$$

$$= \frac{f_0}{(\omega^2 - \alpha^2)^2 + 4\alpha^2 b^2} \left\{ (\omega^2 - \alpha^2) \sin \alpha t - 2b\alpha \cos \alpha t \right\}$$

To obtain the complete solution we must add to this the solution of (17). Hence

$$y = \frac{f_0}{(\omega^2 - \alpha^2)^2 + 4\alpha^2 b^2} \left\{ (\omega^2 - \alpha^2) \sin \alpha t - 2b\alpha \cos \alpha t \right\}$$
$$+ e^{-bt}(c_1 e^{Rt} + c_2 e^{-Rt}) \quad (2\text{-}24)$$

It is seen that the complementary function decays exponentially with t and will be damped out eventually. It is therefore of little interest in physical applications. The amplitude of the oscillations,

$$\frac{f_0}{(\omega^2 - \alpha^2)^2 + 4\alpha^2 b^2}$$

has a maximum when the impressed (angular) frequency has the value

$$\alpha = (\omega^2 - 2b^2)^{1/2}$$

This is said to be the condition of resonance between the impressed force and the vibrating system. If b is zero there occurs what is sometimes referred to as the "resonance catastrophe," for in that case the amplitude is infinite when $\alpha = \omega$.

(α) *Mechanical system.*
The present theory can be applied, for instance, to a mass m held in equilibrium by a spring of stiffness k and damping constant l. We then have, as in sec. 7c,

$$b = \frac{l}{2m}, \quad \omega^2 = \frac{k}{m}, \quad f_0 = \frac{F_0}{m}$$

Resonance occurs when

$$\alpha = \left(\frac{k}{m} - \frac{l^2}{2m^2} \right)^{1/2}$$

(β) *Electrical system.*
For an electrical system with an impressed electromotive force $E_0 \sin \alpha t$ we have (cf. sec. 7d), $b = R/2L$, $\omega = (LC)^{-1/2}$, $f_0 = E_0/L$. Resonance

occurs when

$$\alpha = \left(\frac{1}{LC} - \frac{R^2}{2L^2} \right)^{1/2}$$

The solution (24) represents the charge, q, residing on the condenser at any instant. The current I is obtained by differentiating q with respect to the time. Both terms in braces then become positive, and

$$I = A[(\omega^2 - \alpha^2) \cos \alpha t + 2b\alpha \sin \alpha t]$$

where A stands for $E_0\alpha/L[(\omega^2 - \alpha^2)^2 + 4\alpha^2 b^2]$. The power expended in the circuit is $\int_0^T EI dt$. This integral contains two terms, one with the integrand $\sin \alpha t \cos \alpha t$, the other with the integrand $\sin^2 \alpha t$. The first of these is 0 provided T is taken large enough to include a great number of cycles $2\pi/\alpha$, the last gives $\int_{0_-}^T \sin^2 \alpha t dt = T/2$. Hence the power expended is

$$Ab\alpha T$$

The part of the current proportional to $\cos \alpha t$ causes no power consumption; it is a "wattless" current which is always out of phase with the impressed electromotive force.

b. *Electrical polarization.*

An equation like (23) also describes the response of ordinary matter to an impinging electromagnetic wave. A light wave, for instance, which is polarized in such a way that its electric vector is along y, when incident upon an electron inside a refracting medium, will exert a force equal to $eE_0 \sin \alpha t$ upon this electron. Here E_0 is the amplitude of the electric vector of the light wave, e the charge on an electron, α the frequency of the light (assumed monochromatic). f_0 in (23) is then $(e/m) E_0$, m being the electron mass. The solution is given by (24). y represents the displacement of the electron under consideration at the time t. This gives rise to a dipole of moment ey. By "polarization" is meant the dipole moment per unit volume of the material, and this is obtained on multiplying the dipole moment due to one electron by the number of displaceable electrons per unit volume. If this number is N, then the polarization

$$P = \frac{Ne^2 E_0}{m} \frac{\left\{ (\omega^2 - \alpha^2) \sin \alpha t - 2b\alpha \cos \alpha t \right\}}{(\omega^2 - \alpha^2)^2 + 4\alpha^2 b^2}$$

Further considerations of a physical nature[5] show how the index of refrac-

---

[5] See, for instance, Page, L., "Introduction to Theoretical Physics," Third Edition, D. Van Nostrand Co., 1952, p. 582 et seq.

tion and the conductivity of the substance may be deduced very easily from this expression for P.

2.9. Other Special Forms of Second Order Differential Equations.—
a. An equation of the type

$$\frac{d^2y}{dx^2} = f(x) \tag{2-25}$$

can be integrated by the method of sec. 8. If this is done, only formula (21) is applicable, for the second formula (22) involves the quantity

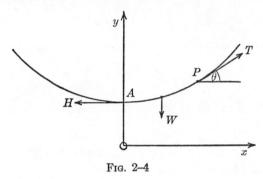

FIG. 2–4

$r_1 - r_2$, which is zero, the auxiliary equation corresponding to (23) having equal roots: $r_1 = r_2 = 0$. The solution is

$$y = \int \varphi(x)dx + c_1 + c_2 x = \int \left[\int f(x)dx \right] dx + c_1 + c_2 x$$

This procedure is here very artificial, of course, for this result could have been obtained directly by integrating (25) twice.

Example. *Suspension bridge.*
Consider the part of the cable between A and the variable point P. It is in equilibrium under the action of three forces: the horizontal force, H, the tension, T, at P, and the weight W of, or supported by, AP, which of course need not act at the middle of the segment. Hence we have

$$T \sin \theta = W; \quad T \cos \theta = H \quad \therefore \tan \theta = \frac{dy}{dx}\bigg|_P = \frac{W}{H}$$

This relation is true for every point P, provided W is the load between A and P. It is generally more convenient to write the equation in terms of $w = dW/dx$, i.e., the load per unit horizontal distance; $w = w(x)$:

$$\frac{d^2y}{dx^2} = \frac{w(x)}{H} \tag{2-26}$$

where H is, of course, a constant.

In the case of the suspension bridge, the load is uniform along x, hence $w = $ const.

Solution:

$$y = \frac{wx^2}{2H} + c_1 x + c_2, \quad \text{a parabola}$$

b. *Equations not containing y.*

If the equation to be solved is

$$\frac{d^2y}{dx^2} = f\left(x, \frac{dy}{dx}\right)$$

introduce the new variable $p = dy/dx$. The resulting equation

$$\frac{dp}{dx} = f(x,p)$$

can then be solved by one of the methods already discussed.

Example. *Cable hanging under its own weight.*

The equation describing the cable is (26), but w is not constant. In this case it is dW/ds, the weight per unit length of cable, which is constant, provided the latter is uniform. Put $dW/ds = \lambda$. Then

$$\frac{d^2y}{dx^2} = \frac{\lambda}{H}\frac{ds}{dx} = \frac{\lambda}{H}\sqrt{1 + \left(\frac{dy}{dx}\right)^2}$$

From this $dp/\sqrt{1 + p^2} = (\lambda/H)dx$, so that

$$\sinh^{-1} p = \frac{\lambda}{H} x + c_1.$$

If the origin is chosen at the lowest point of the cable, $c_1 = 0$, and

$$\frac{dy}{dx} = \sinh\frac{\lambda}{H} x; \quad y = \frac{H}{\lambda}\cosh\frac{\lambda}{H} x + c = \frac{H}{\lambda}\left(\cosh\frac{\lambda}{H} x - 1\right)$$

This curve is known as a catenary.

c. *Equations not containing x.*

$$\frac{d^2y}{dx^2} = f\left(y, \frac{dy}{dx}\right)$$

Again we put $dy/dx = p$, but now we write

$$\frac{d^2y}{dx^2} = \frac{dp}{dx} = \frac{dp}{dy}\cdot\frac{dy}{dx} = p\frac{dp}{dy}$$

The resulting equation

$$p \frac{dp}{dy} = f(y,p)$$

is solved for p, then integrated once more.

All linear homogeneous equations of the second order with constant coefficients discussed in sec. 2.7 can be solved by this method, but the treatment of sec. 2.7 is usually simpler.

Example. *Anharmonic oscillator.*

Differential equation:

$$\frac{d^2 y}{dt^2} + \omega^2 y + \lambda y^2 = 0$$

Solution:

$$pdp = -(\omega^2 y + \lambda y^2)dy, \quad p = \frac{dy}{dt} = (c_1 - \omega^2 y^2 - \tfrac{2}{3}\lambda y^3)^{1/2}$$

The integration of this equation leads to an elliptic function.[6]

Problem. Solve the equation for the anharmonic oscillator by successive approximation, assuming that $\lambda y \ll \omega^2$.

Ans.

$$y = a \cos (\omega t + \epsilon) - \frac{\lambda a^2}{2\omega^2} [1 - \tfrac{1}{3} \cos 2(\omega t + \epsilon)]$$

INTEGRATION IN SERIES

A type of differential equation occurring very commonly in physics has the form

$$y'' + X_1 y' + X_2 y = 0 \tag{2-27}$$

where X_1 and X_2 are functions of x, the independent variable. Here and in the following, primes denote differentiations with respect to x. The methods developed in the preceding sections of this chapter are suitable for solving (27) when X_1 and X_2 have special forms, but are far from yielding solutions of that equation in general. In fact, such solutions are frequently not available in closed or finite form. For certain regions of x, however, they may be found in the form of convergent series by a procedure to be studied presently.

---

[6] See Peirce, B. O., " Short Table of Integrals," Third Revised Edition, Ginn and Co., New York, 1929. Introductory treatments of elliptic integrals may be found in " Higher Mathematics," by R. S. Burington and C. C. Torrance, McGraw-Hill Book Co., New York, 1939, " Higher Mathematics for Engineers and Physicists," by I. S. and E. S. Sokolnikoff, McGraw-Hill Book Co., New York, Second Edition, 1941.

2.10. Qualitative Considerations Regarding Eq. 27.—Before turning to the consideration of exact solutions of (27), a few remarks concerning their qualitative behavior in limited domains of x may be of value. To survey their behavior, it is often advisable to remove the first derivative occurring in (27), which is always possible by means of a simple transformation of the dependent variable. Instead of y, we introduce v, related to y by

$$y = ve^{-\frac{1}{2}\int X_1 dx}$$

When this is substituted into (27) and the exponential factor is then cancelled, there results an equation for v:

$$v'' + (X_2 - \tfrac{1}{2}X_1' - \tfrac{1}{4}X_1^2)v = 0$$

from which the first derivative is absent. This represents essentially a relation between v and the curvature of v and may be written

$$v'' = f(x)v$$

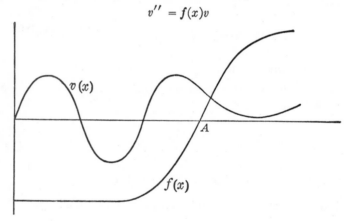

Fig. 2-5

One fact is at once apparent: provided v is finite, it has a point of inflexion wherever $f(x) = 0$. Furthermore, in regions where $f(x) > 0$ two facts are to be noted: If v is positive and has a positive slope, the slope will continually increase as x increases, causing v to grow rapidly; if v is positive and has a negative slope, the positive v'' will continually diminish its steepness, causing v to approach the x-axis and then in general to turn upwards again. For negative v the words " positive " and " negative " in the preceding sentence should be interchanged. This qualitative behavior is most easily remembered if we think of the special case in which $f(x) = $ const. $= \omega^2 > 0$. The solution is then

$$v = c_1 e^{\omega x} + c_2 e^{-\omega x}$$

which typifies the foregoing remarks.

If, however, we consider a region in which $f(x) < 0$, the slope of positive v will be continually diminished. Thus if v starts out with positive slope this will soon be zero and then decrease until $v = 0$; as v then becomes negative its negative slope will increase until it is horizontal and v turns back toward zero. In short, v is oscillatory. This again is easily remembered if we consider the special case in which $f(x) = -\omega^2 < 0$ for it has the solution $v = c \sin (\omega x + \delta)$.

Fig. 5 illustrates these facts. To the left of A, v oscillates; at A it has a point of inflexion; to the right of A it is of exponential behavior.

2.11. Example of Integration in Series. Legendre's Equation.— To illustrate the method of series integration, let us postpone fundamental matters and start by studying a specific example. An equation of considerable interest is Legendre's; it has the form

$$(1 - x^2)y'' - 2xy' + l(l + 1)y = 0 \tag{2-28}$$

in which l is a constant. We attempt to find a solution which is a series in positive powers of x. If the lowest power occurring is κ, this solution will have the general form

$$y = \sum_{\lambda=0}^{\infty} a_\lambda x^{\kappa+\lambda} \tag{2-29}$$

Solving the differential equation then amounts to determining the coefficients a_λ. Whether the series converges can be tested after this has been achieved. At present it will be assumed that this is the case, and that (29) may be differentiated term by term. When (29) is substituted in (28) the result is

$$\sum_\lambda a_\lambda(\kappa + \lambda)(\kappa + \lambda - 1)x^{\kappa+\lambda-2} - \sum_\lambda a_\lambda[(\kappa + \lambda)(\kappa + \lambda - 1)$$
$$+ 2(\kappa + \lambda) - l(l + 1)]x^{\kappa+\lambda} = 0 \tag{2-30}$$

This equation must hold for every value of x, and this can be true only if the coefficient of *every* power of x is identically zero. Since λ cannot, by hypothesis, be negative, the lowest power of x occurring in (30) is $x^{\kappa-2}$, and it is present only in the first summation of (30). Thus we find, putting $\lambda = 0$ to obtain the term in question,

$$a_0\kappa(\kappa - 1) = 0 \tag{2-31}$$

a_0 is the lowest coefficient in our summation and hence not zero. Equation (31) therefore determines κ. It is often called the *indicial equation*. Clearly, two values of κ are permissible:

$$\kappa = 0, 1$$

Next, we see what further information eq. (30) will give. According to the foregoing, the coefficient of $x^{\kappa+j}$ must vanish for every positive integer j. Now the term corresponding to the $(\kappa + j)$-th power of x is obtained in the first summation by putting $\lambda = j + 2$, in the second by putting $\lambda = j$. Hence

$$a_{j+2}(\kappa + j + 2)(\kappa + j + 1) = a_j[(\kappa + j)(\kappa + j + 1) - l(l + 1)]$$

or

$$a_{j+2} = \frac{(\kappa + j)(\kappa + j + 1) - l(l + 1)}{(\kappa + j + 1)(\kappa + j + 2)} a_j \qquad (2\text{–}32)$$

Thus, if a_j is given, a_{j+2} can be computed from this relation. Starting with a_0, (32) permits us to obtain, successively, a_2, a_4, etc.; a_0, however, is arbitrary; it is one of the two arbitrary constants appearing in the general solution of a second order differential equation. On the other hand, if a_1 is assigned arbitrarily, all coefficients with odd subscripts are deducible from (32).

Choice 1. Let us take $\kappa = 0$. Eq. (32) then reads

$$a_{j+2} = \frac{j(j + 1) - l(l + 1)}{(j + 1)(j + 2)} a_j \qquad (2\text{–}33)$$

On taking a_0 and a_1 as arbitrary constants, the solution becomes

$$y = \left(1 - \frac{l(l + 1)}{2} x^2 - \frac{6 - l(l + 1)}{12} \cdot \frac{l(l + 1)}{2} x^4 + \cdots \right) a_0$$

$$+ \left(x + \frac{2 - l(l + 1)}{6} x^3 + \frac{2 - l(l + 1)}{6} \cdot \frac{12 - l(l + 1)}{20} x^5 + \cdots \right) a_1$$

$$= \left(1 - \frac{l(l + 1)}{2!} x^2 + \frac{l(l - 2)(l + 1)(l + 3)}{4!} x^4 + \cdots \right.$$

$$+ (-1)^r \frac{l(l - 2) \cdots (l - 2r + 2)(l + 1) \cdots (l + 2r - 1)}{(2r)!} x^{2r}$$

$$\left. + \cdots \right) a_0$$

$$+ \left(x - \frac{(l - 1)(l + 2)}{3!} x^3 + \frac{(l - 1)(l - 3)(l + 2)(l + 4)}{5!} x^5 + \cdots \right.$$

$$+ (-1)^r \frac{(l - 1)(l - 3) \cdots (l - 2r + 1)(l + 2) \cdots (l + 2r)}{(2r + 1)!} x^{2r+1}$$

$$\left. + \cdots \right) a_1 \qquad (2\text{–}34')$$

Choice 2. Let us take $\kappa = 1$. Eq. (32) then reads

$$a_{j+2} = \frac{(j + 1)(j + 2) - l(l + 1)}{(j + 2)(j + 3)} a_j$$

If now we take again a_0 and a_1 as arbitrary constants, we find

$$y = x\left(1 + \frac{2 - l(l+1)}{6}\, x^2 + \frac{2 - l(l+1)}{6} \cdot \frac{12 - l(l+1)}{20}\, x^4 + \cdots\right) a_0$$

$$+ x\left(x + \frac{6 - l(l+1)}{12}\, x^3 + \frac{6 - l(l+1)}{12} \cdot \frac{20 - l(l+1)}{30}\, x^5 + \cdots\right) a_1$$

$$= x\left(1 - \frac{(l-1)(l+2)}{3!}\, x^2 + \frac{(l-1)(l-3)(l+2)(l+4)}{5!}\, x^4 + \cdots\right) a_0$$

$$+ x\left(x - \frac{(l-2)(l+3)}{12}\, x^3 + \frac{(l-2)(l-4)(l+3)(l+5)}{360}\, x^5 + \cdots\right) a_1$$

$$(2\text{--}35')$$

The terms multiplying a_0 in $(35')$ are seen to be identical with those multiplying a_1 in $(34')$; hence these two particular solutions are the same. The second part of $(35')$, however, does not agree with the first of $(34')$, both of which represent series in even powers of x. It might seem, therefore, as if we had obtained altogether *three* independent solutions, which is, of course, impossible. But closer inspection would show that the second part of $(35')$ is not a solution at all. This is seen at once if, after assuming any specific value for l, we substitute it back into the differential equation. The trouble is that, putting $\kappa = 1$ and $a_0 = 0$, we have carelessly discarded any constant term which might appear in the sequence. The present example indicates clearly that the solution of a differential equation is not an altogether mechanical matter and that caution must be used at every step. Summarizing, we observe that the significant parts of $(34')$ and $(35')$ are:

$$y = \left[1 - \frac{l(l+1)}{2!}\, x^2 + \frac{l(l-2)(l+1)(l+3)}{4!}\, x^4 + \cdots + (-1)^r \right.$$

$$\left. \frac{l(l-2) \cdots (l-2r+2)(l+1) \cdots (l+2r-1)}{(2r)!}\, x^{2r} \right.$$

$$\left. + \cdots \right] a_0 \qquad\qquad\qquad (2\text{--}34)$$

$$y = \left[x - \frac{(l-1)(l+2)}{3!}\, x^3 + \frac{(l-1)(l-3)(l+2)(l+4)}{5!}\, x^5 + \cdots \right.$$

$$\left. + (-1)^r \frac{(l-1)(l-3) \cdots (l-2r+1)(l+2) \cdots (l+2r)}{(2r+1)!}\, x^{2r+1} \right.$$

$$\left. + \cdots \right] a_1 \qquad\qquad\qquad (2\text{--}35)$$

Problem. Show that the equation $y'' + y = 0$, if integrated in series, has two particular solutions. one of which may be identified with the cosine series, the other with the sine series.

One further point should be observed. When any one term in (34) is zero, all succeeding terms vanish also and the series becomes a polynomial. The conditions under which infinite series like (34) reduce to polynomials are of great importance in many physical problems and will be discussed more fully later.

The work thus far has only established the fact that the series (34) and (35) are formal solutions of Legendre's equation, that is, they would satisfy (28) if substituted in it. Whether the solutions are of any interest depends on their convergence properties. A series converges if the ratio of the absolute values of two successive terms,

$$\frac{|u_{j+2}|}{|u_j|}$$

is smaller than unity for large j. Now this ratio is clearly

$$\frac{|a_{j+2}|}{|a_j|} x^2$$

But

$$\frac{|a_{j+2}|}{|a_j|}$$

is immediately obtainable from (33). As $j \to \infty$ it becomes 1. Hence the condition that (34) and (35) converge is that $x^2 < 1$, and this is true as long as $|x| < 1$. For values of x in the range $-1 < x < 1$ our solution is a significant one; for other values it fails. Is it possible to construct a solution valid for $|x^2| > 1$? This is indeed not difficult.

Let us suppose that y, instead of being given by (29), has the form $y = \sum_{\lambda} a_\lambda x^{\kappa - \lambda}$. Eq. (30) will then read

$$\sum_{\lambda} a_\lambda (\kappa - \lambda)(\kappa - \lambda - 1) x^{\kappa - \lambda - 2}$$

$$- \sum_{\lambda} a_\lambda [(\kappa - \lambda)(\kappa - \lambda + 1) - l(l + 1)] x^{\kappa - \lambda} = 0$$

κ now denotes the *highest* power occurring in the series. The indicial equation is obtained by putting the coefficient of the highest power of x equal to zero. Thus

$$\kappa(\kappa + 1) - l(l + 1) = 0$$

whence

$$\kappa = l \quad \text{or} \quad -l - 1$$

As before, the coefficient of $x^{\kappa - j}$ must vanish for every positive integer j This implies

$$a_{j-2}(\kappa - j + 2)(\kappa - j + 1) = a_j[(\kappa - j)(\kappa - j + 1) - l(l + 1)]$$

or, on replacing j by $j + 2$,

$$a_{j+2} = \frac{(\kappa - j)(\kappa - j - 1)}{(\kappa - j - 2)(\kappa - j - 1) - l(l + 1)} a_j \qquad (2\text{-}36)$$

Choice 1.   Let us take $\kappa = l$.   Eq. (36) then reads

$$a_{j+2} = \frac{(l - j)(l - j - 1)}{(j + 2)(j - 2l + 1)} a_j$$

If a_0 is chosen arbitrarily, the series becomes

$$y = x^l \left(1 - \frac{l(l - 1)}{2(2l - 1)} x^{-2} + \frac{l(l - 1)(l - 2)(l - 3)}{8(2l - 1)(2l - 3)} x^{-4} - \cdots \right.$$

$$\left. (-1)^r \frac{(l - 2r + 1)(l - 2r + 2) \cdots (l - 1)l}{2r \cdots 2(2l - 2r + 1) \cdots (2l - 1)} x^{-2r} + \cdots \right) a_0 \qquad (2\text{-}37)$$

The series formally obtained by putting $a_0 = 0$ is of no interest since it violates the assumption, previously made, that κ, i.e., l, represents the highest power of the sequence.   We shall therefore omit it at once.

Choice 2.   Let us take $\kappa = -l - 1$.   Then

$$a_{j+2} = \frac{(j + l + 1)(j + l + 2)}{(j + 2)(2l + j + 3)} a_j$$

If again we put $a_1 = 0$, there results the particular solution

$$y = x^{-l-1} \left(1 + \frac{(l + 1)(l + 2)}{2(2l + 3)} x^{-2} + \frac{(l + 1)(l + 2)(l + 3)(l + 4)}{2 \cdot 4 \,(2l + 3)(2l + 5)} x^{-4} + \cdots \right.$$

$$\left. + \frac{(l + 1) \cdots (l + 2r)}{2 \cdot 4 \cdots 2r(2l + 3) \cdots (2l + 2r + 1)} x^{-2r} + \cdots \right) a_0 \qquad (2\text{-}38)$$

The two solutions (37) and (38) are independent, hence their sum represents the general solution of Legendre's equation.   It is easily seen to converge if $|x| > 1$, *unless* l has such a value that the denominator of one of the coefficients in the series vanishes.   This case will be studied shortly.

We are now in possession of two forms of solution of eq. (28).   The first (eqs. 34 and 35) converges when $|x| < 1$, the second (eqs. 37 and 38) when $|x| > 1$.   Under special circumstances, however, (34) or (35) as well as (37) or (38) may become polynomials, which remain finite for every finite value of x.   It is interesting to see what happens to the various particular solutions when this contingency arises.

Eq. (34) reduces to a polynomial when l is an even positive or an odd negative integer (or zero).

a.   Let l be even and positive; $l = 2k$.   (34) then becomes

$$y = a \left(1 - \frac{l(l + 1)}{2!} x^2 + \cdots (-1)^{l/2} \frac{l\,(l - 2) \cdots 2(l + 1) \cdots (2l - 1)}{l!} x^l \right)$$

On the other hand, (37) becomes under these conditions

$$y = ax^l \left(1 - \frac{l(l-1)}{2(2l-1)} x^{-2} + \cdots \right.$$

$$\left. (-1)^{l/2} \frac{l!}{l(l-2)\cdots 2(l+1)\cdots(2l-1)} x^{-l} \right)$$

These two solutions become identical if the second is multiplied by the constant factor

$$(-1)^{l/2} \frac{l(l-2)\cdots 2(l+1)\cdots(2l-1)}{l!}$$

Hence the particular solution (34) coalesces with (37).

b. Let l be odd and negative. Inspection shows that (34) now becomes identical with (38).

Eq. (35) reduces to a polynomial when l is an odd positive or an even negative integer.

c. If l is odd and positive, (35) reads

$$y = a \left(x - \frac{(l-1)(l+2)}{3!} x^3 + \cdots \right.$$

$$\left. + (-1)^{(l-1)/2} \frac{(l-1)(l-3)\cdots 2(l+2)\cdots(2l-1)}{l!} x^l \right),$$

while (37) becomes

$$y = ax^l \left(1 - \frac{l(l-1)}{2(2l-1)} x^{-2} + \cdots \right.$$

$$\left. + (-1)^{(l-1)/2} \frac{l!}{2 \cdot 4 \cdots (l-1)(l+2)\cdots(2l-1)} x^{-l+1} \right)$$

These two expressions become identical when the second is multiplied by the coefficient of its last term in parenthesis.

d. If l is an even and negative integer (35) turns into (38).

Having established these important relations between solutions (34)–(38) we now return to the consideration of (37) and (38). Solutions (37) and (38) for integral values of l are of great importance in mathematical physics. If the constant a_0 in (37) is chosen to be

$$\frac{(2l)!}{2^l(l!)^2} \equiv \frac{(2l-1)(2l-3)\cdots 1}{l!}$$

the resulting polynomial of degree l is called a *Legendre polynomial* (or Legendre coefficient or " zonal harmonic "). It is usually denoted by P_l.

For purposes of reference we write it down again:

$$P_l(x) = \frac{1 \cdot 3 \cdot 5 \cdots (2l - 1)}{l!}$$

$$\left\{ x^l - \frac{l(l-1)}{2(2l-1)} x^{l-2} + \frac{l(l-1)(l-2)(l-3)}{2 \cdot 4(2l-1)(2l-3)} x^{l-4} - \cdots \right\} \quad (2\text{-}39)$$

The series here is to be continued down to the constant term. On the other hand, (38) with the constant a_0 chosen to be $2^l(l!)^2/(2l+1)!$, l being a positive integer, is often denoted by Q_l. It is an infinite series:

$$Q_l = \frac{l!}{1 \cdot 3 \cdots (2l+1)} \left\{ x^{-l-1} + \frac{(l+1)(l+2)}{2(2l+3)} x^{-l-3} + \cdots \right.$$

$$\left. + \frac{(l+1) \cdots (l+2r)}{2 \cdot 4 \cdots 2r(2l+3) \cdots (2l+2r+1)} x^{-l-2r-1} + \cdots \right\} \quad (2\text{-}40)$$

The following facts will be noted:

When l is a positive integer, (37) is a polynomial, but (38) is an infinite series. The general solution of (28) is a linear combination of (37) and (38).

When l is a negative integer, (37) is an infinite series, and (38) is a polynomial. The general solution of (28) is a linear combination of (37) and (38).

When $2l$ is equal to some *positive odd* integer, solution (37) degenerates into (38). To see this, suppose $2l = 2n - 1$. There will then appear a vanishing denominator in the coefficient of x^{l-2n} and in every subsequent term of (37). To remove these infinities one may multiply the entire series by $(n - r)$, which causes all terms of order higher than $l - 2n$ to vanish while the others remain finite. Hence the series begins with the power $x^{l-2n} = x^{-l-1}$, and inspection shows it then to be identical with (38). In this case, our method has yielded but *one* particular solution, and this is an infinite series. Procedures leading to a general solution are discussed in treatises on Differential Equations.[7]

When $2l$ is equal to an *odd negative* integer, (38) degenerates into (37) in a manner similar to the above. In that case also no general solution can be obtained by the present method.

Having now given a fairly complete mathematical analysis of the solutions of Legendre's equation, we state some conclusions of practical importance. In almost all applications (cf. Chapters 7, 8, 11) the independent variable x appearing in eq. (28) is the cosine of an angle. *The functions of interest are therefore those which remain finite for all values which $x = \cos \theta$ can assume;* these values include $x = \pm 1$. Such functions exist only when

---

[7] See Forsyth, A. R., "Differential Equations," Macmillan Co., London, 1914.

ORDINARY DIFFERENTIAL EQUATIONS **68**

l is a positive or a negative integer, as we have shown. But when l is an integer, consideration may be limited to solutions (37) and (38), because the others reduce to these. Moreover, inspection shows that solution (38) with l replaced by $-(l+1)$ is the same as solution (37). Hence we may further limit our consideration to positive values of l (including 0) and retain only (37) as a significant solution. Finally we note that (37) is identical with (39). Hence:

In physio-chemical problems, where $x = \cos\theta$, the only solution of Legendre's equation which is of practical interest is $P_l(\cos\theta)$.

Problems.

a. Prove that, when l is an even negative integer, the expressions (35) and (38) become identical.

b. Prove that, when $2l$ is an odd negative integer, expressions (37) and (38) become identical.

Differential Equation for Associated Legendre Functions, or Associated Spherical Harmonics.

An equation similar to Legendre's plays a considerable rôle in mathematical physics. It is[8]

$$(1 - x^2)y'' - 2xy' + \left[l(l+1) - \frac{m^2}{1-x^2}\right]y = 0 \qquad (2\text{--}41)$$

where l and m are both integers, and has a particular solution:

$$y = (1 - x^2)^{m/2}\frac{d^m}{dx^m}P_l(x) \qquad (2\text{--}42)$$

The other particular solution is related to Q_n and is of lesser interest in applications. To construct (42) by the method of series integration is perfectly feasible, but we shall here use a simpler method based on the foregoing results. If $P_l(x)$ is a solution of

$$(1 - x^2)y'' - 2xy' + l(l+1)y = 0$$

then

$$\frac{d^m}{dx^m}P_l(x)$$

---

[8] The equation occurs more commonly in the equivalent forms

$$\frac{d^2y}{d\theta^2} + \cot\theta\frac{dy}{d\theta} + \left[l(l+1) - \frac{m^2}{\sin^2\theta}\right]y = 0$$

or

$$\frac{1}{\sin\theta}\left[\frac{d}{d\theta}\left(\sin\theta\frac{dy}{d\theta}\right)\right] + \left[l(l+1) - \frac{m^2}{\sin^2\theta}\right]y = 0$$

which reduce to (41) on substitution of $\cos\theta = x$.

for which we shall write $P_l^{(m)}(x)$, satisfies the equation

$$(1 - x^2)P_l^{(m)''} - 2(m + 1)xP_l^{(m)'} + [l(l + 1) - m(m + 1)]P_l^{(m)} = 0$$

$$(2\text{-}43)$$

as is seen when Legendre's equation is differentiated m times. Now let

$$P_l^{(m)}(x) = (1 - x^2)^r y \qquad (2\text{-}44)$$

and determine, by substituting this into (43), what differential equation y will satisfy. After substitution, (43) will read

$$(1 - x^2)^{r-1}\{(4r^2x^2 - 2r - 2rx^2)y - 4r(1 - x^2)xy' + (1 - x^2)^2 y''$$
$$- 2(m + 1)(1 - x^2)xy' + 4r(m + 1)x^2 y + [l(l + 1)$$
$$- m(m + 1)](1 - x^2)y\} = 0$$

If here the special value $r = -m/2$ is chosen, this equation reduces to (41). We have shown, therefore, that (44) is true with $r = -m/2$ and hence that

$$y = (1 - x^2)^{m/2}P_l^{(m)}(x)$$

as was asserted. The function $P_l^{(m)}$, which is a polynomial of degree $l - m$ and which satisfies eq. (43), is sometimes referred to by physicists as *Helmholtz' function*. The function (42) is known as an associated Legendre function, or more frequently, an *associated spherical harmonic*.

2.12. General Considerations Regarding Series Integration. Fuchs' Theorem.—Before continuing, the reader will wish to know the limits of applicability of the method applied in sec. 2.11, and in particular what properties of the solution one may read directly from the differential equation. First, then, let us ask the question: Will the method described in sec. 2.11 always work? In preparation for the answer, we consider the differential equation

$$y'' + y/x^3 = 0$$

On putting $y = \sum a_\lambda x^{\kappa+\lambda}$ it is seen that

$$\sum_\lambda a_\lambda(\kappa + \lambda)(\kappa + \lambda - 1)x^{\kappa+\lambda-2} = -\sum_\lambda a_\lambda x^{\kappa+\lambda-3}$$

The indicial equation, obtained by putting the coefficient of the lowest power of x equal to zero, simply reads

$$a_0 = 0$$

and does not determine κ. Furthermore,

$$a_{j+1} = -(\kappa + j)(\kappa + j - 1)a_j$$

so that $a_1 = -a_0(\kappa - 1)\kappa$. Since $a_0 = 0$, this means that either $\kappa = \infty$ or a_1 is also zero. In neither case do we get any solution at all.

Equally instructive is the equation

$$y'' + \frac{y'}{x^2} = 0$$

Its indicial equation yields $\kappa = 0$. The recurrence relation between coefficients is

$$a_{j+1} = -\frac{j(j-1)}{j+1}\,a_j$$

Thus we have apparently determined a solution. But let us apply a convergence test. Denoting again the terms of the series by u_r one sees that

$$\lim_{n \to \infty} \frac{|u_{n+1}|}{|u_n|} = \lim_{n \to \infty} \frac{|a_{n+1}|\,|x^{n+1}|}{|a_n|\,|x^n|} = \lim_{n \to \infty} \frac{(n-1)n}{n+1} \cdot x = nx$$

This is greater than 1 as $n \to \infty$ for every finite value of x, so that there is no range of x at all in which the series converges. Again, the method fails.

To enlarge our outlook, let us now return to the general form of the equation we wish to solve, that is, to eq. (27). As a rule there will be values of x for which one or both of the functions X_1 and X_2 become infinite. If $x = x_0$ is such a value, then x_0 is said to be a *singular point* of the equation. It is at such singular points that the method of integration in series may break down. To be more specific, a solution of the form $y = \sum_{\lambda} a_\lambda (x - x_0)^{\kappa+\lambda}$ may not exist at singular points x_0.

In dealing with Legendre's equation, a power series development was attempted about the point $x_0 = 0$. It succeeded because, after writing the equation in the form (27), neither $X_1 = -2x/(1 - x^2)$ nor $X_2 = l(l + 1)/(1 - x^2)$ becomes infinite at $x = 0$. But the points $x = \pm 1$ are singular points of the equation, and it is for this reason that the general solution obtained breaks down at these two points. Again, the two equations just considered, $y'' + x^{-3}y = 0$ and $y'' + x^{-2}y' = 0$ possess a singular point at $x = 0$, and this is the cause of the failure of the present method.

But while the method *often* fails if the differential equation has a singular point at the place where the power series development is attempted, it does not always do so. For instance, the equation

$$y'' + x^{-1}y' - x^{-2}y = 0$$

may be developed in the form $y = \sum_{\lambda} a_\lambda x^{\kappa+\lambda}$ despite its singularities at $x = 0$. The indicial equation yields $\kappa = \pm 1$. When the positive sign is chosen, the coefficients must satisfy the equation

$$[(j + 1)^2 - 1]a_j = 0$$

which is no longer a recurrence relation but serves to determine the coefficients just as well.   For it says that every $a_j = 0$, except for $j = 0$.
The corresponding solution is $y = a_0 x$.   For $\kappa = -1$ we have

$$[(j - 1)^2 - 1]a_j = 0$$

and this indicates that all coefficients must be zero except that corresponding to $j = 0$ and to $j = 2$.   Hence the solution is

$$y = x^{-1}(a_0 + a_2 x^2)$$

The constants a_0 and a_2 are arbitrary, which implies that the solution is a general one, including $y = $ const. x as a special case.   Obviously, then, it is important to settle what kind of singularities do, and what kind do not, permit an integration in series about the singular point.

This issue is settled by an important theorem due to Fuchs, which states the following:

If the differential equation

$$y'' + X_1 y' + X_2 y = 0$$

possesses a singular point at $x = x_0$, then a convergent development of the solution in a power series about the point $x = x_0$ having only a finite number of terms with negative exponents is nevertheless possible provided that $(x - x_0)X_1(x_0)$ and $(x - x_0)^2 X_2(x_0)$ remain finite.

This clearly is true for the equation

$$y'' + x^{-1}y' - x^{-2}y = 0$$

at $x_0 = 0$, but not for

$$y'' + x^{-2}y' = 0$$

Thus the results just obtained are accounted for.   The proof of Fuchs' theorem is a matter of some length and will not be undertaken here.[9]
In conformity with the theorem singularities in X_1 and X_2 occurring at $x = x_i$ which are removable by multiplication by the factors $(x - x_i)$ and $(x - x_i)^2$ respectively are called *non-essential* singularities of the differential equation; all others are essential ones.[10]   All regular and non-essentially singular points are sometimes referred to as regular points of the differential equations (German: " Stellen der Bestimmtheit ").   An equation which has no essential singularities in the entire infinite complex plane is said to belong to the Fuchsian class of differential equations.

[9] See, for instance, Schmidt, H., " Theorie der Wellengleichung," Leipzig, 1931.
[10] Whether the point at infinity is an essentially singular one cannot at once be seen in this way.   To examine it the transformation $\xi = 1/x$ must be made.   One may then show that the point at infinity is *essentially singular* if $X_1 x$ or $X_2 x^2$ become infinite there; it is *non-essentially singular* if $2x - X_1 x^2 \to \infty$ or $X_2 x^4 \to \infty$; otherwise it is regular.

A final remark on the nature of the solutions obtained by the method of integration in series is in order. Even if the point at which the development is made satisfies the Fuchs conditions it may not be possible to obtain *two independent* solutions which, when combined linearly with the use of two arbitrary constants, will yield the general solution. If this process is to produce a general solution, further conditions must be met. Since general solutions are not often required in physical and chemical applications, this matter will not be considered in detail here.[11] We note, however, that two independent solutions in the form $y_1 = \sum a_\lambda (x - x_0)^{\kappa_1 + \lambda}$ and $y_2 = \sum a_\lambda (x - x_0)^{\kappa_2 + \lambda}$ can always be obtained when the two roots of the indicial equation, κ_1 and κ_2, do not differ by an integer or by zero.

SPECIAL EQUATIONS SOLVABLE BY SERIES INTEGRATION

2.13. Gauss' (Hypergeometric) Differential Equation.—

$$(x^2 - x)y'' + [(1 + \alpha + \beta)x - \gamma]y' + \alpha\beta y = 0 \qquad (2\text{-}45)$$

The parameters α, β, γ are constants, and it will be assumed that γ is not an integer. Eq. (45) has singularities at 0, 1, and ∞, but they are all nonessential. On development about $x = 0$, the indicial equation reads

$$\kappa(\kappa - 1) + \kappa\gamma = 0$$

hence $\kappa = 0, 1 - \gamma$. Choosing $\kappa = 0$, we obtain the recurrence formula

$$a_{j+1} = \frac{(\alpha + j)(\beta + j)}{(j + 1)(j + \gamma)} a_j \qquad (2\text{-}46)$$

and hence the particular solution

$$y = a \left\{ 1 + \frac{\alpha\beta}{1 \cdot \gamma} x + \frac{\alpha(\alpha + 1)\,\beta(\beta + 1)}{1 \cdot 2 \cdot \gamma(\gamma + 1)} x^2 + \cdots \right.$$
$$\left. + \frac{\alpha(\alpha + 1) \cdots (\alpha + r - 1) \cdot \beta(\beta + 1) \cdots (\beta + r - 1)}{r!\, \gamma(\gamma + 1) \cdots (\gamma + r - 1)} x^r + \cdots \right\} \qquad (2\text{-}47)$$

The series in $\{\}$ is known as the *hypergeometric series*. It converges if $|x| < 1$. For $\alpha = 1$, $\beta = \gamma$ it reduces to the ordinary geometric series; hence its name. It is customary to denote the hypergeometric series by $F(\alpha, \beta, \gamma; x)$. With this abbreviation, then, this particular solution is

$$y = aF(\alpha, \beta, \gamma; x)$$

Next, we take $\kappa = 1 - \gamma$. The recurrence relation reads

$$a_{j+1} = \frac{(\alpha - \gamma + j + 1)(\beta - \gamma + j + 1)}{(j + 1)(j + 2 - \gamma)} a_j \qquad (2\text{-}48)$$

---

[11] For particulars, see Bôcher, M., "Regular Points of Linear Differential Equations of the Second Order," Harvard University Press.

When the new constants: $\alpha' = \alpha - \gamma + 1$, $\beta' = \beta - \gamma + 1$, $\gamma' = 2 - \gamma$, are introduced in (48) it becomes

$$a_{j+1} = \frac{(\alpha' + j)(\beta' + j)}{(j + 1)(j + \gamma')} a_j$$

that is, it takes the same form as (46). The particular solution corresponding to (48) may therefore be written

$$ax^{1-\gamma}F(\alpha - \gamma + 1, \beta - \gamma + 1, 2 - \gamma; x)$$

We have thus arrived at the following general solution of (45):

$$y = AF(\alpha,\beta,\gamma;x) + Bx^{1-\gamma}F(\alpha - \gamma + 1, \beta - \gamma + 1, 2 - \gamma; x) \quad (2\text{-}49)$$

whose range of convergence is $|x| < 1$.

There is an interesting and sometimes useful relation between the solutions of Gauss' and those of Legendre's equation. Let us introduce in (45) the new independent variable ξ, given by

$$x = \tfrac{1}{2}(1 - \xi)$$

so that it takes the form

$$(1 - \xi^2)\frac{d^2y}{d\xi^2} + [1 + \alpha + \beta - 2\gamma - (\alpha + \beta + 1)\xi]\frac{dy}{d\xi} - \alpha\beta y = 0 \quad (2\text{-}50)$$

This reduces to Legendre's equation (28) if we specify the constants to be

$$\alpha = l + 1, \quad \beta = -l, \quad \gamma = 1$$

One particular solution of Legendre's equation is therefore

$$y = aF\left(l + 1, -l, 1; \frac{1 - \xi}{2}\right)$$

From the fact that this solution, expanded in powers of ξ, starts with a constant term it is clear that it must be identical (aside from a constant factor) with (34). In particular, if l is a positive integer, it must be P_l. This happens to be true, as the reader may verify, even with respect to the constant factor if P_l is defined as in (39). Thus

$$P_l(\xi) = F\left(l + 1, -l, 1; \frac{1 - \xi}{2}\right) \quad (2\text{-}51)$$

An equation known to mathematicians as *Tschebyscheff's* results when in (50) we specialize the constants as follows:

$$\alpha = -\beta = n, \quad \text{an integer}; \quad \gamma = \tfrac{1}{2}$$

The equation then reads:

$$(1 - \xi^2)\frac{d^2y}{d\xi^2} - \xi\frac{dy}{d\xi} + n^2y = 0 \quad (2\text{-}52)$$

Its solution is clearly

$$y(\xi) = AF\left(n, -n, \tfrac{1}{2}; \frac{1-\xi}{2}\right) + B\left(\frac{1-\xi}{2}\right)^{1/2} F\left(n+\tfrac{1}{2}, -n+\tfrac{1}{2}, \tfrac{3}{2}; \frac{1-\xi}{2}\right)$$

(2–53)

The first particular solution here written is a polynomial known as the *Tschebyscheff polynomial*, of degree n. If multiplied by the proper factor it has the alternative form:

$$T_n(x) = 2^{n-1}\left(x^n - \frac{n}{1! \, 2^2} x^{n-2} + \frac{n(n-3)}{2! \, 2^4} x^{n-4} - \right.$$

$$\left. \frac{n(n-4)(n-5)}{3! \, 2^6} x^{n-6} + \cdots\right)$$

(2–54)

This development stops with a constant or a term proportional to x.

The function $F(\alpha,\beta,\gamma;x)$ reduces to a polynomial when $\alpha = -n$, n being a positive integer, as may be seen from its definition (47). The resulting polynomial, which is of degree n, is known as a *Jacobi polynomial*, defined as follows:

$$J_n(p,q;x) \equiv F(-n, p + n, q; x)$$

(2–55)

It satisfies the differential equation

$$(x^2 - x)y'' + [(1 + p)x - q]y' - n(p + n)y = 0$$

(2–56)

in which q must satisfy $q > 0$. Substitution of $\alpha = -n$, $\beta = p + n$, $\gamma = q$ into (47) shows that[12]

$$J_n(p,q;x) = 1 +$$

$$\sum_{\lambda=1}^{n} (-1)^\lambda \binom{n}{\lambda} \frac{(p + n)(p + n + 1) \cdots (p + n + \lambda - 1)}{q(q + 1) \cdots (q + \lambda - 1)} x^\lambda$$

Problem. Find the solution of (45) about the point $x = 1$; i.e., find solutions of the form

$$y = \sum_\lambda a_\lambda (x - 1)^{\kappa+\lambda}$$

Ans.

$$y = AF(\alpha,\beta,\alpha+\beta-\gamma+1; 1-x) + B(1-x)^{\gamma-\alpha-\beta}F(\gamma-\beta, \gamma-\alpha, 1-\alpha-\beta+\gamma; 1-x)$$

2.14. Bessel's Equation.—

$$x^2y'' + xy' + (x^2 - n^2)y = 0$$

(2–57)

n is a constant. Since the equation is regular at $x = 0$, its solution may be developed as a power series about that point. The indicial equation

$$(\kappa^2 - n^2)a_0 = 0$$

---

[12] Cf. eq. 12–2 for the definition of $\binom{n}{\lambda}$.

has the two roots $\kappa = \pm n$. According to the remarks at the end of sec. 2.12 we can obtain two independent particular solutions if $2n$ is not an integer; if it is, the method may allow us to determine only one. Taking $\kappa = n$ one finds

$$y = a_0 x^n \left\{ 1 - \frac{x^2}{2(2n+2)} + \frac{x^4}{2\cdot 4(2n+2)(2n+4)} - \cdots \right.$$

$$\left. + (-1)^r \frac{x^{2r}}{2\cdot 4\cdots 2r(2n+2)(2n+4)\cdots(2n+2r)} + \cdots \right\} \quad (2\text{-}58)$$

For $\kappa = -n$

$$y = a_0 x^{-n} \left\{ 1 + \frac{x^2}{2(2n-2)} + \frac{x^4}{2\cdot 4(2n-2)(2n-4)} + \cdots \right.$$

$$\left. + \frac{x^{2r}}{2\cdot 4\cdots 2r(2n-2)(2n-4)\cdots(2n-2r)} + \cdots \right\} \quad (2\text{-}59)$$

When the constant a_0 in (58) is chosen to be[13] $1/[2^n \Gamma(n+1)]$, the resulting expression

$$y = J_n(x) \equiv \sum_{\lambda=0}^{\infty} \frac{(-1)^\lambda}{\Gamma(\lambda+1)\Gamma(\lambda+n+1)} \left(\frac{x}{2}\right)^{n+2\lambda} \quad (2\text{-}60)$$

is called a *Bessel function* of order n.

When (59) is multiplied by the same factor it becomes $J_{-n}(x)$. Hence the complete solution of Bessel's equation (when n is not an integer) is

$$y = A J_n(x) + B J_{-n}(x) \quad (2\text{-}61)$$

Inspection of (58) and (59) shows that no difficulty arises when n is half-integral, although the difference of the roots of the indicial equation is an integer. But if n is an integer, J_{-n} is no longer independent of J_n. For in that case the coefficient of x^n in (59) has a vanishing term in the denominator, and every subsequent coefficient likewise becomes infinite. Multiplication by the vanishing term makes every term preceding the n-th zero. The series then starts with x^n and is seen to be identical (except for a constant multiplier) with (58). For integral n, therefore, we have obtained only one solution, namely $J_n(x)$.[14] By choosing the constants A and B of

---

[13] The Gamma function appearing here is a generalization of the factorial $n!$ which is defined only for integers (and zero). If n is an integer, $\Gamma(n+1) = n!$. In general, $\Gamma(x) = \int_0^\infty e^{-t} t^{x-1} dt$; it is easily seen to reduce to $n!$ when $x = n$. Moreover, this integral defines the " smoothest " function which takes on the values $n!$ at the integers. Cf. sec. 3.2.

[14] The second particular solution for integral n is derived in Forsyth, " Differential Equations," Macmillan, p. 182.

(61) suitably, several particular solutions of Bessel's equation (such as Neumann's and Hankel's functions) having useful properties may be constructed. They will be discussed in sec. 3.9.

2.15. Hermite's Differential Equation.—

$$y'' - 2xy' + 2\alpha y = 0; \quad \alpha = \text{constant} \tag{2-62}$$

The roots of the indicial equation are $\kappa = 0, 1$; the recurrence relations between the coefficients

$$a_{j+2} = \frac{2(\kappa + j) - 2\alpha}{(\kappa + j + 2)(\kappa + j + 1)} a_j$$

For $\kappa = 0$ we find the solution

$$y = a_0 \left(1 - \frac{2\alpha}{2!} x^2 + \frac{2^2 \alpha(\alpha - 2)}{4!} x^4 - \frac{2^3 \alpha(\alpha - 2)(\alpha - 4)}{6!} x^6 + \cdots \right.$$
$$\left. + (-2)^r \frac{\alpha(\alpha - 2) \cdots (\alpha - 2r + 2)}{(2r)!} x^{2r} + \cdots \right) \tag{2-63}$$

while for $\kappa = 1$

$$y = a_0 x \left(1 - \frac{2(\alpha - 1)}{3!} x^2 + \frac{2^2(\alpha - 1)(\alpha - 3)}{5!} x^4 - \cdots \right.$$
$$\left. + (-2)^r \frac{(\alpha - 1)(\alpha - 3) \cdots (\alpha - 2r + 1)}{(2r + 1)!} x^{2r} + \cdots \right) \tag{2-64}$$

The general solution of Hermite's equation is a superposition of these. If α is an even integer n, (63) reduces to an even polynomial of degree n. On choosing for a_0 the value

$$(-1)^{n/2} \frac{n!}{\left(\dfrac{n}{2} \right)!}$$

this polynomial becomes

$$H_n(x) = (2x)^n - \frac{n(n - 1)}{1!} (2x)^{n-2}$$
$$+ \frac{n(n - 1)(n - 2)(n - 3)}{2!} (2x)^{n-4} - \cdots \tag{2-65}$$

and this is known as the *Hermite polynomial* of degree n. If α is an odd integer, n, (64) reduces to an odd polynomial of degree n. In fact if we choose for a_0 the value

$$(-1)^{(n-1)/2} \frac{2 \cdot n!}{\left(\dfrac{n - 1}{2} \right)!}$$

that particular solution also takes on the form $H_n(x)$.

An equation very similar to that of Hermite is

$$y'' + (1 - x^2 + 2\alpha)y = 0 \qquad (2\text{--}66)$$

For if we put $y = e^{-x^2/2}v$, so that $y'' = \{(x^2 - 1)v - 2xv' + v''\}e^{-x^2/2}$, the equation turns into

$$v'' - 2xv' + 2\alpha v = 0$$

which is identical with (62). Hence the solution of (66) is simply any solution of Hermite's equation, multiplied by $e^{-x^2/2}$.

2.16. Laguerre's Differential Equation.—

$$xy'' + (1 - x)y' + \alpha y = 0; \quad \alpha = \text{constant} \qquad (2\text{--}67)$$

has a non-essential singularity at the origin. Developing about $x = 0$, the indicial equation has the single root $\kappa = 0$. Only one solution will be obtained, this being of considerable importance in physics. The recurrence relation reads:

$$a_{j+1} = \frac{j - \alpha}{(j + 1)^2} a_j$$

hence

$$y = a_0 \left(1 - \alpha x + \frac{\alpha(\alpha - 1)}{(2\,!)^2} x^2 - \cdots \right.$$
$$\left. + (-1)^r \frac{\alpha(\alpha - 1) \cdots (\alpha - r + 1)}{(r\,!)^2} x^r + \cdots \right) \qquad (2\text{--}68)$$

This expression becomes a polynomial when $\alpha = n$, a positive integer. On putting

$$a_0 = (-1)^n n\,!$$

and for integral n, y becomes the *Laguerre polynomial* of degree n:

$$L_n(x) = (-1)^n \left(x^n - \frac{n^2}{1\,!} x^{n-1} + \frac{n^2(n - 1)^2}{2\,!} x^{n-2} + \cdots \right.$$
$$\left. + (-1)^n n\,! \right) \qquad (2\text{--}69)$$

A differential equation at once reducible to Laguerre's is

$$xy'' + (k + 1 - x)y' + (\alpha - k)y = 0, \quad k \text{ an integer} \geq 0 \qquad (2\text{--}70)$$

It results when (67) is differentiated k times and y is replaced by its k-th derivative. Hence a solution of (70) for integral and positive α and k is

$$y = \frac{d^k}{dx^k} L_n(x) \equiv L_n^k(x)$$

This is sometimes called the *associated Laguerre polynomial* of degree $n - k$.

A third function closely related to the Laguerre polynomials satisfies the differential equation

$$xy'' + 2y' + \left[n - \frac{k-1}{2} - \frac{x}{4} - \frac{k^2 - 1}{4x} \right] y = 0 \qquad (2\text{--}71)$$

If we substitute in this equation $y = e^{-x/2}x^{(k-1)/2}v$, then v is seen to be a solution of

$$xv'' + (k + 1 - x)v' + (n - k)v = 0$$

Comparison with (70) shows, therefore, that $v = L_n^k(x)$. Hence a particular solution of (71) is

$$y = e^{-x/2}x^{(k-1)/2}L_n^k(x) \qquad (2\text{--}72)$$

This function is known as an *associated Laguerre function;* it is of great importance in the theory of the hydrogen atom. We observe that if n in (71) were not an integer but any constant α, the corresponding solution of (71) would be

$$y = e^{-x/2}x^{(k-1)/2} \frac{d^k}{dx^k} L_\alpha(x)$$

where L_α is written for the series (68); provided, of course, that k is a positive integer. This solution would no longer be a polynomial in x multiplied by $e^{-x/2}$, but an infinite sequence.

2.17. Mathieu's Equation.—In the previous sections attention has been given to differential equations in which X_1 and X_2[15] were algebraic functions of x. Equations sometimes arise in which these functions are periodic. The simplest instance of these is *Mathieu's equation*, usually written in the form

$$\frac{d^2y}{dx^2} + (a + 16b \cos 2x)y = 0 \qquad (2\text{--}73)$$

where a and b are constants. Its general solution may be obtained by the method of integration in series if the substitution

$$\xi = \cos^2 x$$

is made. (73) then reads

$$4\xi(1 - \xi) \frac{d^2y}{d\xi^2} + 2(1 - 2\xi) \frac{dy}{d\xi} + (a - 16b + 32b\xi)y = 0 \qquad (2\text{--}74)$$

---

[15] Defined by eq. (27).

This equation has a non-essential singularity at $\xi = 0$ and can therefore be developed as a power series about the origin. On inserting

$$y = \sum_\lambda a_\lambda \xi^{\kappa+\lambda}$$

in (74) we obtain

$$2\sum_\lambda (\kappa + \lambda)(2\kappa + 2\lambda - 1)a_\lambda \xi^{\kappa+\lambda-1} - \sum_\lambda [4(\kappa + \lambda)^2 - a + 16b]a_\lambda \xi^{\kappa+\lambda}$$
$$+ 32b\sum_\lambda a_\lambda \xi^{\kappa+\lambda+1} = 0$$

Here a feature arises which was not encountered before; the equation contains *three* different summations instead of two and will therefore lead to a three-term recurrence relation between the coefficients a_λ instead of the two-term relations that occurred in the former instances. This, however, requires no modification of procedure, except that it will force us to advance step by step in the computation of the coefficients. Only the first summation can contribute to the coefficient of $\xi^{\kappa-1}$, which must be zero. Hence the indicial equation is formed as before:

$$\kappa(2\kappa - 1) = 0$$

whence we obtain the two choices: $\kappa = 0, \frac{1}{2}$. Next, we equate to zero the coefficients of ξ^κ, to which the first and second summations contribute. This leads to

$$2(\kappa + 1)(2\kappa + 1)a_1 = (4\kappa^2 - a + 16b)a_0$$

so that

$$a_1 = \frac{1}{2} \frac{4\kappa^2 - a + 16b}{(\kappa + 1)(2\kappa + 1)} a_0$$

from which a_1 may be determined when the arbitrary constant a_0 is assumed. On equating to zero the coefficient of $\xi^{\kappa+1}$ to which all three summations contribute, one gets

$$2(\kappa + 2)(2\kappa + 3)a_2 - [4(\kappa + 1)^2 - a + 16b]a_1 + 32ba_0 = 0$$

a relation permitting the calculation of a_2, etc. In this way two series can be constructed, one for $\kappa = 0$, the other for $\kappa = \frac{1}{2}$, linear composition of which yields the general solution of (74) and hence of (73). Investigation shows that this solution converges if $|\xi| < 1$.

This general solution, however, is rarely of interest in physics and chemistry, for it is not periodic in x. In most problems leading to Mathieu's equation, x is an angle, so that there is no significant distinction between x and $x + 2n\pi$, where n is an integer. Thus the solutions usually sought must have the property that $y(x + 2\pi n) = y(x)$. The general solution here found, which is of the form

$$\sum_\lambda a_\lambda \xi^\lambda + \xi^{1/2}\sum_\lambda b_\lambda \xi^\lambda \qquad (2\text{-}75)$$

does not possess this periodicity, as closer investigation would show. Qualitatively this defect is apparent from the failure of the solution to converge for $\xi = \pm 1$, which excludes the values $x = n\pi$ from consideration altogether, as well as from the existence of a branch point of (75) at $\xi = 0$ (arising from the factor $\xi^{1/2}$).

In fact it is impossible to obtain solutions of Mathieu's equation which are periodic and of period 2π in x, unless definite restrictions are placed upon the constant a. It turns out that the latter must be a complicated function of b if the solution is to be periodic.[16]

Floquet's Theorem. An important theorem concerning the general solution of Mathieu's equation, or indeed of any linear differential equation with periodic coefficients which are one-valued functions of x, will now be established. Suppose that $y_1(x)$ and $y_2(x)$ are two linearly independent solutions of (73), so that any particular solution y may be compounded from them by means of two constants A_1 and A_2 as follows:

$$y = A_1 y_1 + A_2 y_2 \qquad (2\text{-}76)$$

Now it is clear that, if $y_1(x)$ and $y_2(x)$ are solutions of (73), $y_1(x + 2\pi)$ and $y_2(x + 2\pi)$ will also be solutions, for the substitution of $x + 2\pi$ in place of x causes no change in the differential equation. This must, of course, not be interpreted as implying that $y_1(x + 2\pi) = y_1(x)$ and $y_2(x) = y_2(x + 2\pi)$; but it does mean that

$$y_1(x + 2\pi) = \alpha_{11}y_1(x) + \alpha_{12}y_2(x); \quad y_2(x + 2\pi) = \alpha_{21}y_1(x) + \alpha_{22}y_2(x)$$

the α's being constants. Similarly, using (76)

$$y(x + 2\pi) = A_1 y_1(x + 2\pi) + A_2 y_2(x + 2\pi)$$
$$= (A_1\alpha_{11} + A_2\alpha_{21})y_1(x) + (A_1\alpha_{12} + A_2\alpha_{22})y_2(x)$$

We observe that the constants α are fixed by the choice of y_1 and y_2, but A_1 and A_2 may be chosen at will and still leave y a particular solution of the equation. It is possible to choose them so as to satisfy the equations

$$A_1\alpha_{11} + A_2\alpha_{21} = kA_1; \quad A_1\alpha_{12} + A_2\alpha_{22} = kA_2 \qquad (2\text{-}77)$$

where k is a constant not within our control, for if eqs. (77) are to be satis-

---

[16] Cf. Whittaker, E. T., and Watson, G. N., " A Course of Modern Analysis," Fourth Edition, Cambridge Press, 1940, for further details regarding periodic solutions.

fied then k must be subject to the equation

$$\begin{vmatrix} \alpha_{11} - k & \alpha_{21} \\ \alpha_{12} & \alpha_{22} - k \end{vmatrix} = 0 \qquad (2\text{-}78)$$

But if (77) holds then

$$y(x + 2\pi) = k[A_1 y_1(x) + A_2 y_2(x)] = ky(x) \qquad (2\text{-}79)$$

In other words, there exists a particular solution $y(x)$ such that, when x is increased by 2π, the solution itself is multiplied by the constant k. If k were unity, this solution would be periodic.

This result may be expressed in a different way. On putting

$$k = e^{2\pi\mu}, \quad y(x) = e^{\mu x}P(x)$$

eq. (79) reads

$$e^{\mu(x+2\pi)}P(x + 2\pi) = e^{2\pi\mu + \mu x}P(x)$$

so that $P(x)$ turns out to be a periodic function. Thus it is seen that there exists a particular solution of Mathieu's equation of the form

$$y = e^{\mu x}P(x) \qquad (2\text{-}80)$$

where P is periodic. From here it is only a simple step to obtain a general solution of (73). The differential equation is insensitive to the substitution of $-x$ for x. Hence $e^{-\mu x}P(-x)$ must also be a solution. Moreover, it is an independent solution, since it is not a constant multiple of (80). The complete solution is, therefore, a linear combination of these two:

$$y = c_1 e^{\mu x}P(x) + c_2 e^{-\mu x}P(-x) \qquad (2\text{-}81)$$

This result, known as Floquet's theorem, is of interest in some astronomical applications and chiefly in the quantum theory of metals.[17]

Problem. Show that the Schrödinger equation

$$\frac{d^2\psi}{dx^2} + [A + V(x)]\psi = 0,$$

in which A is a constant, and V is a periodic function of x such that $V(x + l) = V(x)$, has solutions of the form

$$\psi = e^{ikx}v(x),$$

where v is also periodic: $v(x + l) = v(x)$.

This is sometimes called Bloch's theorem.[18]

---

[17] See Seitz, F., "Modern Theory of Solids," McGraw-Hill Book Co., New York, 1940, Chap. VIII.

[18] Bloch, F., Z. *Physik* **52**, 555 (1928).

2.18.   Pfaff Differential Expressions and Equations.—The equations of thermodynamics are peculiar inasmuch as they usually occur in the form

$$dW = \sum_{\lambda=1}^{n} X_\lambda dx_\lambda \qquad (2\text{--}82)$$

where the X_λ are functions of some or all the independent variables x_λ. While (82), which is known as a *Pfaff expression*, is not a differential equation of the customary kind, its importance in chemistry and physics requires consideration.   It is for lack of a more adequate place that this material is inserted in the chapter on differential equations.   Some of the material which will be developed from a mathematical point of view in this section has already been used in Chapter 1, to which reference should be made for further applications.   The equation

$$\sum_{\lambda=1}^{n} X_\lambda dx_\lambda = 0$$

is sometimes called a *total* differential equation or, more generally, a *Pfaff* equation.

Clearly, the expression dW, eq. (82), can be integrated along any path in n-dimensional space, but the integral will *in general* depend on the path of integration.   (See Prob. *a*, p. 87;   also the example in sec. 1.8.)

When $\int dW$ depends on the path of integration, dW is said to be *incomplete* or *inexact*.

The condition that (82) be a complete differential is

$$dW = df(x_1 x_2 \cdots x_n) \qquad (2\text{--}83)$$

for then $\int_{r_1}^{r_2} dW = f(r_2) - f(r_1)$, independently of path.   Now

$$df = \sum_\lambda \frac{\partial f}{\partial x_\lambda} dx_\lambda$$

Comparing with (82), we find

$$X_\lambda = \frac{\partial f}{\partial x_\lambda}$$

To state this relation without explicitly introducing the function f, we differentiate it with respect to x_μ, $\mu \neq \lambda$.

$$\frac{\partial X_\lambda}{\partial x_\mu} = \frac{\partial^2 f}{\partial x_\lambda \partial x_\mu}$$

But also

$$\frac{\partial X_\mu}{\partial x_\lambda} = \frac{\partial^2 f}{\partial x_\mu \partial x_\lambda}$$

Hence the necessary condition of " exactness " may be written in the form

$$\frac{\partial X_\lambda}{\partial x_\mu} = \frac{\partial X_\mu}{\partial x_\lambda}, \quad \lambda, \mu = 1 \cdots n \tag{2-84}$$

The reader who is already familiar with vector analysis will note that, if the X_λ are interpreted as components of a *vector* $\mathbf{R}$, (82) may be written

$$dW = \mathbf{R} \cdot d\mathbf{r} \tag{2-82'}$$

and the condition of " exactness " becomes

$$\frac{\partial X}{\partial y} - \frac{\partial Y}{\partial x} = \frac{\partial X}{\partial z} - \frac{\partial Z}{\partial x} = \frac{\partial Y}{\partial z} - \frac{\partial Z}{\partial y} = 0$$

or

$$\nabla \times \mathbf{R} = 0 \tag{2-84'}$$

These results are of importance in vector analysis where they are usually expressed as follows: The condition that the line integral of $\mathbf{R}$ (expression 82′) around any closed curve shall vanish is that $\mathbf{R}$ be the gradient of some scalar function, and this is equivalent to condition (84′). (Cf. sec. 4.17.)

We return now to the general situation:

dW is not exact

and distinguish two cases:

    A The equation $dW = 0$ has a solution.
    B The equation $dW = 0$ does not have a solution.

A. The equation dW = 0 possesses a solution. Leaving aside for the moment all considerations as to when such solutions may be found, we shall first sketch the consequences of the existence of solutions. The equation $dW = 0$ assigns to every point a *direction*, or, what amounts to the same thing, an *element of surface*. (From the point of view of vector analysis this is immediately clear because the relation $\mathbf{R} \cdot d\mathbf{r}$ specifies at every point $(x_1 \cdots x_n)$ the direction $d\mathbf{r}$ which is perpendicular to the vector $\mathbf{R}$.)

When integrated, the equation $dW = 0$ leads to

$$\phi(x_1 x_2 \cdots x_n) = c \tag{2-85}$$

which represents a one-parameter family of surfaces in n-dimensional space. These surfaces consist of the elements specified by $dW = 0$.

We now wish to show that there exists an integrating denominator, $t(x_1 \cdots x_n)$, such that dW/t is an exact differential. The proof is as follows.

Along the surface $\phi(x_1 \cdots x_n) = c$ (cf. Fig. 2–6), we have both $d\phi = 0$ and $dW = 0$. The same is true along a neighboring surface $\phi = c + dc$. Suppose we wish to go from A to C. The change occurring in ϕ is dc, no matter whether the crossing is made at B_1 or at B_2. But the change dW will depend on the path. The important point to note is that no change occurs in W as we pass along either curve; a change can occur only at the *crossing:* dW = function of the point at which the crossing is made. (If $dW \neq 0$ along the two curves, then it would depend on the whole path, not merely on the point of crossing!)

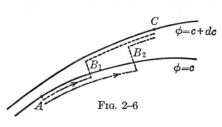

FIG. 2–6

Hence $dW = t(B)d\phi$, where B is the *point* of crossing. Hence $dW = t(x_1 \cdots x_n)d\phi$, or

$$d\phi = \frac{dW}{t}$$

But $d\phi$ is an exact differential.

Along the curves ϕ = const., the equation $F(\phi)$ = const. will likewise be satisfied if F represents a unique, single-valued function. If, then, we use $F(\phi)$ in place of ϕ in the preceding analysis, we are led to

$$dF = \frac{dW}{T} \quad \text{instead of} \quad d\phi = \frac{dW}{t}$$

Since, however, $dF = (dF/d\phi)d\phi$, we see that $T = t/(dF/d\phi)$ is also an integrating denominator. It is clear that, if there exists one integrating denominator t for a Pfaff expression, an infinite number of others can be formed by the above rule.

Only the points on the surface $\phi = c$ are connected with A by paths along which $dW = 0$. It is clear that in the neighborhood of A there is an infinite number of points *not* connected with A by such paths. Hence the fact, important in thermodynamics (though somewhat trivial geometrically!):

If the inexact differential dW possesses an integrating denominator t, then there exist, in the neighborhood of every point P, innumerable points which cannot be reached from P along paths for which dW = 0.

We now consider the question of *how to find the integrating denominator.*

1. *Case of two variables.* First solve the equation

$$dW = 0; \quad Xdx + Ydy = 0 \tag{2–86}$$

The solution is

$$y = f(x,c), \quad \text{or} \quad \phi(x,y) = c \tag{2–87}$$

Along the curves (87), $\phi_x dx + \phi_y dy = 0$, hence

$$\frac{dy}{dx} = -\frac{\phi_x}{\phi_y} \tag{2–88}$$

But from (86)

$$\frac{dy}{dx} = -\frac{X}{Y}$$

so that

$$\frac{\phi_x}{\phi_y} = \frac{X}{Y}, \quad \text{or} \quad \frac{\phi_x}{X} = \frac{\phi_y}{Y} = u(x,y) \tag{2-89}$$

Now

$$d\phi = \frac{dW}{t} = \phi_x dx + \phi_y dy = uX dx + uY dy = u dW$$

Hence

$$t = \frac{1}{u} = \frac{X}{\phi_x} = \frac{Y}{\phi_y} \cdots \tag{2-90}$$

by (89).

2. *Case of three variables.* First solve

$$dW = 0; \quad X dx + Y dy + Z dz = 0 \tag{2-91}$$

The solution is

$$\phi(x,y,z) = c$$

Along these surfaces, $\phi_x dx + \phi_y dy + \phi_z dz = 0$, hence

$$\left.\frac{dy}{dx}\right|_z = -\frac{\phi_x}{\phi_y}, \quad \left.\frac{dz}{dx}\right|_y = -\frac{\phi_x}{\phi_z}, \quad \left.\frac{dz}{dy}\right|_x = -\frac{\phi_y}{\phi_z}$$

But from (91)

$$\left.\frac{dy}{dx}\right|_z = -\frac{X}{Y}, \quad \left.\frac{dz}{dx}\right|_y = -\frac{X}{Z}, \quad \left.\frac{dz}{dy}\right|_x = -\frac{Y}{Z}$$

Hence

$$\frac{\phi_x}{\phi_y} = \frac{X}{Y}, \quad \frac{\phi_x}{\phi_z} = \frac{X}{Z}, \quad \frac{\phi_y}{\phi_z} = \frac{Y}{Z}$$

or

$$\frac{\phi_x}{X} = \frac{\phi_y}{Y} = \frac{\phi_z}{Z} = u(x,y,z)$$

Now

$$d\phi = \frac{dW}{t} = \phi_x dx + \phi_y dy + \phi_z dz = u(X dx + Y dy + Z dz)$$

Therefore

$$t = \frac{1}{u} = \frac{X}{\phi_x} = \frac{Y}{\phi_y} = \frac{Z}{\phi_z}$$

Similarly for more than three variables.

We now consider the *condition that the equation*

$$dW = 0$$

shall have a solution. (*Condition of integrability.*)

Suppose a solution of $\sum_\lambda X_\lambda dx_\lambda = 0$ exists in the form

$$\phi(x_1 \cdots x_n) = c$$

Then

$$u(x_1 \cdots x_n)X_i = \frac{\partial \phi}{\partial x_i}, \quad i = 1, 2, \cdots n \qquad (2\text{-}92)$$

Let i, j, k, be different indices. It follows from (92) that

$$\frac{\partial}{\partial x_i}(uX_j) = \frac{\partial^2 \phi}{\partial x_i \partial x_j} = \frac{\partial}{\partial x_j}(uX_i)$$

whence

$$u\left(\frac{\partial X_j}{\partial x_i} - \frac{\partial X_i}{\partial x_j}\right) = X_i \frac{\partial u}{\partial x_j} - X_j \frac{\partial u}{\partial x_i}$$

Similarly,

$$u\left(\frac{\partial X_i}{\partial x_k} - \frac{\partial X_k}{\partial x_i}\right) = X_k \frac{\partial u}{\partial x_i} - X_i \frac{\partial u}{\partial x_k}$$

$$u\left(\frac{\partial X_k}{\partial x_j} - \frac{\partial X_j}{\partial x_k}\right) = X_j \frac{\partial u}{\partial x_k} - X_k \frac{\partial u}{\partial x_j}$$

Multiply the last three equations by X_k, X_j, and X_i, respectively, and add·

$$X_k\left(\frac{\partial X_j}{\partial x_i} - \frac{\partial X_i}{\partial x_j}\right) + X_j\left(\frac{\partial X_i}{\partial x_k} - \frac{\partial X_k}{\partial x_i}\right) + X_i\left(\frac{\partial X_k}{\partial x_j} - \frac{\partial X_j}{\partial x_k}\right) = 0 \qquad (2\text{-}93)$$

By closer analysis, this equation may be shown to be both necessary and sufficient; it represents the condition of integrability for the Pfaff equation $dW = 0$. In three variables, eq. (93) takes the form

$$\mathbf{R} \cdot \nabla \times \mathbf{R} = 0$$

provided $\mathbf{R}$ is interpreted as the vector having components X_1, X_2, X_3. The total number of equations of the form (93) is equal to the number of triangles that can be formed with n given points as corners; it is therefore $\frac{1}{6}n(n-1)(n-2)$. These equations are therefore not independent.

It is to be observed that, in the case of two variables, eq. (93) is *always* satisfied. Hence every Pfaff equation of the form

$$Xdx + Ydy = 0$$

possesses a solution.

B. The equation $dW = 0$ does not possess a proper solution, i.e., eq. (93) is not satisfied. For simplicity, we consider only the case of *three* variables, where the solutions can be visualized easily in ordinary space. Generalization to more variables introduces no complications. It will be seen that " improper " solutions of eq. (82) are still possible, but that they represent a greater variety of functions than the proper solutions considered in the preceding paragraphs.

We now choose an *arbitrary* relation

$$\psi(x,y,z) = 0 \qquad (2\text{–}94)$$

and impose this upon eq. (82), thereby effectively eliminating one degree of freedom. From (94) and its differential form

$$\psi_x dx + \psi_y dy + \psi_z dz = 0$$

the variables z and dz are obtained in terms of x, y, dx, dy, and these solutions are substituted in eq. (82). It will then be of the form

$$X dx + Y dy = 0$$

and this has a solution

$$\phi(x,y) = 0 \qquad (2\text{–}95)$$

The improper solutions of (82) are said to be those curves which satisfy (94) and (95) simultaneously. They represent, therefore, prescribed curves upon *arbitrary* surfaces. Further investigation would show that every point in the neighborhood of a given point can be reached by a continuous curve satisfying (94) and (95) from the given point, the state of affairs being quite different from that described under **A**.

Problem a. Let $dW = x(dx + dy)$. Compute the integral $\int_{x_1 y_1}^{x_2 y_2} dW$ along two paths:

1. $x_1 y_1 \rightarrow x_2 y_1 \rightarrow x_2 y_2$.
2. $x_1 y_1 \rightarrow x_1 y_2 \rightarrow x_2 y_2$.

Show that the two results differ by the area enclosed by the two paths of integration.

Problem b. Show that the expression

$$dW = -y dx + x dy + k dz = 0$$

where k is a constant, does not possess an integral.[19]

[19] See Born, M., *Physik. Z.* **22**, 250 (1921).

REFERENCES

Rainville, Earl D., "Intermediate Course in Differential Equations," John Wiley and Sons, Inc., New York, 1943. Clear and understandable discussion, particularly of equations of Fuchsian type.

Ince, E. L., "Ordinary Differential Equations," Dover Publications, New York, 1944. More advanced than Rainville but much easier to follow than Forsyth, footnote 7.

Kamke, E., "Differentialgleichungen, Lösungsmethoden und Lösungen, Band 1, Gewöhnliche Differentialgleichungen," Third Edition, Chelsea Publishing Co., New York, 1948. Remarkable reference book, with more than 1500 differential equations, together with their solutions and some comment, arranged so that one can readily find the solution of a given equation.

CHAPTER 3

SPECIAL FUNCTIONS

3.1. Elements of Complex Integration. Theorems of Cauchy.—Some acquaintance with the calculus of complex variables facilitates this work; hence the present section and the next will outline the elements of this useful subject.

As to notation, $i^2 = -1$, and the symbols x and y are used for single real variables. Furthermore,

$$z = x + iy = \rho e^{i\theta}.$$

Through the *Argand diagram*, which consists of a real axis along x, an imaginary axis along y, and presents z as the point with rectangular co-ordinates x and y, this last relation is at once made clear: $\rho^2 = x^2 + y^2$, $\theta = \tan^{-1}y/x$.

Now let $f(z)$ be a single-valued function and analytic in the sense that it has a unique derivative with respect to both x and y at every point of the Argand plane. We may then write, in terms of two new analytic functions u and v,

$$f(z) = u(x,y) + iv(x,y)$$

Hence follows an important result. Since

$$\frac{\partial f}{\partial x} = \frac{df}{dz} = \frac{\partial u}{\partial x} + i\frac{\partial v}{\partial x}$$

and

$$\frac{\partial f}{\partial y} = i\frac{df}{dz} = \frac{\partial u}{\partial y} + i\frac{\partial v}{\partial y}$$

one finds on equating real and imaginary parts of df/dz that

$$\frac{\partial u}{\partial x} = \frac{\partial v}{\partial y}, \quad \frac{\partial v}{\partial x} = -\frac{\partial u}{\partial y}$$

These are the famous *Cauchy-Riemann* conditions which the components u and v of an analytic $f(z)$ must satisfy. Further differentiation yields another important set of relations, which we shall often use hereafter:

$$\frac{\partial^2 u}{\partial x^2} + \frac{\partial^2 u}{\partial y^2} = 0, \quad \frac{\partial^2 v}{\partial x^2} + \frac{\partial^2 v}{\partial y^2} = 0$$

Cauchy's theorem asserts that the integral

$$\int_C f(z)dz = 0$$

provided it is taken along a closed curve C on and within which $f(z)$ is analytic. A simple proof is as follows.

$$\int_C f(z)dz = \int_C [u(dx + idy) + iv(dx + idy)]$$

$$= \int_C (udx - vdy) + i \int_C (vdx + udy)$$

By virtue of the Cauchy-Riemann conditions, both of the final integrands are exact in the sense of sec. 1.7, and the line integral around the closed contour vanishes.

Equally important is an extension of Cauchy's theorem to which we now turn. Suppose again that $f(z)$ is analytic within and on a closed curve C in the Argand plane, and denote by z_0 a fixed point within C. The function $f(z)/(z - z_0)$ will then have a singularity at z_0 and its line integral along C will not be zero. But the value of this integral will remain unchanged if we alter C, so long as the contour does not cross z_0. This follows at once from Cauchy's theorem, for the difference between the old and the new value of the integral will itself be a line integral around a region over which $f(z)/(z - z_0)$ is analytic, and will therefore vanish. Let us denote the infinitesimally small circle of radius ρ surrounding the point z_0 by Γ.

$$\int_C \frac{f(z)}{z - z_0} dz = \int_\Gamma \frac{f(z)}{z - z_0} dz = f(z_0) \int_\Gamma \frac{dz}{z - z_0} = f(z_0) \int_\Gamma \frac{d(\rho e^{i\theta})}{\rho e^{i\theta}}$$

$$= f(z_0)i \int_0^{2\pi} d\theta = 2\pi i f(z_0)$$

provided we traverse C in *counter-clockwise* fashion. Hence the theorem

$$\frac{1}{2\pi i} \int_C \frac{f(z)}{z - z_0} dz = f(z_0) \tag{3-1}$$

Henceforth we shall understand that $\int_C$ denotes counter-clockwise (so-called positive) integration along a closed curve C.

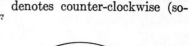

3.1a. Theorem of Laurent. Residues.—Let $f(z)$ be analytic on the ring formed by two concentric circles C_1 and C_2 including these boundaries. (See Fig. 1.) Apply eq. (1) to the point $z_0 = \zeta$, choosing a contour which goes from A along C_2 to B, thence inside to C_1, around C_1 in a negative sense, and finally back to A. The two horizontal portions of the path make equal and opposite contributions to the integral and therefore cancel. Hence

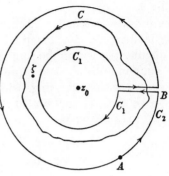

Fig. 3–1

$$f(\zeta) = \frac{1}{2\pi i} \int_{C_2} \frac{f(z)}{z - \zeta}\,dz - \frac{1}{2\pi i} \int_{C_1} \frac{f(z)}{z - \zeta}\,dz \tag{a}$$

In the first integral we may write (z_0 now denotes the common center of C_1 and C_2)

$$\frac{1}{z - \zeta} = \frac{1}{z - z_0} \cdot \frac{1}{1 - \dfrac{\zeta - z_0}{z - z_0}} = \frac{1}{z - z_0} \sum_{\lambda=0}^{\infty} \left(\frac{\zeta - z_0}{z - z_0}\right)^{\lambda}$$

and obtain a series which converges on C_2. Therefore

$$\frac{1}{2\pi i} \int_{C_2} \frac{f(z)}{z - \zeta}\,dz = \sum_{\lambda=0}^{\infty} a_\lambda (\zeta - z_0)^{\lambda} \tag{b}$$

provided we define the coefficients

$$a_\lambda = \frac{1}{2\pi i} \int_{C_2} \frac{f(z)\,dz}{(z - z_0)^{\lambda+1}} \tag{c}$$

In the second integral of (a) we use the convergent expansion

$$\frac{1}{\zeta - z} = \frac{1}{\zeta - z_0} \cdot \frac{1}{1 - \dfrac{z - z_0}{\zeta - z_0}} = \frac{1}{\zeta - z_0} \sum_{\lambda} \left(\frac{z - z_0}{\zeta - z_0}\right)^{\lambda}$$

so that

$$\frac{1}{2\pi i} \int_{C_1} \frac{f(z)}{z - \zeta}\,dz = -\sum_{\lambda=0}^{\infty} b_\lambda (\zeta - z_0)^{-\lambda-1} \tag{d}$$

provided

$$b_\lambda = \frac{1}{2\pi i} \int_{C_1} f(z)(z - z_0)^\lambda dz. \qquad \text{(e)}$$

Now put $\lambda = -\mu - 1$. Then b_λ becomes a_μ, except that the integration in b_λ is along C_1, in a_μ along C_2. But the integrands of (c) and (e) have no singularities within the ring, hence the path of integration may be taken the same in both of these expressions. We may indeed take it to be any curve, C, within the ring, which encloses the point ζ. Thus, because the sense of C_2 is opposite to that of C_1, $b_{-\mu-1} = -a_\mu$.

Eq. (d) now reads

$$\frac{1}{2\pi i} \int_{C_1} \frac{f(z)}{z - \zeta} dz = \sum_{\mu=-\infty}^{-1} a_\mu(\zeta - z_0)^\mu \qquad \text{(f)}$$

When we add (b) and (f) to form (a), and replace ζ by z in the final formula, we find

$$f(z) = \sum_{\lambda=-\infty}^{\infty} a_\lambda(z - z_0)^\lambda \qquad \text{a)}$$

with

$$a_\lambda = \frac{1}{2\pi i} \int_C \frac{f(z)dz}{(z - z_0)^{\lambda+1}} \qquad \text{b)}$$

$$(3\text{--}2)$$

Eq. (2) is called *Laurent's theorem*. It shows that a function $f(z)$, free from singularities on a circular ring, can be expanded as a *Laurent series* (eq. 2a), which contains negative as well as positive powers of $z - z_0$.

The term a_{-1}, formed by means of eq. (2b), is especially interesting. It is

$$a_{-1} = \frac{1}{2\pi i} \int_C f(z)dz \qquad (3\text{--}3)$$

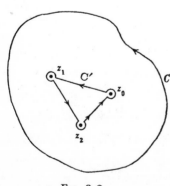

FIG. 3–2

As to the contour C, we noted that it includes the point z_0. If $f(z)$ is analytic at z_0, $a_{-1} = 0$. The theorem is useful, therefore, when z_0 is a singular point of $f(z)$. The constant a_{-1} is called the *residue* of the function $f(z)$ at z_0, and eq. (3) is known as the *theorem of residues*.

If $f(z)$ has a number of singularities z_0, z_1, z_2, etc., within a contour C, this path may be distorted in the manner shown in Fig. 2 to form C' Since the triangular path, exclusive of the circles, lies in a region free from singularities, it contributes nothing to $\int f dz$. The remainder comes from the singularities and is equal to the sum of their residues. Hence

$$\int_C f(z)dz = 2\pi i \text{ (sum of residues within } C).$$

Example. To evaluate the integral

$$I = \int_{-\pi}^{\pi} \frac{d\phi}{a + b \cos \phi + c \sin \phi}$$

Let $z = e^{i\phi}$; $\phi = -i \log z$, $d\phi = -i(dz/z)$. Then $\cos \phi = \frac{1}{2}(z + z^{-1})$, $\sin \phi = (1/2i)(z - z^{-1})$.

$$I = -i \oint \frac{dz}{az + \dfrac{b}{2}(z^2 + 1) + \dfrac{c}{2i}(z^2 - 1)}$$

the contour being the unit circle about 0. The denominator of the integrand may be written

$$\tfrac{1}{2}(b - ic)z^2 + az + \tfrac{1}{2}(b + ic) = \tfrac{1}{2}(b - ic)\left[z - \frac{1}{b - ic}(-a + R) \right]$$

$$\times \left[z - \frac{1}{b - ic}(-a - R) \right]$$

provided we put

$$\sqrt{a^2 - b^2 - c^2} \equiv R$$

If $a^2 - b^2 - c^2 > 0$ then

$$\left| \frac{1}{b - ic}(-a + R) \right| < 1$$

The other root > 1 and lies outside the unit circle. The residue of the integrand at $z = (-a + R)/(b - ic)$ is

$$\frac{1}{\tfrac{1}{2}(b - ic)\left[\dfrac{1}{b - ic}(-a + R) - \dfrac{1}{b - ic}(-a - R) \right]} = \frac{1}{R}$$

Therefore

$$I = -i \cdot 2\pi i (a^2 - b^2 - c^2)^{-1/2}$$

$$= \frac{2\pi}{\sqrt{a^2 - b^2 - c^2}}$$

3.2. Gamma Function.—The gamma function is a generalization of the factorial $n!$ for non-integral values of n; more specifically, $\Gamma(z)$ is so chosen that, if n is an integer, $\Gamma(n) = (n - 1)!$. A fundamental definition, due to Euler, states

$$\Gamma(z) = \lim_{n \to \infty} \frac{1 \cdot 2 \cdot 3 \cdots (n-1)}{z(z+1) \cdots (z+n-1)} n^z \qquad (3\text{--}4)$$

Several important properties of the Γ-function follow at once from this definition. Since from (4)

$$\Gamma(z+1) = \lim_{n \to \infty} \frac{1 \cdot 2 \cdots (n-1)}{(z+1)(z+2) \cdots (z+n)} n^{z+1}$$

$$\Gamma(z+1) = \lim_{n \to \infty} \frac{zn}{(z+n)} \cdot \frac{1 \cdot 2 \cdots (n-1)}{z(z+1) \cdots (z+n-1)} n^z = z\Gamma(z) \quad (3\text{--}5)$$

On the other hand, (4) also shows that

$$\Gamma(1) = \lim_{n \to \infty} \frac{n!}{n!} = 1 \qquad (3\text{--}6)$$

From (5) and (6) it is at once apparent that, **if** n **is a positive integer,**

$$\Gamma(n) = (n-1)! \qquad (3\text{--}7)$$

as was stated above. It is also evident from the definition (4) that $\Gamma(z)$ becomes infinite at $z = 0,\ -1,\ -2$, etc., and that it is an analytic function everywhere else.

It is often useful to represent $\Gamma(z)$ by means of a definite integral. To achieve this, we consider the function

$$F(z,n) \equiv \int_0^n \left(1 - \frac{t}{n}\right)^n t^{z-1} dt \qquad (3\text{--}8)$$

wherein n stands for a positive integer, and the real part of z is taken to be greater than zero in order to insure convergence of the integral. The transformation $\tau = t/n$ converts F into

$$F(z,n) = n^z \int_0^1 (1 - \tau)^n \tau^{z-1} d\tau$$

The integral appearing here may be evaluated by repeated partial integrations:

$$\int_0^1 (1-\tau)^n \tau^{z-1} d\tau = \left[(1-\tau)^n \frac{\tau^z}{z}\right]_0^1 + \frac{n}{z} \int_0^1 (1-\tau)^{n-1} \tau^z d\tau$$

The integrated part here vanishes at both limits, and the remainder may again be subjected to a partial integration, yielding

$$\frac{n}{z} \left\{ \left[(1-\tau)^{n-1} \frac{\tau^{z+1}}{z+1}\right]_0^1 + \frac{n-1}{z+1} \int_0^1 (1-\tau)^{n-2} \tau^{z+1} d\tau \right\}$$

The integrated part is again zero.   By continuing this process we find

$$F(z,n) = \frac{n(n-1)\cdots 1}{z(z+1)\cdots(z+n-1)} n^z \int_0^1 \tau^{z+n-1}d\tau = \frac{1\cdot 2\cdots n}{z(z+1)\cdots(z+n)} n^z$$

As n approaches infinity, this expression becomes identical with (4); hence

$$\lim_{n\to\infty} F(z,n) = \Gamma(z) \qquad (3\text{-}9)$$

On the other hand, since $e = \lim\limits_{p\to\infty} (1 + 1/p)^p$ and therefore

$$e^x = \lim_{px\to\infty} (1 + 1/p)^{px} = \lim_{n\to\infty} (1 + x/n)^n$$

the quantity $(1 - t/n)^n$ appearing in (8) approaches the limit e^{-t}.   We conclude, therefore, that in view of (8) and (9)

$$\int_0^\infty e^{-t}t^{z-1}dt = \Gamma(z) \qquad (3\text{-}10)$$

This result is valid, we recall, when the real part of z is greater than zero.

A definition of the Γ-function, or rather its reciprocal, by means of an infinite product has been given by *Weierstrass*.   Since it is a useful one, we shall here derive it by simple steps (the rigor of which is not always obvious) from Euler's definition (4).   We first note that the product

$$\frac{1}{z}\cdot\frac{1}{z+1}\cdot\frac{2}{z+2}\cdots\frac{n-1}{z+n-1}$$

which appears in (4), may be written $\dfrac{1}{z}\displaystyle\prod_{m=1}^{n-1}(1 + z/m)^{-1}$, so that (4) becomes

$$\Gamma(z) = \frac{1}{z} \lim_{n\to\infty} n^z\prod_1^n\left(1 + \frac{z}{m}\right)^{-1}$$

or

$$\frac{1}{\Gamma(z)} = z \lim_{n\to\infty} n^{-z}\prod_1^n\left(1 + \frac{z}{m}\right)$$

If we multiply the right-hand side of this equation by unity in the form of

$$\left[\lim_{n\to\infty} e^{(1+1/2+\cdots 1/n)z}\right]\left[\lim_{n\to\infty} \prod_1^n e^{-z/m}\right]$$

we obtain

$$\frac{1}{\Gamma(z)} = z\left[\lim_{n\to\infty} e^{(1+1/2+\cdots 1/n-\log n)z}\right]\left[\lim_{n\to\infty} \prod_1^n\left(1 + \frac{z}{m}\right)e^{-z/m}\right]$$

Now the infinite series: $\lim_{n \to \infty} (1 + \frac{1}{2} + \cdots 1/n - \log n) = C$ converges; it has the value $C = 0.5772 \cdots$, known as the *Euler-Mascheroni* constant. Hence

$$\frac{1}{\Gamma(z)} = ze^{Cz}\prod_{1}^{\infty}\left(1 + \frac{z}{n}\right)e^{-z/n} \tag{3-11}$$

which is the Weierstrass definition. It shows, again, that $\Gamma(z)$ has poles at $z = 0, -1, -2$, etc.

A further important property of Γ-functions, namely the relation

$$\Gamma(z)\Gamma(1 - z) = \frac{\pi}{\sin \pi z} \tag{3-12}$$

is readily derived from the Weierstrass definition. First, we recall the theorem:

$$\frac{\sin \pi z}{\pi z} = \prod_{1}^{\infty}\left(1 - \frac{z^2}{n^2}\right) \tag{3-13}$$

which may be proved by an expansion of the infinite product as a sum of powers of z^2. (The details are left as an exercise for the reader.) From (11),

$$\Gamma(z)\Gamma(-z) = -\frac{1}{z^2}\prod_{1}^{\infty}\left(1 + \frac{z}{n}\right)^{-1}\left(1 - \frac{z}{n}\right)^{-1}$$

$$= -\frac{1}{z^2}\prod_{1}^{\infty}\left(1 - \frac{z^2}{n^2}\right)^{-1}$$

$$= -\frac{\pi}{z \sin \pi z} \tag{3-14}$$

the last step because of (13). But in view of (5)

$$\Gamma(-z) = -\frac{1}{z}\Gamma(1 - z)$$

and this, when inserted in (14), yields (12).

Several other formulas for the derivation of which the reader should refer to mathematical treatises,[1] will now be listed without proof.

$$\Gamma(z)\Gamma(z + \tfrac{1}{2}) = 2^{1-2z}\pi^{1/2}\Gamma(2z) \tag{3-15}$$

An infinite product of the form

$$\frac{1-a}{1-b} \cdot \frac{2-a}{2-b} \cdot \frac{3-a}{3-b}$$

may be expressed in terms of Γ-functions:

$$\prod_{1}^{\infty}\frac{n - a}{n - b} = \frac{\Gamma(1 - b)}{\Gamma(1 - a)} \tag{3-16}$$

[1] Cf., for instance, Whittaker and Watson, p. 235.

Also,

$$\prod_1^\infty \frac{n(a + b + n)}{(a + n)(b + n)} = \frac{\Gamma(a + 1)\Gamma(b + 1)}{\Gamma(a + b + 1)} \tag{3-16a}$$

If m and n are positive constants, not necessarily integral, we have

$$2 \int_0^{\pi/2} \cos^{m-1} x \sin^{n-1} x\, dx = \frac{\Gamma\left(\dfrac{m}{2}\right) \Gamma\left(\dfrac{n}{2}\right)}{\Gamma\left(\dfrac{m + n}{2}\right)} \tag{3-17}$$

This relation may be modified as follows. Put $m = 2r$, $n = 2s$, and introduce the new variable of integration $\cos^2 x = u$ on the left. The integral will then be converted into

$$\int_0^1 u^{r-1} (1 - u)^{s-1}\, du$$

which is a function of r and s known as the *Eulerian integral of the first kind*, or simply the *B*-function, and denoted by $B(r,s)$. Eq. (17) may therefore be put in the form

$$B(r,s) = \frac{\Gamma(r)\Gamma(s)}{\Gamma(r + s)} \tag{3-17'}$$

The logarithmic derivative of the Γ-function is given by

$$\frac{d}{dz} \ln \Gamma(z) = \int_0^\infty \left(\frac{e^{-t}}{t} - \frac{e^{-zt}}{1 - e^{-t}} \right) dt \tag{3-18}$$

if x = real part of $z > 0$, as was shown by Gauss.

From this result it is possible to obtain an expression for $\ln \Gamma(z)$ which is useful in evaluating $\Gamma(z)$ for large values of z:

$$\ln \Gamma(z) = (z - \tfrac{1}{2}) \ln z - z + \tfrac{1}{2} \ln (2\pi) + O\left(\frac{1}{x}\right) \tag{3-19}$$

where $O(1/x)$ represents a series of terms which vanish for large z at least as strongly as $1/x$. For real z, (19) takes the form of *Stirling's* series, when written for Γ instead of its logarithm:

$$\Gamma(x) =$$
$$e^{-x}x^{x-1/2}(2\pi)^{1/2} \left\{ 1 + \frac{1}{12x} + \frac{1}{288x^2} - \frac{139}{51840x^3} - \frac{571}{2488320x^4} + \cdots \right\} \tag{3-20}$$

It is valid when x is large. This expansion may be used for the approximate evaluation of factorials of large numbers:

$$N! = N\Gamma(N) = e^{-N}N^N(2\pi N)^{1/2}(1 + \cdots) \tag{3-21}$$

In concluding, let us compute a few numerical values of the Γ-function. It has already been noted that

$$\Gamma(0) = \infty, \quad \Gamma(1) = 1, \quad \Gamma(2) = 1, \quad \Gamma(3) = 2!, \quad \Gamma(4) = 3! \quad \text{etc.}$$

If the values of $\Gamma(x)$ in the interval $0 < x < 1$ are known, $\Gamma(x)$ can be computed for all real positive x by means of (5). $\Gamma(\frac{1}{2})$ is easily obtained from eq. (10):

$$\Gamma(\tfrac{1}{2}) = \int_0^\infty e^{-t}t^{-1/2}dt = 2\int_0^\infty e^{-x^2}dx$$

if x^2 is written for t. Hence $\Gamma(\frac{1}{2}) = \sqrt{\pi}$. The same result could have been obtained by putting $z = \frac{1}{2}$ in (12). Thus we find: $\Gamma(\frac{1}{2}) = \pi^{1/2}$, $\Gamma(\frac{3}{2}) = \frac{1}{2}\pi^{1/2}$, $\Gamma(\frac{5}{2}) = \frac{3}{4}\pi^{1/2}$, $\Gamma(\frac{7}{2}) = \frac{15}{8}\pi^{1/2}$, etc.[2] The qualitative behavior of $\Gamma(x)$ is plotted in Fig. 3.

Problems.

a. Prove eq. (13) by expanding the infinite product.

b. Prove eq. (17′) directly. *Hint:* Express $\Gamma(r)\Gamma(s)$ as a double integral in accordance with eq. (10). Next, put the two variables of integration, respectively, equal to x^2 and y^2 and then transform to polar coordinates. The radial integral will be $\Gamma(r + s)$, the remainder $B(r,s)$.

c. Show that $B(r,s) = B(r + 1,s) + B(r,s + 1)$.

3.3. Legendre Polynomials.—Of the solutions of Legendre's equation (2.28), the functions denoted by $P_l(x)$ in sec. 2.11 are of greatest interest because they remain finite at $x = \pm 1$. In physical problems, the argument of P_l is usually the cosine of an angle and has therefore the range $-1 \leqq x \leqq 1$. P_l is definite at the endpoints of that range; the other solutions are not. Hence the present discussion will be restricted to the polynomials P_l. We repeat their definition:

$$P_l(x) = \frac{(2l)!}{2^l(l!)^2}\left\{x^l - \frac{l(l-1)}{2(2l-1)}x^{l-2} + \frac{l(l-1)(l-2)(l-3)}{2 \cdot 4(2l-1)(2l-3)}x^{l-4} - \cdots\right\} \quad (3\text{-}22)$$

Specifically,

$$P_0 = 1, P_1 = x, P_2 = \tfrac{1}{2}(3x^2 - 1), P_3 = \tfrac{1}{2}(5x^3 - 3x), P_4 = \tfrac{1}{8}(35x^4 - 30x^2 + 3),$$
$$\text{etc.}$$

An interesting representation of P_l is easily established. When the function

$$F(x,y) = (1 - 2xy + y^2)^{-1/2} \quad (3\text{-}23)$$

is differentiated n times with respect to y and y is then put equal to zero, the result is seen to be $l!P_l(x)$. Hence if $F(x,y)$ is expanded in a

---

[2] Tabulated values of the Γ-function for real arguments may be found in Jahnke, E., and Emde, F., "Tables of Functions," Dover Publications, New York, 1943.

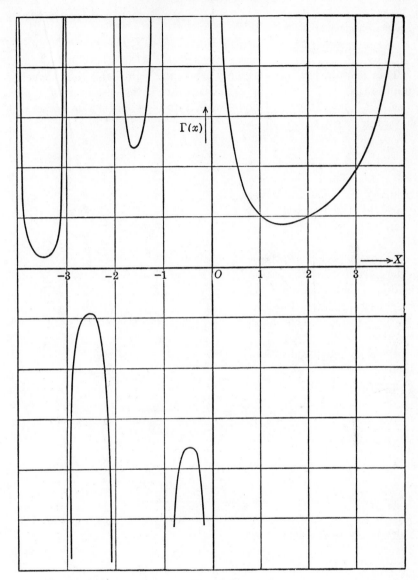

Fig. 3-3

MacLaurin series about the value $F(x,0)$ the result

$$F(x,y) = F(x,0) + y \left.\frac{\partial F(x,y)}{\partial y}\right|_{y=0} + \frac{y^2}{2!}\left.\frac{\partial^2 F(x,y)}{\partial y^2}\right|_{y=0} + \cdots + \frac{y^l}{l!}\left.\frac{\partial^l F(x,y)}{\partial y^l}\right|_{y=0}$$
$$+ \cdots$$

becomes

$$F(x,y) \equiv (1 - 2xy + y^2)^{-1/2} = \sum_{l=0}^{\infty} P_l(x)y^l \qquad (3\text{--}24)$$

This relation has meaning only when the right-hand side converges. Suppose that $|x| \leq 1$ which, as pointed out, is the case in most applications. $P_l(x)$ will then also lie between 1 and -1, for the definition of P shows that every

$$P_l(1) = 1$$

and that $|P_l(x)| < 1$ for $|x| < 1$. Thus the coefficients of y^l in (24) are never greater, in absolute value, than 1, and the series converges when $y < 1$.

Theorem (24) is of interest in the calculation of the potential due to a static distribution of electrical charges. In terms of Fig. 4, which depicts a distribution of charges $q_1 \cdots q_4$ of different magnitudes and possibly of different signs, the potential at P is

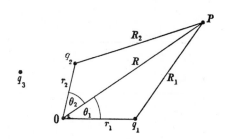

FIG. 3–4

$$V = \sum_i \frac{q_i}{R_i} = \frac{1}{R} \sum_i \frac{q_i}{\left[1 - 2\dfrac{r_i}{R}\cos\theta_i + \left(\dfrac{r_i}{R}\right)^2\right]^{\frac{1}{2}}}$$

On identifying $\cos\theta_i$ with x and (r_i/R) with y in (24) one obtains

$$V = \sum_{l=0}^{\infty} \frac{\sum_i q_i r_i^l P_l(\cos\theta_i)}{R^{l+1}}$$

With the use of the definition

$$Q_l = \sum_i q_i r_i^l P_l(\cos\theta_i) \qquad (3\text{--}25)$$

this result becomes the *multipole expansion* of the potential arising from an electric charge distribution:

$$V = \sum_l \frac{Q_l}{R^{l+1}} \qquad (3\text{--}26)$$

The monopole strength Q_0 is $\sum_i q_i$; the dipole strength Q_1 is $\sum_i q_i r_i \cos\theta_i$, and it represents the component of what is called the *dipole moment* of the charge distribution, $\sum_i q_i \mathbf{r}_i$, in the direction toward P. The quadrupole strength, $Q_2 = \sum_i q_i r_i^2 P_2(\cos\theta_i)$, is a scalar quantity constructible from the components of a tensor called the *quadrupole moment*,[3] and so on.

Q_l is called the strength of the 2^l-pole of the charge distribution. Its value depends on the choice of origin. If all charges have the same sign, Q_1 can be made to vanish by a suitable choice of origin. Furthermore, Q_2 can be given an especially simple form by choice of origin and axes, etc. Similar remarks are true about multipole moments.

The reader might find it interesting to verify the following statements.

(1) Two equal charges of opposite sign produce a dipole moment which is independent of the choice of origin. Their quadrupole moment can be made to vanish by taking the origin midway between them, in which case all Q with even subscripts vanish.

(2) Four equal charges disposed with alternating signs about the corners of a parallelogram produce a zero dipole moment and hence a vanishing Q_1; the quadrupole moment for a given orientation of axes is finite and independent of the choice of origin. Q_2 depends on the angle of orientation.

(3) A continuous spherical distribution of charge has a finite quadrupole-moment tensor, but vanishing Q_2.

The entire analysis leading to eq. (25) presupposes, of course, that every charge is closer to the origin than the point P, since the requirement $y = (r_i/R) < 1$ must be obeyed.

From the foregoing results one can derive a useful expression for P_l. Let $\mathbf{r}$ be a vector extending from the origin, Δz an increment in the z-direction, and $\mathbf{R} = \mathbf{r} + \Delta \mathbf{z}$. (Cf. Fig. 5.) If then we express

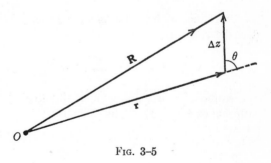

Fig. 3-5

$$\frac{1}{R} = [r^2 + (\Delta z)^2 + 2r\Delta z \cos\theta]^{-1/2} = \frac{1}{r}\left[1 + \frac{2\Delta z}{r}\cos\theta + \left(\frac{\Delta z}{r}\right)^2\right]^{-1/2}$$

---

[3] If we label the Cartesian components of r_i by $r_1^i = x_i$, $r_2^i = y_i$, $r_3^i = z_i$, the tensor in question is $T_{lm} = \sum_i q_i r_l^i r_m^i$. In the physical literature the terminology is sometimes confused, the multipole strength being identified with the multipole moment. It is correct to say that the *multipole strength*, i.e., the term appearing in the expansion, is a scalar form involving the components of the *multipole moment*, a tensor of rank l.

by means of (24), putting $-\cos\theta = x$ and $\Delta z/r = y$, we have

$$\frac{1}{R} = \frac{1}{r}\sum_{l=0}^{\infty} P_l(-\cos\theta)\left(\frac{\Delta z}{r}\right)^l \tag{3-27}$$

On the other hand, if $1/R = 1/|\,\mathbf{r}+\Delta\mathbf{z}\,|$ is expanded in a Taylor series about $\mathbf{R} = \mathbf{r}$ the result is

$$\frac{1}{R} = \frac{1}{r} + \Delta z\,\frac{\partial}{\partial z}\left(\frac{1}{r}\right) + \cdots + \frac{(\Delta z)^l}{l!}\frac{\partial^l}{\partial z^l}\left(\frac{1}{r}\right) + \cdots \tag{3-28}$$

On comparing the coefficients of $(\Delta z)^l$ in (27) and (28) it is seen that

$$\frac{1}{r^{l+1}}P_l(-\cos\theta) = \frac{1}{l!}\frac{\partial^l}{\partial z^l}\left(\frac{1}{r}\right)$$

Since $P_l(-x) = (-1)^l P_l(x)$, this is equivalent to

$$P_l(\cos\theta) = \frac{(-1)^l}{l!}r^{l+1}\frac{\partial^l}{\partial z^l}\left(\frac{1}{r}\right) \tag{3-29}$$

In using this relation it is understood that $\cos\theta = z/r$, and $r^2 = x^2 + y^2 + z^2$.

Another expression for P_l involving an l-th derivative, and in some sense simpler than (29), is known as *Rodrigues' formula*. To obtain it we observe, first of all, that

$$(x^2 - 1)^l = \sum_{\lambda=0}^{l}(-1)^\lambda\frac{l!}{\lambda!(l-\lambda)!}\,x^{2(l-\lambda)}$$

in accordance with the binomial theorem. When this expression is differentiated l times, there results:

$$\frac{d^l}{dx^l}(x^2 - 1)^l = \sum_{\lambda}(-1)^\lambda\frac{l!}{\lambda!(l-\lambda)!}\,\frac{(2l-2\lambda)!}{(l-2\lambda)!}\,x^{l-2\lambda}$$

the summation extending over all integers λ including 0 until λ equals either $\frac{1}{2}l$ or $\frac{1}{2}(l-1)$. The right-hand side of this equation may be written

$$\frac{(2l)!}{l!}\left\{x^l - \frac{l(l-1)}{2(2l-1)}\,x^{l-2} + \cdots\right\}$$

Hence, from the definition of P_l, eq. (22),

$$P_l(x) = \frac{1}{2^l l!}\frac{d^l}{dx^l}(x^2 - 1)^l \tag{3-30}$$

which is the formula of Rodrigues.

An integral representation of P_l is due to *Schlaefli*. It can be derived by combining eq. (30) with the fact expressed by eq. (2). If the latter equation is differentiated l times with respect to the argument of $f(a)$, the result may be written

$$\frac{d^l}{dx^l} f(x) = \frac{l!}{2\pi i} \oint \frac{f(z)}{(z-x)^{l+1}} \, dz$$

Now choose $f(x)$ to be $(x^2 - 1)^l$, so that

$$\frac{d^l}{dx^l} (x^2 - 1)^l = \frac{l!}{2\pi i} \oint \frac{(z^2 - 1)^l}{(z-x)^{l+1}} \, dz$$

On comparing this with (30), it is seen that

$$P_l(x) = \frac{2^{-l}}{2\pi i} \oint \frac{(z^2 - 1)^l}{(z-x)^{l+1}} \, dz \tag{3-31}$$

The path of integration here is understood to be some contour enclosing the point x in a counter-clockwise sense. From this result, which is known as Schlaefli's formula, it is possible to derive the formula:

$$P_l(x) = \frac{1}{\pi} \int_0^\pi [x + \sqrt{x^2 - 1} \cos \varphi]^l \, d\varphi \tag{3-32}$$

To do this, one may take the contour to be a circle of radius $\sqrt{|x^2 - 1|}$, so that z in (31) is $x + \sqrt{x^2 - 1} \, e^{i\varphi}$ and φ varies from $-\pi$ to $+\pi$. The integral then becomes

$$P_l(x) = \frac{2^{-l}}{2\pi i} \int_{-\pi}^\pi \frac{[x^2 - 1 + 2x\sqrt{x^2 - 1}\, e^{i\varphi} + (x^2 - 1)e^{2i\varphi}]^l}{[\sqrt{x^2 - 1}\, e^{i\varphi}]^{l+1}} \sqrt{x^2 - 1}\, e^{i\varphi} i \, d\varphi$$

$$= \frac{1}{2\pi} \int_{-\pi}^\pi [x + \sqrt{x^2 - 1} \cos \varphi]^l \, d\varphi$$

This result is equivalent to (32) because the integral from $-\pi$ to zero is equal to that from zero to π.[4]

---

[4] In general, the definite integral $\int_{-\alpha}^0 I(x)dx = \int_0^\alpha I(x)dx$ if the integrand is an even function of x, that is, if $I(-x) = I(x)$. To see this, change the variable of integration to $-x$, and make a corresponding change in the limits:

$$\int_{-\alpha}^0 I(x)dx = -\int_\alpha^0 I(-x)dx = -\int_\alpha^0 I(x)dx = \int_0^\alpha I(x)dx$$

If $I(x)$ is odd, that is if $I(-x) = -I(x)$,

$$\int_{-\alpha}^0 I(x)dx = -\int_0^\alpha I(x)dx, \quad \text{so that} \quad \int_{-\alpha}^\alpha I(x)dx = 0$$

3.4. Integral Properties of Legendre Polynomials.—Integrals over products of Legendre polynomials, which are needed in many quantum mechanical problems, are best obtained with the use of Rodrigues' formula. We wish to calculate

$$\int_{-1}^{1} P_l(x)P_{l'}(x)dx$$

First, we suppose that $l' > l$. Substituting in accordance with eq. (30) this integral becomes

$$\frac{1}{2^{l+l'}l!\,l'!} \int_{-1}^{1} \frac{d^l}{dx^l}(x^2-1)^l \cdot \frac{d^{l'}}{dx^{l'}}(x^2-1)^{l'}dx \qquad (3\text{-}33)$$

After l' successive partial integrations, in which all the integrated parts vanish because every derivative of $(x^2-1)^l$ is zero at $x = \pm 1$, the remaining integral reads

$$\frac{(-1)^{l'}}{2^{l+l'}l!\,l'!} \int_{-1}^{1} \frac{d^{l+l'}}{dx^{l+l'}}(x^2-1)^l \cdot (x^2-1)^{l'}dx \qquad (3\text{-}34)$$

But the $(l+l')$-th derivative of $(x^2-1)^l$ is certainly zero because the highest power of x in $(x^2-1)^l$ is x^{2l}, and $l+l'$ is, by hypothesis, greater than $2l$. Therefore the integral vanishes. This is clearly true whenever $l' \neq l$; for if l should be greater than l' we need only " unpeel " the derivatives appearing in (33) in the reverse manner by partial integrations, and we are left with an expression like (34) but with l and l' interchanged.

Next, suppose that $l = l'$. The integral in (34) then reads

$$\int_{-1}^{1}(x^2-1)^l \frac{d^{2l}}{dx^{2l}}(x^2-1)^l dx = (2l)! \int_{-1}^{1}(x^2-1)^l dx$$

the latter because the only term of $(x^2-1)^l$ which will not vanish after $2l$ differentiations is x^{2l}. But on putting $x = \cos\theta$ it is seen that

$$\int_{-1}^{1}(x^2-1)^l dx = (-1)^l \cdot 2 \int_{0}^{\pi/2} \sin^{2l+1}\theta\,d\theta = \frac{(-1)^l \cdot 2^{l+1}l!}{3 \cdot 5 \cdots (2l+1)}$$

Collecting coefficients, we find in place of (34)

$$\frac{(-1)^l}{2^{2l}(l!)^2} \cdot (2l)! \cdot \frac{(-1)^l \cdot 2^{l+1}l!}{3 \cdot 5 \cdots (2l+1)} = \frac{2}{2l+1}$$

Our results may be combined in the formula

$$\int_{-1}^{1} P_l(x)P_{l'}(x)dx = \frac{2}{2l+1}\delta_{l,l'} \qquad (3\text{-}35)$$

The symbol $\delta_{l,l'}$ here employed is freely used in mathematics and physics;

it is called the " Kronecker " δ and represents a discontinuous factor which is taken to be unity when the two subscripts have the same value $(l = l')$ but is zero when they are not equal.

3.5. Recurrence Relations between Legendre Polynomials.—Relations between Legendre polynomials are most simply derived from Schlaefli's representation of $P_l(x)$, eq. (31). We first observe that

$$\oint \frac{(z^2 - 1)^l}{(z - x)^l}\, dz = \oint \frac{z(z^2 - 1)^l}{(z - x)^{l+1}}\, dz - x \oint \frac{(z^2 - 1)^l}{(z - x)^{l+1}}\, dz \quad (3\text{--}36)$$

The first term on the right of this equation, however, may be transformed as follows. Since

$$\frac{d}{dz}\left(\frac{z^2 - 1}{z - x}\right)^{l+1} = (l + 1)\left[2z\,\frac{(z^2 - 1)^l}{(z - x)^{l+1}} - \frac{(z^2 - 1)^{l+1}}{(z - x)^{l+2}}\right]$$

and since the integral of the left-hand side around a closed contour must vanish, we find

$$\oint \frac{z(z^2 - 1)^l}{(z - x)^{l+1}}\, dz = \tfrac{1}{2} \oint \frac{(z^2 - 1)^{l+1}}{(z - x)^{l+2}}\, dz$$

Equation (36) thus reads

$$\oint \frac{(z^2 - 1)^l}{(z - x)^l}\, dz = \tfrac{1}{2} \oint \frac{(z^2 - 1)^{l+1}}{(z - x)^{l+2}}\, dz - x \oint \frac{(z^2 - 1)^l}{(z - x)^{l+1}}\, dz$$

Reference to equation (31) allows the two terms on the right to be identified, after multiplication by $(2^{l+1}\pi i)^{-1}$, with $P_{l+1}(x) - xP_l(x)$. Hence

$$P_{l+1}(x) - xP_l(x) = \frac{2^{-l}}{2\pi i} \oint \frac{(z^2 - 1)^l}{(z - x)^l}\, dz \quad (3\text{--}37)$$

When this is differentiated with respect to x, there results

$$P'_{l+1}(x) - xP'_l(x) = (l + 1)P_l(x) \quad (3\text{--}38)$$

which is the first important relation to be derived. It connects Legendre functions, and their derivatives, of degrees l and $l + 1$.

A relation connecting Legendre polynomials of three different degrees may be deduced in a similar way. Clearly, since

$$\oint \frac{d}{dz}\left[\frac{z(z^2 - 1)^l}{(z - x)^l}\right] dz = 0$$

$$\oint \frac{(z^2 - 1)^l}{(z - x)^l}\, dz + 2l \oint \frac{z^2(z^2 - 1)^{l-1}}{(z - x)^l}\, dz - l \oint \frac{z(z^2 - 1)^l}{(z - x)^{l+1}}\, dz = 0$$

In the second integrand, we introduce $z^2 = (z^2 - 1) + 1$, and in the third

$z = (z - x) + x$, so that

$$(l + 1) \oint \frac{(z^2 - 1)^l}{(z - x)^l} \, dz + 2l \oint \frac{(z^2 - 1)^{l-1}}{(z - x)^l} \, dz - lx \oint \frac{(z^2 - 1)^l}{(z - x)^{l+1}} \, dz = 0$$

Now the first term appearing here may be identified by means of (37), the others by (31). Thus, after simple rearrangement,

$$(l + 1)P_{l+1}(x) - (2l + 1)xP_l(x) + lP_{l-1}(x) = 0 \qquad (3\text{--}39)$$

The remaining relations are derived by differentiation and elimination among formulas (38) and (39). Thus, when (39) is differentiated with respect to x and P'_{l+1} is eliminated by means of (38),

$$xP'_l(x) - P'_{l-1}(x) = lP_l(x) \qquad (3\text{--}40)$$

Finally, the reader will have no difficulty in proving, by eliminations among (38), (39), and (40), that

$$P'_{l+1}(x) - P'_{l-1}(x) = (2l + 1)P_l(x) \qquad (3\text{--}41)$$

and

$$(x^2 - 1)P'_l(x) = lxP_l(x) - lP_{l-1}(x) \qquad (3\text{--}42)$$

It may be remarked that the recurrence relations here derived are also correct for Legendre functions having non-integral indices l, although the above proof does not indicate this fact explicitly.

3.6. Associated Legendre Polynomials.—The associated Legendre polynomial has been defined in sec. 2.11 as

$$P_l^m(x) = (1 - x^2)^{m/2} \frac{d^m}{dx^m} P_l(x) \qquad (3\text{--}43)$$

This definition is meaningless except when m is an integer not smaller than zero. In the present discussion, this will always be understood to be the case.

Recurrence Relations. We first derive the more important recurrence relations between these functions, which, as was shown in 2.11, satisfy the differential equation

$$(1 - x^2) \frac{d^2 P_l^m}{dx^2} - 2x \frac{dP_l^m}{dx} + \left\{ l(l + 1) - \frac{m^2}{1 - x^2} \right\} P_l^m = 0$$

The function $P_l^{(m)} = (d^m/dx^m)P_l$ was seen to be a solution of

$$(1 - x^2) \frac{d^2}{dx^2} P_l^{(m)} - 2(m + 1)x \frac{d}{dx} P_l^{(m)}$$

$$+ [l(l + 1) - m(m + 1)]P_l^{(m)} = 0 \qquad (3\text{--}44)$$

When this equation is multiplied by $(1 - x^2)^{m/2}$ and the definition (43) is used, it may be written

$$P_l^{m+2} - 2(m + 1) \frac{x}{\sqrt{1 - x^2}} P_l^{m+1} + [l(l + 1) - m(m + 1)]P_l^m = 0$$

or, on replacing m by $m - 1$

$$P_l^{m+1} - 2m \frac{x}{\sqrt{1 - x^2}} P_l^m + [l(l + 1) - m(m - 1)]P_l^{m-1} = 0 \qquad (3\text{-}45)$$

This represents the fundamental relation between three associated Legendre functions with equal l but consecutive values of m.

To get a similar relation for equal m but consecutive l we return to eqs. (39) and (41). Differentiating the first of these m times we have

$$(l + 1)P_{l+1}^{(m)} - (2l + 1)xP_l^{(m)} - (2l + 1)mP_l^{(m-1)} + lP_{l-1}^{(m)} = 0 \qquad (3\text{-}46)$$

When (41) is differentiated $m - 1$ times, the result is

$$P_{l+1}^{(m)} - P_{l-1}^{(m)} = (2l + 1)P_l^{(m-1)} \qquad (3\text{-}47)$$

On eliminating $P_l^{(m-1)}$ between (46) and (47) we find

$$(2l + 1)xP_l^{(m)} = (l + m)P_{l-1}^{(m)} + (l - m + 1)P_{l+1}^{(m)}$$

When this is multiplied by $(1 - x^2)^{m/2}$, there results the desired relation:

$$xP_l^m = (2l + 1)^{-1}[(l + m)P_{l-1}^m + (l - m + 1)P_{l+1}^m] \qquad (3\text{-}48)$$

Two " mixed " relations, in which both l and m have different values, are often useful (cf. Chapter 11) and will now be derived. One is at once obtained from (47) when that equation is written with m replaced by $m + 1$ and then multiplied by $(1 - x^2)^{(m+1)/2}$:

$$(1 - x^2)^{1/2}P_l^m = (2l + 1)^{-1}[P_{l+1}^{m+1} - P_{l-1}^{m+1}] \qquad (3\text{-}49)$$

The other can be deduced from eq. (45). When xP_l^m in (45) is eliminated by means of (48) it reads:

$$P_l^{m+1} = \frac{2m}{(1 - x^2)^{1/2}(2l + 1)} [(l + m)P_{l-1}^m + (l - m + 1)P_{l+1}^m]$$
$$- [l(l + 1) - m(m - 1)]P_l^{m-1}$$

Here P_l^{m-1} can be expressed in terms of P_{l+1}^m and P_{l-1}^m by means of (49) (written for $m - 1$ instead of m). When this is done and the terms are collected, we find

$$(1 - x^2)^{1/2}P_l^{m+1} = (2l + 1)^{-1}[(l + m)(l + m + 1)P_{l-1}^m$$
$$- (l - m)(l - m + 1)P_{l+1}^m] \qquad (3\text{-}50a)$$

For convenience in later work this may be written in a form similar to (49) if only m is replaced by $m - 1$. Thus

$$(1 - x^2)^{1/2}P_l^m = (2l + 1)^{-1}[(l + m)(l + m - 1)P_{l-1}^{m-1}$$
$$- (l - m + 1)(l - m + 2)P_{l+1}^{m-1}] \quad (3\text{-}50)$$

It is seen from (49) and (50) that $\sqrt{1 - x^2}P_l^m$ can be expressed in terms of P_{l-1}^{m+1} and P_{l+1}^{m+1}, as well as P_{l-1}^{m-1} and P_{l+1}^{m-1}. Relations (48), (49), and (50) are used in calculating quantum mechanical matrix elements in central field problems (cf. sec. 11.13).

Integral Properties. It is desired to evaluate the integral over the product of two associated Legendre functions having the same index m. If we use the definition (43) together with Rodrigues' formula (30) we have

$$\int_{-1}^{1} P_l^m(x)P_{l'}^m(x)dx = \frac{(-1)^m}{2^{l+l'}l!\,l'!} \int_{-1}^{1} X^m \frac{d^{l+m}}{dx^{l+m}} X^l \cdot \frac{d^{l'+m}}{dx^{l'+m}} X^{l'}dx \quad (3\text{-}51)$$

where $X = x^2 - 1$. As was done in connection with the Legendre polynomials, we again carry out a sequence of partial integrations, $l' + m$ in number, in which all the integrated parts are zero. The integral in (51) then reads

$$(-1)^{l'+m} \int_{-1}^{1} \frac{d^{l'+m}}{dx^{l'+m}} \left[X^m \frac{d^{l+m}}{dx^{l+m}} X^l \right] X^{l'}dx$$

$$= (-1)^{l'+m} \int_{-1}^{1} X^{l'} \sum_\lambda \binom{l' + m}{\lambda} \frac{d^{l'+m-\lambda}}{dx^{l'+m-\lambda}} (X^m) \frac{d^{l+m+\lambda}}{dx^{l+m+\lambda}} (X^l)dx$$

Now the term of highest power in X^m is x^{2m}, that in X^l is x^{2l}. Therefore every term in the summation over λ will be zero unless, simultaneously,

$$l' + m - \lambda \leqq 2m \quad \text{and} \quad l + m + \lambda \leqq 2l \quad (3\text{-}52)$$

The first of these implies: $\lambda \geqq l' - m$, the second $\lambda \leqq l - m$. Let us suppose that $l < l'$. Since m is positive, these two relations are incompatible, and the summation contains *no* term which is different from zero. Hence the integral (51) vanishes. If $l > l'$, it must also be zero because the integrand is perfectly symmetrical with respect to l and l'. To show this result explicitly, the partial integrations must be performed in the reverse manner.

If $l' = l$ the two relations (52) are indeed compatible, but only for the single value $\lambda = l - m$. Hence the sum over λ contains only one term, and the integral becomes

$$(-1)^{l+m} \int_{-1}^{1} X^l \binom{l + m}{l - m} \frac{d^{2m}}{dx^{2m}} (X^m) \frac{d^{2l}}{dx^{2l}} (X^l)dx$$

$$= (-1)^{l+m} \binom{l + m}{l - m} (2l)!\,(2m)! \int_{-1}^{1} X^l dx$$

The remaining integral has already been computed (preceding eq. 35); it was found to be

$$\frac{(-1)^l 2^{2l+1}(l!)^2}{(2l+1)!}$$

On the other hand, $\binom{l+m}{l-m} = \dfrac{(l+m)!}{(l-m)!\,(2m)!}$. Thus we find, collecting the various factors,

$$\int_{-1}^{1} [P_l^m(x)]^2 dx = \frac{(-1)^m}{2^{2l}(l!)^2} \cdot \frac{(l+m)!\,(-1)^{l+m}}{(l-m)!\,(2m)!} \cdot (2l)!\,(2m)! \cdot \frac{(-1)^l 2^{2l+1}(l!)^2}{(2l+1)!}$$

$$= \frac{(l+m)!}{(l-m)!} \cdot \frac{2}{2l+1}$$

It has thus been proved that

$$\int_{-1}^{1} P_l^m(x)P_{l'}^m(x)dx = \frac{(l+m)!}{(l-m)!} \frac{2}{2l+1}\delta_{l'l} \tag{3-53}$$

If x is taken to be $\cos\theta$, this result may be written in the equivalent form:

$$\int_{0}^{\pi} P_l^m(\cos\theta)P_{l'}^m(\cos\theta)\sin\theta d\theta = \frac{(l+m)!}{(l-m)!} \frac{2}{2l+1}\delta_{l'l} \tag{3-53a}$$

3.7. Addition Theorem for Legendre Polynomials.—To prove the famous addition theorem for Legendre polynomials (eq. 61) it is necessary first to establish a formula due to *Heine*. If we substitute the Schlaefli integral, eq. (31), for P_l in the definition of P_l^m (eq. 43) and carry out the differentiations with respect to x under the integral sign, we have

$$P_l^m(x) = \frac{2^{-l}}{2\pi i}(l+1)(l+2)\cdots(l+m)(1-x^2)^{m/2} \times$$

$$\oint (z^2-1)^l(z-x)^{-l-m-1}dz$$

Now let $z = x + \sqrt{x^2-1}\,e^{i\phi}$ and integrate over ϕ from $-\pi$ to π in accordance with the meaning of the contour (cf. eq. 31 et seq.). Then

$$P_l^m(x) = \frac{(l+1)(l+2)\cdots(l+m)}{2\pi}(1-x^2)^{m/2} \times$$

$$\int_{-\pi}^{\pi} \frac{[x+\sqrt{x^2-1}\cos\phi]^l}{[\sqrt{x^2-1}\,e^{i\phi}]^m}d\phi$$

$$= \frac{(l+1)(l+2)\cdots(l+m)}{2\pi}(-1)^{m/2} \times$$

$$\int_{-\pi}^{\pi}[x+\sqrt{x^2-1}\cos\phi]^l e^{-im\phi}d\phi$$

$$P_l^m(x) = \frac{(l+1)(l+2)\cdots(l+m)(-1)^{m/2}}{\pi} \times$$

$$\int_0^\pi [x + \sqrt{x^2-1}\cos\phi]^l \cos m\phi\, d\phi \quad (3\text{-}54a)$$

In taking the last step we observe that, of the two constituents of $e^{im\phi} = \cos m\phi + i \sin m\phi$, the first is an even, the second an odd function of ϕ. Since the other remaining factor of the integrand is even, only the cosine part of $e^{im\phi}$ will give a finite integral, and this has twice the value of the integral between the limits 0 and π. Eq. (54a) is Heine's formula.

If in the differential equation for $P_l^m(x)$ we substitute $-l-1$ for l, the equation remains unaltered. Therefore $P_{-l-1}^m = P_l^m$. In view of this, Heine's formula may also be written

$$P_l^m(x) = P_{-l-1}^m = \frac{(-l)(-l+1)\cdots(-l-1+m)(-1)^{m/2}}{\pi} \times$$

$$\int_0^\pi [x + \sqrt{x^2-1}\cos\phi]^{-l-1} \cos m\phi\, d\phi$$

$$= \frac{l(l-1)\cdots(l-m+1)(-1)^{3m/2}}{\pi} \times$$

$$\int_0^\pi \frac{\cos m\phi\, d\phi}{[x + \sqrt{x^2-1}\cos\phi]^{l+1}} \quad (3\text{-}54b)$$

To prove the addition theorem we consider the equation

$$\sum_{l=0}^\infty p^l \frac{[x_1 + (x_1^2-1)^{1/2}\cos(\omega-\alpha)]^l}{[x_2 + (x_2^2-1)^{1/2}\cos\alpha]^{l+1}}$$
$$= \{x_2 + (x_2^2-1)^{1/2}\cos\alpha - p[x_1 + (x_1^2-1)^{1/2}\cos(\omega-\alpha)]\}^{-1} \quad (3\text{-}55)$$

which is an identity for sufficiently small values of the parameter p. All other quantities appearing in (55) are supposed to be real, but otherwise unrestricted. This relation is simply an application of the expansion

$$\sum_{l=0}^\infty x^l = (1-x)^{-1}, \quad |x| < 1$$

Let us integrate eq. (55) over $d\alpha$ from $-\pi$ to π. The integral on the right may be evaluated by means of the formula

$$\int_{-\pi}^\pi \frac{d\alpha}{a + b\cos\alpha + c\sin\alpha} = 2\pi(a^2 - b^2 - c^2)^{-1/2}$$

which was proved as an example on page 93.   Here

$$a = x_2 - px_1$$
$$b = (x_2^2 - 1)^{1/2} - p(x_1^2 - 1)^{1/2} \cos \omega$$
$$c = -p(x_1^2 - 1)^{1/2} \sin \omega$$

Hence the right-hand side of (55) becomes after integration

$$2\pi \{1 - 2p[x_1 x_2 - (x_1^2 - 1)^{1/2}(x_2^2 - 1)^{1/2} \cos \omega] + p^2\}^{-1/2}$$

As will be seen forthwith, the expression in [ ] appearing here has a very simple geometrical meaning.   For the present, let us designate it by x:

$$x \equiv x_1 x_2 - (x_1^2 - 1)^{1/2}(x_2^2 - 1)^{1/2} \cos \omega \qquad (3\text{--}56)$$

The result of the integration may therefore be written

$$2\pi(1 - 2px + p^2)^{-1/2}$$

But by the theorem on Legendre functions, eq. (24), this is

$$2\pi \sum_{l=0}^{\infty} p^l P_l(x)$$

The left hand side of (55) may be integrated term by term because the expression is assumed to converge.   On comparing coefficients of p^l we see, therefore, that

$$P_l(x) = \frac{1}{2\pi} \int_{-\pi}^{\pi} \frac{[x_1 + (x_1^2 - 1)^{1/2} \cos (\omega - \alpha)]^l}{[x_2 + (x_2^2 - 1)^{1/2} \cos \alpha]^{l+1}} \, d\alpha \qquad (3\text{--}57)$$

The last step of the proof involves an expansion of $P_l(x)$ in a Fourier series.[5]   Clearly, $P_l(x)$, being a polynomial of degree l in $\cos \omega$, can be expressed in the form

$$P_l(x) = \tfrac{1}{2}c_0 + \sum_{m=1}^{l} c_m \cos m\omega \qquad (3\text{--}58)$$

The coefficients c_m are given by

$$c_m = \frac{1}{\pi} \int_{-\pi}^{\pi} P_l(x) \cos m\omega \, d\omega$$

$$= \frac{1}{2\pi^2} \int_{-\pi}^{\pi} d\alpha \int_{-\pi}^{\pi} d\omega \, \frac{[x_1 + (x_1^2 - 1)^{1/2} \cos (\omega - \alpha)]^l}{[x_2 + (x_2^2 - 1)^{1/2} \cos \alpha]^{l+1}} \cos m\omega$$

when $P_l(x)$ is replaced in accordance with (57).   In this integral, we may

[5] See sec. 8.2.

introduce the variable $\omega - \alpha \equiv \beta$ in place of ω, so that

$$c_m = \frac{1}{2\pi^2} \int_{-\pi}^{\pi} \frac{d\alpha}{[x_2 + (x_2^2 - 1)^{1/2} \cos \alpha]^{l+1}} \int_{-\pi}^{\pi} [x_1 + (x_1^2 - 1)^{1/2} \cos \beta]^l \times$$
$$(\cos m\alpha \cos m\beta - \sin m\alpha \sin m\beta) d\beta$$

The integration with respect to β over the term containing $\sin m\beta$ is obviously zero because the integrand is odd. The other term can be evaluated by means of eq. (54a). The result is

$$c_m =$$

$$\frac{1}{2\pi^2} \int_{-\pi}^{\pi} \frac{\cos m\alpha \, d\alpha}{[x_2 + (x_2^2 - 1)^{1/2} \cos \alpha]^{l+1}} \cdot \frac{2\pi(-1)^{-m/2}}{(l+1)(l+2)\cdots(l+m)} P_l^m(x_1)$$

In the remaining integration over α we use (54b), obtaining

$$c_m = 2 \frac{(l-m)!}{(l+m)!} P_l^m(x_1) P_l^m(x_2)$$

Hence from (58):

$$P_l(x) = P_l(x_1)P_l(x_2) + 2 \sum_{m=1}^{l} \frac{(l-m)!}{(l+m)!} P_l^m(x_1)P_l^m(x_2) \cos m\omega \qquad (3\text{-}59)$$

which is the desired addition theorem.

Finally, let us investigate the meaning of x defined in eq. (56). If θ_1, φ_1 and θ_2, φ_2 denote, respectively, the polar and azimuthal angles of two lines passing through the origin, then Θ, the angle between these two lines, is given by

$$\cos \Theta = \cos \theta_1 \cos \theta_2 + \sin \theta_1 \sin \theta_2 \cos (\varphi_1 - \varphi_2) \qquad (3\text{-}60)$$

Thus, if in (56) the following identifications are made:

$$x = \cos \Theta, \quad x_1 = \cos \theta_1, \quad x_2 = \cos \theta_2, \quad \omega = \varphi_1 - \varphi_2$$

then (59) becomes

$$P_l(\cos \Theta) = P_l(\cos \theta_1) P_l(\cos \theta_2) + 2 \sum_{m=1}^{l} \frac{(l-m)!}{(l+m)!} \times$$

$$P_l^m(\cos \theta_1) P_l^m(\cos \theta_2) \cos m(\varphi_1 - \varphi_2) \qquad (3\text{-}61)$$

In quantum mechanics (cf. sec. 11.12) it is convenient to use associated Legendre functions which differ from $P_l^m(x)$ by factors depending on l and m, but constant with respect to x and so chosen that the integral over the square of the functions is unity. These functions are called "normalized" associated Legendre functions. They will here be denoted by Π_l^m. Let us put $\Pi_l^m(x) = N_{l,m}P_l^m(x)$. Then if we wish $\int_0^{\pi} [\Pi_l^m(\cos \theta)]^2 \sin \theta d\theta$ to be equal to unity we must, in view of (53), put

$$N_{l,m} = \left[\frac{2l + 1}{2} \frac{(l - m)!}{(l + m)!} \right]^{1/2}$$

It is also customary to permit the index m of Π_l^m to be negative and to define

$$\Pi_l^m(x) = \left[\frac{2l + 1}{2} \frac{(l - |m|)!}{(l + |m|)!} \right]^{1/2} P_l^{|m|}(x) \tag{3-62}$$

The index m may then take on all integral values including zero from $-l$ to $+l$, while l is always a positive integer. In terms of the functions Π_l^m, the addition theorem takes a particularly simple form:

$$\sqrt{\frac{2l + 1}{2}} \, \Pi_l^0(\cos \Theta) = \sum_{m = -l}^{l} \Pi_l^m(\cos \theta_1) \Pi_l^m(\cos \theta_2) \cos m(\varphi_1 - \varphi_2) \tag{3-61a}$$

One may also replace the factor $\cos m(\varphi_1 - \varphi_2)$ in each term of this summation by $e^{im(\varphi_1 - \varphi_2)}$, because each pair of terms corresponding to $+m$ and $-m$ then yields a cosine function.

Problem: Express eqs. (48) to (50) in terms of Π-functions.

3.8. Bessel Functions.—In sec. 2.14 we have shown that a particular solution of Bessel's differential equation, (eq. 2-57), is the " Bessel function of order n," defined as (cf. eq. 2-60)

$$J_n(x) = \sum_{\lambda=0}^{\infty} \frac{(-1)^\lambda}{\Gamma(\lambda + 1)\Gamma(\lambda + n + 1)} \left(\frac{x}{2}\right)^{n+2\lambda} \tag{3-63}$$

It is of interest, first, to note that for integral n, $J_n(x)$ is the coefficient of u^n in the expansion of $\exp [(x/2)(u - 1/u)]$. In fact $J_n(x)$ for integral n, called Bessel's coefficient, was originally defined by means of this relation. To prove it we merely expand the exponential, using the binomial theorem[6] to express $(u - 1/u)^\nu$:

$$\exp \left[\frac{x}{2}\left(u - \frac{1}{u}\right) \right] = \sum_{\nu=0}^{\infty} \frac{1}{\nu!} \left(\frac{x}{2}\right)^\nu \left(u - \frac{1}{u}\right)^\nu = \sum_{\nu, \lambda} \frac{1}{\nu!} \left(\frac{x}{2}\right)^\nu \binom{\nu}{\lambda} u^{\nu - \lambda}(-u)^{-\lambda}$$

$$= \sum_{\nu, \lambda} \frac{(-1)^\lambda \left(\frac{x}{2}\right)^\nu}{(\nu - \lambda)! \, \lambda!} u^{\nu - 2\lambda}$$

If we now put $\nu - 2\lambda = n$, this becomes

$$\exp \left[\frac{x}{2}\left(u - \frac{1}{u}\right) \right] = \sum_n u^n \left[\sum_{\lambda=0}^{\infty} \frac{(-1)^\lambda}{(n + \lambda)! \, \lambda!} \left(\frac{x}{2}\right)^{n+2\lambda} \right] \tag{3-64}$$

[6] The binomial theorem states:

$$(a + b)^n = \sum_\lambda \binom{n}{\lambda} a^{n - \lambda} b^\lambda$$

For integral n, the bracket appearing here is identical with the expansion (63); hence the above-mentioned theorem is proved.

From (64) an integral representation may be derived quite simply. By the theorem of residues (eq. 3) the coefficient of z^{-1} in an expansion of $f(z)$ is given by

$$a_{-1} = \frac{1}{2\pi i} \oint f(z)\, dz$$

the integral being taken in a counter-clockwise sense about $z = 0$. Similarly, the coefficient of z^n in the expansion of $f(z)$ will be:

$$a_n = \frac{1}{2\pi i} \oint \frac{f(z)}{z^{n+1}}\, dz$$

The theorem just proved is therefore tantamount to the relation

$$J_n(x) = \frac{1}{2\pi i} \oint u^{-n-1} e^{(x/2)(u-1/u)} du \qquad (3\text{-}65)$$

It is customary to write this result in a slightly different form, obtainable on replacement of the variable u by $2t/x$. Eq. (65) then reads

$$J_n(x) = \frac{1}{2\pi i} \left(\frac{x}{2}\right)^n \oint t^{-n-1} \exp\left[t - \frac{x^2}{4t}\right] dt \qquad (3\text{-}66)$$

While this integral has been shown here to be identical with the convergent sum of eq. (63) only if n is an integer, a more special consideration would indeed establish the equivalence of (63) and (66) for non-integral n also. A simple way to prove this fact is to show, by substitution, that (66) satisfies Bessel's differential equation. On performing the differentiations indicated on the left of eq. (2-57)[7] and substituting therein, we find

$$\frac{x^{n+2}}{2^{n+1}\pi i} \oint t^{-n-1} \left[1 - \frac{n+1}{t} + \frac{x^2}{4t^2}\right] \exp\left(t - \frac{x^2}{4t}\right) dt$$

$$= \frac{x^{n+2}}{2^{n+1}\pi i} \oint \frac{d}{dt}\left[t^{-n-1} \exp\left(t - \frac{x^2}{4t}\right)\right] dt = 0$$

because the integral around a closed loop of an exact differential is zero. We may therefore regard either (63) or (66) as a definition of $J_n(x)$ for both integral and non-integral values of n. For non-integral n, however, caution is required in the choice of the contour of integration in (66). This must clearly enclose the origin. But if we were to take, for example, a circle about the origin as center we should encounter a difficulty. For non-inte-

---

[7] The differentiations may be carried out in (66) without regard to the fixed path of integration, that is, " under the integral sign."

gral n the integrand is a many-valued function of t. Thus if the amplitude of t should vary, say, from $-\pi$ to $+\pi$, the integrand will not have performed a closed loop[8] and the last equation above would not be true. It is necessary, therefore, to select a path of integration which (a) encloses the origin and no other singularity of the integrand; (b) starts and ends at a point in the t-plane which will cause the integrand to perform a closed loop also.

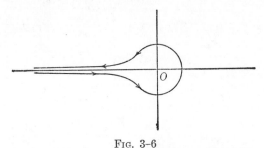

FIG. 3-6

Such a path is that illustrated in Fig. 6. Whenever eq. (65) or (66) is used, we shall understand that the contour integral is taken along this path. (The reader familiar with the theory of many-valued functions will observe that this path confines the integrand entirely to one of its branches provided that the argument of t is given its principal value.)

Recurrence Relations. From (66) one may show by direct differentiation that

$$\frac{d}{dx}[x^{-n}J_n(x)] = -x^{-n}J_{n+1}(x) \qquad (3\text{-}67)$$

or, when the differentiation on the left is carried out,

$$J_n'(x) = \frac{n}{x}J_n(x) - J_{n+1}(x) \qquad (3\text{-}68)$$

To obtain the other fundamental recurrence formula we perform the differentiations in the equation

$$\oint \frac{d}{dt}\left[t^{-n}\exp\left(t - \frac{x^2}{4t}\right)\right]dt = 0$$

The result is

$$\oint\left(t^{-n} + \frac{x^2}{4}t^{-n-2} - nt^{-n-1}\right)\exp\left(t - \frac{x^2}{4t}\right)dt = 0$$

[8] As an example of many-valued functions, consider $\sqrt{t} = \sqrt{\rho e^{i\theta}} = \rho^{1/2}e^{i\theta/2}$. When t moves along a circle of unit radius from $-\pi$ to $+\pi$, initial and final points are identical. But $\sqrt{t}$ has the initial point $e^{-i\pi/2} = -i$ and the final point $e^{i\pi/2} = +i$.

When use is now made of the definition (66) this reads

$$2\pi i \left(\frac{2}{x}\right)^{n-1} \left\{ J_{n-1} + J_{n+1} - \frac{2n}{x} J_n \right\} = 0$$

hence

$$J_{n-1}(x) + J_{n+1}(x) = \frac{2n}{x} J_n(x) \tag{3-69}$$

On eliminating J_{n+1} from (68) by means of (69) there results

$$J_n'(x) = J_{n-1}(x) - \frac{n}{x} J_n(x) \tag{3-70}$$

and from this and (68)

$$J_n'(x) = \tfrac{1}{2}[J_{n-1}(x) - J_{n+1}(x)] \tag{3-71}$$

Bessel's Integral. Let us consider $J_n(x)$ as defined by eq. (65):

$$J_n(x) = \frac{1}{2\pi i} \oint u^{-n-1} e^{x/2\,(u-1/u)} du$$

We may free ourselves from the condition that n be an integer by choosing a path like that of Fig. 6. More specifically, the contour will be taken to start at $-\infty$, pass to the right below the real axis ($u = te^{-i\pi}$, $\infty > t \geq 1$) up to the point -1, then perform a circle of unit radius in a counter-clockwise sense about the origin ($u = e^{i\theta}$, $-\pi < \theta \leq \pi$) and finally to return to $-\infty$ above the real axis ($u = te^{i\pi}$, $+1 \leq t < +\infty$). The contour integral then becomes

$$J_n(x) = \frac{1}{2\pi i} e^{(n+1)i\pi} \int_1^\infty t^{-n-1} \exp \frac{x}{2}\left(-t + \frac{1}{t}\right) dt +$$
$$\frac{1}{2\pi} \int_{-\pi}^{\pi} e^{-ni\theta + ix\sin\theta} d\theta - \frac{1}{2\pi i} e^{-(n+1)i\pi} \int_{+1}^{+\infty} t^{-n-1} \exp \frac{x}{2}\left(-t + \frac{1}{t}\right) dt$$

The second of these integrals may be written

$$\frac{1}{\pi} \int_0^\pi \cos\,(n\theta - x\sin\theta)d\theta$$

because the odd part of the exponential, $\sin\,(n\theta - x\sin\theta)$, vanishes on integration between $-\pi$ and $+\pi$. The first and last may be transformed by putting $t = e^\theta$ and noting that

$$e^\theta - e^{-\theta} = 2\sinh\theta$$

When they are combined, the result is

$$\frac{1}{2\pi i}[e^{(n+1)i\pi} - e^{-(n+1)i\pi}] \int_0^\infty e^{-n\theta - x\sinh\theta}\,d\theta = -\frac{1}{\pi}\sin n\pi \int_0^\infty e^{-n\theta - x\sinh\theta}\,d\theta$$

Hence

$$J_n(x) = \frac{1}{\pi} \left\{ \int_0^\pi \cos(n\theta - x\sin\theta)d\theta - \right.$$

$$\left. \sin n\pi \int_0^\infty \exp(-n\theta - x\sinh\theta)d\theta \right\} \quad (3\text{--}72)$$

This is a generalized form of *Bessel's integral,* derived by Bessel for integral values of n. In that special case the second integral vanishes and

$$J_n(x) = \frac{1}{\pi} \int_0^\pi \cos(n\theta - x\sin\theta)d\theta, \quad n = \text{integer} \quad (3\text{--}72a)$$

Bessel Functions of Half-Odd Order. When n is half an odd integer, for instance, $p + \frac{1}{2}$, $J_n(x)$ takes a particularly simple form and is related closely to the trigonometric functions. To show this, let us first compute $J_{1/2}(x)$ by the expansion (63). We may then use the recurrence formulas to obtain $J_{3/2}$, etc. Thus,

$$J_{1/2}(x) = \left(\frac{x}{2}\right)^{1/2} \sum_\lambda \frac{(-1)^\lambda x^{2\lambda}}{2^{2\lambda}\lambda! \, \Gamma(\frac{3}{2} + \lambda)}$$

But in view of eq. (5), etc., $[\Gamma(x + 1) = x\Gamma(x)]$

$$\Gamma(\tfrac{3}{2} + \lambda) = \frac{2\lambda + 1}{2} \cdot \frac{2\lambda - 1}{2} \cdots \tfrac{3}{2}\Gamma(\tfrac{3}{2})$$

$$= \frac{(2\lambda + 1)!}{2^{2\lambda}\lambda!} \Gamma(\tfrac{3}{2}) = \frac{(2\lambda + 1)!}{2^{2\lambda+1}\lambda!} \sqrt{\pi}$$

When this is substituted in the series for $J_{1/2}(x)$, there results

$$J_{1/2}(x) = \left(\frac{x}{2}\right)^{1/2} \sum_\lambda \frac{2(-1)^\lambda x^{2\lambda}}{(2\lambda + 1)! \, \sqrt{\pi}} = \left(\frac{2}{\pi x}\right)^{1/2} \sum_\lambda \frac{(-1)^\lambda x^{2\lambda+1}}{(2\lambda + 1)!}$$

$$= \left(\frac{2}{\pi x}\right)^{1/2} \sin x \quad (3\text{--}73)$$

From (67),

$$J_{3/2}(x) = -x^{1/2}\frac{d}{dx}[x^{-1/2}J_{1/2}(x)] = \left(\frac{2}{\pi x}\right)^{1/2}\left(\frac{\sin x}{x} - \cos x\right)$$

This process may be continued if the explicit form of the functions $J_{p+1/2}(x)$ is desired. A general formula is readily obtainable as follows. Eq. (67) may be written

$$x^{-n-1}J_{n+1}(x) = -\frac{1}{x}\frac{d}{dx}[x^{-n}J_n(x)] = -2\frac{d}{d(x^2)}[x^{-n}J_n(x)]$$

or, by repeated application of (67)

$$x^{-n-p}J_{n+p}(x) = (-2)^p \frac{d^p}{d(x^2)^p}[x^{-n}J_n(x)] \qquad (3\text{-}74)$$

Hence, on putting $n = \frac{1}{2}$,

$$J_{p+1/2}(x) = x^{p+1/2}(-2)^p \frac{d^p}{d(x^2)^p}[x^{-1/2}J_{1/2}(x)] =$$

$$\frac{(-1)^p(2x)^{p+1/2}}{\pi^{1/2}} \frac{d^p}{d(x^2)^p}\left(\frac{\sin x}{x}\right) \qquad (3\text{-}75)$$

The first few functions of half-odd order are given below.

p	$\sqrt{\dfrac{\pi x}{2}}\, J_{p+1/2}(x)$	$\sqrt{\dfrac{\pi x}{2}}\, J_{-p-1/2}(x)$
0	$\sin x$	$\cos x$
1	$\dfrac{\sin x}{x} - \cos x$	$-\sin x - \dfrac{\cos x}{x}$
2	$\left(\dfrac{3}{x^2} - 1\right)\sin x - \dfrac{3}{x}\cos x$	$\dfrac{3}{x}\sin x + \left(\dfrac{3}{x^2} - 1\right)\cos x$
3	$\left(\dfrac{15}{x^3} - \dfrac{6}{x}\right)\sin x - \left(\dfrac{15}{x^2} - 1\right)\cos x$	$-\left(\dfrac{15}{x^2} - 1\right)\sin x - \left(\dfrac{15}{x^3} - \dfrac{6}{x}\right)\cos x$
4	$\left(\dfrac{105}{x^4} - \dfrac{45}{x^2} + 1\right)\sin x$	$\left(\dfrac{105}{x^3} - \dfrac{10}{x}\right)\sin x$
	$-\left(\dfrac{105}{x^3} - \dfrac{10}{x}\right)\cos x$	$+\left(\dfrac{105}{x^4} - \dfrac{45}{x^2} + 1\right)\cos x$

When the differentiations in (75) are carried out it is easily established that the asymptotic form of $J_{p+1/2}$ is given, for all p, by

$$\lim_{x \to \infty} J_{p+1/2}(x) = \left(\frac{2}{\pi x}\right)^{1/2}\sin\left(x - p\frac{\pi}{2}\right) \qquad (3\text{-}76)$$

3.9. Hankel Functions and Summary on Bessel Functions.—The Bessel function $J_n(x)$ is only one particular solution of Bessel's differential equation. However, as was noted in sec. 2.14, $J_{-n}(x)$ is also a particular solution, for the differential equation is insensitive to the substitution of $-n$ for n. Hence a general solution of the form

$$y = aJ_n(x) + bJ_{-n}(x) \qquad (3\text{-}77)$$

is at hand provided J_n and J_{-n} are different functions, i.e., are linearly independent.   As was also shown in sec. 2.14, this is true as long as n is not an integer.   The *Hankel function*, frequently used in physical problems,[9] is of interest only in connection with non-integral n and the following remarks are restricted to that case.   This function is a solution of Bessel's equation of the form (77) with the constants a and b suitably chosen.   We distinguish two kinds of Hankel function, generally denoted by $H_n^{(1)}$ and $H_n^{(2)}$:

$$\left.\begin{aligned} H_n^{(1)} &= \frac{i}{\sin n\pi}\,[e^{-n\pi i}J_n(x) - J_{-n}(x)] \\[2mm] H_n^{(2)} &= \frac{-i}{\sin n\pi}\,[e^{n\pi i}J_n(x) - J_{-n}(x)] \end{aligned}\right\} \tag{3-78}$$

Hence, conversely,

$$\left.\begin{aligned} J_n(x) &= \tfrac{1}{2}[H_n^{(1)}(x) + H_n^{(2)}(x)] \\[1mm] J_{-n}(x) &= \tfrac{1}{2}[e^{n\pi i}H_n^{(1)}(x) + e^{-n\pi i}H_n^{(2)}(x)] \end{aligned}\right\} \tag{3-79}$$

These definitions hold, of course, for complex as well as for real values of the argument. Hankel functions are particularly useful for complex arguments, for they vanish strongly when the modulus of the argument approaches infinity, which is a requirement in many physical problems.

The qualitative properties of Bessel and Hankel functions may be summarized in the following brief survey.

A. $J_n(z)$ is real if z is real, complex if z is complex.

1. At $x = 0$,

$$J_n(x) = \begin{cases} 1 \text{ if } n = 0 \\ 0 \text{ if } n > 0 \\ 0 \text{ if } n < 0, \text{ and } n \text{ is an integer} \\ \infty \text{ if } n < 0, \text{ and } n \text{ is not an integer} \end{cases}$$

2. At $x \to \infty$, all $J_n(x)$ oscillate, but with ever-decreasing amplitude (provided x is real).

$$\lim_{x \to \infty} J_n(x) = \begin{cases} \pm\sqrt{\dfrac{2}{\pi x}}\,\cos\left(x - \dfrac{\pi}{4}\right) \text{ if } n \text{ is even} \\[4mm] \pm\sqrt{\dfrac{2}{\pi x}}\,\sin\left(x - \dfrac{\pi}{4}\right) \text{ if } n \text{ is odd} \end{cases}$$

---

[9] See, for instance, Stratton, J., " Electromagnetic Theory," McGraw-Hill Book Co., 1941. For applications in: propagation of radio waves, cf. Sommerfeld, A., *Ann. der Phys.* **28**, 692 (1909); theory of optical diffraction, cf. Wolfsohn, G., *Handb. d. Phys.*, **XX**, p. 282; quantum mechanics, cf. Margenau, H., *Phys. Rev.* **46**. 613 (1934).

B. $H_n(z)$ is complex if z is real, but $i^{n+1}H_n^{(1)}(ix)$ and $i^{-(n+1)}H_n^{(2)}(-ix)$ are always real if x is real and > 0.

1. At $z = 0$, both $H_n^{(1)}$ and $H_n^{(2)}$ become infinite. In fact

$$\lim_{x \to 0} i^{n+1}H_n^{(1)}(ix) = \lim_{x \to 0} i^{-(n+1)}H_n^{(2)}(-ix) = \frac{(n-1)!}{\pi}\left(\frac{2}{x}\right)^n$$

2. At $z \to \infty$, either $H_n^{(1)}(z)$ or $H_n^{(2)}(z)$ vanishes exponentially.

$$\lim_{|z| \to \infty} H_n^{(1)}(z) = \begin{cases} 0 \text{ if the imaginary part of } z > 0 \\ \\ \infty \text{ if the imaginary part of } z < 0 \end{cases}$$

$$\lim_{|z| \to \infty} H_n^{(2)}(z) = \begin{cases} \infty \text{ if the imaginary part of } z > 0 \\ \\ 0 \text{ if the imaginary part of } z < 0 \end{cases}$$

The behavior at infinity of both J_n and H_n is most easily remembered by noting the general similarity between

$$H_n^{(1)}(z) \quad \text{and} \quad e^{iz}$$

$$H_n^{(2)}(z) \quad \text{and} \quad e^{-iz}$$

$$J_n(z) \quad \text{and} \quad \tfrac{1}{2}(e^{iz} + e^{-iz}) = \cos z$$

The important difference between the Bessel functions and the circular functions is in the fact that the former have neither constant amplitude nor constant wave length.

Useful Formulas Involving Bessel Functions. We conclude the discussion of Bessel functions by appending here a list of formulas involving Bessel functions. Some of these are easily proved with the use of the theory here developed; to establish others reference should be made to more comprehensive treatises, such as that of Nielsen[10] and that of Gray and Mathews.[11] An extensive table of differential equations having Bessel functions as solutions is given in Jahnke and Emde.[12]

$$\left| J_0(x) \right| \leq 1, \quad \left| J_n(x) \right| \leq \sqrt{\tfrac{1}{2}} \quad \text{for} \quad n \geq 1; \quad x \text{ real}$$

$$[J_0(z)]^2 + 2\sum_{n}^{\infty} [J_n(z)]^2 = 1$$

$$J_{-n}(z)J_{n-1}(z) + J_{-n+1}(z)J_n(z) = \frac{2 \sin n\pi}{\pi z}$$

[10] Nielsen, N., "Handbuch der Theorie der Cylinderfunktionen," Teubner, 1904.

[11] Gray, A., and Mathews, G. B., "A Treatise on Bessel Functions," Macmillan. London, 1922.

[12] Jahnke, E., and Emde, F., "Tables of Functions With Formulae and Curves," Dover Publications, New York, 1943, pp. 146–147. See also Kamke, E., "Differentialgleichungen, Lösungsmethoden und Lösungen," Bd. 1. Third Edition, Chelsea Publishing Co., New York, 1948.

$$\int_0^x x^m J_{n+1}(x) J_{n-1}(x) dx = \frac{x^{m+1}}{m-1} \left[J_{n+1}(x) J_{n-1}(x) - (J_n(x))^2 \right]$$

$$+ \frac{m+1}{m-1} \int_0^x x^m [J_n(x)]^2 dx$$

provided n is a positive integer and $m + 1 > 0$

$$J_{n-1}(x) H_n^{(1)}(x) - J_n(x) H_{n-1}^{(1)}(x) = H_{n-1}^{(2)}(x) J_n(x) - H_n^{(2)}(x) J_{n-1}(x)$$

$$= \frac{2}{\pi i x}$$

$$J_n(x_1 + x_2) = \left(1 + \frac{x_2}{x_1} \right)^n \sum_{\lambda=0}^{\infty} \frac{(-1)^\lambda x_2^\lambda}{\lambda!} \left(1 + \frac{1}{2} \frac{x_2}{x_1} \right)^\lambda J_{n+\lambda}(x)$$

$$\int_0^x J_n(x) dx = 2 \sum_{\lambda=0}^{\infty} J_{n+2\lambda+1}(x)$$

$$\int_0^x x[J_n(\alpha x)]^2 dx = \frac{x^2}{2} \left\{ [J_n(\alpha x)]^2 - J_{n-1}(\alpha x) J_{n+1}(\alpha x) \right\}$$

This formula is also valid when all J's are replaced by $H^{(1)}$ or $H^{(2)}$.

$$\int_0^\infty x^{-n+m} J_n(\alpha x) dx = 2^{-n+m} \alpha^{n-m-1} \frac{\Gamma\left(\dfrac{m+1}{2} \right)}{\Gamma\left(n - \dfrac{m-1}{2} \right)}$$

if $2n + 1 > m > -1$.

$$\lim_{|z| \to \infty} J_n(z) = \left(\frac{2}{\pi z} \right)^{1/2} \times$$
$$\left[\cos\left(z - \frac{n\pi}{2} - \frac{\pi}{4} \right) - \frac{1}{8} \frac{4n^2 - 1}{z} \sin\left(z - \frac{n\pi}{2} - \frac{\pi}{4} \right) + O\left(\frac{1}{z^2} \right) \right]^{13}$$

$$\int_0^x x J_n(\alpha x) J_n(\beta x) dx = \frac{\beta x J_n(\alpha x) J_{n-1}(\beta x) - \alpha x J_{n-1}(\alpha x) J_n(\beta x)}{\alpha^2 - \beta^2}$$

3.10. Hermite Polynomials and Functions.—In sec. 2.15 the Hermite polynomial of degree n has been defined as the polynomial solution of Hermite's differential equation

$$y'' - 2xy' + 2ny = 0 \tag{3–80}$$

---

[13] The notation $O(1/z^2)$ is to be read " terms of the order $1/z^2$."

Such solutions were seen to exist when n is an integer. Explicitly, $y =$

$$H_n(x) = (2x)^n - \frac{n(n-1)}{1!}(2x)^{n-2} +$$
$$\frac{n(n-1)(n-2)(n-3)}{2!}(2x)^{n-4} - \cdots \quad (3\text{-}81)$$

We shall now find an equivalent expression for H_n in terms of a definite integral. If we put

$$y_n = \frac{1}{2\pi i} \oint z^{-n-1} e^{x^2 - (z-x)^2} dz \quad (3\text{-}82)$$

and take the contour around a circle which has the origin as its center, then

$$\frac{dy_n}{dx} = \frac{1}{2\pi i} \oint 2z^{-n} e^{x^2 - (z-x)^2} dz \quad (3\text{-}82a)$$

and

$$\frac{d^2 y_n}{dx^2} = \frac{1}{2\pi i} \oint 4z^{-n+1} e^{x^2 - (z-x)^2} dz$$

The differentiations here may be performed under the integral sign. When these derivatives are substituted on the left of the differential equation (80), it is found that

$$y_n'' - 2xy_n' + 2ny_n = \frac{1}{2\pi i} \oint (4z^2 - 4xz + 2n) e^{x^2 - (z-x)^2} z^{-n-1} dz$$
$$= -\frac{2}{2\pi i} \oint \frac{d}{dz} (z^{-n} e^{x^2 - (z-x)^2}) dz = 0$$

The last step follows because the contents of the parenthesis, being a single-valued function of z, if n is an integer, takes the same value at the initial and final points of the contour integration. It has thus been shown that expression (82) is also a solution of Hermite's equation.[14] Since it represents a polynomial in x it must be identical with $H_n(x)$ except for a constant multiplier. This constant may be found by computing, for example, $H_n(0)$ from (81) and $y_n(0)$ from (82) for even n (since otherwise $H_n(0)$ would vanish). Eq. (81) gives

$$H_n(0) = \frac{(-1)^{n/2} n!}{\left(\dfrac{n}{2}\right)!}$$

---

[14] The function y_n defined by (82) is a solution of Hermite's differential equation even when n is non-integral, but in that case the contour must be specified differently: to make the integrand return to its original value, the path must start at $+\infty$, go in toward the origin, encircle it in a counter-clockwise sense and return to $+\infty$.

while from (82) we obtain

$$y_n(0) = \frac{1}{2\pi i} \oint z^{-n-1} e^{-z^2} dz = \frac{(-1)^{n/2}}{\dfrac{n}{2}\,!}$$

by the theorem of residues (eq. 3).   Hence we see that H_n is $n!$ times y_n:

$$H_n(x) = \frac{n!}{2\pi i} \oint z^{-n-1} e^{x^2-(z-x)^2} dz \qquad (3\text{–}83)$$

This result may be expressed in a different way.   On examining (82) in the light of the theorem of residues it is apparent that y_n is the coefficient of z^n in the expansion of $e^{x^2-(z-x)^2}$ as a power series in z.   Hence

$$e^{x^2-(z-x)^2} = \sum_{n=0}^{\infty} y_n z^n = \sum_{n} \frac{H_n(x)}{n!} z^n \qquad (3\text{–}84)$$

Recurrence relations between Hermite polynomials of different degree are easily derived.   The first is implicit in eq. (82a), which may now be written $y_n'(x) = 2y_{n-1}(x)$, or

$$H_n'(x) = 2nH_{n-1}(x) \qquad (3\text{–}85)$$

The second follows from the differential equation

$$H_n''(x) - 2xH_n' + 2nH_n = 0 \qquad (3\text{–}86)$$

Others may be derived ad libitum by repeated application of (85):

$$H_n'' = 4n(n-1)H_{n-2} \quad \text{etc.}$$

Thus far two representations of H_n have been obtained, the series form (81) and the integral form (83).   A third may be deduced from (84).   Let us take the n-th derivative with respect to z on both sides.   The left becomes

$$e^{x^2} \frac{\partial^n}{\partial z^n} e^{-(z-x)^2} = e^{x^2} (-1)^n \frac{\partial^n}{\partial x^n} e^{-(z-x)^2}$$

and the right simply

$$H_n(x) + H_{n+1}(x)z + \cdots$$

These two expressions are equal for all values of z.   On putting $z = 0$, there results

$$H_n(x) = (-1)^n e^{x^2} \frac{d^n}{dx^n} e^{-x^2} \qquad (3\text{–}87)$$

A function closely related to the polynomial $H_n(x)$ was introduced in sec. 2.15. It is

$$y = e^{-x^2/2}H_n(x) \tag{3-88}$$

and satisfies the differential equation

$$y'' + (1 - x^2 + 2n)y = 0 \tag{3-89}$$

The function defined by (88) is called the *Hermite (orthogonal) function*. It is of interest because it appears (cf. sec. 11.11) as an eigenfunction in the quantum mechanical problem of the simple harmonic oscillator. We shall here derive a few integrals involving this function which will be found useful later.

The first is the integral over the product of two Hermite functions,

$$\int_{-\infty}^{\infty} e^{-x^2}H_n(x)H_m(x)dx$$

In view of (84)

$$e^{x^2-(z_1-x)^2} \cdot e^{x^2-(z_2-x)^2} = \left(\sum_{\lambda} \frac{H_\lambda(x)}{\lambda!} z_1^\lambda\right)\left(\sum_{\mu} \frac{H_\mu(x)}{\mu!} z_2^\mu\right)$$

Hence, multiplying each side by e^{-x^2} and integrating

$$\sum_{\lambda\mu} \left[\int_{-\infty}^{\infty} e^{-x^2}H_\lambda(x)H_\mu(x)dx\right] \frac{z_1^\lambda z_2^\mu}{\lambda!\mu!} = \int_{-\infty}^{\infty} e^{x^2-(z_1-x)^2-(z_2-x)^2}dx \tag{3-90}$$

The integral on the right has the value[15] $\sqrt{\pi}e^{2z_1z_2}$.

This may be expanded to read

$$\sqrt{\pi}e^{2z_1z_2} = \sqrt{\pi} \sum_{\lambda} \frac{(2z_1z_2)^\lambda}{\lambda!} = \sqrt{\pi} \sum_{\lambda\mu} \frac{2^\lambda}{\lambda!} z_1^\lambda z_2^\mu \delta_{\lambda,\mu} \tag{3-91}$$

where the single summation over λ has been changed into a double summation over λ and μ by the artificial use of the Kronecker δ-symbol, defined on page 104. Since (90) is true for every value of z_1 and z_2, the individual coefficients of every power of z_1 and z_2 in both expansions must be equal. On comparing (91) with the left side of (90) we see that

$$\int_{-\infty}^{\infty} \frac{e^{-x^2}H_\lambda H_\mu dx}{\lambda!\mu!} = \sqrt{\pi} \frac{2^\lambda}{\lambda!} \delta_{\lambda,\mu}$$

or

$$\int_{-\infty}^{\infty} e^{-x^2}H_n(x)H_m(x)dx = 2^n n! \sqrt{\pi}\delta_{n,m} \tag{3-92}$$

[15] In evaluating it, use is made of the formula:

$$\int_{-\infty}^{\infty} e^{-ax^2+2bx}dx = \sqrt{\frac{\pi}{a}} e^{b^2/a}$$

The integral

$$\int_{-\infty}^{\infty} xe^{-x^2}H_n(x)H_m(x)dx$$

can be evaluated in a similar way.   In place of (90) we now write

$$\sum_{\lambda\mu}\left[\int_{-\infty}^{\infty} xe^{-x^2}H_\lambda(x)H_\mu(x)dx\right]\frac{z_1^\lambda z_2^\mu}{\lambda!\mu!} = \int_{-\infty}^{\infty} xe^{x^2-(z_1-x)^2-(z_2-x)^2}dx$$

$$= \sqrt{\pi}(z_1 + z_2)e^{2z_1z_2}$$

The last result is, on expansion,

$$\sqrt{\pi}\left(\sum_\lambda \frac{2^\lambda z_1^{\lambda+1}z_2^\lambda}{\lambda!} + \sum_\lambda \frac{2^\lambda z_1^\lambda z_2^{\lambda+1}}{\lambda!}\right)$$

$$= \sqrt{\pi}\left(\sum_{\lambda\mu} \frac{2^{\lambda-1}z_1^\lambda z_2^\mu}{(\lambda - 1)!}\delta_{\mu,\lambda-1} + \sum_{\lambda\mu} \frac{2^\lambda z_1^\lambda z_2^\mu}{\lambda!}\delta_{\mu,\lambda+1}\right)$$

Equating coefficients of $z_1^n z_2^m$ then yields

$$\int_{-\infty}^{\infty} xe^{-x^2}H_n(x)H_m(x)dx = \sqrt{\pi}(2^{n-1}n!\delta_{m,n-1} + 2^n(n + 1)!\delta_{m,n+1})$$

$$(3-93)$$

The integral vanishes when $n = m$ and also when n and m differ by more than unity.   The same method may be used to calculate other integrals of the type

$$\int_{-\infty}^{\infty} x^r e^{-x^2}H_n(x)H_m(x)dx$$

Later, however, we shall learn of simpler ways, involving matrix algebra, for deriving these from the result established in eq. (93).   (See problem at the end of sec. 11.17.)

Example.   A simple harmonic oscillator, if treated by the methods of quantum mechanics, has a distribution of mass about the attracting center which is given by

$$P(\xi) = ce^{-\xi^2}[H_n(\xi)]^2$$

where $\xi = \sqrt{\beta}x$, β being a quantity characteristic of the oscillator, and n is a quantum number which depends on the total energy possessed by the vibrating point.   The moment of inertia of this mass distribution is given by

$$\overline{mx^2} = m \int_{-\infty}^{\infty} x^2 e^{-\xi^2} [H_n(\xi)]^2 dx \Big/ \int_{-\infty}^{\infty} e^{-\xi^2} [H(\xi)]^2 dx$$

$$= \frac{m}{\beta} \int_{-\infty}^{\infty} \xi^2 e^{-\xi^2} [H_n(\xi)]^2 d\xi \Big/ \int_{-\infty}^{\infty} e^{-\xi^2} [H_n(\xi)]^2 d\xi$$

The integral in the denominator has already been calculated (cf. 92) and is equal to $2^n n! \sqrt{\pi}$.   The integral in the numerator may be computed by the same method and is found to be

$$\frac{2n + 1}{2} 2^n n! \sqrt{\pi}$$

Hence

$$\overline{mx^2} = \frac{m}{\beta} \frac{2n + 1}{2}$$

Later (cf. sec. 11.11) it will be shown that $\beta = 4\pi^2 m \nu_0 / h$ so that

$$\overline{mx^2} = \frac{2n + 1}{2} \frac{h}{4\pi^2 \nu_0}$$

where ν_0 is the " classical " frequency of the oscillator.

LIST OF HERMITE POLYNOMIALS

$H_0(\xi) = 1$
$H_1(\xi) = 2\xi$
$H_2(\xi) = 4\xi^2 - 2$
$H_3(\xi) = 8\xi^3 - 12\xi$
$H_4(\xi) = 16\xi^4 - 48\xi^2 + 12$
$H_5(\xi) = 32\xi^5 - 160\xi^3 + 120\xi$
$H_6(\xi) = 64\xi^6 - 480\xi^4 + 720\xi^2 - 120$
$H_7(\xi) = 128\xi^7 - 1344\xi^5 + 3360\xi^3 - 1680\xi$
$H_8(\xi) = 256\xi^8 - 3584\xi^6 + 13440\xi^4 - 13440\xi^2 + 1680$
$H_9(\xi) = 512\xi^9 - 9216\xi^7 + 48384\xi^5 - 80640\xi^3 + 30240\xi$
$H_{10}(\xi) = 1024\xi^{10} - 23040\xi^8 + 161280\xi^6 - 403200\xi^4 + 302400\xi^2 - 30240$

3.11. Laguerre Polynomials and Functions.—The theory of Laguerre polynomials may be developed along lines very similar to those of the last section.   A Laguerre polynomial $L_n(x)$ has been defined in sec. 2.16 as the polynomial solution of Laguerre's differential eq. (2–67):

$$xy'' + (1 - x)y' + ny = 0 \qquad (3\text{–}94)$$

It exists whenever n is an integer and was found to be

$$L_n(x) = (-1)^n \left(x^n - \frac{n^2}{1!} x^{n-1} + \frac{n^2(n-1)^2}{2!} x^{n-2} + \cdots (-1)^n n! \right)$$

$$(3\text{–}95)$$

We first establish a representation of L_n in the form of a definite integral. Consider

$$y_n = \frac{1}{2\pi i} \oint \frac{z^{-n-1}}{1-z} \exp\left(\frac{-xz}{1-z}\right) dz \qquad (3\text{-}96)$$

where the contour is taken to include the origin. Differentiations with respect to x may be performed under the integral sign; hence

$$y_n' = -\frac{1}{2\pi i} \oint \frac{z^{-n}}{(1-z)^2} \exp\left(\frac{-xz}{1-z}\right) dz$$

and

$$y_n'' = \frac{1}{2\pi i} \oint \frac{z^{-n+1}}{(1-z)^3} \exp\left(\frac{-xz}{1-z}\right) dz$$

On substituting in the left-hand side of (94) we find

$$\frac{1}{2\pi i} \oint \left[\frac{xz^2}{(1-z)^2} - \frac{(1-x)z}{1-z} + n\right] \frac{z^{-n-1}}{1-z} \exp\left(\frac{-xz}{1-z}\right) dz$$

But this is easily seen to be

$$-\frac{1}{2\pi i} \oint \frac{d}{dz}\left[\frac{z^{-n}}{1-z} \exp\left(\frac{-xz}{1-z}\right)\right] dz$$

an expression which vanishes because the quantity in brackets takes on the same value at the initial and final point of the contour. Hence (96) is a solution of Laguerre's differential equation. Moreover, it is a polynomial, as an analysis in the light of the theorem of residues will show. Its relation to $L_n(x)$ may be established by computing both $y_n(x)$ and $L_n(x)$ for a particular value of x, say zero. From (95)

$$L_n(0) = n!$$

from (96)

$$y_n(0) = \frac{1}{2\pi i} \oint \frac{z^{-n-1}}{1-z} dz = \frac{1}{2\pi i} \oint z^{-n-1}(1 + z + z^2 + \cdots) \, dz = 1$$

Therefore

$$L_n = n!\, y_n$$

Again, using the theorem of residues (eq. 3), we find, since

$$L_n = \frac{n!}{2\pi i} \oint \frac{z^{-n-1}}{1-z} \exp\left(\frac{-xz}{1-z}\right) dz \qquad (3\text{-}97)$$

that

$$(1-z)^{-1} \exp\left(\frac{-xz}{1-z}\right) = \sum_{n=0}^{\infty} y_n z^n = \sum_{n=0}^{\infty} \frac{L_n(x)}{n!} z^n \qquad (3\text{-}98)$$

This result is quite similar to (84).

Next, we turn our attention to the recurrence relations existing between Laguerre polynomials of different degrees, and between the derivatives of a given polynomial. A relation of the latter type follows at once from the differential equation:

$$xL_n'' + (1 - x)L_n' + nL_n = 0 \tag{3-99}$$

The former relation may be obtained by differentiating (98) with respect to z:

$$\frac{1 - x - z}{(1 - z)^3} \exp\left(\frac{-xz}{1 - z}\right) = \sum_{\lambda=0}^{\infty} \frac{L_\lambda(x)z^{\lambda-1}}{(\lambda - 1)!}$$

When the left-hand side of this equation is again expressed in terms of Laguerre polynomials with the use of (98), the result may be written

$$(1 - x - z) \sum_\lambda \frac{L_\lambda(x)z^\lambda}{\lambda!} = (1 - 2z + z^2) \sum_\lambda \frac{L_\lambda(x)}{(\lambda - 1)!} z^{\lambda-1}$$

On equating the coefficients of z^n, there results

$$(1 - x)\frac{L_n}{n!} - \frac{L_{n-1}}{(n - 1)!} = \frac{L_{n+1}}{n!} - \frac{2L_n}{(n - 1)!} + \frac{L_{n-1}}{(n - 2)!}$$

whence,

$$(1 + 2n - x)L_n - n^2 L_{n-1} - L_{n+1} = 0 \tag{3-100}$$

which is the relation here sought.

For some purposes it is convenient to have L_n in the form of a derivative. To find it we differentiate (98) n times with respect to z and afterwards put $z = 0$, thus obtaining

$$e^x \lim_{z \to 0} \frac{\partial^n}{\partial z^n}\left[(1 - z)^{-1} \exp\left(\frac{-x}{1 - z}\right)\right] = L_n(x)$$

The reader will be able to show without difficulty that

$$\lim_{z \to 0} \frac{\partial^n}{\partial z^n}\left[(1 - z)^{-1} \exp\left(\frac{-x}{1 - z}\right)\right] = \frac{d^n}{dx^n}(x^n e^{-x})$$

Hence

$$L_n(x) = e^x \frac{d^n}{dx^n}(x^n e^{-x}) \tag{3-101}$$

The *associated Laguerre polynomial*, of degree $n - k$, was shown in sec. 2.16 to satisfy the differential equation (2-70)

$$xy'' + (k + 1 - x)y' + (n - k)y = 0$$

and is given by

$$y = L_n^k = \frac{d^k}{dx^k} L_n(x) \tag{3-102}$$

On differentiating (98) k times with respect to x, it is seen at once that

$$(-1)^k (1 - z)^{-1} \left(\frac{z}{1 - z}\right)^k \exp\left(\frac{-xz}{1 - z}\right) = \sum_{\lambda=k}^{\infty} \frac{L_\lambda^k(x)}{\lambda!} z^\lambda \tag{3-103}$$

A function of great importance in quantum mechanics is the *associated Laguerre function,* for it describes, in a sense to be discussed fully in Chapter 11, the motion of the electron in the hydrogen atom. It satisfies differential eq. (71) in sec. 2.16, and was there shown to be represented by

$$y_{n,k} = e^{-x/2} x^{(k-1)/2} L_n^k(x) \tag{3-104}$$

Certain integrals involving this function are often used and will here be calculated. They are of the form

$$I_{n,m} = \int_0^\infty e^{-x} x^{k-1} L_n^k(x) L_m^k(x) \cdot x^p dx$$

where p is another integer which we shall take in this work to be either 1, 2, or 3. Furthermore, our interest will be confined to $I_{n,n}$. If we multiply eq. (103) in which z_1 is written for z, by a similar one in which z is replaced by z_2, there results

$$\sum_{\lambda,\mu=k}^{\infty} \frac{z_1^\lambda z_2^\mu}{\lambda!\mu!} L_\lambda^k(x) L_\mu^k(x)$$

$$= (z_1 z_2)^k (1 - z_1)^{-k-1} (1 - z_2)^{-k-1} \exp\left(\frac{-xz_1}{1 - z_1} - \frac{xz_2}{1 - z_2}\right)$$

Let us now multiply each side of this equation by $e^{-x} x^{k+p-1}$ and then integrate with respect to x. In view of the definition of $I_{n,m}$, the result may be written

$$\sum_{\lambda,\mu=k}^{\infty} \frac{z_1^\lambda z_2^\mu}{\lambda!\mu!} I_{\lambda,\mu} = (z_1 z_2)^k [(1 - z_1)(1 - z_2)]^{-k-1}$$

$$\int_0^\infty x^{k+p-1} \exp\left[x\left(1 - \frac{1}{1 - z_1} - \frac{1}{1 - z_2}\right)\right] dx$$

Now

$$\int_0^\infty e^{-ax} x^r dx = \alpha^{-r-1} r!$$

as may be shown by r-fold partial integration or from eq. (10). If we put

$$\alpha = (1 - z_1)^{-1} + (1 - z_2)^{-1} - 1 = (1 - z_1 z_2)(1 - z_1)^{-1}(1 - z_2)^{-1}$$

we obtain, therefore,

$$\sum_{\lambda,\mu} \frac{z_1^\lambda z_2^\mu}{\lambda!\mu!} I_{\lambda,\mu} = \frac{(z_1 z_2)^k (1 - z_1)^{p-1}(1 - z_2)^{p-1}}{(1 - z_1 z_2)^{k+p}} (k + p - 1)! \qquad (3\text{-}105)$$

When the denominator on the right is expanded by the binomial theorem

$$(1 - z_1 z_2)^{-k-p} = \sum_\lambda \binom{k + p + \lambda - 1}{\lambda} (z_1 z_2)^\lambda$$

$$= \sum_\lambda \frac{(k + p + \lambda - 1)!}{(k + p - 1)!\lambda!} (z_1 z_2)^\lambda$$

the right-hand side of (105) becomes

$$(1 - z_1)^{p-1}(1 - z_2)^{p-1} \sum_\lambda \frac{(k + p + \lambda - 1)!}{\lambda!} (z_1 z_2)^{k+\lambda} \qquad (3\text{-}106)$$

Thus, in view of (105), $I_{n,n}$ is simply $(n!)^2$ times the coefficient of $(z_1 z_2)^n$ of this expression.

a. When $p = 1$, this is obtained by choosing that term of the summation in (106) for which $k + \lambda = n$, that is $\lambda = n - k$.

$$I_{n,n} = (n!)^3/(n - k)! \qquad (3\text{-}107a)$$

b. When $p = 2$, (106) becomes

$$\sum_\lambda \frac{(k + \lambda + 1)!}{\lambda!} [(z_1 z_2)^{k+\lambda} - z_1^{k+\lambda+1} z_2^{k+\lambda} - z_1^{k+\lambda} z_2^{k+\lambda+1} + (z_1 z_2)^{k+\lambda+1}]$$

The second and third terms in the bracket, in which z_1 and z_2 appear with different exponents, cannot contribute to $I_{n,n}$; the first terms contribute when $\lambda = n - k$, the last when $\lambda = n - k - 1$. Hence

$$I_{n,n} = (n!)^2 \left[\frac{(n + 1)!}{(n - k)!} + \frac{n!}{(n - k - 1)!} \right]$$

$$= \frac{(n!)^3}{(n - k)!} (2n - k + 1) \qquad (3\text{-}107b)$$

c. When $p = 3$, the significant parts of (106) are

$$\sum_\lambda \frac{(k + \lambda + 2)!}{\lambda!} [(z_1 z_2)^{k+\lambda} + 4(z_1 z_2)^{k+\lambda+1} + (z_1 z_2)^{k+\lambda+2}]$$

terms with different exponents of z_1 and z_2 having been omitted. Consequently,

$$I_{n,n} = (n!)^2 \left[\frac{(n + 2)!}{(n - k)!} + \frac{4(n + 1)!}{(n - k - 1)!} + \frac{n!}{(n - k - 2)!} \right]$$

$$= \frac{(n!)^3}{(n - k)!} (6n^2 - 6nk + k^2 + 6n - 3k + 2) \qquad (3\text{-}107c)$$

Obviously, the same process permits the evaluation of $I_{n,n}$ for any value of p. The quantities $I_{n,m}$ for $n \neq m$ are rarely needed, but can be obtained by this method also.

Example. The electronic charge of the hydrogen atom is distributed about the proton as origin in accordance with the distribution function

$$P(\rho) = c\rho^{2l+2}e^{-\rho}[L_{n+l}^{2l+1}(\rho)]^2$$

as will be shown in sec. 11.13. In this expression n and l stand for the " total " and " angular momentum " quantum numbers which designate the state of the atom; c is a constant which is different for different states, and ρ is proportional to r, the radius vector: $\rho = (2/na_0)r$. The proportionality constant depends on the quantum number n and differs for different states or the atom, a_0 is the fundamental constant known as the first " Bohr radius." $P(\rho)$, finally, represents the charge to be found within the spherical shell enclosed between ρ and $\rho + d\rho$.

Let it be desired to find the mean value of $1/r$ and r for this distribution. Clearly

$$\overline{r^{-1}} = \frac{\int_0^\infty \frac{P(\rho)}{r}dr}{\int_0^\infty P(\rho)dr} = \frac{\frac{2}{na_0}\int_0^\infty \frac{P(\rho)d\rho}{\rho}}{\int_0^\infty P(\rho)d\rho}$$

The integral in the numerator is simply $I_{n+l,\ n+l}$, with $k = 2l + 1$ and $p = 1$, that in the denominator is also $I_{n+l,\ n+l}$ with the same k, but with $p = 2$. Hence, using (107a and b)

$$\overline{r^{-1}} = \frac{2}{na_0} \cdot \frac{1}{2n} = \frac{1}{n^2a_0}$$

Similarly,

$$\overline{r} = \frac{\int_0^\infty P(\rho)rdr}{\int_0^\infty P(\rho)dr} = \frac{\frac{na_0}{2}\int_0^\infty P(\rho)\rho d\rho}{\int_0^\infty P(\rho)d\rho}$$

$$= \frac{na_0}{2} \cdot \frac{I_{n+l,n+l}(p = 3)}{I_{n+l,n+l}(p = 2)}, \quad \text{with } k = 2l + 1$$

In view of (107c and b)

$$\overline{r} = \frac{na_0}{2} \cdot \frac{6n^2 - 2l(l + 1)}{2n} = \frac{a_0}{2}[3n^2 - l(l + 1)]$$

For the ground state of the hydrogen atom $(n = 1, l = 0)$

$$\overline{r^{-1}} = a_0^{-1} \quad \text{and} \quad \overline{r} = \tfrac{3}{2}a_0$$

3.12. Generating Functions.—A simple and powerful way of representing functions of the more unfamiliar types is by means of *generating functions*, that is, functions of two arguments which, when expanded in a power series with respect to one argument, contain the functions to be generated as coefficients involving the other argument parametrically. Examples of generating functions have occurred in the preceding sections; they will here be exhibited once more for easy reference.

1. *Legendre Polynomials.*

$$(1 - 2xy + y^2)^{-1/2} = \sum_{l=0}^{\infty} P_l(x)y^l \quad \text{(Cf. eq. 24)}$$

2. *Associated Legendre Polynomials.*

$$\frac{(2m)!(1 - x^2)^{m/2}y^m}{2^m m!(1 - 2xy + y^2)^{m+1/2}} = \sum_{l=m}^{\infty} P_l^m(x)y^l$$

This was not used in the text, but is easily derived by differentiation from (24) on the basis of (43).

3. *Bessel Functions (of integral order).*

$$\exp\left[\frac{x}{2}\left(u - \frac{1}{u}\right)\right] = \sum_{n=0}^{\infty} J_n(x)u^n \quad \text{(Cf. 64)}$$

4. *Hermite Polynomials.*

$$\exp[x^2 - (z - x)^2] = \sum_{n=0}^{\infty} \frac{H_n(x)}{n!} z^n \quad \text{(Cf. 84)}$$

5. *Laguerre Polynomials.*

$$(1 - z)^{-1} \exp\left(\frac{-xz}{1 - z}\right) = \sum_{n=0}^{\infty} \frac{L_n(x)}{n!} z^n \quad \text{(Cf. 98)}$$

6. *Associated Laguerre Polynomials.*

$$(-1)^k (1 - z)^{-1} \left(\frac{z}{1 - z}\right)^k \exp\left(\frac{-xz}{1 - z}\right) = \sum_{n=k}^{\infty} \frac{L_n^k(x)}{n!} z^n \quad \text{(cf. 103)}$$

7. *Tschebyscheff Polynomials.*

$$\frac{1 - xy}{1 - 2xy + y^2} = \sum_{n=0}^{\infty} T_n(x)y^n$$

(Not proved in text, but Cf. eq. 2–54.)

3.13. Linear Dependence.—A set of functions $\varphi_1, \varphi_2 \cdots \varphi_n$ is said to be linearly dependent when a set of constants, $k_1, k_2 \cdots k_n$, not all zero, exists such that

$$\sum_1^n k_\lambda \varphi_\lambda = 0$$

If this relation can be satisfied only by putting all k_λ equal to zero, the functions are *linearly independent*.

A criterion for linear dependence is easily derived. We observe that the integral

$$I(k_1 \cdots k_n) = \int \left| \left(\sum_1^n k_\lambda \varphi_\lambda \right) \right|^2 dx \qquad (3\text{--}108)$$

taken over the range of x in which the functions φ_λ are considered, cannot be smaller than zero. It will attain the minimum, zero, for specific values of the parameters k_λ. Now it will first be shown that, if I has a stationary value at all, this value must be zero.

For this purpose, let us vary I, replacing every k_λ by $k_\lambda(1 + \delta k)$. The result is

$$I + \delta I = (1 + \delta k)^2 I$$

and

$$\delta I = [2\delta k + (\delta k)^2]I$$

Where I has a stationary value, δI must vanish; but it is seen that δI cannot vanish unless I itself is zero. Therefore the stationary value of (108) is zero, and we may say that the conditions

$$\frac{\partial I}{\partial k_\lambda} = \frac{\partial I}{\partial k_\lambda^*} = 0, \quad \lambda = 1, 2, \cdots n \qquad (3\text{--}109)$$

are both sufficient and necessary for the vanishing of I or, what amounts to the same thing, for the validity of

$$\sum_1^n k_\lambda \varphi_\lambda = 0$$

If, therefore, eqs. (109) have a solution other than the trivial one in which all k_λ are zero, the functions φ_λ are linearly dependent.

But (108) may be written in a different way. If we define the coefficients

$$a_{\lambda\mu} = \int \varphi_\lambda^* \varphi_\mu dx$$

we have

$$I = \sum_{\lambda\mu} a_{\lambda\mu} k_\lambda^* k_\mu$$

and eqs. (109) now read

$$\sum_\lambda a_{\lambda\mu} k_\lambda^* = 0, \quad \sum_\lambda a_{\mu\lambda} k_\lambda = 0 \qquad (3\text{--}110)$$

These are identical because $a_{\mu\lambda} = a^*_{\lambda\mu}$, so that one is merely the complex form of the other. Now the condition that (110) shall have a *non-vanishing* solution $k_1, k_2, \cdots k_n$ is that the determinant

$$\left| \, a_{\lambda\mu} \, \right| = 0$$

This, therefore, is the condition for linear dependence of the functions φ_λ. Conversely, if $\left| \, a_{\lambda\mu} \, \right| \neq 0$, the set of functions is linearly independent. The determinant $\left| \, a_{\lambda\mu} \, \right|$ is named after the mathematician *Gram*.

A simpler test, applicable when the functions $\varphi_1, \cdots \varphi_n$ are differentiable $n - 1$ times within their range of definition, may be conducted as follows. If the functions are linearly dependent,

$$\left. \begin{aligned} \sum_1^n k_\lambda \varphi_\lambda &= 0 \\[6pt] \sum_1^n k_\lambda \varphi_\lambda' &= 0 \\[2pt] &\cdots\cdots\cdots \\[2pt] \sum_1^n k_\lambda \varphi_\lambda^{(n-1)} &= 0 \end{aligned} \right\}$$

These n homogeneous equations may be regarded as determining the set of constants k_λ. It will be shown in section 10.9 that they possess solutions other than $k_1 = k_2 = k_3 \cdots k_n = 0$ only if the determinant of the coefficients of k_λ, called the *Wronskian*,

$$\begin{vmatrix} \varphi_1 & \varphi_2 & \cdots & \varphi_n \\ \varphi_1' & \varphi_2' & \cdots & \varphi_n' \\ \multicolumn{4}{c}{\cdots\cdots\cdots\cdots\cdots\cdots} \\ \varphi_1^{(n-1)} & \varphi_2^{(n-1)} & \cdots & \varphi_n^{(n-1)} \end{vmatrix} = 0$$

For independence of the solutions, then, the Wronskian must not vanish. It should be stated, however, that the vanishing of this determinant is not a sufficient condition for linear dependence of the functions.

3.14. Schwarz' Inequality.—Let f and g be any two functions of x such that the integrals

$$A = \int f^* f\, dx, \quad B = \int f^* g\, dx, \quad C = \int g^* g\, dx \qquad (3\text{–}111)$$

exist. The integrations extend over any definite range of the variable x. Certainly the integral

$$\int [\lambda f^*(x) + g^*(x)][\lambda f(x) + g(x)]\, dx = A\lambda^2 + (B^* + B)\lambda + C$$

in which λ is to be considered as a *real* variable, independent of x, is always positive or zero (zero only when g is directly proportional to f) and hence has no real roots in λ. But the roots of $A\lambda^2 + (B^* + B)\lambda + C$ are given by

$$\lambda = -\frac{B^* + B}{2A} \pm \frac{1}{2A}\sqrt{(B^* + B)^2 - 4AC}$$

They are real unless

$$4AC \geq (B^* + B)^2 \tag{3-112}$$

The equality sign here holds only when $g = \text{const.} \times f$.

The right-hand side of (112) is twice the real part of B. Hence, if f and g are real functions, the inequality becomes

$$\int f^2 dx \cdot \int g^2 dx \geq \left(\int fg\,dx\right)^2 \tag{3-113}$$

which is one form of Schwarz' inequality.

For complex functions f and g, (112) may be modified. Write f and g in polar form:

$$f(x) = \rho_1(x)e^{i\theta_1(x)} \; ; \; g(x) = \rho_2(x)e^{i\theta_2(x)}$$

Then $B = \displaystyle\int \rho_1\rho_2 e^{i(\theta_2 - \theta_1)}dx$. Since (112) holds for every pair of functions f and g (which have integrable squares), it must also be true when g is replaced by $g' = ge^{i(\theta_1 - \theta_2)}$. But the substitution of g' for g leaves the values of A and C unchanged while it converts both B^* and B into $\displaystyle\int \rho_1\rho_2 dx = |B|$, which is the modulus of B. Hence

$$\int f^* f\,dx \int g^* g\,dx \geq \left| \int f^* g\,dx \right|^2 \tag{3-114}$$

This is the more general form of the Schwarz inequality. Further generalization to functions of more than one real variable is obvious.

A relation like (114) is also valid for sums:

$$\left(\sum_1^n f_i^* f_i\right)\left(\sum_1^n g_i^* g_i\right) \geq \left| \sum_1^n f_i^* g_i \right|^2 \tag{3-115}$$

For ordinary vectors $\mathbf{U}$ and $\mathbf{V}$ this is equivalent to

$$U^2 V^2 \geq (\mathbf{U} \cdot \mathbf{V})^2$$

Problem. Prove inequality (115).

REFERENCES

Magnus, W., and Oberhettinger, F., " Formulas and Theorems for the Special Functions of Mathematical Physics," Chelsea Publishing Co., New York, 1949.

Churchill, P. V., " Introduction to Complex Variables and Applications," McGraw-Hill Book Co., New York, 1948.

Forsythe, A. R., " Theory of Functions of a Complex Variable," Cambridge, 1893.

McLachlan, N. W., " Complex Variables and Operational Calculus," Cambridge, New York, 1939.

Cambi, E., " Bessel Functions," Dover Publications, New York, 1948.

Gray, A., Mathews, G. B., and MacRobert, T. M., " Treatise on Bessel Functions," Macmillan, London, 1922.

McLachlan, N. W., " Bessel Functions for Engineers," Oxford, New York, 1934.

Watson, G. N., " Treatise on the Theory of Bessel Functions," Cambridge, New York, 1944.

NBS Mathematical Tables Project, " Tables of J_0 and J_1, N_0 and N_1 for Complex Argument," Columbia University Press, New York, 1947–1950.

Nielsen, N., " Theorie der Gammafunktion," B. C. Teubner, Leipzig, 1906.

MacRobert, T. M., " Spherical Harmonics," Methuen, London, 1927; Reprint, Dover Publications, New York, 1948.

Tallquist, H. J., " Tafeln der 32 Ersten Kugelfunktionen P_n $(cos\ \theta)$," *Acta Soc. Sci. Fennicae*, Ser. A, Tom. II, No. 11, 1938.

Hobson, E. W., " Theory of Spherical and Ellipsoidal Harmonics," Cambridge, New York, 1931.

NBS Mathematical Tables Project, " Tables of Associated Legendre Functions," Columbia University Press, New York, 1945.

McLachlan, N. W., " Theory and Application of Mathieu Functions," Oxford, New York, 1947.

NBS Mathematical Tables Project, " Tables Relating to Mathieu Functions," Columbia University Press, New York, 1951.

CHAPTER 4

VECTOR ANALYSIS

4.1. Definition of a Vector.—A physical quantity possessing both magnitude and direction is called a *vector;* typical examples are velocity, acceleration, force and angular momentum; other quantities such as mass, volume, temperature and time, having magnitude only, are called *scalars.* It is customary to represent vectors by letters in **bold-face type** and scalars in *italics*, so that **A** stands for a vector whose magnitude is A. This custom will here be followed. A vector may be indicated graphically by an arrow drawn between two points, tail and head of the arrow being its *origin* and *terminus*, respectively; the scalar part of the vector equals (or is proportional to) the length of the arrow and the direction of arrow and vector coincide.

It is often necessary to locate a vector relative to a coordinate system, which may be done by giving the coordinates of origin and terminus. Let the selected coordinate system be the usual right-handed[1] Cartesian one with three mutually perpendicular axes X, Y and Z so oriented that if the positive X-axis points towards the reader's right and the positive Y-axis towards the top of the page, the positive Z-axis will point up from the page towards the reader. Let the coordinates of origin and terminus of **A** be (x_1, y_1, z_1) and (x_2, y_2, z_2), respectively; then the three rectangular Cartesian components of **A**, relative to the axes X, Y, Z, are defined to be

$$A_x = x_2 - x_1; \quad A_y = y_2 - y_1; \quad A_2 = z_2 - z_1$$

The length of the vector is the distance between the two points:

$$A = \sqrt{A_x^2 + A_y^2 + A_z^2}$$

The two points might be located relative to many other coordinate systems, one of which could be obtained from the previous one by rotation of the axes to X', Y', Z' and translation of the origin from O to O' as shown in Fig. 1. Suppose the coordinates of O' in the first system are (x_0, y_0, z_0), then the position of the second system is determined with respect to the first

---

[1] Left-handed systems are sometimes used. They may be obtained from right-handed ones by changing the direction of one of the axes, or by interchanging the names (or directions) of three of the axes. If two axes (or directions) are exchanged, the system remains unchanged.

when the angles between $O'X'$, $O'Y'$, $O'Z'$ and OX, OY, OZ are known. The cosines of these nine angles are given in Table 1, where for the present purpose the first row and column are to be used; for example, a_{13} is the cosine of the angle between the two straight lines $O'X'$ and OZ.

TABLE 1

		OX	OY	OZ
		A_x	A_y	A_z
$O'X'$	A'_x	a_{11}	a_{12}	a_{13}
$O'Y'$	A'_y	a_{21}	a_{22}	a_{23}
$O'Z'$	A'_z	a_{31}	a_{32}	a_{33}

In order to locate the vector in the system $O'X'Y'Z'$, it is necessary first to obtain relations between the nine direction cosines. From a well-known

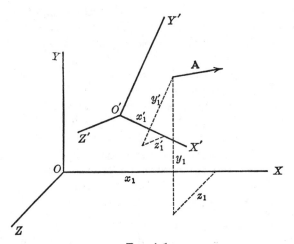

FIG. 4–1

formula of solid analytic geometry, if θ is the angle between two straight lines, whose angles with the coordinate axes are α_1, β_1, γ_1, α_2, β_2, γ_2,

$$\cos \theta = \cos \alpha_1 \cos \alpha_2 + \cos \beta_1 \cos \beta_2 + \cos \gamma_1 \cos \gamma_2 \qquad (4\text{–}1)$$

If the lines are perpendicular to each other, as is true for the axes OY and OZ, $\cos \theta = 0$, hence

$$a_{12}a_{13} + a_{22}a_{23} + a_{32}a_{33} = 0 \qquad (4\text{–}2)$$

Five similar equations result for the other mutually perpendicular axes. Six further relations are obtained from the fact that the sum of the squares of the direction cosines of any line is unity; for example, for the line OX relative to $O'X'Y'Z'$

$$a_{11}^2 + a_{21}^2 + a_{31}^2 = 1 \qquad (4\text{-}3)$$

There are yet ten more relations, in addition to the twelve expressed in eqs. (2) and (3). Nine of them are of the type $a_{11} = a_{22}a_{33} - a_{32}a_{23}$ and the tenth is the determinant of the cosines, which equals unity. It is evident that the nine cosines are not linearly independent.

Now let (x_1, y_1, z_1) and (x_1', y_1', z_1') be the coordinates of the same point P in $OXYZ$ and $O'X'Y'Z'$ and let α_1, β_1, γ_1, α_2 be the direction angles of $O'P$ with OX, OY, OZ, $O'X'$. Then from (1) and Table 1,

$$x_1' = O'P \cos \alpha_2 = O'P(a_{11} \cos \alpha_1 + a_{12} \cos \beta_1 + a_{13} \cos \gamma_1)$$

$$= a_{11}(x_1 - x_0) + a_{12}(y_1 - y_0) + a_{13}(z_1 - z_0)$$

In like manner,

$$y_1' = a_{21}(x_1 - x_0) + a_{22}(y_1 - y_0) + a_{23}(z_1 - z_0)$$

$$z_1' = a_{31}(x_1 - x_0) + a_{32}(y_1 - y_0) + a_{33}(z_1 - z_0)$$

Similarly, if the components of **A** in $O'X'Y'Z'$ are

$$A_x' = x_2' - x_1'; \quad A_y' = y_2' - y_1'; \quad A_z' = z_2' - z_1'$$

then,

$$A_x' = a_{11}A_x + a_{12}A_y + a_{13}A_z \qquad (4\text{-}4)$$

and two other expressions for A_y' and A_z' may be derived in the same way. These three equations may be solved for the unprimed quantities in terms of the primed ones, or the same method may be continued to give three relations like

$$A_x = a_{11}A_x' + a_{21}A_y' + a_{31}A_z' \qquad (4\text{-}5)$$

All of them are symbolized, in self-explanatory fashion, in Table 1 if the second row and column are used. While it is usually true that the components of a vector are different in different reference frames, certain properties such as the length and the angle between two vectors are equal in all frames. It is readily shown, using (2), (3) and (5), that

$$A = A' = \sqrt{A_x'^2 + A_y'^2 + A_z'^2}$$

Considerable simplification often results in expressing physical laws in vector notation, without reference to a selected coordinate system. The transformation properties just described, however, show that it is always possible to list the components of a vector in any given reference frame when so desired. In accordance with these ideas, a vector is sometimes defined as a set of numbers (A_x, A_y, A_z) referred to a reference frame, so that if the numbers are then referred to a second frame, they will become (A_x', A_y', A_z') with relations as given by Table 1. Provided these conditions are met the vector is said to be a *proper vector*. This analytical definition is more restrictive than the intuitive conception of a vector as a quantity possessing magnitude and direction, but it leads naturally to the more general idea of the tensor and it may be readily extended to the vector in n-dimensional space as used in many branches of modern analysis. Moreover, the analytical definition is more precise than the usual one, which offers no explanation of the words "magnitude and direction." The three components of a vector define these words provided they are the same in all reference frames. Further comments on this matter will be found in sec. 4.21.

4.2. Unit Vectors.—Vectors of unit length, drawn along the axes OX, OY, and OZ, respectively, are called *unit vectors* (cf. Fig. 2); they are designated by **i**, **j**, and **k**, respectively. Any directed line along either of the

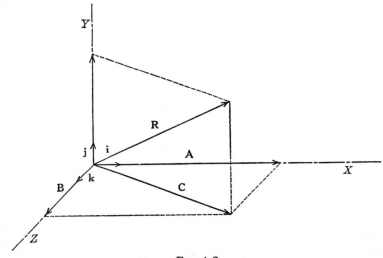

Fɪɢ. 4–2

three axes is also a vector, for if its length is A_x units along the X-axis, the scalar magnitude is thereby given and its direction is specified by the unit vector **i**, the whole vector being designated by $A_x\mathbf{i}$. Similar vectors could be drawn along the Y- or Z-axes, $A_y\mathbf{j}$ or $A_z\mathbf{k}$.

4.3. Addition and Subtraction of Vectors.—Referring again to Fig. 2, it is seen that the diagonal of the parallelogram, whose two unequal sides are the vectors $A_x\mathbf{i}$ and $A_z\mathbf{k}$, is also a vector, its origin being taken as coincident with the coordinate origin. From the previous discussion, it follows that the reference frame $OXYZ$ is superfluous, so that the symbols $A_x\mathbf{i}$

and $A_z\mathbf{k}$ may be replaced by more general symbols, $\mathbf{A}$ and $\mathbf{B}$. The resultant vector $\mathbf{C}$ representing the diagonal is called the *vector sum* of $\mathbf{A}$ and $\mathbf{B}$,

$$\mathbf{A} + \mathbf{B} = \mathbf{C}$$

The addition of vectors thus obeys the familiar rule for composition of forces in mechanics. To obtain the *difference* of two vectors, $\mathbf{A} - \mathbf{B}$, it is only necessary to define the negative of a vector. This is taken to mean a vector whose length is equal and whose direction is opposite to that of the original vector. Thus $\mathbf{A} - \mathbf{B} = \mathbf{A} + (-\mathbf{B})$. Hence the rule: To form the difference of two vectors graphically, reverse the direction of the minuend and complete the parallelogram as before.

From the parallelogram law, it is seen that any vector in a plane may be resolved in numerous ways into two components in the same plane, and that a vector in space may be resolved in numerous ways into three components, not in the same plane. If the resolution is made along the rectangular axes, the result may be symbolized in terms of unit vectors,

$$\mathbf{C} = A_x\mathbf{i} + A_z\mathbf{k}$$

and

$$\mathbf{R} = A_x\mathbf{i} + A_y\mathbf{j} + A_z\mathbf{k} \tag{4-6}$$

From the geometry of Fig. 2, the lengths of $\mathbf{C}$ and $\mathbf{R}$ are

$$C = (A_x^2 + A_z^2)^{1/2}$$
$$R = (A_x^2 + A_y^2 + A_z^2)^{1/2} \tag{4-7}$$

The laws which govern addition and subtraction of vectors are easily seen to be associative, commutative, and distributive. Multiplication of a vector by a scalar is understood to mean multiplication of its length by the scalar factor, without change in its direction. Vector algebra thus developed enables one to demonstrate many geometrical theorems in a simple way.[2]

Problem a. Prove that the diagonals of a parallelogram bisect each other.

Problem b. Prove that the line that joins one corner of a parallelogram to the middle point of an opposite side trisects the diagonal and is trisected by it.

4.4. The Scalar Product of Two Vectors.—The *scalar* (or *inner*) product[3] of two vectors is defined by

$$\mathbf{A} \cdot \mathbf{B} = AB \cos \theta \tag{4-8}$$

where θ is the angle between $\mathbf{A}$ and $\mathbf{B}$. It follows that the scalar product of

---

[2] Numerous examples may be found in books on vector analysis; see for example: Phillips, H. B., " Vector Analysis," John Wiley and Sons, New York, 1933; Gibbs-Wilson, " Vector Analysis," Yale University Press, New Haven, Conn., 1925.

[3] Also called the dot product.

two perpendicular unit vectors must vanish since $\theta = \pi/2$, $\cos \theta = 0$. Similarly the scalar product of a unit vector by itself must equal unity since $\theta = 0$, $\cos \theta = 1$. In vector notation,

$$\mathbf{i} \cdot \mathbf{j} = \mathbf{j} \cdot \mathbf{i} = \mathbf{i} \cdot \mathbf{k} = \mathbf{k} \cdot \mathbf{i} = \mathbf{j} \cdot \mathbf{k} = \mathbf{k} \cdot \mathbf{j} = 0 \qquad (4\text{--}9)$$

$$\mathbf{i} \cdot \mathbf{i} = \mathbf{j} \cdot \mathbf{j} = \mathbf{k} \cdot \mathbf{k} = i^2 = j^2 = k^2 = 1 \qquad (4\text{--}10)$$

If $\mathbf{A} = \mathbf{B}$, $\theta = 0$, so from (8) and (10),

$$\mathbf{A} \cdot \mathbf{A} = A^2 = A_x^2 + A_y^2 + A_z^2$$

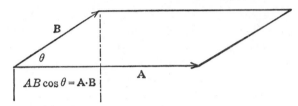

FIG. 4–3

an equation which defines the square of the length of $\mathbf{A}$ (see also Fig. 3). If

$$\mathbf{A} \cdot \mathbf{B} = 0 \qquad (4\text{--}11)$$

for any two vectors, $\mathbf{A}$ and $\mathbf{B}$ are perpendicular to each other, unless one vanishes; if

$$\mathbf{A} \cdot \mathbf{B} = AB$$

then $\mathbf{A}$ and $\mathbf{B}$ are parallel. In a Cartesian system,

$$\mathbf{A} \cdot \mathbf{B} = A_x B_x + A_y B_y + A_z B_z \qquad (4\text{--}12)$$

The scalar product obeys the rules of ordinary multiplication

$$\mathbf{A} \cdot \mathbf{B} = \mathbf{B} \cdot \mathbf{A}$$

$$\mathbf{A} \cdot (\mathbf{B} + \mathbf{C}) = (\mathbf{A} \cdot \mathbf{B}) + (\mathbf{A} \cdot \mathbf{C})$$

From (8), it is seen that any relation involving the cosine of an included angle may be written in terms of the scalar product. For example, the mechanical work W done by a force $\mathbf{F}$ which makes an angle θ with the displacement $\mathbf{D}$ is $W = FD \cos \theta$ or in vector notation, $W = \mathbf{F} \cdot \mathbf{D}$.

Problem. If $\mathbf{A}$ and $\mathbf{B}$ are the sides of a parallelogram and $\mathbf{C}$, $\mathbf{D}$ are the diagonals, show that $C^2 + D^2 = 2(A^2 + B^2)$; $C^2 - D^2 = 4AB \cos \widehat{AB}$.

4.5. The Vector Product of Two Vectors.—Let two arbitrary vectors, $\mathbf{A}$ and $\mathbf{B}$, be drawn from a common origin 0 with an included angle θ, $0 < \theta < \pi$, and let $\mathbf{C}$ be a vector perpendicular to Σ, the plane of $\mathbf{A}$ and $\mathbf{B}$

(cf. Fig. 4).   Then from (11) and (12)

$$\mathbf{C}\cdot\mathbf{A} = C_xA_x + C_yA_y + C_zA_z = 0$$
$$\mathbf{C}\cdot\mathbf{B} = C_xB_x + C_yB_y + C_zB_z = 0$$

Solving, we find

$$C_x = m(A_yB_z - A_zB_y)$$
$$C_y = m(A_zB_x - A_xB_z) \qquad\qquad (4\text{--}13)$$
$$C_z = m(A_xB_y - A_yB_x)$$

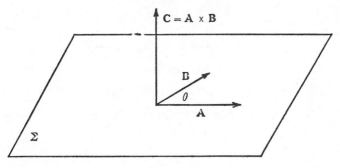

Fɪɢ. 4–4

where m is an arbitrary constant, which is conveniently taken as $+1$. Then from (6) and (13),

$$C^2 = C_x^2 + C_y^2 + C_z^2 = (A_x^2 + A_y^2 + A_z^2)(B_x^2 + B_y^2 + B_z^2)$$
$$- (A_xB_x + A_yB_y + A_zB_z)^2$$

The first member on the right-hand side of this equals A^2B^2 by (6), the second member equals $(\mathbf{A}\cdot\mathbf{B})^2 = (AB\cos\theta)^2$ by (12) and (8), hence

$$C^2 = (A^2B^2 - A^2B^2\cos^2\theta) = (AB\sin\theta)^2$$

The vector $\mathbf{C}$ may thus be described as the product of two other vectors, $\mathbf{A}$ and $\mathbf{B}$; it is called the *vector* (or *skew*) *product*[4] and is written

$$\mathbf{C} = \mathbf{A}\times\mathbf{B}$$

Its length is $C = AB\sin\theta$; its direction is perpendicular to the plane determined by $\mathbf{A}$ and $\mathbf{B}$.   Using (13) and the unit vectors, we may also write

$$\mathbf{C} = \mathbf{A}\times\mathbf{B} = (A_yB_z - A_zB_y)\mathbf{i} + (A_zB_x - A_xB_z)\mathbf{j}$$
$$+ (A_xB_y - A_yB_x)\mathbf{k}$$

[4] Also called the cross product or outer product.

This may be put in the form of a determinant:

$$\mathbf{A} \times \mathbf{B} = \begin{vmatrix} \mathbf{i} & \mathbf{j} & \mathbf{k} \\ A_x & A_y & A_z \\ B_x & B_y & B_z \end{vmatrix} \tag{4-14}$$

As a consequence of (14), vector products of the unit vectors become

$$\mathbf{i} \times \mathbf{j} = -\mathbf{j} \times \mathbf{i} = \begin{vmatrix} \mathbf{i} & \mathbf{j} & \mathbf{k} \\ 1 & 0 & 0 \\ 0 & 1 & 0 \end{vmatrix} = \mathbf{k}$$

$$\mathbf{j} \times \mathbf{k} = -\mathbf{k} \times \mathbf{j} = \mathbf{i}; \quad \mathbf{k} \times \mathbf{i} = -\mathbf{i} \times \mathbf{k} = \mathbf{j}$$

and

$$\mathbf{i} \times \mathbf{i} = \mathbf{j} \times \mathbf{j} = \mathbf{k} \times \mathbf{k} = 0 \tag{4-15}$$

Eq. (14) shows that vector multiplication is not commutative,

$$\mathbf{A} \times \mathbf{B} = -\mathbf{B} \times \mathbf{A}$$

The distributive law of ordinary multiplication, however, is retained:

$$\mathbf{A} \times (\mathbf{B} + \mathbf{C}) = \mathbf{A} \times \mathbf{B} + \mathbf{A} \times \mathbf{C}$$

$$(\mathbf{A} + \mathbf{B}) \times (\mathbf{C} + \mathbf{D}) = \mathbf{A} \times \mathbf{C} + \mathbf{A} \times \mathbf{D} + \mathbf{B} \times \mathbf{C} + \mathbf{B} \times \mathbf{D}$$

Problem. Prove by vector methods the trigonometric relations

$$\cos (x \pm y) = \cos x \cos y \mp \sin x \sin y$$
$$\sin (x \pm y) = \sin x \cos y \pm \cos x \sin y$$

Hint: Take three vectors: $\mathbf{A} = \cos x \mathbf{i} + \sin x \mathbf{j}$
$$\mathbf{B} = \cos y \mathbf{i} + \sin y \mathbf{j}$$
$$\mathbf{C} = \cos y \mathbf{i} - \sin y \mathbf{j}$$

Form the scalar and vector products.

The close connection between the vector $\mathbf{C} = \mathbf{A} \times \mathbf{B}$ and the parallelogram whose sides are $\mathbf{A}$ and $\mathbf{B}$ suggests that it may be useful generally to represent areas by vectors. The convention usually adopted in this connection, with reference to plane areas, is the following. The area is represented by a vector perpendicular to the area; and of length equal to its size. This leaves the direction of the vector undetermined. The latter is fixed relative to the sense in which the contour of the area is described: it is taken to be that direction in which a right-handed screw would advance when turned in the sense in which the contour is to be described. When the sense of the contour is not specified, the direction of the area vector remains undetermined. For closed surfaces, it is customary to draw the vector along the outward normal.

We now consider two important examples of vector products.

a. *Moment of a Force.* In mechanics, the moment of a force about a point O is defined as the product of the force by its perpendicular distance from O. From the geometry of Fig. 5a this product equals twice the area of the triangle OPQ. It may be represented as

$$\mathbf{M} = \mathbf{D} \times \mathbf{F}$$

where $\mathbf{M}$, $\mathbf{D}$, and $\mathbf{F}$ are vectors representing the moment, perpendicular distance and force, respectively. The sign of $\mathbf{M}$, fixed by the previous

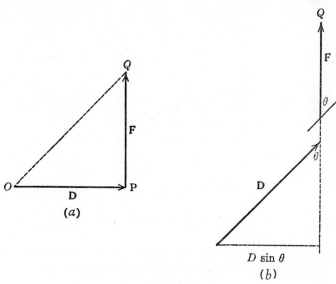

Fig. 4–5

definition of the area as a vector quantity, is positive on that side of the plane passed through O and the line F on which the force tends to produce a rotation about O in the positive direction. If $\mathbf{D}$ be drawn from O to any point in the line of action of $\mathbf{F}$ (cf. Fig. 5b), the perpendicular distance is $D \sin \theta$ and the moment is still given by the vector product, $\mathbf{D} \times \mathbf{F}$. If the force has components F_x, F_y, F_z and $\mathbf{D}$ has components D_x, D_y, D_z, the components of $\mathbf{M}$ are

$$M_x = (D_y F_z - D_z F_y)$$
$$M_y = (D_z F_x - D_x F_z)$$
$$M_z = (D_x F_y - D_y F_x)$$

b. *Angular and Linear Velocity.* Suppose a rigid body is rotating about a fixed axis, with a constant *angular velocity* of ω radians per second. The rotation of the body is then described by the vector **ω** with length equal to the scalar ω and direction parallel to the axis of rotation. Its sign, by the convention, is positive in the same direction in which a right-handed screw would progress under the given rotation. Any point P, not on the axis (cf. Fig. 6), will then describe a circle concentric with, and in a plane perpendicular to the axis, this point being determined by any vector **R** drawn from a point O on the axis of rotation. The *linear velocity* of P is at right angles to both **ω** and **R**, its magnitude being $L = \omega R \sin\theta$, or in vector symbols

$$\mathbf{L} = \mathbf{\omega} \times \mathbf{R} \qquad (4\text{--}16)$$

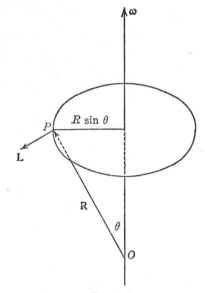

FIG. 4–6

4.6. Products Involving Three Vectors.

—From three arbitrary vectors, **A**, **B** and **C**, the following products may tentatively be formed:

(a) $\mathbf{A}(\mathbf{B} \cdot \mathbf{C})$ (d) $\mathbf{A}(\mathbf{B} \times \mathbf{C})$

(b) $\mathbf{A} \cdot (\mathbf{B} \times \mathbf{C})$ (e) $\mathbf{A} \cdot (\mathbf{B} \cdot \mathbf{C})$

(c) $\mathbf{A} \times (\mathbf{B} \times \mathbf{C})$ (f) $\mathbf{A} \times (\mathbf{B} \cdot \mathbf{C})$

Of these expressions, (e) and (f) are meaningless since vector products have only been defined when vectors stand on both sides of the dot or cross. Furthermore, no meaning has been attached to two vectors standing together in the absence of one of these signs, hence (d) is of no interest here.

a. Since $(\mathbf{B} \cdot \mathbf{C}) = BC \cos\theta$ is a scalar, the triple product $\mathbf{A}(\mathbf{B} \cdot \mathbf{C})$ is a new vector whose direction is the same as that of **A**; its magnitude equals A multiplied by $BC \cos\theta$.

b. The product $\mathbf{A} \cdot (\mathbf{B} \times \mathbf{C})$, called the *scalar triple product*, is a scalar, for $(\mathbf{B} \times \mathbf{C}) = \mathbf{D}$, a new vector. We have

$$\mathbf{A} \cdot (\mathbf{B} \times \mathbf{C}) = (\mathbf{B} \times \mathbf{C}) \cdot \mathbf{A} = \mathbf{A} \cdot \mathbf{D} = \mathbf{D} \cdot \mathbf{A} = \text{a scalar}$$

Moreover, the new vector **D** is perpendicular to both **B** and **C**, or from (11)

$$\mathbf{B} \cdot (\mathbf{B} \times \mathbf{C}) = \mathbf{C} \cdot (\mathbf{B} \times \mathbf{C}) = 0$$

If the three vectors **A**, **B** and **C** are the edges of a parallelepiped, as shown in Fig. 7, then $(\mathbf{B} \times \mathbf{C})$ is a vector whose length equals the area of the parallelogram forming the base of the parallelepiped; its direction is perpendicular to the plane of **B** and **C**. The scalar triple product is thus the area of the base multiplied by the projection of the slant height of **A** on the

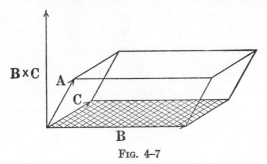

FIG. 4–7

vector (**B** × **C**) or a scalar whose magnitude equals the volume of the parallelepiped, v. By taking various faces in turn, we find from Fig. 7,

$$\mathbf{A} \cdot (\mathbf{B} \times \mathbf{C}) = \mathbf{B} \cdot (\mathbf{C} \times \mathbf{A}) = \mathbf{C} \cdot (\mathbf{A} \times \mathbf{B}) = v$$

Since a change of order in the vector product changes the sign there are other possible relations which may be abbreviated by writing

$$v = [\mathbf{ABC}] = \mathbf{A} \cdot (\mathbf{B} \times \mathbf{C}) = (\mathbf{B} \times \mathbf{C}) \cdot \mathbf{A}, \quad \text{etc.} \qquad (4\text{–}17)$$

and

$$[\mathbf{ABC}] = [\mathbf{BCA}] = [\mathbf{CAB}] = -[\mathbf{ACB}] = -[\mathbf{BAC}] = -[\mathbf{CBA}] \qquad (4\text{–}18)$$

Each term in square brackets stands for the two possible ways of writing the triple product as shown in (17). It also follows from (18) that the cross and dot may be exchanged at will, provided the cyclical order of the three vectors is retained. The parenthesis in a product like **A** · (**B** × **C**) is superfluous but it is often written for clarity.

Because of (15), the scalar triple products of unit vectors all disappear except

$$[\mathbf{ijk}] = -[\mathbf{ikj}] = 1$$

which follows from (9). If the three vectors **A**, **B**, **C** are written in terms of unit vectors and the indicated multiplications performed, the use of (9), (10) and (15) gives

$$[\mathbf{ABC}] = A_x B_y C_z + B_x C_y A_z + C_x A_y B_z - A_x C_y B_z - B_x A_y C_z - C_x B_y A_z$$

$$= \begin{vmatrix} A_x & A_y & A_z \\ B_x & B_y & B_z \\ C_x & C_y & C_z \end{vmatrix} \qquad (4\text{–}19)$$

c. The product, **V** = **A** × (**B** × **C**), called the *vector triple product*, is a vector since it is the vector product of two vectors, **A** and (**B** × **C**). It is therefore perpendicular to both of its components:

$$\mathbf{V} \cdot \mathbf{A} = 0; \quad \mathbf{V} \cdot (\mathbf{B} \times \mathbf{C}) = 0$$

but $\mathbf{V}$ must lie in the plane of $\mathbf{B}$ and $\mathbf{C}$, since it is perpendicular to the vector product of $\mathbf{B}$ and $\mathbf{C}$, which itself is perpendicular to both $\mathbf{B}$ and $\mathbf{C}$.

The most important property of this triple product is that it permits decomposition into two scalar products:

$$\mathbf{A} \times (\mathbf{B} \times \mathbf{C}) = \mathbf{B}(\mathbf{A} \cdot \mathbf{C}) - \mathbf{C}(\mathbf{A} \cdot \mathbf{B}) \qquad (4\text{-}20)$$

a relation which may be proved geometrically or analytically by expanding in Cartesian coordinates. Since the vector product changes its sign when the order of multiplication is changed, the sign of the triple vector product will change when the order of the factors in the parenthesis is changed or when the position of the parenthesis is changed:

$$\mathbf{A} \times (\mathbf{B} \times \mathbf{C}) = -\mathbf{A} \times (\mathbf{C} \times \mathbf{B}) = (\mathbf{C} \times \mathbf{B}) \times \mathbf{A} = -(\mathbf{B} \times \mathbf{C}) \times \mathbf{A}$$

Products of more than three vectors may always be reduced to one of the three preceding types of triple products by successive application of the above rules.

Problem. Verify the relations:

$$(\mathbf{A} \times \mathbf{B}) \cdot (\mathbf{C} \times \mathbf{D}) = (\mathbf{A} \cdot \mathbf{C})(\mathbf{B} \cdot \mathbf{D}) - (\mathbf{A} \cdot \mathbf{D})(\mathbf{B} \cdot \mathbf{C})$$

$$(\mathbf{A} \times \mathbf{B}) \times (\mathbf{C} \times \mathbf{D}) = \mathbf{B}[\mathbf{ACD}] - \mathbf{A}[\mathbf{BCD}] = \mathbf{C}[\mathbf{ABD}] - \mathbf{D}[\mathbf{ABC}]$$

$$[\mathbf{A} \times \mathbf{B} \ \ \mathbf{B} \times \mathbf{C} \ \ \mathbf{C} \times \mathbf{A}] = [\mathbf{ABC}]^2$$

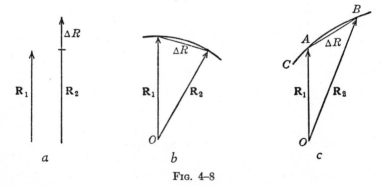

a b c

Fig. 4–8

4.7. Differentiation of Vectors.—If a vector $\mathbf{R}$ is a function of a single scalar t, which for convenience may be assumed to be the time, there are three possible ways in which $\mathbf{R}$ may vary. Let $\mathbf{R}_1$ and $\mathbf{R}_2$ refer to times t_1 and t_2, then $\mathbf{R}_2$ may differ from $\mathbf{R}_1$: (a) in magnitude only; (b) in direction only; (c) in both magnitude and direction as shown in Fig. 8. Since no complication arises from treating the general case, let us assume that a curve is traced by the terminus of a continuously varying vector $\mathbf{R}$, the origin of the latter being kept fixed at the origin of a coordinate system. The vector $\Delta\mathbf{R} = \mathbf{R}_2 - \mathbf{R}_1$, having the direction of the secant AB of Fig. 8, approaches the tangent of the curve C at the point $\mathbf{R}_1$ as $\Delta t = t_2 - t_1$ approaches zero. The quotient $\Delta\mathbf{R}/\Delta t$ is the average rate of change of $\mathbf{R}$ in the time interval between t_1 and t_2. Following the usual methods of differential calculus, the derivative of $\mathbf{R}$ is defined as

$$\lim_{t \to 0} \frac{\Delta\mathbf{R}}{\Delta t} = \frac{d\mathbf{R}}{dt}$$

In terms of unit vectors, and with the use of primes for differentiation

$$\mathbf{R} = iR_x + jR_y + kR_z$$
$$\mathbf{R}' = iR_x' + jR_y' + kR_z'$$
$$\mathbf{R}'' = iR_x'' + jR_y'' + kR_z''$$

For a composite function of two or more vectors, each of which depends on the single scalar t, the usual rules of differentiation hold except that, of course, the order of the vectors must not be changed in cases involving the vector product.

In the special case of Fig. 8a, where $\mathbf{R}$ is constant in direction but variable in magnitude, $\Delta\mathbf{R}$ is parallel to $\mathbf{R}$. Similarly, in case (b), $\Delta\mathbf{R}$ is perpendicular to $\mathbf{R}$, for the fixed length of $\mathbf{R}$ is $\mathbf{R} \cdot \mathbf{R} = R^2$, $d(\mathbf{R} \cdot \mathbf{R})/dt = 0$ and hence $\mathbf{R} \cdot d\mathbf{R}/dt = 0$, the latter being the requirement that $\mathbf{R}$ and $d\mathbf{R}/dt$ be perpendicular.

4.8. Scalar and Vector Fields.—A *scalar field* is defined as a region of space, with each point of which there is associated a *scalar point function* (cf. sec. 1.7). A simple example is the temperature of points in the atmosphere at a given moment. On the other hand, if there is a vector associated with each point in a region of space, the points and vectors constitute a *vector field*, an example being the wind velocity of points in the atmosphere at any instant.

Suppose $\phi(x,y,z)$ is a scalar point function referred to a given coordinate system. It will usually change its form if referred to another system, say $\phi'(x',y',z')$ but its value at any point must be unchanged, or $\phi = \phi'$. For example, the temperature at any point in the atmosphere cannot depend on the coordinate system used to describe the point. Differentiating $\phi = \phi'$ partially, we obtain

$$\frac{\partial\phi'}{\partial x'} = \frac{\partial x}{\partial x'}\frac{\partial\phi}{\partial x} + \frac{\partial y}{\partial x'}\frac{\partial\phi}{\partial y} + \frac{\partial z}{\partial x'}\frac{\partial\phi}{\partial z}$$

$$= a_{11}\frac{\partial\phi}{\partial x} + a_{12}\frac{\partial\phi}{\partial y} + a_{13}\frac{\partial\phi}{\partial z}$$

$$\frac{\partial \phi'}{\partial y'} = a_{21} \frac{\partial \phi}{\partial x} + a_{22} \frac{\partial \phi}{\partial y} + a_{23} \frac{\partial \phi}{\partial z}$$

$$\frac{\partial \phi'}{\partial z'} = a_{31} \frac{\partial \phi}{\partial x} + a_{32} \frac{\partial \phi}{\partial y} + a_{33} \frac{\partial \phi}{\partial z}$$

with three similar equations for $\partial \phi/\partial x$, $\partial \phi/\partial y$, $\partial \phi/\partial z$. Comparing these derivatives with (4) and (5), it follows that $(\partial \phi/\partial x, \partial \phi/\partial y, \partial \phi/\partial z)$ are the three components of a vector since they transform from one reference frame to another in the manner prescribed for vector components.

Using the abbreviation

$$\nabla \equiv \mathbf{i}\partial/\partial x + \mathbf{j}\partial/\partial y + \mathbf{k}\partial/\partial z \tag{4-21}$$

let us study the quantities $\nabla * \psi$, where ψ is either a scalar or a vector and $(*)$ is either to be omitted or replaced by a dot or a cross in order to give products which have meaning. The operator, ∇, called "*del*," is not a vector in the geometrical sense since it has no scalar magnitude, but it does transform properly, so that it may be treated formally as a vector. The possible products are $\nabla \phi$, where ϕ is a scalar point function; $\nabla \cdot \mathbf{V}$ and $\nabla \times \mathbf{V}$, where $\mathbf{V}$ is a vector field.

4.9. The Gradient.—The first of these products, called the **gradient** of the scalar ϕ

$$\nabla \phi = \mathbf{grad}\ \phi = \mathbf{i}\partial \phi/\partial x + \mathbf{j}\partial \phi/\partial y + \mathbf{k}\partial \phi/\partial z \tag{4-22}$$

is a vector, since it is the product of a scalar ϕ and a vector ∇. To perceive its physical significance, let us consider the family of surfaces, $\phi(x,y,z) = $ constant, or the equivalent of this relation

$$d\phi = (\partial \phi/\partial x)dx + (\partial \phi/\partial y)dy + (\partial \phi/\partial z)dz = 0$$

At any point P with coordinates (x,y,z), on one of these surfaces $d\mathbf{R} = \mathbf{i}dx + \mathbf{j}dy + \mathbf{k}dz$ is a vector, tangent to P, provided dx, dy, dz satisfy the preceding equation. Since $\nabla \phi \cdot d\mathbf{R} = d\phi = 0$, $d\mathbf{R}$ and $\nabla \phi$ are perpendicular to each other, or $\nabla \phi$ is perpendicular to that surface of the family which passes through P. By the convention of signs previously established, the direction of $\nabla \phi$ is that in which ϕ is increasing. For any other direction determined by the unit vector $\mathbf{s}$ with direction cosines (l,m,n) through P, the component of $\nabla \phi$ in the direction $\mathbf{s}$ is

$$l\partial \phi/\partial x + m\partial \phi/\partial y + n\partial \phi/\partial z$$

which may be written $\mathbf{s} \cdot \nabla \phi$. This is the *directional derivative* of ϕ in the direction $\mathbf{s}$. In going from P on one of the surfaces (cf. Fig. 9) to any point Q on the surface $\phi + d\phi$, the increase in ϕ is the same wherever the point Q is chosen, but the distance PQ will be smallest and hence $\mathbf{s} \cdot \nabla \phi$

greatest when **s** is in the direction of the normal **N**. Therefore, since $\nabla\phi$ is normal to the surface ϕ = const., at the point P its direction and magnitude give the maximum space rate of increase of the scalar ϕ.

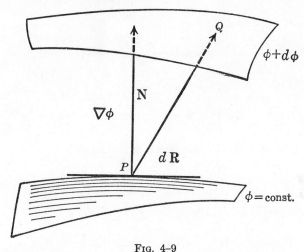

FIG. 4-9

4.10. The Divergence.—The scalar product of the vector operator ∇ and a vector **V** gives a scalar which is called the divergence of **V**.

$$\nabla \cdot \mathbf{V} = \operatorname{div} \mathbf{V} = \left\{ \mathbf{i}\,\frac{\partial}{\partial x} + \mathbf{j}\,\frac{\partial}{\partial y} + \mathbf{k}\,\frac{\partial}{\partial z} \right\} \cdot \left\{ \mathbf{i}V_x + \mathbf{j}V_y + \mathbf{k}V_z \right\}$$

$$= \partial V_x/\partial x + \partial V_y/\partial y + \partial V_z/\partial z \qquad (4\text{-}23)$$

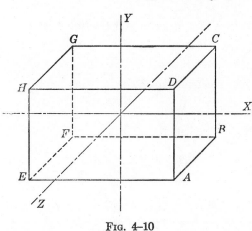

FIG. 4-10

If **V** is a vector field, the derivative $\partial V_x/\partial x$ transforms, when a change of coordinate system is made, like the product A_xB_x of the x-components of two vectors **A** and **B**, hence the divergence of **V** is a scalar point function. Suppose that **V** represents at each point in space the direction and magnitude of flow (density times velocity) of some fluid such as water or a gas, or that it represents thermal or electrical flux. Consider, for example (Fig. 10), a small parallelepiped of volume $dxdydz = d\tau$, through which a fluid is passing.

The loss of fluid mass through face $ABCD$ per unit time is

$$\mathbf{i} \cdot \left\{ \mathbf{V}(x,y,z) + \frac{\partial \mathbf{V}}{\partial x}\frac{dx}{2} \right\} dydz$$

while the gain through $EFGH$ is

$$\mathbf{i} \cdot \left\{ \mathbf{V}(x,y,z) - \frac{\partial \mathbf{V}}{\partial x}\frac{dx}{2} \right\} dydz$$

Therefore the net loss through these two faces is

$$\mathbf{i} \cdot \frac{\partial \mathbf{V}}{\partial x} dxdydz$$

The losses through the other two pairs of faces are

$$\mathbf{j} \cdot \frac{\partial \mathbf{V}}{\partial y} dxdydz \quad \text{and} \quad \mathbf{k} \cdot \frac{\partial \mathbf{V}}{\partial z} dxdydz$$

The total loss from the parallelepiped is therefore

$$\left\{ \mathbf{i} \cdot \frac{\partial \mathbf{V}}{\partial x} + \mathbf{j} \cdot \frac{\partial \mathbf{V}}{\partial y} + \mathbf{k} \cdot \frac{\partial \mathbf{V}}{\partial z} \right\} dxdydz = \nabla \cdot \mathbf{V}d\tau$$

If $\mathbf{v}$ is the velocity of the fluid of density ρ, $\mathbf{V} = \rho\mathbf{v}$ is called the *flux density* and represents the total flow of fluid per unit cross section in unit time. Then if no fluid is created or destroyed within the parallelepiped, this loss of mass must equal $-(\partial\rho/\partial t)d\tau$,

$$\nabla \cdot \mathbf{V} = \frac{-\partial\rho}{\partial t}$$

a relation usually called the *equation of continuity*. If the liquid is incompressible, $\partial\rho/\partial t = 0$, hence

$$\nabla \cdot \mathbf{V} = 0$$

A similar relation holds for $\mathbf{D}$, the electric displacement,

$$\nabla \cdot \mathbf{D} = 0$$

4.11. The Curl.—The vector product of ∇ and $\mathbf{V}$ is called the **curl** or rotation of $\mathbf{V}$

$$\operatorname{curl} \mathbf{V} = \nabla \times \mathbf{V} = \mathbf{i}\left\{\frac{\partial V_z}{\partial y} - \frac{\partial V_y}{\partial z}\right\} + \mathbf{j}\left\{\frac{\partial V_x}{\partial z} - \frac{\partial V_z}{\partial x}\right\}$$

$$+ \mathbf{k}\left\{\frac{\partial V_y}{\partial x} - \frac{\partial V_x}{\partial y}\right\} = \begin{vmatrix} \mathbf{i} & \mathbf{j} & \mathbf{k} \\ \frac{\partial}{\partial x} & \frac{\partial}{\partial y} & \frac{\partial}{\partial z} \\ V_x & V_y & V_z \end{vmatrix} \tag{4-24}$$

This function may be used to describe the motion of a rigid body rotating about an axis with uniform angular velocity, ω. The linear velocity of any point P in the body with radius vector $\mathbf{R}$ is (cf. 16) $\mathbf{L} = \omega \times \mathbf{R}$ and

$$\text{curl } \mathbf{L} = \nabla \times (\omega \times \mathbf{R}) \tag{4-25}$$

Expanding (25), by (20)

$$\text{curl } \mathbf{L} = \omega (\nabla \cdot \mathbf{R}) - (\nabla \cdot \omega)\mathbf{R}$$

Since $\mathbf{R} = \mathbf{i}R_x + \mathbf{j}R_y + \mathbf{k}R_z = \mathbf{i}x + \mathbf{j}y + \mathbf{k}z$, $\nabla \cdot \mathbf{R} = 3$. The angular velocity is a constant vector, hence $\nabla \cdot \omega = \omega \cdot \nabla$ and we may write the last member of the above equation in the form $(\omega \cdot \nabla)\mathbf{R}$, which is to be interpreted as the product of a scalar $(\omega \cdot \nabla)$ and a vector $\mathbf{R}$. Expanding,

$$(\omega \cdot \nabla)\mathbf{R} = \left\{\omega_x \frac{\partial}{\partial x} + \omega_y \frac{\partial}{\partial y} + \omega_z \frac{\partial}{\partial z}\right\}\mathbf{R} = \mathbf{i}\omega_x + \mathbf{j}\omega_y + \mathbf{k}\omega_z = \omega$$

Hence, $\text{curl } \mathbf{L} = 3\omega - \omega = 2\omega$

or the curl of the linear velocity of any point of a rigid body equals twice the angular velocity, for magnitude, not direction changes.

4.12. Composite Functions Involving ∇.—The following relations involving ∇ may be verified by expanding the vectors in terms of their components along three unit vectors, $\mathbf{i}$, $\mathbf{j}$ and $\mathbf{k}$.

$$\nabla * (A + B) = \nabla * A + \nabla * B$$
$$\nabla * (\phi A) = \nabla\phi * A + \phi\nabla * A \tag{4-26}$$
$$\nabla (\mathbf{U} \cdot \mathbf{V}) = (\mathbf{V} \cdot \nabla)\mathbf{U} + (\mathbf{U} \cdot \nabla)\mathbf{V} + \mathbf{V} \times (\nabla \times \mathbf{U}) + \mathbf{U} \times (\nabla \times \mathbf{V})$$
$$\nabla \cdot (\mathbf{U} \times \mathbf{V}) = \mathbf{V} \cdot \nabla \times \mathbf{U} - \mathbf{U} \cdot \nabla \times \mathbf{V}$$
$$\nabla \times (\mathbf{U} \times \mathbf{V}) = (\mathbf{V} \cdot \nabla)\mathbf{U} - \mathbf{V}(\nabla \cdot \mathbf{U}) - (\mathbf{U} \cdot \nabla)\mathbf{V} + \mathbf{U}(\nabla \cdot \mathbf{V})$$

In these equations A and B are either scalars or vectors depending on the choice of $(*)$, ϕ is a scalar and $\mathbf{U}$, $\mathbf{V}$ are vectors. If $\mathbf{R} = \mathbf{i}x + \mathbf{j}y + \mathbf{k}z$,

$$\nabla \cdot \mathbf{R} = 3$$
$$\nabla \times \mathbf{R} = 0$$
$$\mathbf{U} \cdot \nabla\mathbf{R} = \mathbf{U}$$

Problem. Prove eqs. (26).

4.13. Successive Applications of ∇.—There are six possible combinations in which ∇ occurs twice. The following relations may be proved as above by expansion in terms of $\mathbf{i}$, $\mathbf{j}$ and $\mathbf{k}$.

a. $\nabla \cdot \nabla\phi = \nabla^2\phi = \nabla \cdot \text{grad } \phi = \text{div grad } \phi$

$$= \frac{\partial^2\phi}{\partial x^2} + \frac{\partial^2\phi}{\partial y^2} + \frac{\partial^2\phi}{\partial z^2} \tag{4-27}$$

The operator ∇^2 is generally called the *Laplacian*.

b. Since ∇^2 is a scalar, it may also be applied to a vector, the result being a new vector

$$(\nabla \cdot \nabla)\mathbf{V} = \nabla^2\mathbf{V} = \frac{\partial^2 \mathbf{V}}{\partial x^2} + \frac{\partial^2 \mathbf{V}}{\partial y^2} + \frac{\partial^2 \mathbf{V}}{\partial z^2} \tag{4-28}$$

c. $$\nabla(\nabla \cdot \mathbf{V}) = \text{grad div } \mathbf{V}$$

$$= \mathbf{i}\frac{\partial^2 V_x}{\partial x^2} + \mathbf{j}\frac{\partial^2 V_y}{\partial y^2} + \mathbf{k}\frac{\partial^2 V_z}{\partial z^2} + \mathbf{i}\left\{\frac{\partial^2 V_y}{\partial x \partial y} + \frac{\partial^2 V_z}{\partial x \partial z}\right\}$$

$$+ \mathbf{j}\left\{\frac{\partial^2 V_x}{\partial x \partial y} + \frac{\partial^2 V_z}{\partial y \partial z}\right\} + \mathbf{k}\left\{\frac{\partial^2 V_x}{\partial x \partial z} + \frac{\partial^2 V_y}{\partial y \partial z}\right\} \tag{4-29}$$

d. $\nabla \times \nabla\phi = \text{curl grad } \phi = \begin{vmatrix} \mathbf{i} & \mathbf{j} & \mathbf{k} \\ \partial/\partial x & \partial/\partial y & \partial/\partial z \\ \partial\phi/\partial x & \partial\phi/\partial y & \partial\phi/\partial z \end{vmatrix} = 0$

$$\tag{4-30}$$

This is an identity. If for some vector $\mathbf{V}$, $\nabla \times \mathbf{V} = 0$, then $\mathbf{V} = \nabla\phi$, where ϕ is some scalar function. Under these conditions, $\mathbf{V}$ is said to be *irrotational*. Expansion also yields

e. $$\nabla \cdot \nabla \times \mathbf{V} = \text{div curl } \mathbf{V} = 0 \tag{4-31}$$

Thus if for any vector $\mathbf{W}$, $\nabla \cdot \mathbf{W} = 0$ then $\mathbf{W} = \nabla \times \mathbf{V}$ and $\mathbf{W}$ is said to be *solenoidal*.

Finally, the reader will easily check by expansion in rectangular components, the relation

f. $\nabla \times (\nabla \times \mathbf{V}) = \text{curl curl } \mathbf{V} =$

$$\text{grad div } \mathbf{V} - \nabla^2\mathbf{V} = \nabla(\nabla \cdot \mathbf{V}) - \nabla \cdot \nabla\mathbf{V} \tag{4-32}$$

Problem. Show by expansion that eqs. (4–27, 28, 29, 30, 31, 32) are correct.

4.14. Vector Integration.—As a simple example of vector integration, we consider the motion of a particle under the constant acceleration of gravity. The equation of motion is

$$d^2\mathbf{R}/dt^2 = \mathbf{G}$$

where $\mathbf{G}$ is a constant vector. Integration results in $d\mathbf{R}/dt = \mathbf{G}t + \mathbf{V}_0$; $\mathbf{R} = \mathbf{G}t^2/2 + \mathbf{V}_0 t + \mathbf{C}_0$, where $\mathbf{V}_0$ and $\mathbf{C}_0$ are the constants of integration which are vectors not necessarily collinear with $\mathbf{G}$. They are determined by the values of $d\mathbf{R}/dt$ and $\mathbf{R}$, respectively, when $t = 0$.

More complicated cases may arise, however, for in the general case, the integral is $\int \psi * dr$, where ψ and dr may be scalars or vectors, (*) has the same meaning as before and the integrals may be multiple.

4.15. Line Integrals.—Suppose dr is the vector ds, where $\mathbf{s} = \mathbf{s}(t)$ is the equation for a curve. It is then possible to form the integrals:

$$(a) \int_C \phi d\mathbf{s}; \qquad (b) \int_C \mathbf{V} \cdot d\mathbf{s}; \qquad (c) \int_C \mathbf{V} \times d\mathbf{s} \qquad (4\text{-}33)$$

each of these being called the *line integral* along the curve C. The results of integration are respectively, a vector, a scalar and a vector.

Since

$$d\mathbf{s} = \mathbf{i}dx + \mathbf{j}dy + \mathbf{k}dz \qquad (4\text{-}34)$$

the first integral in (33) becomes:

$$\int_C \phi d\mathbf{s} = \int_A^B \phi(x,y,z)(\mathbf{i}dx + \mathbf{j}dy + \mathbf{k}dz) = \int_{x_1}^{x_2} \phi(x,y,z)\mathbf{i}dx$$
$$+ \int_{y_1}^{y_2} \phi(x,y,z)\mathbf{j}dy + \int_{z_1}^{z_2} \phi(x,y,z)\mathbf{k}dz$$

where A and B are initial and final points of the curve, with coordinates (x_1,y_1,z_1) and (x_2,y_2,z_2). The first integral on the right may be evaluated when y and z are known in terms of x for points on the curve C. The remaining integrals are determined in a similar fashion. The problem thus reduces to the usual line integral in scalar calculus except that it is necessary to specify the direction in which the radius vector $\mathbf{s}$ describes the curve during integration, for if the direction A to B is taken as positive, then

$$\int_A^B \phi d\mathbf{s} = -\int_B^A \phi d\mathbf{s}$$

In case C is a closed curve, the direction is always taken so that the enclosed curve appears positive (cf. sec. 4.5).

No difficulty is experienced in the interpretation of (b) and (c) of (33) as the following example shows. Let $\mathbf{V} = xy\mathbf{i} - z^2\mathbf{j} + xyz\mathbf{k}$; evaluate $\int \mathbf{V} \cdot d\mathbf{s}$ from the point $A = (0,0,0)$ to $B = (1,1,1)$ along the curve $\mathbf{s} = \mathbf{i}t + \mathbf{j}t^2 + \mathbf{k}t^3$.

$$\int_A^B \mathbf{V} \cdot d\mathbf{s} = \int_A^B (xy\mathbf{i} - z^2\mathbf{j} + xyz\mathbf{k}) \cdot (\mathbf{i}dx + \mathbf{j}dy + \mathbf{k}dz)$$
$$= \int_A^B (xydx - z^2dy + xyzdz)$$

Since $d\mathbf{s}$ is the position vector of points on the curve, the coordinates of any point in terms of t are

$$x = t; \quad y = t^2; \quad z = t^3$$

Hence,

$$\int_A^B \mathbf{V} \cdot d\mathbf{s} = \int_0^1 x^3 dx - \int_0^1 y^3 dy + \int_0^1 z^2 dz$$
$$= \tfrac{1}{3}$$

An important special case arises in scalar calculus when the function to be integrated is an exact differential, where the value of the integral is independent of the path. In vector calculus, suppose

$$\mathbf{V} = \mathbf{grad}\ \phi = \nabla \phi \qquad (4\text{-}35)$$

with ϕ a scalar point function. Then using (22) and (34)

$$\int_A^B \mathbf{V} \cdot d\mathbf{s} = \int_A^B \nabla \phi \cdot d\mathbf{s} = \int_A^B \left\{ \frac{\partial \phi}{\partial x}\, dx + \frac{\partial \phi}{\partial y}\, dy + \frac{\partial \phi}{\partial z}\, dz \right\}$$
$$= \int_A^B d\phi = \phi_B - \phi_A \qquad (4\text{-}36)$$

If the integration is taken around a closed curve, $B = A$, then

$$\int_A^A \nabla \phi \cdot d\mathbf{s} = \oint \nabla \phi \cdot d\mathbf{s} = 0$$

Conversely, if $\oint \mathbf{V} \cdot d\mathbf{s} = 0$, then (35) must hold, i.e., $\mathbf{V}$ is the **gradient** of some scalar point function ϕ. We have therefore shown that if $\mathbf{V} = \mathbf{grad}\ \phi$, the line integral $\int_A^B \mathbf{V} \cdot d\mathbf{s}$ depends only on the initial and final values of ϕ and is independent of the path.

4.16. Surface and Volume Integrals.—Let Σ be any surface, divided into infinitesimal elements each of which may be considered as a vector, $d\mathbf{S}$. The surface integral may then be described as in ordinary analysis, but again there are three cases:

$$(a) \iint_S \phi d\mathbf{S}; \qquad (b) \iint_S \mathbf{V} \cdot d\mathbf{S}; \qquad (c) \iint_S \mathbf{V} \times d\mathbf{S}$$

giving a vector, a scalar and a vector. As before, it is important to specify the side of the surface over which the integration is performed, for although $d\mathbf{S}$ is normal to the surface, the signs of the normals on opposite sides are opposite. The sign of the normal is uniquely determined by the previous conventions except for the case of a one-sided surface[5] such as the *Möbius strip*. If the surface encloses a portion of space, $d\mathbf{S}$ is taken as the outward

---

[5] See, for example, Burington, R. S., and Torrance, C. C., "Higher Mathematics," McGraw-Hill Book Co., New York, 1939, pp. 250ff.

pointing normal. The surface integral $\iint_S \mathbf{V} \cdot d\mathbf{S}$ is called the *flux* of $\mathbf{V}$ through the surface, for if $\mathbf{V}$ is the product of density and the velocity of a fluid, the integral is the amount of fluid flowing through a surface in unit time. The vector $\mathbf{V}$ may also refer to electric, magnetic or gravitational force, flow of heat and so on.

Let $d\tau = dxdydz$ be an element of volume. Since this is a scalar, there are only two possible volume integrals

$$(a) \quad \iiint_\tau \phi d\tau; \quad (b) \quad \iiint_\tau \mathbf{V} d\tau$$

the first being a scalar and the second a vector.

It is often convenient to convert multiple integrals into others with fewer integral signs. One possibility has previously been presented in (36), namely that the line integral $\int_C \mathbf{V} \cdot d\mathbf{s}$ may be reduced to the difference between two scalar quantities, provided $\mathbf{V} = \nabla\phi$. A line integral may also be converted into a double or surface integral by *Stokes' theorem*, or conversely, the double integral may be reduced to a single integral.

4.17. Stokes' Theorem.—This theorem may be stated in the form

$$\oint \mathbf{V} \cdot d\mathbf{s} = \iint_S \nabla \times \mathbf{V} \cdot d\mathbf{S} \tag{4-37}$$

Conversely, if $\mathbf{W} = \nabla \times \mathbf{V}$, where $\mathbf{V}$ is another vector, then the value of the surface integral $\iint_S \mathbf{W} \cdot d\mathbf{S}$ depends only upon values of $\mathbf{V}$ at points on the boundary of the surface,

$$\iint_S \mathbf{W} \cdot d\mathbf{S} = \oint \mathbf{V} \cdot d\mathbf{s}$$

The vector $\mathbf{V}$ may be taken as flux density of a fluid or as the field of a mechanical or electrical force. In the latter case, the line integral represents the work done on a particle moving along a curve C. If the curve is closed, forming the boundary of a region Σ, then according to the theorem the work done equals the surface integral of the **curl** of the force field. In the special case where the work done is independent of the path, the line integral vanishes so that a requirement for independence of the path is that $\nabla \times \mathbf{V} = 0$.

A proof of Stokes' theorem follows. Consider a surface Σ bounded by the closed contour C. Let C' be the projection of Σ on the XY-plane; we are thus associating a point $P(x,y)$ on the plane with every point

$P'(x,y,z)$ of the surface. This means that on the surface, where z is a function of x and y, a function $u(x,y,z)$ becomes

$$u(x,y,z) = \phi(x,y) \tag{4-38}$$

since the value of ϕ on C' must equal the value of u on C. Similarly with other functions

$$v(x,y,z) = \chi(x,z); \quad w(x,y,z) = \psi(y,z)$$

when projections are made on the XZ- and YZ-planes. We may write for the vector defined at each point on the surface,

$$\mathbf{V} = u\mathbf{i} + v\mathbf{j} + w\mathbf{k}$$

If we furthermore take a unit vector $\mathbf{n}$, perpendicular to the surface at any point, the right-hand side of (37) becomes after expansion

$$\iint_S \mathbf{n} \cdot \nabla \times \mathbf{V} dS = \iint_S \mathbf{n} \cdot (\nabla \times u\mathbf{i} + \nabla \times v\mathbf{j} + \nabla \times w\mathbf{k}) dS \tag{4-39}$$

A typical term of (39) may be transformed as follows

$$\mathbf{n} \cdot \nabla \times u\mathbf{i} = \mathbf{n} \cdot \left\{ \mathbf{j}\frac{\partial u}{\partial z} - \mathbf{k}\frac{\partial u}{\partial y} \right\} = -\mathbf{n} \cdot \mathbf{k}\frac{\partial \phi}{\partial y} \tag{4-40}$$

the second expression coming from (24). The last member of (40) is obtained as follows. The partial derivative

$$\frac{\partial \mathbf{s}}{\partial y} = \mathbf{j} + \mathbf{k}\frac{\partial z}{\partial y}$$

of $\mathbf{s} = x\mathbf{i} + y\mathbf{j} + z\mathbf{k}$ is a vector, tangent to the curve cut from Σ by a plane perpendicular to the X-axis. It is perpendicular to $\mathbf{n}$, hence

$$\mathbf{n} \cdot \left\{ \mathbf{j} + \mathbf{k}\frac{\partial z}{\partial y} \right\} = 0 \tag{4-41}$$

Substitution of (41) and the partial derivative of (38),

$$\frac{\partial \phi}{\partial y} = \frac{\partial u}{\partial y} + \frac{\partial u}{\partial z}\frac{\partial z}{\partial y}$$

into (40), gives the last term of that equation. Since $\mathbf{n} \cdot \mathbf{k} dS = dxdy$, we may write

$$\iint_S \mathbf{n} \cdot \nabla \times u\mathbf{i} dS = -\iint \frac{\partial \phi}{\partial y} dxdy \tag{4-42}$$

The integral on the right of (42) may be written

$$\iint \frac{\partial \phi}{\partial y} dxdy = \oint (\phi_2 - \phi_1) dx$$

where ϕ_2 and ϕ_1 are the values of ϕ at the maximum and minimum values of y, y_2 and y_1, respectively.   If $d\sigma$ is a line element of the contour C', we may write $dx = \pm (\partial x/\partial \sigma)d\sigma$, choosing the sign in accordance with the position of $d\sigma$ on the contour.   Since it is negative at y_2 and positive at y_1, the integral becomes

$$-\oint (\phi_2 + \phi_1) \frac{\partial x}{\partial \sigma} d\sigma = -\oint_{C'} \phi dx$$

Remembering (38) and the fact that between two points on C' the change in x is the same as that between the equivalent points on C,

$$\oint_{C'} \phi dx = \oint_C u dx$$

so that we finally have

$$\int\int_S \mathbf{n} \cdot \nabla \times u i dS = \oint_C u dx$$

Similar equations are obtained from consideration of projections of Σ on the XZ- and YZ-planes.   When they are added together, Stokes' theorem results.

4.18. Theorem of the Divergence.—A method of reducing triple integrals to double integrals is offered in the *theorem of the divergence*, which may be written

$$\int\int\int_\tau \nabla \cdot \mathbf{V} d\tau = \int\int_S \mathbf{V} \cdot d\mathbf{S} \qquad (4\text{--}43)$$

The Cartesian form of this equation

$$\int\int\int_\tau \left\{ \frac{\partial V_x}{\partial x} + \frac{\partial V_y}{\partial y} + \frac{\partial V_z}{\partial z} \right\} d\tau = \int\int_S (V_x dydz + V_y dxdz + V_z dxdy)$$

is often called *Gauss' theorem*.   Suppose $\mathbf{V}$ represents the flux density of an incompressible fluid.   Then, as we have shown, $\nabla \cdot \mathbf{V}$ is the total amount of fluid flowing out of a volume $d\tau$ per second.   The total flow from a large volume is $\int\int\int \nabla \cdot \mathbf{V} d\tau$, which must equal the rate of flow across all of the surfaces of the volume $\int\int \mathbf{V} \cdot d\mathbf{S}$.   This proves the theorem.[6]   If we assume a steady state, the total amount of flow neither increases nor decreases in time and hence must be maintained constant by sources or sinks within the region, unless the density of the fluid is continually chang-

---

[6] An analytical proof, which does not depend on the flow of a liquid and is similar to the one given here for Stokes' theorem, may be found in books on vector analysis.

ing (which is contrary to the initial assumption). In view of Gauss' theorem the divergence of the field takes on an interesting meaning. Since

$$\operatorname{div} \mathbf{V} = \mathbf{\nabla} \cdot \mathbf{V} = \lim_{d\tau \to 0} \frac{1}{d\tau} \int \int_S \mathbf{V} \ d\mathbf{S} \qquad (4\text{-}44)$$

the divergence is the same as the intensity of the steady flow at a given point. This argument may be continued to derive the equation of continuity which has been obtained in another way in sec. 4.10.

A further application of the divergence theorem arises in the problem of heat flow. Consider the flow of heat into a thermally isotropic solid body, the temperature of which is not the same at all points. The rate of flow of heat into the body is*

$$-\int_S \mathbf{V} \cdot d\mathbf{S}$$

where $\mathbf{V}$ is the flux of heat, the amount of heat which crosses unit area drawn perpendicular to the lines of flow per unit time. By Fourier's law, heat flows in the direction of most rapid decrease in temperature, U, with a rate proportional to the thermal conductivity κ, of the solid or

$$\mathbf{V} = -\kappa \mathbf{\nabla} U \qquad (4\text{-}45)$$

If there are no sources or sinks of heat within the body, and if ρ is the density of the solid and s its specific heat, the amount of heat entering unit volume in unit time is

$$s\rho \frac{\partial U}{\partial t}$$

For the whole body, the heat gained must equal that passing through the surface

$$\int_\tau s\rho \frac{\partial U}{\partial t} \ d\tau = -\int_S \mathbf{V} \cdot d\mathbf{S}$$

and this becomes in view of eq. (43)

$$\int_\tau \left\{ s\rho \frac{\partial U}{\partial t} + \mathbf{\nabla} \cdot \mathbf{V} \right\} \ d\tau = 0$$

This equation must hold for every surface, hence

$$s\rho \frac{\partial U}{\partial t} = -\mathbf{\nabla} \cdot \mathbf{V}$$

---

* Henceforth, single integral signs will be written in multiple integrals when the meaning is otherwise clear.

Thus because of (45)

$$sp \frac{\partial U}{\partial t} = \nabla \cdot (\kappa \nabla U)$$

or

$$\frac{\partial U}{\partial t} = \mu^2 \nabla^2 U$$

with $\mu^2 = \kappa/sp$ and κ assumed constant. For a stationary state, $\nabla^2 U = 0$; this is Laplace's equation, the same law holds for the distribution of temperature as for the distribution of potential in charge-free space.

4.19. Green's Theorems.—The three fundamental relations (36), (37) and (43) may be used to obtain a large number of formulas for the transformation of integrals, the results corresponding to integration by parts in scalar calculus. The two most important such formulas are known as *Green's theorems*, when given in Cartesian form. In vector notation, these are

$$\int_\tau \nabla\phi \cdot \nabla\psi \, d\tau = \int_S \phi\nabla\psi \cdot dS - \int_\tau \phi\nabla^2\psi \, d\tau$$

$$= \int_S \psi\nabla\phi \cdot dS - \int_\tau \psi\nabla^2\phi \, d\tau \qquad (4\text{-}46)$$

$$\int_\tau (\phi\nabla^2\psi - \psi\nabla^2\phi) d\tau = \int_S (\phi\nabla\psi - \psi\nabla\phi) \cdot dS \qquad (4\text{-}47)$$

Green's first theorem is easily found by substituting $V = \phi\nabla\psi$ in (43). The second theorem is obtained by interchanging ϕ and ψ in (46) and subtracting the result from (46).

Problem. Verify eqs. (46) and (47).

4.20. Tensors.—In many physical problems, the notion of a vector is too restricted. For example, in an isotropic medium, stress **S** and strain **X** are related by the vector equation $\mathbf{S} = k\mathbf{X}$, **X** and **S** having the same direction. If the medium is not isotropic, **S** and **X** are not in general in the same direction; it is then necessary to replace the scalar k by a more general mathematical construct capable, when acting on the vector **X**, of changing its direction as well as its magnitude. Such a construct is a *tensor*. A similar generalization has to be made in the vector equations

$$\mathbf{P} = \epsilon\mathbf{E}$$

where **P** and **E** represent electric polarization and field strength,

$$\mathbf{I} = \mu\mathbf{H}$$

where **I** and **H** represent intensity of magnetization and field strength; for anisotropic media, the susceptibilities ϵ and μ must be replaced by tensors.

Again, if it is desired to represent the displacements $\delta \mathbf{v}$ of the points in a strained elastic medium as functions of their position vectors **v**, a *tensor* equation of the form $\delta \mathbf{v} = t\mathbf{v}$ is needed, for $\delta \mathbf{v}$ and **v** differ in direction, and the tensor t must effect this difference. This example will be treated in detail in sec. 4.23; but first we shall discuss the analytical properties of tensors.

Let us consider for complete generality a space of ν dimensions and assume that two different reference frames are given so that a point whose coordinates[7] in the first one are $(x^1, x^2, \cdots, x^\nu)$ has the coordinates $(\bar{x}^1, \bar{x}^2, \cdots, \bar{x}^\nu)$ in the second system. Further let there be relations

$$\bar{x}^m = f^m(x^1, x^2, \cdots, x^\nu); \quad x^m = g^m(\bar{x}^1, \bar{x}^2, \cdots, \bar{x}^\nu) \tag{4-48}$$

$$m = 1, 2, 3, \cdots, \nu$$

so that we may transform from one system to the other. Then if ν quantities $(A^1, A^2, \cdots, A^\nu)$ are related to ν other quantities $(\bar{A}^1, \bar{A}^2, \cdots, \bar{A}^\nu)$ by the equations

$$\bar{A}^m = \sum_{i=1}^{\nu} \frac{\partial \bar{x}^m}{\partial x^i} A^i; \quad m = 1, 2, \cdots, \nu \tag{4-49}$$

they are said to be the components of a *contravariant vector* or a *tensor of the first rank*. To simplify the notation, it is customary to omit the summation sign and sum over indices which are repeated on the same side of the equation. An index which is not repeated is understood to take successively the values $1, 2, \cdots, \nu$, so that there are altogether ν different equations. With these conventions, we may rewrite (49) as

$$\bar{A}^m = \frac{\partial \bar{x}^m}{\partial x^i} A^i \tag{4-50}$$

A further word about notation should be added. Since a repeated index (it is often called a *dummy* or *umbral index*) indicates summation, another letter may be substituted for it at will. Thus (50) may also be written

$$\bar{A}^m = \frac{\partial \bar{x}^m}{\partial x^j} A^j = \frac{\partial \bar{x}^m}{\partial x^n} A^n, \quad \text{etc.}$$

We will often use the same symbol such as A^i to indicate both the tensor and the i-th component of a tensor. No confusion should result from this arrangement.

---

[7] The upper suffix is not an exponent. Its position has an important meaning as the subsequent discussion will show.

A *covariant vector* with components A_m in one system and $\bar{A}_m$ in another is defined by the relation

$$\bar{A}_m = \frac{\partial x^i}{\partial \bar{x}^m} A_i \tag{4-51}$$

If (48) is differentiated we obtain

$$d\bar{x}^m = \frac{\partial \bar{x}^m}{\partial x^i} dx^i \tag{4-51a}$$

hence we see that the components of an ordinary vector in ν-dimensional space are actually the components of a contravariant tensor of rank one. To find an example of a covariant vector consider a scalar point function $\varphi(x^m) = \bar{\varphi}(\bar{x}^m)$. The components of the gradient of φ will be $\partial\varphi/\partial x^m$ and

$$\frac{\partial \bar{\varphi}}{\partial \bar{x}^m} = \frac{\partial \varphi}{\partial x^i} \frac{\partial x^i}{\partial \bar{x}^m}$$

Thus the gradient of such a function is a covariant vector. The reader should not conclude, however, that a covariant vector is necessarily the gradient of a scalar.

These ideas may be extended easily to define tensors of any rank. If $\varphi(x^m) = \bar{\varphi}(\bar{x}^m)$, we speak of φ as a tensor of zero rank or a *scalar* or *invariant*. There are three varieties of second rank tensors[8] defined by the transformations

$$\bar{A}^{mn} = \frac{\partial \bar{x}^m}{\partial x^i} \frac{\partial \bar{x}^n}{\partial x^j} A^{ij} \tag{4-52}$$

$$\bar{A}_{mn} = \frac{\partial x^i}{\partial \bar{x}^m} \frac{\partial x^j}{\partial \bar{x}^n} A_{ij} \tag{4-53}$$

$$\bar{A}^m_n = \frac{\partial \bar{x}^m}{\partial x^i} \frac{\partial x^j}{\partial \bar{x}^n} A^i_j \tag{4-54}$$

They are called contravariant, covariant and mixed, respectively. A useful mixed tensor of the second rank is the Kronecker delta

$$\delta^m_n = 1; \quad m = n$$
$$= 0; \quad m \neq n \tag{4-55}$$

This is seen as follows. Suppose δ^i_j is this tensor in the coordinate system x^i; then from (54)

$$\bar{\delta}^m_n = \frac{\partial \bar{x}^m}{\partial x^i} \frac{\partial x^j}{\partial \bar{x}^n} \delta^i_j = \frac{\partial \bar{x}^m}{\partial x^i} \frac{\partial x^i}{\partial \bar{x}^n}$$
$$= \frac{\partial \bar{x}^m}{\partial \bar{x}^n} = \delta^m_n$$

[8] Tensors of the second rank are also called *dyadics;* see Gibbs-Wilson, loc. cit.

We thus see that δ_n^m has the same components in all coordinate systems.

Tensors of higher rank are defined by similar laws, for example, a mixed tensor of rank four is

$$\bar{A}_{npq}^m = \frac{\partial \bar{x}^m}{\partial x^i} \frac{\partial x^j}{\partial \bar{x}^n} \frac{\partial x^k}{\partial \bar{x}^p} \frac{\partial x^h}{\partial \bar{x}^q} A_{jkh}^i \tag{4-56}$$

It should be noted that if ν is the number of dimensions of the coordinate system, then a tensor of rank α has ν^α components.

4.21. Addition, Multiplication and Contraction.—The sum or difference of two or more tensors of the same rank and type is a tensor of the same rank and type. For example, if

$$A^{mn} + B^{mn} = C^{mn}$$

it follows from (52) that C^{mn} is a tensor. It frequently happens that the components of a tensor satisfy the relation

$$A^{mn} = A^{nm}$$

such a tensor being called *symmetric*. On the other hand, if $A^{mn} = -A^{nm}$, the tensor is *skew-symmetric*. When neither of these relations holds, a given tensor may always be written as the sum of a symmetric and a skew-symmetric tensor. To see this let us take

$$S^{mn} = \tfrac{1}{2}(A^{mn} + A^{nm}); \quad T^{mn} = \tfrac{1}{2}(A^{mn} - A^{nm}) \tag{4-57}$$

where A^{mn} is neither symmetric nor skew-symmetric. Then

$$A^{mn} = S^{mn} + T^{mn}$$

The property of being symmetric or skew-symmetric is unaltered when a tensor is transformed from one reference frame to another.

An important relation exists between vectors and skew-symmetric tensors. Suppose $\mathbf{C} = \mathbf{A} \times \mathbf{B}$, where the components of $\mathbf{C}$ are given by (13). But the components of $\mathbf{A}$ (or $\mathbf{B}$) form a skew-symmetric tensor, $a_{ij} = -a_{ji}$, $a_{ii} = 0$, where $a_{13} = A_y$, $a_{21} = A_z$, $a_{32} = A_x$. We note, however, that if the vectors $\mathbf{A}$ and $\mathbf{B}$ were drawn in a left-handed coordinate system, their directions would both be opposite to those in a right-handed system while $\mathbf{C}$, their vector product, would have the same direction in both coordinate systems.

The more common type of vector, such as that representing translation or a mechanical force, is often called a *polar vector* to distinguish it from a vector $\mathbf{C}$ which has the unusual behavior just described. The latter, called an *axial vector* or a *pseudovector*,[9] requires the idea not merely of a displacement, but of some basic direction such as that implicit in a right-handed (or left-handed) coordinate system. A typical example of it is the vector product of two polar vectors, like angular momentum or the moment of a force. A pseudovector in a three-dimensional Cartesian coordinate system, as we have seen, behaves in most respects like a proper vector but in the more general case it transforms like a skew-symmetric tensor.

The scalar product of a polar vector and a pseudovector is called a *pseudoscalar*.[9] It differs from a true scalar, which must have the same magnitude in all coordinate systems, since it will change its sign if the direction of its coordinate system is changed.

Problem. Show that in two dimensions a skew-symmetric tensor of second rank is a pseudoscalar and that one of third rank is impossible; in three dimensions, that a second rank skew-symmetric tensor is a pseudovector and a third rank tensor is a pseudoscalar.

If we write a tensor in matrix form and compare it with eq. (10–16) it is clear that the components of the tensor are also the elements of a matrix. The only difference lies in the fact that tensors may always be written in matrix form if so desired, but the elements of a matrix do not need to transform in the same manner as tensors.

If we multiply A^m by B_n we obtain the mixed tensor $A^m B_n = C_n^m$. It is easily seen that C_n^m tranforms like (54). This type of product, called the outer product, may be obtained with tensors of any rank or type; thus $A_n^m B_{pq} = C_{npq}^m$. It should not be inferred, however, that every tensor can be written as a product in this way. Neither should we conclude that the outer product is the same as the vector product of sec. 4.5.

Let us set $m = q$ in the mixed tensor of (56) and write $B_{np} = A_{npm}^m$. To show that our notation, which indicates that A_{npm}^m is a covariant tensor of rank two, is justified we use the transformation law (56),

$$\bar{B}_{np} = \bar{A}_{npm}^m = \frac{\partial \bar{x}^m}{\partial x^i} \frac{\partial x^j}{\partial \bar{x}^n} \frac{\partial x^k}{\partial \bar{x}^p} \frac{\partial x^h}{\partial \bar{x}^m} A_{jkh}^i$$

$$= \frac{\partial x^j}{\partial \bar{x}^n} \frac{\partial x^k}{\partial \bar{x}^p} \delta_i^h A_{jkh}^i = \frac{\partial x^j}{\partial \bar{x}^n} \frac{\partial x^k}{\partial \bar{x}^p} A_{jki}^i$$

Comparison with (53) convinces us that A_{jki}^i is indeed a covariant vector of rank two since it transforms in the required way. This process of summing over a pair of contravariant and covariant indices is called

---

[9] For further properties of them, see Herbert Goldstein, "Classical Mechanics," Addison-Wesley Press, Inc., Cambridge, Mass., 1951.

contraction. It always reduces the rank of a mixed tensor by two, thus when it is applied to a mixed tensor of rank two the result is a scalar:

$$\bar{A}_m^m = \frac{\partial \bar{x}^m}{\partial x^i} \frac{\partial x^j}{\partial \bar{x}^m} A_j^i = A_i^i = A_m^m$$

When two tensors are multiplied together and then contracted, we speak of *inner multiplication*, thus

$$A^{mn} B_{npq} = C_{pq}^m; \quad A^m B_m = \text{a scalar}$$

The last example is clearly equivalent to the scalar product in rectangular coordinates (cf. sec. 4.4), hence in tensor analysis, we say that if l is the length of A^m or A_m

$$l^2 = A^m A_m \tag{4-58}$$

From (8) we conclude that the angle θ between two vectors A_m and B_m is defined by

$$\cos \theta = \frac{A_m B^m}{[(A_m A^m)(B_m B^m)]^{1/2}}$$

and if A_m and B_m are perpendicular to each other,

$$A_m B^m = 0$$

We have just shown how new tensors may be obtained by addition, multiplication and contraction. We now inquire whether it is possible to change contravariant tensors to covariant ones or the reverse. Let g_{mn} be any symmetric covariant tensor and g be the determinant of the components of g_{mn}. Also let G^{mn} be the co-factor[10] of g_{mn} in g, then if we define

$$g^{mn} = \frac{G^{mn}}{g} \tag{4-59}$$

it follows from the rules for the expansion of determinants that

$$g_{mn} g^{pn} = \delta_m^p \tag{4-60}$$

We would like to justify our notation and prove that g^{mn} is actually a tensor. Let A^n be a vector, then $B_m = g_{mn} A^n$ is also a vector, moreover

$$g^{mn} B_m = g^{mn} g_{mp} A^p = \delta_p^n A^p = A^n$$

so that g^{mn} changes a covariant vector into a contravariant one; hence it must itself be a tensor.

Two vectors related by the equations

$$A^m = g^{mn} A_n$$

or

$$A_m = g_{mn} A^n$$

---

[10] Note that G^{mn} is not a tensor. See sec. 10.3 for discussion of determinants and sec. 5.16 for further properties of these tensors.

are called *associated*. It is often said that both are the same vector, A^m being the contravariant components and A_m the covariant ones. Tensors of any rank may be treated in the same way, thus

$$A^{mn} = g^{mp}g^{nq}A_{pq}$$

It should be clear that

$$A_{mn}B^{mn} = A^{mn}B_{mn}; \quad A_{mn}B^{pn} = A_m{}^nB^p{}_n$$

Because of the fact that dummy indices may be changed from one letter to another at will it follows that they enjoy a certain freedom of motion. They may be raised in one place if they are lowered in another. We have indicated this procedure in the last equation by spacing the indices. Such information is needed, for it is not true that

$$A_m{}^n = g_{mp}A^{pn} \quad \text{and} \quad A^n{}_m = g_{pm}A^{np}$$

are identical unless A is a symmetrical tensor.

4.22. Differentiation of Tensors.—It has been shown in sec. 4.20 that the derivative of a scalar point function is a covariant vector. The derivative of a covariant vector is not a tensor, however, for if

$$\bar{A}_m = \frac{\partial x^h}{\partial \bar{x}^m} A_h$$

$$\frac{\partial \bar{A}_m}{\partial \bar{x}^n} = \frac{\partial^2 x^h}{\partial \bar{x}^n \partial \bar{x}^m} A_h + \frac{\partial x^h}{\partial \bar{x}^m} \frac{\partial A_h}{\partial \bar{x}^n}$$

$$= \frac{\partial^2 x^h}{\partial \bar{x}^n \partial \bar{x}^m} A_h + \frac{\partial x^h}{\partial \bar{x}^m} \frac{\partial x^p}{\partial \bar{x}^n} \frac{\partial A_h}{\partial x^p} \qquad (4\text{-}61)$$

and the presence of the second derivative shows that $\partial \bar{A}_m/\partial \bar{x}^n$ does not transform like a tensor. In order to find a " *derivative* " of the proper tensor character we first rewrite the second derivative in terms of first derivatives. To do this let us use the two tensors g_{ij} and g^{ij} defined previously. Let us further introduce the following quantities (they are *not* tensors) called the *Christoffel three-index symbols*

$$[mn,q] = \frac{1}{2}\left(\frac{\partial g_{mq}}{\partial x^n} + \frac{\partial g_{nq}}{\partial x^m} - \frac{\partial g_{mn}}{\partial x^q}\right) \qquad (4\text{-}62)$$

$$\{mn,q\} = \tfrac{1}{2}g^{qs}\left(\frac{\partial g_{ms}}{\partial x^n} + \frac{\partial g_{ns}}{\partial x^m} - \frac{\partial g_{mn}}{\partial x^s}\right) \qquad (4\text{-}63)$$

the significance of which will soon be evident. From these definitions we see that

$$\{mn,q\} = g^{qs}[mn,s] \qquad (4\text{-}64)$$

According to (53), we have

$$\bar{g}_{mn} = \frac{\partial x^i}{\partial \bar{x}^m} \frac{\partial x^j}{\partial \bar{x}^n} g_{ij} \tag{4-65}$$

and it is also true that

$$\frac{\partial g_{ij}}{\partial \bar{x}^q} = \frac{\partial g_{ij}}{\partial x^k} \frac{\partial x^k}{\partial \bar{x}^q} \tag{4-66}$$

Differentiating (65) and using (66) we get

$$\frac{\partial \bar{g}_{mn}}{\partial \bar{x}^q} = g_{ij}\left(\frac{\partial^2 x^i}{\partial \bar{x}^q \partial \bar{x}^m} \frac{\partial x^j}{\partial \bar{x}^n} + \frac{\partial x^i}{\partial \bar{x}^m} \frac{\partial^2 x^j}{\partial \bar{x}^q \partial \bar{x}^n}\right) + \frac{\partial x^i}{\partial \bar{x}^m} \frac{\partial x^j}{\partial \bar{x}^n} \frac{\partial x^k}{\partial \bar{x}^q} \frac{\partial g_{ij}}{\partial x^k} \tag{4-67}$$

In the same way if we differentiate $\bar{g}_{nq}$ and $\bar{g}_{mq}$ we obtain

$$\frac{\partial \bar{g}_{nq}}{\partial \bar{x}^m} = g_{ij}\left(\frac{\partial^2 x^i}{\partial \bar{x}^m \partial \bar{x}^n} \frac{\partial x^j}{\partial \bar{x}^q} + \frac{\partial x^i}{\partial \bar{x}^n} \frac{\partial^2 x^j}{\partial \bar{x}^m \partial \bar{x}^q}\right) + \frac{\partial x^i}{\partial \bar{x}^m} \frac{\partial x^j}{\partial \bar{x}^n} \frac{\partial x^k}{\partial \bar{x}^q} \frac{\partial g_{jk}}{\partial x^i} \tag{4-68}$$

$$\frac{\partial \bar{g}_{mq}}{\partial \bar{x}^n} = g_{ij}\left(\frac{\partial^2 x^i}{\partial \bar{x}^n \partial \bar{x}^m} \frac{\partial x^j}{\partial \bar{x}^q} + \frac{\partial x^i}{\partial \bar{x}^m} \frac{\partial^2 x^j}{\partial \bar{x}^n \partial \bar{x}^q}\right) + \frac{\partial x^i}{\partial \bar{x}^m} \frac{\partial x^j}{\partial \bar{x}^n} \frac{\partial x^k}{\partial \bar{x}^q} \frac{\partial g_{ik}}{\partial x^j} \tag{4-69}$$

We may exchange i and j in the second term on the right of these expressions. If we add (68) and (69), subtract (67) and use eq. (62) we obtain

$$\overline{[mn,q]} = g_{ij} \frac{\partial^2 x^i}{\partial \bar{x}^m \partial \bar{x}^n} \frac{\partial x^j}{\partial \bar{x}^q} + \frac{\partial x^i}{\partial \bar{x}^m} \frac{\partial x^j}{\partial \bar{x}^n} \frac{\partial x^k}{\partial \bar{x}^q} [ij,k]$$

where the bar over the Christoffel symbol indicates that it refers to the coordinate system $\bar{x}^m$. Now multiply this equation by $\bar{g}^{qr}(\partial x^h/\partial \bar{x}^r)$ and use (64), which gives

$$\overline{\{mn,r\}} \frac{\partial x^h}{\partial \bar{x}^r} = g_{ij} \frac{\partial^2 x^i}{\partial \bar{x}^m \partial \bar{x}^n} g^{qr} \frac{\partial x^j}{\partial \bar{x}^q} \frac{\partial x^h}{\partial \bar{x}^r}$$
$$+ \bar{g}^{qr} \frac{\partial x^k}{\partial \bar{x}^q} \frac{\partial x^h}{\partial \bar{x}^r} \frac{\partial x^i}{\partial \bar{x}^m} \frac{\partial x^j}{\partial \bar{x}^n} [ij,k]$$

By means of (52) we may eliminate $\bar{g}^{qr}$ from the right-hand side of this equation to obtain

$$g_{ij}g^{jh} \frac{\partial^2 x^i}{\partial \bar{x}^m \partial \bar{x}^n} + \frac{\partial x^i}{\partial \bar{x}^m} \frac{\partial x^j}{\partial \bar{x}^n} g^{kh}[ij,k]$$

Finally, remembering that $g_{ij}g^{jh} = g_i^h = \delta_i^h$, we see that

$$\frac{\partial^2 x^h}{\partial \bar{x}^m \partial \bar{x}^n} = \overline{\{mn,r\}} \frac{\partial x^h}{\partial \bar{x}^r} - \frac{\partial x^i}{\partial \bar{x}^m} \frac{\partial x^j}{\partial \bar{x}^n} \{ij,h\}$$

Let us put this result into (61) which then becomes

$$\frac{\partial \bar{A}_m}{\partial \bar{x}^n} = \left[\overline{\{mn,r\}} \frac{\partial x^h}{\partial \bar{x}^r} - \frac{\partial x^i}{\partial \bar{x}^m} \frac{\partial x^j}{\partial \bar{x}^n} \{ij,h\}\right] A_h + \frac{\partial x^i}{\partial \bar{x}^m} \frac{\partial x^j}{\partial \bar{x}^n} \frac{\partial A_i}{\partial x^j} \tag{4-70}$$

where we have changed the dummy indices in the last term from h and p to i and j. Finally we see from (51) that we have

$$\bar{A}_r = \frac{\partial x^h}{\partial \bar{x}^r} A_h$$

so that (70) may be written

$$\frac{\partial \bar{A}_m}{\partial \bar{x}^n} - \{\overline{mn,r}\} \bar{A}_r = \frac{\partial x^i}{\partial \bar{x}^m} \frac{\partial x^j}{\partial \bar{x}^n} \left[\frac{\partial A_i}{\partial x^j} - \{ij,h\} A_h \right]$$

Now if we use the comma abbreviation

$$A_{i,j} \equiv \frac{\partial A_i}{\partial x^j} - \{ij,h\} A_h$$

it follows that

$$\bar{A}_{m,n} = \frac{\partial x^i}{\partial \bar{x}^m} \frac{\partial x^j}{\partial \bar{x}^n} A_{i,j}$$

hence this quantity is a covariant tensor of the second rank. It is called the *covariant derivative* of A_i with respect to g_{ij}.

In a similar way it may be shown that the covariant derivative of A^i with respect to g_{ij} is

$$A^i_{,j} = \frac{\partial A^i}{\partial x^j} + \{jh,i\} A^h \qquad (4\text{–}71)$$

Problem a. Prove that $[mn,p] = g_{pq}\{mn,q\}$.

Problem b. Show that second derivatives of tensors may be derived in the form

$$A_{ij,k} = \frac{\partial A_{ij}}{\partial x^k} - A_{ih}\{jk,h\} - A_{hj}\{ik,h\}$$

$$A^{ij,k} = \frac{\partial A^{ij}}{\partial x^k} + A^{ih}\{hk,j\} + A^{hj}\{hk,i\}$$

$$A^i_{j,k} = \frac{\partial A^i_j}{\partial x^k} + A^h_j\{hk,i\} - A^i_h\{jk,h\}$$

4.23. Tensors and the Elastic Body.—As an example of the use of the tensor in a physical problem let us consider a deformable body subjected to an infinitely small deformation or strain. Let P_0 be a point of the medium in the unstrained state and let P be its deformed position. If the coordinates of P_0 and P are x_0^r and x^r then the components u_0^r of the displacement vector will be

$$u_0^r = x^r - x_0^r = u_0^r(x_0^1, x_0^2, x_0^3) \qquad (4\text{–}72)$$

Suppose Q_0 is a neighboring point as shown in Fig. 11 which is deformed to the position Q. Now if the components of the vectors P_0Q_0 and PQ

are v_0^r and v^r, the coordinates of Q_0 and Q will be $x_0^r + v_0^r$ and $x^r + v^r$. It follows[11] that

$$(x^r + v^r) - (x_0^r + v_0^r) = u_0^r + \left(\frac{\partial u^r}{\partial x^s}\right)_0 v_0^s \qquad (4\text{--}73)$$

and, on using (72), that[12]

$$\delta v^r = v^r - v_0^r = \left(\frac{\partial u^r}{\partial x^s}\right)_0 v_0^{'s} \qquad (4\text{--}74)$$

The coefficients $(\partial u^r/\partial x^s)_0$ which relate the two vectors δv^r and v_0^s are the components of a tensor. The terms $(\partial u^1/\partial x^1)$, $(\partial u^2/\partial x^2)$, $(\partial u^3/\partial x^3)$

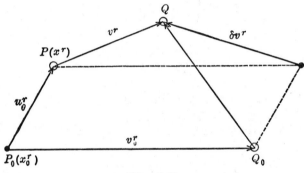

Fig. 4–11

are *tension strains* parallel to the axes x^1, x^2, x^3, respectively. The remaining terms are *shearing strains* about these axes; for example, $(\partial u^2/\partial x^1 + \partial u^1/\partial x^2)$ is the shearing strain about the axis perpendicular to x^1 and x^2.

If the nine components of the tensor are written out it will be seen that it is not in general symmetric. However, it can be made so as shown in sec. 4.21. Dropping the zero subscripts from (74) we write

$$\delta v^r = t_s^r v^s = e_s^r v^s + \omega_s^r v^s \qquad (4\text{--}75)$$

where

$$t_s^r = (\partial u^r/\partial x^s); \quad e_s^r = \tfrac{1}{2}(t_s^r + t_r^s)$$
$$\omega_s^r = \tfrac{1}{2}(t_s^r - t_r^s) \qquad (4\text{--}76)$$

The coefficients e_s^r are now the components of a symmetrical tensor, which is called a *pure strain*. It may be shown (see problem at end of this

---

[11] The zero subscript on the derivative is meant to indicate that it is evaluated at the point P_0.

[12] This result only holds for rectangular coordinates. If (74) is to hold in generalized coordinates, we must use the covariant derivative of u^r.

section) that ω_s^r represents a rotation of the neighborhood of P_0 about P. We could also add to (75) a translation by the amount a^r, so that

$$\delta v^r = a^r + e_s^r v^s + \omega_s^r v^s$$

represents the most general displacement of an elastic body, the total motion being composed of: (1) a translation, (2) a pure strain, and (3) a rotation.

This brief discussion of tensors is entirely inadequate to indicate its great value in mathematical physics. The subject has been most frequently employed in the general theory of relativity.[13] It may also be applied with advantage in the study of dynamics, electricity, and hydrodynamics.[14] The material presented here is sufficient for the use which will be made of tensors in this book.

Problem. Show that the tensor ω_s^r represents a rotation.

Hint: Write out the components of (76) and it will be seen that the resulting vector is the vector product of two other vectors.

[13] Eddington, A. S., " The Mathematical Theory of Relativity," Second Edition, Cambridge Press, 1930.

[14] These subjects have been so treated by McConnell, A. J., " Applications of the Absolute Differential Calculus," Blackie and Sons, London, 1931, and more briefly by Thomas, T. Y., " The Elementary Theory of Tensors," McGraw-Hill Book Co., New York, 1931. See also Kron, G., " Short Course in Tensor Analysis for Electrical Engineers," John Wiley and Sons, New York, 1942. Tensor methods have been used to discuss the elastic properties of solids by Partington, J. R., " An Advanced Treatise on Physical Chemistry," Vol. 3, The Properties of Solids, Longmans, Green and Co., New York, 1952.

REFERENCES

In the following books the emphasis is on the mathematical aspects:
Craig, H. V., " Vector and Tensor Analysis," McGraw-Hill Book Co., New York, 1943.
Lass, H., " Vector and Tensor Analysis," McGraw-Hill Book Co., New York, 1950.
Spain, B., " Tensor Calculus," Interscience Publishers, New York, 1953.
Wade, T. L., " Algebra of Vectors and Matrices," Addison-Wesley Press, Inc., Cambridge, 1951.
Weatherburn, C. E., " Elementary and Advanced Vector Analysis," 2 vols., G. Bell, London, 1928.
Vector or tensor methods are applied to physical problems in the following:
Brillouin, L., " Les Tenseurs en Méchanique et en Élasticité," Masson et Cie, Paris, 1938; Dover Publications, Inc., New York.
Milne, E. A., " Vectorial Mechanics," Interscience Publishers, Inc., New York, 1948.
Rutherford, D. E., " Vector Methods," Fifth Edition, Interscience Publishers, Inc., New York, 1948.
Dyadics, which are particularly useful in many cases, are treated by:
Gibbs-Wilson, " Vector Analysis," Yale University Press, New Haven, Conn., 1925.
Morse, P. M., and Feshbach, H., " Methods of Theoretical Physics," Part I, Chapter 1, McGraw-Hill Book Co., New York, 1953.

CHAPTER 5

COORDINATE SYSTEMS

VECTORS AND CURVILINEAR COORDINATES

5.1. Curvilinear Coordinates.—Although the methods of vector analysis prove convenient in the statement of physical laws, it is usually necessary to rewrite the vector equations in terms of suitable coordinates before the final solution of a specific problem can be obtained. It is the purpose of this chapter to show[1] how the components of vectors or vector operators may be formulated in a system of *curvilinear coordinates*, the latter being of so general a nature that it is an easy matter to transform from them to any one of the several kinds of special coordinate systems which have been found useful in physical problems.

In Cartesian coordinates, the position of a point $P(x,y,z)$ is determined by the intersection of three mutually perpendicular planes, x = const., y = const., z = const. When x, y and z are related to three new quantities by the equations

$$x = x(q_1,q_2,q_3)$$
$$y = y(q_1,q_2,q_3)$$
$$z = z(q_1,q_2,q_3)$$

(5-1)

with inverses,

$$q_1 = q_1(x,y,z)$$
$$q_2 = q_2(x,y,z)$$
$$q_3 = q_3(x,y,z)$$

(5-2)

a given point may be described by specifying either x, y, z or q_1, q_2, q_3, for each equation of (2) represents a surface and the intersection of three such surfaces locates the point. The surfaces q_1 = const., q_2 = const., q_3 = const. are called the *coordinate surfaces;* the space curves formed by their intersection in pairs are called the *coordinate lines*. The *coordinate axes* are determined by the tangents to the coordinate lines at the intersection of three surfaces. They are not in general fixed directions in space, as is true for simple Cartesian coordinates. The quantities (q_1,q_2,q_3) are the *curvilinear coordinates* of a point $P(x,y,z)$.

---

[1] The relations which we derive here may be obtained in other ways; see sec. 5.16 and Hobson, E. W., " The Theory of Spherical and Ellipsoidal Harmonics," Cambridge Press, 1931.

From (1),

$$dx = \frac{\partial x}{\partial q_1} dq_1 + \frac{\partial x}{\partial q_2} dq_2 + \frac{\partial x}{\partial q_3} dq_3$$

$$dy = \frac{\partial y}{\partial q_1} dq_1 + \frac{\partial y}{\partial q_2} dq_2 + \frac{\partial y}{\partial q_3} dq_3$$

$$dz = \frac{\partial z}{\partial q_1} dq_1 + \frac{\partial z}{\partial q_2} dq_2 + \frac{\partial z}{\partial q_3} dq_3$$

hence the square of the distance between two adjacent points,

$$ds^2 = dx^2 + dy^2 + dz^2 = Q_{11}^2 dq_1^2 + Q_{22}^2 dq_2^2 + Q_{33}^2 dq_3^2$$
$$+ 2Q_{12} dq_1 dq_2 + 2Q_{13} dq_1 dq_3 + 2Q_{23} dq_2 dq_3 \qquad (5\text{-}3)$$

where,

$$Q_{ij} = \frac{\partial x}{\partial q_i}\frac{\partial x}{\partial q_j} + \frac{\partial y}{\partial q_i}\frac{\partial y}{\partial q_j} + \frac{\partial z}{\partial q_i}\frac{\partial z}{\partial q_j} \qquad (5\text{-}4)$$

$$Q_{ii}^2 = \left(\frac{\partial x}{\partial q_i}\right)^2 + \left(\frac{\partial y}{\partial q_i}\right)^2 + \left(\frac{\partial z}{\partial q_i}\right)^2 \qquad (i,j = 1,2,3;\ i \neq j)$$

For convenience we shall hereafter omit a repeated subscript, writing for instance Q_i instead of Q_{ii}.

The distance between two points on a coordinate line is called the *line element*. It is given by eq. (3) when variation is limited to only one of the q's,

$$ds_i = Q_i dq_i \quad (i = 1,2,3) \qquad (5\text{-}5)$$

The direction cosines between these line elements and dx, dy or dz may be arranged as shown in Table 1 of sec. 4.1; for example, the cosine of the angle between ds_1 and dz is $(\partial z/\partial q_1)(\partial q_1/\partial s_1) = (\partial z/\partial q_1)/Q_1$, and the cosine of the angle θ_{ij} between ds_i and ds_j is

$$\cos \theta_{ij} = Q_{ij}/Q_i Q_j$$

The most useful coordinate systems are orthogonal ones, that is, systems in which surfaces always intersect at right angles. We shall limit ourselves to such systems in secs. 5.2 to 5.15, returning to the more general case of non-orthogonal systems in sec. 5.16. For the present, then, $\cos \theta_{ij} = 0$, $Q_{ij} = 0$, and the cross product terms may be dropped from (3). The three possible *surface elements* in orthogonal systems thus become

$$dS_{ij} = ds_i ds_j = Q_i Q_j dq_i dq_j \quad (i,j = 1,2,3;\ i \neq j) \qquad (5\text{-}6)$$

and the *volume element*,

$$d\tau = ds_1 ds_2 ds_3 = Q_1 Q_2 Q_3 dq_1 dq_2 dq_3 \qquad (5\text{-}7)$$

5.2. Vector Relations in Curvilinear Coordinates.—If ϕ is a scalar point function, $\nabla\phi$ must be the same in all coordinate systems, for $\nabla\phi$ is a vector whose magnitude and direction give the maximum space rate of change of ϕ. A component of $\nabla\phi$ is its directional derivative (see sec. 4.9) in the given direction, thus the component perpendicular to the surface $q_i =$ constant and hence in the direction of s_i is

$$\frac{d\phi}{ds_i} = \frac{1}{Q_i}\frac{\partial\phi}{\partial q_i}$$

in accordance with eq. (5). Since it is also possible to regard ∇ as a vector operator, it may be written in terms of unit vectors, $\mathbf{u}_1$, $\mathbf{u}_2$, $\mathbf{u}_3$ along the curvilinear coordinate axes. Thus,

$$\nabla = \frac{\mathbf{u}_1}{Q_1}\frac{\partial}{\partial q_1} + \frac{\mathbf{u}_2}{Q_2}\frac{\partial}{\partial q_2} + \frac{\mathbf{u}_3}{Q_3}\frac{\partial}{\partial q_3} \tag{5-8}$$

so that

$$\nabla\phi = \frac{\mathbf{u}_1}{Q_1}\frac{\partial\phi}{\partial q_1} + \frac{\mathbf{u}_2}{Q_2}\frac{\partial\phi}{\partial q_2} + \frac{\mathbf{u}_3}{Q_3}\frac{\partial\phi}{\partial q_3} \tag{5-9}$$

Any vector may be written in terms of curvilinear components V_1, V_2, V_3:

$$\mathbf{V} = \mathbf{u}_1 V_1 + \mathbf{u}_2 V_2 + \mathbf{u}_3 V_3 \tag{5-10}$$

but in order to find $\nabla * \mathbf{V}$ (see sec. 4.10) in curvilinear coordinates, we must know the relation between $\mathbf{u}_1$, $\mathbf{u}_2$, $\mathbf{u}_3$ and x, y, z. We proceed by evaluating $\nabla * \mathbf{u}_i$, starting with $\nabla \times \mathbf{u}_i$, since this is needed to obtain $\nabla \cdot \mathbf{u}_i$.

Remembering that $\mathbf{u}_1/Q_1$ is the product of a scalar and a vector, we may write in view of (4-26)

$$\nabla \times \frac{\mathbf{u}_1}{Q_1} = \nabla\left(\frac{1}{Q_1}\right) \times \mathbf{u}_1 + \frac{1}{Q_1}(\nabla \times \mathbf{u}_1)$$

$$= -\mathbf{u}_1 \times \nabla\left(\frac{1}{Q_1}\right) + \frac{1}{Q_1}(\nabla \times \mathbf{u}_1) \tag{5-11}$$

the change of sign coming from the change of order in the vector product. From (9), we note that $\mathbf{u}_1/Q_1 = \nabla q_1$ and from (4-30) that

$$\nabla \times \nabla q_1 = 0$$

hence,

$$\mathbf{u}_1 \times \nabla\left(\frac{1}{Q_1}\right) = \frac{1}{Q_1}(\nabla \times \mathbf{u}_1) \tag{5-12}$$

Now using (8) and performing the differentiation, we find

$$\nabla\left(\frac{1}{Q_1}\right) = \frac{-\mathbf{u}_1}{Q_1^3}\frac{\partial Q_1}{\partial q_1} - \frac{\mathbf{u}_2}{Q_1^2 Q_2}\frac{\partial Q_1}{\partial q_2} - \frac{\mathbf{u}_3}{Q_1^2 Q_3}\frac{\partial Q_1}{\partial q_3} \tag{5-13}$$

When we further recall that

$$\mathbf{u}_i \times \mathbf{u}_i = 0; \quad \mathbf{u}_i \times \mathbf{u}_j = \mathbf{u}_k \tag{5-14}$$

and substitute (13) in (12), we obtain

$$\nabla \times \mathbf{u}_1 = \frac{\mathbf{u}_2}{Q_1 Q_3} \frac{\partial Q_1}{\partial q_3} - \frac{\mathbf{u}_3}{Q_1 Q_2} \frac{\partial Q_1}{\partial q_2} \tag{5-15}$$

The scalar product of ∇ and a unit vector may be written as

$$\nabla \cdot \mathbf{u}_1 = \nabla \cdot (\mathbf{u}_2 \times \mathbf{u}_3) = \mathbf{u}_3 \cdot (\nabla \times \mathbf{u}_2) - \mathbf{u}_2 \cdot (\nabla \times \mathbf{u}_3) \tag{5-16}$$

by using (14) and (4–26). This becomes

$$\nabla \cdot \mathbf{u}_1 = \frac{1}{Q_1 Q_2 Q_3} \frac{\partial (Q_2 Q_3)}{\partial q_1} \tag{5-17}$$

when we expand the vector product by (15) and use the fact that

$$\mathbf{u}_i \cdot \mathbf{u}_i = 1; \quad \mathbf{u}_i \cdot \mathbf{u}_j = 0 \tag{5-18}$$

In order to determine $\nabla \cdot \mathbf{V}$ in curvilinear coordinates, we see from (10), that

$$\nabla \cdot \mathbf{V} = \nabla \cdot (\mathbf{u}_1 V_1) + \nabla \cdot (\mathbf{u}_2 V_2) + \nabla \cdot (\mathbf{u}_3 V_3)$$

a typical term becoming

$$\nabla \cdot (\mathbf{u}_i V_i) = V_i \nabla \cdot \mathbf{u}_i + \mathbf{u}_i \cdot \nabla V_i \tag{5-19}$$

by (4–26). When $\nabla \cdot \mathbf{u}_i$ is written in the form of (17), ∇V_i in the form of (9) and (18) used to eliminate the scalar products of the unit vectors, the three terms of (19) reduce to

$$\nabla \cdot \mathbf{V} = \frac{1}{Q_1 Q_2 Q_3} \left\{ \frac{\partial}{\partial q_1} (V_1 Q_2 Q_3) + \frac{\partial}{\partial q_2} (V_2 Q_1 Q_3) \right.$$
$$\left. + \frac{\partial}{\partial q_3} (V_3 Q_1 Q_2) \right\} \tag{5-20}$$

If $\mathbf{V} = \nabla \phi$,

$$\nabla \cdot \nabla \phi = \nabla^2 \phi = \frac{1}{Q_1 Q_2 Q_3} \left\{ \frac{\partial}{\partial q_1} \left[\frac{Q_2 Q_3}{Q_1} \frac{\partial \phi}{\partial q_1} \right] \right.$$
$$\left. + \frac{\partial}{\partial q_2} \left[\frac{Q_1 Q_3}{Q_2} \frac{\partial \phi}{\partial q_2} \right] + \frac{\partial}{\partial q_3} \left[\frac{Q_1 Q_2}{Q_3} \frac{\partial \phi}{\partial q_3} \right] \right\} \tag{5-21}$$

since the components of $\nabla \phi$ are $V_i = (\partial \phi / \partial q_i)/Q_i$.

The curl of a vector in terms of the unit curvilinear vectors becomes

$$\nabla \times \mathbf{V} = \nabla \times (\mathbf{u}_1 V_1) + \nabla \times (\mathbf{u}_2 V_2) + \nabla \times (\mathbf{u}_3 V_3)$$

which may be expanded by using (11), to give terms like

$$\nabla \times (\mathbf{u}_i V_i) = V_i(\nabla \times \mathbf{u}_i) - \mathbf{u}_i \times (\nabla V_i)$$

When three similar equations are added together, the result in determinantal form is

$$\nabla \times \mathbf{V} = \frac{1}{Q_1 Q_2 Q_3} \begin{vmatrix} Q_1\mathbf{u}_1 & Q_2\mathbf{u}_2 & Q_3\mathbf{u}_3 \\ \partial/\partial q_1 & \partial/\partial q_2 & \partial/\partial q_3 \\ V_1 Q_1 & V_2 Q_2 & V_3 Q_3 \end{vmatrix} \qquad (5\text{-}22)$$

In order to compute $\nabla^2 \mathbf{V}$ in curvilinear coordinates, use is made of the relation (4-32).

$$\nabla^2 \mathbf{V} = \nabla(\nabla \cdot \mathbf{V}) - \nabla \times \nabla \times \mathbf{V}$$

which may be reduced to the desired form by means of (8), (20) and (22). The component of the resulting expression along the $\mathbf{u}_1$ direction is given by

$$\frac{1}{Q_1}\frac{\partial}{\partial q_1}(\nabla \cdot \mathbf{V}) - \frac{1}{Q_2 Q_3}\frac{\partial}{\partial q_2}[Q_3(\nabla \times \mathbf{V})_3] + \frac{1}{Q_2 Q_3}\frac{\partial}{\partial q_3}[Q_2(\nabla \times \mathbf{V})_2] \quad (5\text{-}23)$$

where $(\nabla \times \mathbf{V})_3$ and $(\nabla \times \mathbf{V})_2$ are the components of $\nabla \times \mathbf{V}$ along $\mathbf{u}_3$ and $\mathbf{u}_2$. The two other components of $\nabla^2 \mathbf{V}$ are obtained from (23) by cyclic permutation of the subscripts 1, 2, 3.

The task of computing any of these vector quantities in special coordinate systems is seen to involve calculation of the Q_i which may be done in a straightforward way from (4) provided relations like (1) or (2) are known. In the remainder of this chapter we discuss those special systems which appear to be most useful. We include all those which may be used to solve the three-dimensional Schrödinger wave equation of quantum mechanics. It has been shown[2] that the method of separation of variables (cf. Chapter 7) is applicable to this equation only if the potential energy is of the form

$$V = \Pi f(q_i)/Q_i^2 \qquad (5\text{-}24)$$

and the coordinates have certain special properties. There are eleven such systems; these are the ones described in secs. 5.3–5.9, 5.11–5.13 and the confocal ellipsoidal system of sec. 5.6 expressed in terms of elliptic integrals. We indicate other examples of the use of some of the systems as we proceed. In each case, we describe the geometry, give the relations between the new coordinates and x, y, z and list the resulting Q_i obtained

[2] Robertson, H. P., *Math. Ann.* **98**, 749 (1928); Eisenhart, L. P., *Phys. Rev.* **45**, 427 (1934); **74**, 87 (1948). These coordinate systems have been discussed in considerable detail by Morse, P. M., and Feshbach, H., " Methods of Theoretical Physics," 2 vols., McGraw-Hill Book Co., New York, 1953.

from (4).   Calculation of $\nabla\phi$, $\nabla^2\phi$, $\nabla \times \mathbf{V}$, etc., may be performed as an exercise by the student[3] (see problems in later sections).

SPECIAL ORTHOGONAL COORDINATE SYSTEMS

5.3. Cartesian Coordinates.—These form a trivial case of curvilinear coordinate systems.

$$Q_x^2 = Q_y^2 = Q_z^2 = 1 \tag{5–25}$$

5.4. Spherical Polar Coordinates.—The coordinate surfaces are families of: (1) concentric spheres about the origin (r = const.), (2) right circular cones with apex at the origin and axis along z (θ = const.), (3) half-planes

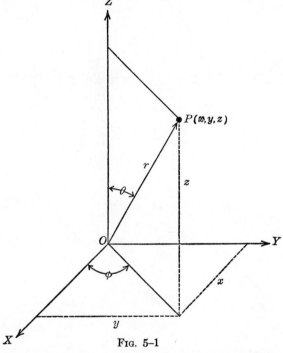

Fig. 5–1

from the Z-axis (ϕ = const.).   A point $P(x,y,z)$ is located by specifying the radius r of the sphere on which it lies, its colatitude θ, and its longitude or azimuth ϕ on the sphere.   From Fig. 1, it follows that

$$\begin{aligned} x &= r \sin\theta \cos\phi \\ y &= r \sin\theta \sin\phi \\ z &= r \cos\theta \end{aligned} \tag{5–26}$$

[3] Some of these quantities for certain of the systems may be found in Pauling and Wilson, " Introduction to Quantum Mechanics," Appendix IV, McGraw-Hill Book Co., New York, 1935; see also Adams, E. P., " Smithsonian Mathematical Formulae," Washington, 1922, and Magnus, W., and Oberhettinger, F., " Formulas and Theorems for the Special Functions of Mathematical Physics," translated by John Wermer, Chelsea Publishing Co., New York, 1949.

Remembering that $ds_i = Q_i dq_i$, values of the Q_i may also be determined by inspection from the figure, thus

$$Q_r^2 = 1; \quad Q_\theta^2 = r^2; \quad Q_\phi^2 = r^2 \sin^2 \theta \qquad (5\text{-}27)$$

5.5. Cylindrical Coordinates.—The coordinate surfaces are: (1) right circular cylinders which form families of concentric circles about the origin in the XY-plane ($\rho = $ const.); (2) half-planes from the Z-axis ($\phi = $ const.); (3) planes parallel to the XY-plane ($z = $ const.). A point

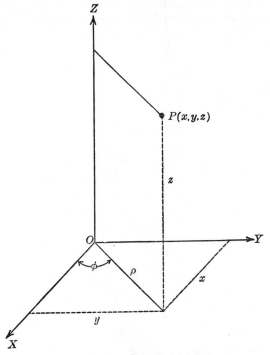

FIG. 5–2

$P(x,y,z)$ is located by giving the distance ρ in the XY-plane from the origin to the cylinder on which the point lies, the angle ϕ in the XY-plane and the distance on the Z-axis from this plane to the point. From Fig. 2

$$x = \rho \cos \phi$$
$$y = \rho \sin \phi \qquad (5\text{-}28)$$
$$z = z$$

$$Q_\rho^2 = Q_z^2 = 1; \quad Q_\phi^2 = \rho^2 \qquad (5\text{-}29)$$

5.6. Confocal Ellipsoidal Coordinates.—In this system, the coordinate surfaces are families of (1) ellipsoids ($\lambda = $ const.); (2) hyperboloids of one sheet (μ = const.); (3) hyperboloids of two sheets (ν = const.) given by the equations

$$\frac{x^2}{a^2 - \lambda} + \frac{y^2}{b^2 - \lambda} + \frac{z^2}{c^2 - \lambda} = 1$$

$$\frac{x^2}{a^2 - \mu} + \frac{y^2}{b^2 - \mu} - \frac{z^2}{\mu - c^2} = 1 \qquad (5\text{-}30)$$

$$\frac{x^2}{a^2 - \nu} - \frac{y^2}{\nu - b^2} - \frac{z^2}{\nu - c^2} = 1$$

where λ, μ, ν are parameters called *ellipsoidal coordinates*; a, b, c are constants; $a^2 > \nu > b^2 > \mu > c^2 > \lambda > -\infty$. It is shown in books on solid analytical geometry that intersections of these three surfaces are orthogonal and that all of them have common foci. Moreover, through any fixed point $P(x,y,z)$ there passes one and only one surface of each type.

The relation between the new and the old coordinates may be found by solving (30) directly. It may be done more easily as follows. · Consider the cubic equation in a parameter q

$$\frac{x^2}{a^2 - q} + \frac{y^2}{b^2 - q} + \frac{z^2}{c^2 - q} - 1 = 0 \qquad (5\text{-}31)$$

with three real roots, λ, μ, ν satisfying the inequalities just stated. As q varies between a^2 and $-\infty$, (31) describes the complete system of confocal surfaces given in (30). On clearing (31) of fractions and equating it to its identity, we have

$$x^2(b^2 - q)(c^2 - q) + y^2(a^2 - q)(c^2 - q) + z^2(a^2 - q)(b^2 - q)$$
$$- (a^2 - q)(b^2 - q)(c^2 - q) \equiv (q - \lambda)(q - \mu)(q - \nu) \equiv 0 \qquad (5\text{-}32)$$

and this must hold for every value of q. Upon setting $q = a^2$, b^2, c^2 in turn, we obtain

$$x^2 = \frac{(a^2 - \lambda)(a^2 - \mu)(a^2 - \nu)}{(b^2 - a^2)(c^2 - a^2)}$$

$$y^2 = \frac{(b^2 - \lambda)(b^2 - \mu)(b^2 - \nu)}{(a^2 - b^2)(c^2 - b^2)} \qquad (5\text{-}33)$$

$$z^2 = \frac{(c^2 - \lambda)(c^2 - \mu)(c^2 - \nu)}{(a^2 - c^2)(b^2 - c^2)}$$

Taking the logarithm of (33), differentiating partially with respect to λ and using (4), we have

$$Q_\lambda^2 = \frac{1}{4}\left\{ \frac{(a^2 - \mu)(a^2 - \nu)}{(a^2 - \lambda)(b^2 - a^2)(c^2 - a^2)} + \frac{(b^2 - \mu)(b^2 - \nu)}{(b^2 - \lambda)(a^2 - b^2)(c^2 - b^2)} \right.$$
$$\left. + \frac{(c^2 - \mu)(c^2 - \nu)}{(c^2 - \lambda)(a^2 - c^2)(b^2 - c^2)} \right\} \qquad (5\text{-}34)$$

Values for Q_μ^2 and Q_ν^2 may be obtained in a similar way or from (34) by cyclic interchange of (λ,μ,ν). Simplification of the resulting expressions yields

$$Q_\lambda^2 = \frac{1}{4}\left\{\frac{(\mu - \lambda)(\nu - \lambda)}{(a^2 - \lambda)(b^2 - \lambda)(c^2 - \lambda)}\right\}$$

$$Q_\mu^2 = \frac{1}{4}\left\{\frac{(\nu - \mu)(\lambda - \mu)}{(a^2 - \mu)(b^2 - \mu)(c^2 - \mu)}\right\} \tag{5-35}$$

$$Q_\nu^2 = \frac{1}{4}\left\{\frac{(\lambda - \nu)(\mu - \nu)}{(a^2 - \nu)(b^2 - \nu)(c^2 - \nu)}\right\}$$

It is somewhat laborious to transform (34) directly into the first equation of (35) but their equivalence may be verified by writing the latter in terms of partial fractions.

Because of the fact that x, y and z appear as squares in (33), a given point $P(x,y,z)$ is not uniquely determined by (λ,μ,ν); in fact, eight points symmetrically located relative to the (XYZ)-axes correspond to the set (λ,μ,ν). This ambiguity may be resolved by adopting some convention concerning the signs of (λ,μ,ν), or in more elegant fashion by the introduction of elliptic functions. The latter procedure may be accomplished either by means of elliptic integrals, Jacobian elliptic functions or Weierstrass p-functions.[4]

The confocal ellipsoidal coordinate system has proved useful in problems of mechanics, potential theory, electrodynamics and hydrodynamics.[5]

5.7. Prolate Spheroidal Coordinates.—Degenerate cases of the preceding system may arise if two or three of the axes in (31) become equal. Additional surfaces are then needed since the resulting equation in q is either quadratic or linear. Instead of following a method similar to that used for ellipsoidal coordinates, it is simpler to proceed by considering the equations of an ellipse and a hyperbola,

$$\frac{z^2}{a^2} + \frac{x^2}{a^2(1 - e_1^2)} = 1$$

$$\frac{z^2}{a^2} - \frac{x^2}{a^2(e_2^2 - 1)} = 1 \tag{5-36}$$

---

[4] Full details concerning these functions may be found in Whittaker, E. T. and Watson, G. N., " A Course of Modern Analysis," Fourth Edition, Cambridge Press, 1927.

[5] Some references to these applications are: MacMillan, W. D., " Statics and Dynamics of a Particle," 1927, " The Theory of the Potential," 1930, McGraw-Hill Book Co.; Kellogg, O. D., " Foundations of Potential Theory," J. Springer, Berlin, 1929; Mason, Max and Weaver, Warren, " The Electromagnetic Field," University of Chicago Press, 1929; Milne-Thomson, L. M., " Theoretical Hydrodynamics," Macmillan and Co.. London, 1938.

where a is the semi-major axis and $e_1 < 1$ is the eccentricity of the ellipse, $e_2 > 1$, the eccentricity of the hyperbola. If we now substitute $a \cosh u$ for a and sech u for e_1 in the ellipse, $a \cos v$ for a and sec v for e_2 in the hyperbola and finally $x^2 + y^2 = r^2$ for x^2, we obtain

$$\frac{z^2}{a^2 \cosh^2 u} + \frac{r^2}{a^2 \sinh^2 u} = 1$$

$$\frac{z^2}{a^2 \cos^2 v} - \frac{r^2}{a^2 \sin^2 v} = 1$$

$$(5\text{-}37)$$

with $0 \le u \le \infty$, $0 \le v \le \pi$. These equations represent the confocal families of: (1) prolate spheroids[6] ($u = $ const.) and (2) hyperboloids (of two sheets) of revolution ($v = $ const.) obtained by rotating the ellipses and hyperbolas of (36) around the Z-axis. The intersection of these surfaces, as shown in Fig. 3, will be a circle of radius r; hence if $0 \le \phi \le 2\pi$, the addition of (3), a family of planes through the Z-axis ($\phi = $ const.), to the spheroids and hyperboloids gives us three suitable coordinate surfaces (u,v,ϕ). We may then solve (37) for z and r and simplify the resulting expressions by means of the relations between trigonometric functions. Finally, we set $x = r \cos \phi$, $y = r \sin \phi$, obtaining

$$x = a \sinh u \sin v \cos \phi$$
$$y = a \sinh u \sin v \sin \phi \qquad (5\text{-}38)$$
$$z = a \cosh u \cos v$$

and from (4),

$$Q_u^2 = Q_v^2 = a^2 (\sinh^2 u + \sin^2 v)$$
$$Q_\phi^2 = a^2 (\sinh^2 u \sin^2 v) \qquad (5\text{-}39)$$

An important property of prolate spheroidal coordinates makes them useful in certain quantum mechanical problems. It is well known from analytical geometry that the sum of the focal radii of an ellipse is a constant, equal to the major axis. Similarly the difference between the focal radii in a hyperbola equals the transverse axis. If r_A and r_B are the distances from the two foci to a point of intersection of the ellipsoids and hyperboloids, we find that

$$r_A + r_B = 2a \cosh u; \quad r_A - r_B = 2a \cos v$$

where we have replaced a by $a \cosh u$ and by $a \cos v$ as before. This procedure thus locates a point relative to any two-center problem such as the diatomic molecule (see sec. 11.21). It is often convenient to introduce the

---

[6] Also called ovary ellipsoids.

coordinates ξ and η in place of $\cosh u$ and $\cos v$, respectively, so that

$$\xi = \frac{r_A + r_B}{2a} \; ; \quad \eta = \frac{r_A - r_B}{2a} \qquad (5\text{-}40)$$

In terms of these variables, the volume element may be seen to take the form

$$d\tau = a^3(\xi^2 - \eta^2)d\xi d\eta d\phi$$

5.8. Oblate Spheroidal Coordinates.—When ellipses are rotated about their minor axis, the resulting surfaces are oblate spheroids.[7] If we rewrite (37) so that the axis of revolution is again the Z-axis, but now[8] the *minor axis* of the ellipse, we have

$$\frac{r^2}{a^2 \cosh^2 u} + \frac{z^2}{a^2 \sinh^2 u} = 1$$

$$\frac{r^2}{a^2 \sin^2 v} - \frac{z^2}{a^2 \cos^2 v} = 1 \qquad (5\text{-}41)$$

with $0 \leq u \leq \infty$, $0 \leq v \leq \pi$, $x = r \cos \phi$, $y = r \sin \phi$, $0 \leq \phi \leq 2\pi$. The coordinate surfaces are thus: (1) oblate spheroids ($u = $ const.); (2) hyperboloids (of one sheet) of revolution ($v = $ const.); (3) planes through the Z-axis ($\phi = $ const.). From (41), we find

$$x = a \cosh u \sin v \cos \phi$$
$$y = a \cosh u \sin v \sin \phi \qquad (5\text{-}42)$$
$$z = a \sinh u \cos v$$

and from (4),

$$Q_u^2 = Q_v^2 = a^2 (\sinh^2 u + \cos^2 v)$$
$$Q_\phi^2 = a^2 \cosh^2 u \sin^2 v \qquad (5\text{-}43)$$

The geometry of the system may be inferred from Fig. 3 by suitable interchange of the X-, Y- and Z-axes.

5.9. Elliptic Cylindrical Coordinates.—If (37) is again rewritten with x^2 in place of z^2 and y^2 in place of r^2, the loci of these equations are cylindrical surfaces, whose elements are parallel to the Z-axis and perpendicular to the XY-plane. Their intersections with this plane are ellipses and hyperbolas. The coordinate surfaces are: (1) elliptic cylinders ($u = $ const.); (2) hyperbolic cylinders ($v = $ const.); (3) planes parallel to the

---

[7] Also called *planetary* ellipsoids. The figures of the earth and of the planet Jupiter are approximately of this form.

[8] At the risk of some confusion, we have interchanged axes in this system and in some of the following ones so that the Z-axis is always the axis of revolution.

XY-plane (z = const.). Proceeding as before,

$$x = a \cosh u \cos v$$
$$y = a \sinh u \sin v \qquad\qquad (5\text{--}44)$$
$$z = z$$

$$Q_u^2 = Q_v^2 = a^2 \,(\sinh^2 u + \sin^2 v); \quad Q_z^2 = 1 \qquad (5\text{--}45)$$

The intersection of these cylinders with the XY-plane may also be inferred from Fig. 3.

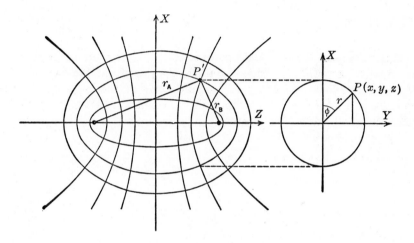

Fig. 5–3

5.10. Conical Coordinates.—A further degenerate case of the system of sec. 5.6 arises when the orthogonal sets of surfaces are: (1) spheres with centers at the origin and radius u (u = const.); (2) cones with apexes at the origin and axes along the Z-axis (v = const.); (3) cones with apexes at the origin and axes along the X-axis (w = const.), their equations being

$$x^2 + y^2 + z^2 = u^2$$
$$\frac{x^2}{v^2} + \frac{y^2}{v^2 - b^2} - \frac{z^2}{c^2 - v^2} = 0 \qquad (5\text{--}46)$$
$$\frac{x^2}{w^2} - \frac{y^2}{b^2 - w^2} - \frac{z^2}{c^2 - w^2} = 0$$

$c^2 > v^2 > b^2 > w^2$. The projections of the surfaces on the XY-plane are families of circles, ellipses and triangles. From (46), we find

$$x^2 = \frac{u^2 v^2 w^2}{b^2 c^2}$$

$$y^2 = \frac{u^2(v^2 - b^2)(w^2 - b^2)}{b^2(b^2 - c^2)} \qquad (5\text{--}47)$$

$$z^2 = \frac{u^2(v^2 - c^2)(w^2 - c^2)}{c^2(c^2 - b^2)}$$

and from (4)

$$Q_u^2 = 1; \quad Q_v^2 = \frac{u^2(v^2 - w^2)}{(v^2 - b^2)(c^2 - v^2)} : \quad Q_w^2 = \frac{u^2(v^2 - w^2)}{(w^2 - b^2)(w^2 - c^2)} \qquad (5\text{--}48)$$

5.11. Confocal Paraboloidal Coordinates.—A system similar to that of sec. 5.6 has coordinate surfaces consisting of confocal families of: (1) elliptic paraboloids extending in the direction of the negative Z-axis (λ = const.); (2) hyperbolic paraboloids (μ = const.); (3) elliptic paraboloids extending along the positive Z-axis (ν = const.). The equations for the surfaces are

$$\frac{x^2}{a^2 - \lambda} + \frac{y^2}{b^2 - \lambda} + 2z + \lambda = 0$$

$$\frac{x^2}{a^2 - \mu} - \frac{y^2}{\mu - b^2} + 2z + \mu = 0 \qquad (5\text{--}49)$$

$$\frac{x^2}{\nu - a^2} + \frac{y^2}{\nu - b^2} - 2z - \nu = 0$$

where $-\infty < \lambda < b^2 < \mu < a^2 < \nu < +\infty$. Proceeding as in the confocal ellipsoidal system, we may write the cubic equation in q,

$$\frac{x^2}{a^2 - q} + \frac{y^2}{b^2 - q} + 2z + q = 0 \qquad (5\text{--}50)$$

with three real roots, λ, μ, ν. As q varies between $-\infty$ and $+\infty$, the complete system of confocal surfaces (49) will be described. On clearing (50) of fractions and equating it to its identity, we have

$$x^2(b^2 - q) + y^2(a^2 - q) + (2z + q)(a^2 - q)(b^2 - q)$$
$$\equiv (q - \lambda)(q - \mu)(q - \nu) \equiv 0 \qquad (5\text{--}51)$$

Expressions for x^2 and y^2 may be obtained from (51) by setting $q = a^2$ and b^2 in turn; the result for z is found by equating the coefficients of q^2 on both sides of (51).   We thus have

$$x^2 = \frac{(a^2 - \lambda)(a^2 - \mu)(a^2 - \nu)}{(b^2 - a^2)}$$

$$y^2 = \frac{(b^2 - \lambda)(b^2 - \mu)(b^2 - \nu)}{(a^2 - b^2)} \tag{5-52}$$

$$z = \tfrac{1}{2}(a^2 + b^2 - \lambda - \mu - \nu)$$

and

$$Q_\lambda^2 = \frac{1}{4}\frac{(\mu - \lambda)(\nu - \lambda)}{(a^2 - \lambda)(b^2 - \lambda)}$$

$$Q_\mu^2 = \frac{1}{4}\frac{(\nu - \mu)(\lambda - \mu)}{(a^2 - \mu)(b^2 - \mu)} \tag{5-53}$$

$$Q_\nu^2 = \frac{1}{4}\frac{(\lambda - \nu)(\mu - \nu)}{(a^2 - \nu)(b^2 - \nu)}$$

Because of the appearance of x and y as squares in (52), a point $P(x,y,z)$ corresponds to four points $P(\lambda,\mu,\nu)$ symmetrically located with respect to the XZ- and YZ-planes.   As in the confocal ellipsoidal system (sec. 5.6) the ambiguity may be removed by the use[9] of elliptic integrals.

5.12. Parabolic Coordinates.—If two roots of (50) become equal, the preceding method fails since there are now only two surfaces.   In this case, consider the families of parabolas

$$x^2 = 2\xi^2(z + \xi^2/2)$$
$$x^2 = -2\eta^2(z - \eta^2/2) \tag{5-54}$$

The vertices of all parabolas lie on the Z-axis at distances $-\xi^2/2$ and $\eta^2/2$, respectively, and all of them have a common focus at the origin of the Cartesian coordinate system.   If we now rotate these parabolas about the Z-axis, the resulting intersections are circles and the paraboloids of revolution are still given by (54) if we replace x^2 by $r^2 = x^2 + y^2$, $x = r \cos \phi$, $y = r \sin \phi$.   We thus obtain

$$x = \xi\eta \cos \phi$$
$$y = \xi\eta \sin \phi \tag{5-55}$$
$$z = (\eta^2 - \xi^2)/2$$

and from (4),

$$Q_\xi^2 = Q_\eta^2 = (\xi^2 + \eta^2)$$
$$Q_\phi^2 = \xi^2\eta^2 \tag{5-56}$$

[9] See, for example, Maxwell, J. C., " A Treatise on Electricity and Magnetism," Vol. I, Third Edition, Oxford Press, 1904, p. 240.   Application of this system is also described there.

The coordinate surfaces are: (1) paraboloids of revolution extending in the direction of the positive Z-axis (ξ = const.); (2) paraboloids of revolution extending toward the negative Z-direction (η = const.); (3) planes through the Z-axis (ϕ = const.). Intersections of these surfaces with the XZ- and XY-planes are shown in Fig. 4. Parabolic coordinates have been used in the treatment of the Stark effect.[10]

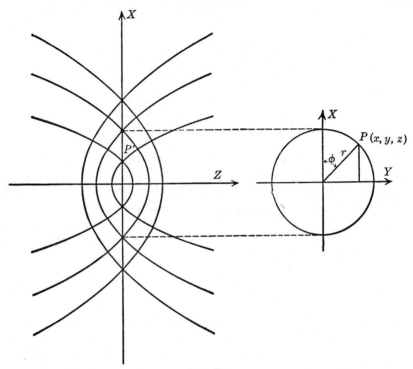

FIG. 5-4

5.13. Parabolic Cylindrical Coordinates.—A system similar to elliptical cylindrical coordinates is obtained by adding planes to the parabolic cylinders represented by (54). If we replace z by y in those equations, we have

$$\begin{aligned} x &= \xi\eta \\ y &= (\eta^2 - \xi^2)/2 \\ z &= z \end{aligned} \tag{5-57}$$

$$\begin{aligned} Q_\xi^2 = Q_\eta^2 &= (\xi^2 + \eta^2) \\ Q_z^2 &= 1 \end{aligned} \tag{5-58}$$

[10] Schrödinger, E., *Ann. Physik* **80.** 457 (1926); Epstein, P. S., *Phys. Rev.* **28.** 695 (1926).

The coordinate surfaces are: (1) parabolic cylinders (ξ = const.); (2) parabolic cylinders (η = const.); (3) planes (z = const.). The intersection of these surfaces with the XY-plane is like the system of confocal parabolas shown in Fig. 4.

5.14. Bipolar Coordinates.—Before considering this system, we list a few relations which are needed in the subsequent discussion. In terms of exponentials, we may write

$$\sin x = \frac{i}{2}(e^{-ix} - e^{ix}); \quad \cos x = \tfrac{1}{2}(e^{-ix} + e^{ix})$$

$$\tan x = \frac{\sin x}{\cos x} = \frac{i(1 - e^{2ix})}{(1 + e^{2ix})} \tag{5-59}$$

Replacing x by ix, we have the corresponding *hyperbolic functions*

$$\sin ix = \frac{i}{2}(e^x - e^{-x}) = i \sinh x$$

$$\cos ix = \tfrac{1}{2}(e^x + e^{-x}) = \cosh x \tag{5-60}$$

$$\tan ix = \frac{i(e^{2x} - 1)}{(e^{2x} + 1)} = i \tanh x$$

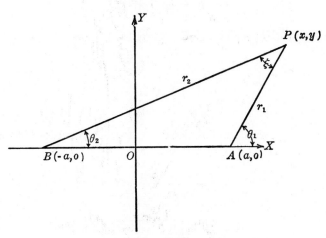

Fig. 5-5

We shall also need the inverse circular function $\tan^{-1} x = u$. Since $x = \tan u$, it follows from (59) that

$$e^{2iu} = \frac{(i - x)}{(i + x)}; \quad 2iu = \ln \frac{(i - x)}{(i + x)}$$

and $(5-61)$

$$u = \tan^{-1} x = \frac{i}{2}\ln\frac{(i + x)}{(i - x)}$$

Suppose a point $P(x,y)$ is located as shown in Fig. 5 by means of two vectors $\mathbf{r}_1$ and $\mathbf{r}_2$ and two angles θ_1 and θ_2. For different positions of the point in the XY-plane, the vectors are always drawn from the fixed points A and B symmetrically located on the X-axis a distance $2a$ apart. If $\rho = x + iy$; $\rho^* = x - iy$, then

$$x = (\rho^* + \rho)/2; \quad y = \frac{i}{2}(\rho^* - \rho) \tag{5-62}$$

The coordinates of the point are

$$\rho - a = r_1 e^{i\theta_1}$$
$$\rho + a = r_2 e^{i\theta_2} \tag{5-63}$$

and from the geometry of Fig. 5, it follows that

$$r_1^2 = (x - a)^2 + y^2; \quad \theta_1 = \tan^{-1} y/(x - a)$$
$$r_2^2 = (x + a)^2 + y^2; \quad \theta_2 = \tan^{-1} y/(x + a) \tag{5-64}$$

Defining new quantities

$$\xi = \theta_1 - \theta_2; \quad \eta = \ln \frac{r_2}{r_1} \tag{5-65}$$

and dividing the two equations of (63) by each other

$$\frac{\rho + a}{\rho - a} = e^{-i\chi}; \quad \frac{\rho}{a} = \frac{e^{-i\chi} + 1}{e^{-i\chi} - 1} \tag{5-66}$$

where

$$\chi = \xi + i\eta \tag{5-67}$$

In order to find x and y as functions of ξ and η, substitute (66) and (67) in (62). When use is made of (59) and (60) the results are

$$x = \frac{a \sinh \eta}{\cosh \eta - \cos}$$
$$y = \frac{a \sin \xi}{\cosh \eta - \cos \xi} \tag{5-68}$$

To find the form of the coordinate surfaces, we start from the definition of ξ and use (61) to obtain

$$\xi = \tfrac{1}{2}i \ln \frac{(ix - ia + y)(ix + ia - y)}{(ix - ia - y)(ix + ia + y)}$$

which may also be written as

$$x^2 + y^2 - a^2 + 2iay \frac{(1 + e^{2i\xi})}{(1 - e^{2i\xi})} = 0$$

We observe from (59) that the last term of this expression equals $-2ay/\tan \xi = -2ay \cot \xi$. Hence

$$x^2 + y^2 - a^2 - 2ay \cot \xi = 0$$

or

$$x^2 + (y - a \cot \xi)^2 = a^2(1 + \cot^2 \xi) = a^2 \csc^2 \xi \qquad (5\text{-}69)$$

In the same way we find

$$e^{2\eta} = \frac{r_2^2}{r_1^2} = \frac{(x + a)^2 + y^2}{(x - a)^2 + y^2}$$

and

$$(x - a \coth \eta)^2 + y^2 = a^2 \operatorname{csch}^2 \eta \qquad (5\text{-}70)$$

We thus see that for $\xi = \text{const.}$, $0 \le \xi \le 2\pi$, we have a family of circles with centers on the Y-axis at the point, $x = 0$, $y = a \cot \xi$, the radii of the circles being $a \csc \xi$. Each member of this family will pass through the

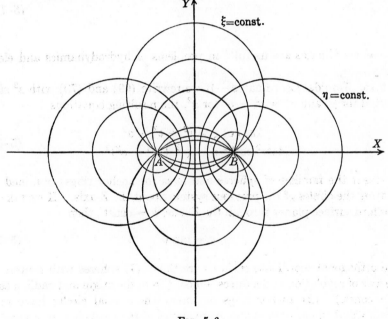

Fɪɢ. 5–6

fixed points A and B as shown in Fig. 6 and will intersect the circles $\eta = \text{const.}$ orthogonally. The members of the second family have radii of length $a \operatorname{csch} \eta$ and are all situated on the X-axis at the points $x = a \coth \eta$, $y = 0$. The point A is obtained when $\eta = +\infty$ and B when

$\eta = -\infty$. When $\eta = 0$, the circles degenerate into points on the Y-axis. The position of a point in the XY-plane is thus fixed when we know in which quadrant it lies and furthermore the constant values of η, ξ of the circles which pass through it. Since the fixed points A and B (that is, the X-axis) divide each circle of the set $\xi = $ const. into two segments, we arbitrarily take $\xi = \xi_0 < \pi$ for the arc above the X-axis and $\xi = \xi_0 + \pi$ for all points below this axis.

In order to use these circles as a coordinate system in space, imagine them to be moved along the Z-axis. Then (69) and (70) represent two families of right circular cylinders with axes parallel to the Z-axis. Suitable coordinate surfaces are then: (1) cylinders with centers on the Y-axis ($\xi = $ const.); (2) cylinders with centers on the X-axis ($\eta = $ const.); (3) planes perpendicular to the Z-axis ($z = $ const.). From (68) and (4),

$$Q_\xi^2 = Q_\eta^2 = \frac{a^2}{(\cosh \eta - \cos \xi)^2}$$

$$Q_z^2 = 1 \tag{5-71}$$

Bipolar coordinates are useful[11] in problems of hydrodynamics and electricity.

5.15. Toroidal Coordinates.—If we rewrite (69) and (70) with z^2 substituted for y^2 and $r^2 = x^2 + y^2$ for x^2, the resulting equations

$$2az \cot \xi = r^2 + z^2 - a^2$$
$$4a^2r^2 \coth^2 \eta = (r^2 + z^2 + a^2)^2 \tag{5-72}$$

represent the families of spheres and *tores* (or anchor rings) obtained by rotating the circles of the previous system about the Z-axis. If we take as the third surface planes through the Z-axis, $\psi = $ const., then

$$y/x = \tan \psi \tag{5-73}$$

The orthogonal coordinate surfaces are thus: (1) spheres with centers on the axis of revolution at distances $\pm a \cot \xi$ from the origin and radii, $a \csc \xi$ ($\xi = $ const.); (2) anchor rings or tores, whose axial circles have radii $a \coth \eta$ and whose cross-sections are circles of radii $a \csch \eta$ ($\eta = $ const.); (3) planes through the Z-axis ($\psi = $ const.). The spheres and anchor rings have a common circle, $r = a$, $z = 0$. With methods similar to those used

---

[11] See, for example, Milne-Thomson (loc. cit.); Maxwell (loc. cit.); Jeans, J. H., "The Mathematical Theory of Electricity and Magnetism," Fifth Edition, Cambridge Press, 1925.

in sec. 5.14,

$$x = r \cos \psi, \quad y = r \sin \psi$$

$$r = \frac{a \sinh \eta}{\cosh \eta - \cos \xi} \tag{5-74}$$

$$z = \frac{a \sin \xi}{\cosh \eta - \cos \xi}$$

$$Q_\xi^2 = Q_\eta^2 = \frac{a^2}{(\cosh \eta - \cos \xi)^2}$$

$$\tag{5-75}$$

$$Q_\psi^2 = \frac{a^2 \sinh^2 \eta}{(\cosh \eta - \cos \xi)^2}$$

This system has found application[12] in certain problems of electricity and of potential theory.

Problem a. Show that in spherical polar coordinates:

$$\nabla \cdot \mathbf{V} = \frac{1}{r^2 \sin \theta} \left\{ \sin \theta \, \frac{\partial}{\partial r} \, (r^2 V_r) + r \, \frac{\partial}{\partial \theta} \, (\sin \theta V_\theta) + r \, \frac{\partial V_\phi}{\partial \phi} \right\}$$

$$\nabla^2 = \frac{1}{r^2 \sin \theta} \left\{ \sin \theta \, \frac{\partial}{\partial r} \left(r^2 \frac{\partial}{\partial r} \right) + \frac{\partial}{\partial \theta} \left(\sin \theta \, \frac{\partial}{\partial \theta} \right) + \frac{1}{\sin \theta} \, \frac{\partial^2}{\partial \phi^2} \right\}$$

$$(\nabla \times \mathbf{V})_r = \frac{1}{r \sin \theta} \left\{ \frac{\partial}{\partial \theta} \, (\sin \theta V_\phi) - \frac{\partial V_\theta}{\partial \phi} \right\}$$

$$(\nabla \times \mathbf{V})_\theta = \frac{1}{r \sin \theta} \left\{ \frac{\partial V_r}{\partial \phi} - \sin \theta \, \frac{\partial (r V_\phi)}{\partial r} \right\}$$

$$(\nabla \times \mathbf{V})_\phi = \frac{1}{r} \left\{ \frac{\partial}{\partial r} \, (r V_\theta) - \frac{\partial V_r}{\partial \theta} \right\}$$

Problem b. Show that in cylindrical coordinates:

$$\nabla^2 = \frac{1}{\rho} \left\{ \frac{\partial}{\partial \rho} \left(\rho \, \frac{\partial}{\partial \rho} \right) + \frac{1}{\rho} \, \frac{\partial^2}{\partial \phi^2} + \rho \, \frac{\partial^2}{\partial z^2} \right\}$$

Problem c. If V is the potential energy and m is the mass of a particle show that Newton's laws of motion become:

(1) in spherical polar coordinates

$$m \{ \ddot{r} - r \dot{\theta}^2 - r \sin^2 \theta \dot{\varphi}^2 \} = -\partial V / \partial r$$

$$m \left\{ \frac{1}{r} \frac{d}{dt} \, (r^2 \dot{\theta}) - r \sin \theta \cos \theta \dot{\varphi}^2 \right\} = -\frac{1}{r} \frac{\partial V}{\partial \theta}$$

$$m \left\{ \frac{1}{r \sin \theta} \frac{d}{dt} \, (r^2 \sin^2 \theta \dot{\varphi}) \right\} = -\frac{1}{r \sin \theta} \frac{\partial V}{\partial \varphi}$$

[12] See Hobson, loc. cit., for references.

(2) in cylindrical coordinates

$$m(\ddot{\rho} - \rho\dot{\varphi}^2) = -\frac{\partial V}{\partial \rho}$$

$$m\left\{\frac{1}{\rho}\frac{d}{dt}(\rho^2\dot{\varphi})\right\} = -\frac{1}{\rho}\frac{\partial V}{\partial \rho}$$

$$m\ddot{z} = -\frac{\partial V}{\partial z}$$

NON-ORTHOGONAL COORDINATE SYSTEMS

5.16. Tensor Relations in Curvilinear Coordinates.—When the coordinate surfaces of a curvilinear system are not orthogonal, the methods of tensor analysis prove convenient (see sec. 4.20 ff.). The relations which we are about to derive are more general than those obtained in the first part of this chapter; in fact, we will show that the two formulations of the problem become equivalent for orthogonal coordinates.

Let (x^1, x^2, x^3) be the usual Cartesian coordinates of a point and (q^1, q^2, q^3) be its curvilinear coordinates, as discussed in sec. 5.1. Then in tensor notation,[13] eq. (3) becomes

$$ds^2 = g_{ij}dq^i dq^j \tag{5-3a}$$

where

$$g_{ij} = \frac{\partial x^m}{\partial q^i}\frac{\partial x^m}{\partial q^j} = g_{ji} \tag{5-4a}$$

is identical with Q_{ij} of eq. (4). The line element is clearly

$$ds_i = \sqrt{g_{ii}}dq^i; \quad (i \text{ not summed}) \tag{5-5a}$$

In order to find the surface element, we recall from sec. 4.5 that a surface may be represented as the vector product of two other vectors. Thus let ds_2 be an infinitesimal displacement at the point (q^1, q^2, q^3) along the coordinate line q^2 and ds_3 be a similar displacement along the line q^3. Then the vector $d\mathbf{S}_1 = d\mathbf{s}_2 \times d\mathbf{s}_3$ is perpendicular to the plane $q^1 =$ const. and its magnitude dS_1 is the desired surface element in that plane. Before we can obtain the appropriate expression for it in terms of the tensor g_{ij} we must digress in order to consider two important systems of vectors in curvilinear coordinates. Suppose $\mathbf{r} = \mathbf{r}(q^1, q^2, q^3)$ is a vector and

$$d\mathbf{r} = \frac{\partial \mathbf{r}}{\partial q^1}dq^1 + \frac{\partial \mathbf{r}}{\partial q^2}dq^2 + \frac{\partial \mathbf{r}}{\partial q^3}dq^3$$

[13] We generally use the summation convention throughout the rest of this chapter. In certain cases, repeated indices are not to be summed; such exceptions should be obvious to the reader.

is a small displacement. If we define three vectors

$$e_i = \frac{\partial r}{\partial q^i}$$

then it is clear that we may write

$$dr = e_i dq^i \tag{5-76}$$

These vectors, **e**, which we call *base vectors*,[14] are directed tangentially along the coordinate curves but they are not necessarily of unit length. While it is usually more convenient to resolve an arbitrary vector **A** into components which are multiples of a unit vector we may also write

$$\mathbf{A} = a^i e_i \tag{5-77}$$

and the three scalars a^i are the contravariant components of **A**. Let us define another set of base vectors

$$e^1 = \frac{e_2 \times e_3}{v} \; ; \quad e^2 = \frac{e_3 \times e_1}{v} \; ; \quad e^3 = \frac{e_1 \times e_2}{v} \tag{5-78}$$

where v is the scalar triple product $[e_1 e_2 e_3]$ of sec. 4.6b. These vectors are perpendicular to the planes of e_2, e_3; e_3, e_1 and e_1, e_2, respectively, and it is easily seen that

$$e^m \cdot e_n = \delta_n^m \tag{5-79}$$

Furthermore, it is true that

$$e_1 = \frac{e^2 \times e^3}{v'} \; ; \quad e_2 = \frac{e^3 \times e^1}{v'} \; ;$$

$$e_3 = \frac{e^1 \times e^2}{v'} \quad \text{where } v' = [e^1 e^2 e^3] \text{ and } vv' = 1; \tag{5-80}$$

hence the two sets of vectors e^m and e_n are said to be *reciprocal* to each other. In terms of the reciprocal set[15] (76) becomes

$$dr = e^i dq_i \tag{5-81}$$

and (77) becomes

$$\mathbf{A} = a_i e^i \tag{5-82}$$

where the a_i are the covariant components of **A**.

[14] The systems of base vectors introduced here are treated by matrix methods in sec. 10.10.

[15] Many interesting properties of reciprocal systems are presented by Gibbs-Wilson, loc. cit., pp. 81–92.

If we equate (76) and (81) we obtain

$$dr = e_i dq^i = e^j dq_j \tag{5-83}$$

and if we multiply by e^i or e_j we find, because of (79), that

$$dq^i = e^i \cdot e^j dq_j; \quad dq_j = e_j \cdot e_i dq^i \tag{5-84}$$

Since the square of the distance between two points is given by $ds^2 = dr \cdot dr$ we see from (83) that

$$ds^2 = e_i \cdot e_j dq^i dq^j = e^i \cdot e^j dq_i dq_j \tag{5-85}$$

We may therefore identify the scalar products of the base vectors with the tensors g^{ij} and g_{ij}

$$g_{ij} = e_i \cdot e_j; \quad g^{ij} = e^i \cdot e^j \tag{5-86}$$

For later use, we also note that we may equate (77) to (82)

$$A = a_i e^i = a^j e_j \tag{5-87}$$

and use (79) and (86) to write

$$a_i = g_{ij} a^j; \quad a^j = g^{ij} a_i \tag{5-88}$$

We also have from (87) the equivalent expressions

$$a_i = A \cdot e_i; \quad a^j = A \cdot e^j$$

hence (87) may be stated in the alternative form

$$A = (A \cdot e_i) e^i = (A \cdot e^j) e_j \tag{5-89}$$

We now have several relations by means of which we may find either the contravariant or covariant components of an arbitrary vector A. If we wish to know the components in terms of unit vectors, tangent to the q^i coordinate lines, we recall the equation defining the length of a vector (sec. 4.4) and see that the appropriate unit vectors are

$$u_i = \frac{e_i}{\sqrt{e_i \cdot e_i}} = \frac{e_i}{\sqrt{g_{ii}}}$$

Therefore, any vector A may also be written as

$$A = A_i u_i$$

where

$$A_i = \sqrt{g_{ii}} a^i$$

If needed, similar equations could be given in the reciprocal system.

Let us now return to the problem of the surface element in curvilinear coordinates. Since $ds_i = e_i dq^i$, we have

$$dS_1 = ds_2 \times ds_3 = (e_2 \times e_3) dq^2 dq^3$$

and

$$dS_1 = [(e_2 \times e_3) \cdot (e_2 \times e_3)]^{1/2} dq^2 dq^3$$

It is easy to show (see Problem a, sec. 4.6) that the scalar product inside the brackets becomes

$$(e_2 \cdot e_2)(e_3 \cdot e_3) - (e_2 \cdot e_3)(e_3 \cdot e_2)$$

Thus when we use (86) we obtain

$$dS_1 = \sqrt{g_{22}g_{33} - g_{23}^2} dq^2 dq^3 \tag{5-6a}$$

Similarly, surface elements on the planes $q^2 = $ const. and $q^3 = $ const. are

$$dS_2 = \sqrt{g_{11}g_{33} - g_{13}^2} dq^1 dq^3$$

$$dS_3 = \sqrt{g_{11}g_{22} - g_{12}^2} dq^1 dq^2$$

The volume element,

$$d\tau = d\mathbf{s}_1 \cdot d\mathbf{s}_2 \times d\mathbf{s}_3 = [e_1 e_2 e_3] dq^1 dq^2 dq^3$$

If we place $\mathbf{A} = e_2 \times e_3$ and use (89) we get

$$\mathbf{A} = e_2 \times e_3 = [e^1 e_2 e_3]e_1 + [e^2 e_2 e_3]e_2 + [e^3 e_2 e_3]e_3$$

Now by means of (4-18) and (78) we eliminate e^i to obtain

$$[e_1 e_2 e_3] = e_1 \cdot \mathbf{A} = \frac{e_1}{[e_1 e_2 e_3]} \{ (e_2 \times e_3 \cdot e_2 \times e_3)e_1$$

$$+ (e_3 \times e_1 \cdot e_2 \times e_3)e_2 + (e_1 \times e_2 \cdot e_2 \times e_3)e_3 \}$$

Finally we expand the scalar products within the brackets using again the result of Problem a, sec. 4.6 and getting

$$[e_1 e_2 e_3]^2 = e_1 \cdot e_1[(e_2 \cdot e_2)(e_3 \cdot e_3) - (e_2 \cdot e_3)(e_3 \cdot e_2)]$$

$$+ e_1 \cdot e_2[(e_2 \cdot e_3)(e_3 \cdot e_1) - (e_2 \cdot e_1)(e_3 \cdot e_3)]$$

$$+ e_1 \cdot e_3[(e_2 \cdot e_1)(e_3 \cdot e_2) - (e_2 \cdot e_2)(e_3 \cdot e_1)]$$

By means of (86) we may replace the scalar products in this equation by the g_{ij}, finding that $[e_1 e_2 e_3] = \sqrt{g}$ where g is the determinant of the components of g_{ij} and the volume element becomes

$$d\tau = \sqrt{g}\, dq^1 dq^2 dq^3 \tag{5-7a}$$

5.17. The Differential Operators in Tensor Notation.—We have seen in sec. 4.20 that the components of the gradient of a scalar point function φ are $\partial\varphi/\partial q^i$. The direction of dq^i is determined by the vector $d\mathbf{s}^i = e_i dq^i = $

$dq^i/\mathbf{e}^i$ or, since $\mathbf{e}^i = g^{ij}\mathbf{e}_j$, we have in curvilinear coordinates

$$\nabla\varphi = \mathbf{e}^i \frac{\partial\varphi}{\partial q^i} = g^{ij}\mathbf{e}_j \frac{\partial\varphi}{\partial q^i} \tag{5-9a}$$

The divergence of a vector $\mathbf{V}$ in terms of its contravariant components is the covariant derivative. Thus, from (4–71)

$$\nabla \cdot \mathbf{V} = V^i,_i = \frac{\partial V^i}{\partial q^i} + V^j\{ij,i\} \tag{5-90}$$

Now according to (4.63)

$$\{ij,i\} = \tfrac{1}{2}g^{ik}\left\{\frac{\partial g_{ik}}{\partial q^j} + \frac{\partial g_{jk}}{\partial q^i} - \frac{\partial g_{ij}}{\partial q^k}\right\}$$

but

$$g^{ik}\frac{\partial g_{jk}}{\partial q^i} = g^{ki}\frac{\partial g_{ji}}{\partial q^k}$$

since we may exchange the dummy indices i and k. Moreover, $g^{ik} = g^{ki}$ and $g_{ij} = g_{ji}$ so we may cancel the second and third terms in $\{ij,i\}$. Finally, we refer to (4–59) and the rule for differentiating determinants (see sec. 10.4) to prove that

$$\frac{\partial g}{\partial g_{ij}} = G^{ij} = gg^{ij}$$

The Christoffel symbol therefore takes the form

$$\{ij,i\} = \tfrac{1}{2}g^{ik}\frac{\partial g_{ik}}{\partial q^j} = \frac{1}{2g}\frac{\partial g}{\partial q^j} = \frac{1}{\sqrt{g}}\frac{\partial(\sqrt{g})}{\partial q^j}$$

Eq. (90) may thus be written as

$$\nabla \cdot \mathbf{V} = \frac{\partial V^i}{\partial q^i} + \frac{V^j}{\sqrt{g}}\frac{\partial(\sqrt{g})}{\partial q^j}$$

$$= \frac{1}{\sqrt{g}}\frac{\partial}{\partial q^i}[V^i\sqrt{g}] \tag{5-20a}$$

A similar expression may be obtained in terms of covariant components of V.

If $\mathbf{V} = \nabla\phi$, the contravariant components of the **gradient** are $V^i = \mathbf{V} \cdot \mathbf{e}^i$ by (88) and by (9a)

$$V^i = \mathbf{e}^i \cdot \mathbf{e}_j g^{kj}\frac{\partial\phi}{\partial q^k} = g^{ij}\frac{\partial\phi}{\partial q^j}$$

Substituting this result in (20a) we find for the Laplacian

$$\nabla^2 \phi = \frac{1}{\sqrt{g}} \frac{\partial}{\partial q^i} \left[g^{ij} \sqrt{g} \frac{\partial \phi}{\partial q^j} \right] \tag{5-21a}$$

The final expression we wish to derive here is the **curl**. Define the covariant tensor of rank two

$$V_{ij} = \frac{\partial V_i}{\partial q^j} - \frac{\partial V_j}{\partial q^i}$$

If we transform to a new coordinate system we see from eq. (4-53) that

$$\bar{V}_{mn} = \frac{\partial q^i}{\partial \bar{q}^m} \frac{\partial q^j}{\partial \bar{q}^n} V_{ij}$$

This tensor which is invariant to such a transformation is the **curl** in curvilinear coordinates. According to its definition, it is skew symmetric, hence the only non-vanishing components are V_{12}, V_{23} and V_{31}. In terms of the base vectors we write

$$\nabla \times V = V_{12}(e^1 \times e^2) + V_{23}(e^2 \times e^3) + V_{31}(e^3 \times e^1)$$

We have shown in (78) how to convert the e^i into the reciprocal base vectors and we have also proved that $v = [e_1 e_2 e_3] = \sqrt{g}$. With these changes, the **curl** of a vector V is*

$$-\nabla \times V = \frac{1}{\sqrt{g}} \left[\left\{ \frac{\partial V_3}{\partial q^2} - \frac{\partial V_2}{\partial q^3} \right\} e_1 + \left\{ \frac{\partial V_1}{\partial q^3} - \frac{\partial V_3}{\partial q^1} \right\} e_2 \right.$$
$$\left. + \left\{ \frac{\partial V_2}{\partial q^1} - \frac{\partial V_1}{\partial q^2} \right\} e_3 \right] \tag{5-22a}$$

It is a simple matter to see what happens when the coordinate surfaces are mutually orthogonal. In that case, the vectors e_1, e_2, e_3 are also perpendicular to each other and e^i is parallel to e_i. Moreover,

$$e^i = \frac{e_i}{e_i \cdot e_i} = \frac{e_i}{g_{ii}}$$

and $e_1 \cdot e_2 = e_2 \cdot e_3 = e_3 \cdot e_1 = 0$. It thus follows that $g_{ij} = 0$ unless $i = j$; in the latter case,

$$g_{ii} = 1/g^{ii}$$

Remembering that g_{ii} is then identical with Q_i^2 as used in the first parts of this chapter, equations such as (3a), (4a), etc., in secs. 5.16 and 5.17 will reduce to the corresponding equations which appeared earlier in this chapter without the letter a.

Problem. Derive by the tensor method the results of Problems a, b, c, of sec. 5.15.

* Note that this tensor differs in sign from the conventional curl of vector analysis.

CHAPTER 6

CALCULUS OF VARIATIONS

One of the elementary problems of the differential calculus is to find the maxima and minima, that is, the stationary values, of a function $y(x)$. The necessary condition for the occurrence of a stationary value at $x = a$ is that $y'(a) = 0$. Sufficient conditions that it shall be a minimum or a maximum are, respectively, $y''(a) > 0$ and $y''(a) < 0$. The calculus of variations deals with a similar, but a more complicated problem, that of finding a function $y(x)$ such that a definite integral, taken over a function of this function, shall be a maximum or a minimum. The simpler parts of this calculus, to which this chapter will be primarily devoted, deal with the necessary conditions that the integral shall be either a maximum or a minimum; in other words, that it shall have a stationary value; sufficiency considerations as well as criteria for establishing the maximum or minimum character of the solutions are not important in many physical applications. For these, the reader should consult the more comprehensive treatises on the subject listed at the end of this chapter.

6.1. Single Independent and Single Dependent Variable.—Let it be desired, then, to find that function $y(x)$ which will cause the integral

$$\int_{x_1}^{x_2} I(x,y,y_x)dx$$

to have a stationary value. The integrand I is taken to be a function of the dependent variable y as well as the independent variable x and $y_x = dy/dx$. The limits x_1 and x_2 are fixed and at each of them, y has a fixed value. The integral over I takes on different values along different paths connecting the points (x_1,y_1) and (x_2,y_2); one of these paths is labeled $Y(x)$ in Fig. 1. We assume that it is either largest or smallest along $y(x)$, for example. The paths $Y(x)$ which are admitted for comparison shall be " adjacent " paths covering a small neighborhood of the *stationary* path $y(x)$, that is, $Y(x) - y(x)$ shall be infinitesimal for all values of x between x_1 and x_2.

We define:

$$\delta y(x) \equiv Y(x) - y(x) \tag{6-1}$$

$$\delta I \equiv I(x,Y,Y_x) - I(x,y,y_x) \tag{6-2}$$

The symbol δ is called *variation;* it represents the increase in the quantity to which it is applied as we pass from the stationary path to the comparison path at the same value of x. Thus, clearly $\delta x = 0$. Furthermore,

$$\delta \frac{dy}{dx} = \frac{dY}{dx} - \frac{dy}{dx} = \frac{d}{dx}(Y - y) = \frac{d}{dx}\delta y \qquad (6\text{–}3)$$

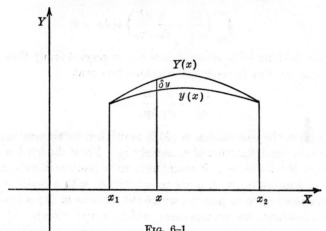

FIG. 6–1

This shows that the symbols δ and d/dx " commute." Since Y and y are adjacent, it follows from (2) that

$$\delta I = I(x, y + \delta y, y_x + \delta y_x) - I(x,y,y_x)$$

$$= \frac{\partial I}{\partial y}\delta y + \frac{\partial I}{\partial y_x}\delta y_x \qquad (6\text{–}4)$$

In words, the formal rules for computing variations are the same as those for computing differentials.

In terms of this notation, the condition that $\displaystyle\int_{x_1}^{x_2} I dx$ be stationary is easily written down. It is simply that the integral along y shall yield the same value as that along $y + \delta y$,

$$\int_{x_1}^{x_2} \delta I dx = 0 \qquad (6\text{–}5)$$

This is of course the analogue of the condition in the ordinary calculus that $y(x)$ be stationary, i.e., $dy = 0$. With the use of (3) and (4), eq. (5) becomes

$$\int_{x_1}^{x_2} \left[\frac{\partial I}{\partial y}\delta y + \frac{\partial I}{\partial y_x}\frac{d}{dx}(\delta y) \right] dx = 0$$

The second term of the integrand yields after partial integration

$$-\int_{x_1}^{x_2} \left(\frac{d}{dx}\frac{\partial I}{\partial y_x}\right)\delta y\, dx + \left[\frac{\partial I}{\partial y_x}\,\delta y\right]_{x_1}^{x_2}$$

But the integrated part vanishes at both limits because $\delta y_1 = \delta y_2 = 0$. Hence the stationarity condition becomes

$$\int_{x_1}^{x_2} \left(\frac{\partial I}{\partial y} - \frac{d}{dx}\frac{\partial I}{\partial y_x}\right)\delta y\, dx = 0 \qquad (6\text{-}6)$$

While the vanishing of an integral does not in general imply that the integrand is zero, we may nevertheless conclude here that

$$\frac{\partial I}{\partial y} - \frac{d}{dx}\frac{\partial I}{\partial y_x} = 0 \qquad (6\text{-}7)$$

This is because the parenthesis in (6) is multiplied by an *arbitrary* though infinitesimally small function of x, namely δy. For if the left-hand side of (7) were not zero for every x, it would have to be positive in some range and negative in another range in order to satisfy (6) with a positive δy. We may then choose δy to be positive where the left side of (7) is positive and negative elsewhere, an arrangement which would violate (6). Hence eq. (7) follows and is the condition we are seeking. A function y which satisfies that differential equation is called an *extremal*. Among these extremals the minimizing or maximizing curve y will be found, provided it exists.

Eq. (7) was first derived by Euler; it is called the *Euler equation* associated with the variation problem. It may be written in a different form:

$$\frac{\partial I}{\partial x} - \frac{d}{dx}\left(I - y_x \frac{\partial I}{\partial y_x}\right) = 0 \qquad (6\text{-}7a)$$

which is useful when I does not depend explicitly on x, for then (7a) shows that

$$I - y_x \frac{\partial I}{\partial y_x} = \text{const.}$$

represents an extremal. The identity of (7) and (7a) is at once established by noting:

$$\frac{dI}{dx} = \frac{\partial I}{\partial x} + y_x \frac{\partial I}{\partial y} + y_{xx} \frac{\partial I}{\partial y_x}$$

Examples.

a. *Geodesics.* It is usually taken for granted that a straight line is the shortest distance between two points in a plane. The calculus of variations provides a formal proof of this assertion. The element of distance in Cartesian coordinates is given by $ds^2 = dx^2 + dy^2$. Hence

$$s = \int_{x_1}^{x_2} ds = \int_{x_1}^{x_2} (1 + y_x^2)^{1/2} dx$$

If this is to be a minimum, Euler's equation (7), with $I = (1 + y_x^2)^{1/2}$ must be satisfied. Hence

$$\frac{d}{dx}\left[\frac{\partial}{\partial y_x}(1 + y_x^2)^{1/2}\right] = 0$$

or

$$\frac{y_x}{\sqrt{1 + y_x^2}} = \text{const.}$$

which means $dy/dx = \text{const.}$

The minimizing curve is the straight line passing through the points y_1 and y_2. Had we chosen polar coordinates, the problem would have been to find r as a function of φ such that

$$s = \int_{\varphi_1}^{\varphi_2} (dr^2 + r^2 d\varphi^2)^{1/2} = \int (r^2 + r_\varphi^2)^{1/2} d\varphi$$

is stationary. The Euler equation then reads

$$\frac{r}{(r^2 + r_\varphi^2)^{1/2}} - \frac{d}{d\varphi}\frac{r_\varphi}{(r^2 + r_\varphi^2)^{1/2}} = 0$$

This reduces to

$$\frac{rr_{\varphi\varphi} - 2r_\varphi^2 - r^2}{(r^2 + r_\varphi^2)^{3/2}} = 0$$

The expression on the left is simply the curvature of the curve in polar coordinates; hence the result is the same as before.

The element of distance on the surface of a sphere of radius a is given by

$$ds = a(d\theta^2 + \sin^2\theta d\varphi^2)^{1/2}$$

If we wish to find φ as a function of θ such that s is stationary, we must solve (7) with $I = (1 + \sin^2\theta\varphi_\theta^2)^{1/2}$:

$$\frac{d}{d\theta}\left[\frac{\sin^2\theta\varphi_\theta}{(1 + \sin^2\theta\varphi_\theta^2)^{1/2}}\right] = 0$$

When the bracket is put equal to a constant, c, we get

$$\varphi_\theta = \frac{c\,\text{cosec}^2\,\theta}{(1 - c^2 - c^2\cot^2\theta)^{1/2}}$$

and on integrating

$$\varphi = \alpha - \sin^{-1}(k\cot\theta)$$

α and k being new constants. To interpret this result we write it in Cartesian coordinates, using $z = a \cos \theta$. We have $ak \cot \theta = a \sin (\alpha - \varphi)$, or, on multiplying by $\sin \theta$,

$$kz = x \sin \alpha - y \cos \alpha$$

This represents a plane passing through the origin and hence cutting the surface of the sphere in a great circle. The shortest (and also the longest) distance between two points on the surface of the sphere is the arc of the great circle connecting them!

b. *The Brachistochrone.*

A problem which held the fascination of mathematicians for several decades of the 17th and 18th centuries is that of finding the path on which an object, in the absence of friction, will slide from one given point to another in the shortest (brachistos) time (chronos). John Bernoulli proposed the problem in 1696; both he and his brother James, and also Newton and Leibnitz, found the correct solution. The path, which happens to be a cycloid, is known as the brachistochrone.

Let the particle start from rest at the origin; the terminal point of the motion is $(x_2 y_2)$. In working this problem it is convenient to extend the Y-axis to the right and to measure x downward. Then from the principle of conservation of energy,

$$\tfrac{1}{2}mv^2 = mgx$$

where v is the velocity of the particle at any point of its path, m its mass and g the acceleration of gravity. Hence, since

$$v = \frac{ds}{dt} = \frac{\sqrt{dx^2 + dy^2}}{dt},$$

$$dt = \frac{(1 + y_x^2)^{1/2}}{(2gx)^{1/2}} \, dx$$

The integral to be minimized is therefore

$$\sqrt{2g}\,t = \int_0^{x_2} \left(\frac{1 + y_x^2}{x}\right)^{1/2} dx$$

Euler's equation reads

$$\frac{d}{dx} \frac{y_x}{[x(1 + y_x^2)]^{1/2}} = 0$$

Hence

$$\frac{y_x^2}{x(1 + y_x^2)} = c, \quad y_x = \frac{x}{\left(\dfrac{x}{c} - x^2\right)^{1/2}}$$

If we introduce the constant $2a = 1/c$, integration leads to

$$y = a \cos^{-1}\left(1 - \frac{x}{a}\right) - (2ax - x^2)^{1/2} + c' \qquad (6\text{-}8)$$

But the new constant of integration, c', must be zero in order to make y vanish at $x = 0$.   Eq. (8) represents the equation of an inverted cycloid with its base along Y and its cusp at the origin.   (Cf. Fig. 2.)   The constant a must be so adjusted that the cycloid passes through the point (x_2, y_2).   The path will also be a cycloid if we allow the particle to fall with a finite initial velocity, as the reader may verify.

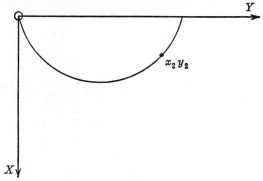

FIG. 6-2

c. *Minimum Surface of Revolution.*

The soap film problem discussed in sec. 2.2i may also be solved by the method outlined above.   Whatever the function y, the surface generated by revolving y about the X-axis has an area

$$2\pi \int_{x_1}^{x_2} y\, ds = 2\pi \int y(1 + y_x^2)^{1/2} dx$$

If this is to be a minimum, eq. (7a) requires that

$$y(1 + y_x^2)^{1/2} - yy_x^2(1 + y_x^2)^{-1/2} = \frac{y}{(1 + y_x^2)^{1/2}} = a$$

$$\frac{dy}{dx} = \sqrt{\frac{y^2}{a^2} - 1}, \quad y = a \cosh\left(\frac{x}{a} + b\right)$$

an expression which is identical with our former solution of the soap film problem.

   Problem.   Solve Example c with the use of equation (7).

   6.2. Several Dependent Variables.—The foregoing simple considerations may be generalized in several obvious ways.   In the present section we shall suppose that the integrand I occurring in the integral to be minimized or maximized is a function of one independent but several *dependent* variables. In almost all examples relevant to this situation the independent variable is the time, while the coordinates are dependent. In view of this fact we shall modify our notation, using t in place of the former x, and x, y, z, etc., in place of the former y. We wish to find the functions $x(t)$, $y(t)$, $z(t)$, $\cdots$ which make the integral

$$\int_{t_1}^{t_2} I(t,x,y,z, \cdots x_t, y_t, z_t, \cdots) \, dt$$

stationary. The Euler condition is desired as before; we must require that

$$\int_{t_1}^{t_2} \delta I \, dt = 0 \qquad (6\text{--}9)$$

But in this case

$$\delta I = \frac{\partial I}{\partial x} \delta x + \frac{\partial I}{\partial y} \delta y + \frac{\partial I}{\partial z} \delta z + \cdots + \frac{\partial I}{\partial x_t} \delta x_t + \frac{\partial I}{\partial y_t} \delta y_t + \frac{\partial I}{\partial z_t} \delta z_t + \cdots$$

In computing the integral (9) we again perform partial integrations in the second group of terms; for example:

$$\int_{t_1}^{t_2} \frac{\partial I}{\partial x_t} \delta x_t dt = \int_{t_1}^{t_2} \frac{\partial I}{\partial x_t} \frac{d}{dt}(\delta x) dt = \left[\frac{\partial I}{\partial x_t} \delta x\right]_{t_1}^{t_2} - \int_{t_1}^{t_2} \frac{d}{dt}\left(\frac{\partial I}{\partial x_t}\right) \delta x \, dt$$

As before, δx vanishes at both limits. Hence (9) becomes

$$\int_{t_1}^{t_2} \left[\left(\frac{\partial I}{\partial x} - \frac{d}{dt}\frac{\partial I}{\partial x_t}\right) \delta x + \left(\frac{\partial I}{\partial y} - \frac{d}{dt}\frac{\partial I}{\partial y_t}\right) \delta y \right. $$
$$\left. + \left(\frac{\partial I}{\partial z} - \frac{d}{dt}\frac{\partial I}{\partial z_t}\right) \delta z + \cdots \right] dt = 0$$

If δx, δy, δz are entirely arbitrary and independent functions of t, each of the parentheses occurring here must vanish separately. Hence we obtain, in place of the one Euler equation (7), as many as there are dependent variables:

$$\left. \begin{array}{l} \dfrac{\partial I}{\partial x} - \dfrac{d}{dt}\dfrac{\partial I}{\partial x_t} = 0; \quad \dfrac{\partial I}{\partial y} - \dfrac{d}{dt}\dfrac{\partial I}{\partial y_t} = 0; \\[3mm] \dfrac{\partial I}{\partial z} - \dfrac{d}{dt}\dfrac{\partial I}{\partial z_t} = 0; \cdots \end{array} \right\} \qquad (6\text{--}10)$$

6.3. Example: Hamilton's Principle.—The elementary formulation of the laws of mechanics is Newton's; it involves in an essential manner the concept of force. Numerous other formulations based on different fundamental ideas, particularly the energy concept, have been proposed through-out the history of the subject.   The most important of these is Hamilton's principle.   It should be regarded not as a consequence of Newton's laws of force (although it can be shown to be consistent with them) but as a parallel fundamental postulate of mechanics which may be useful in cases where Newton's laws are cumbersome in their application.   The principle takes for granted a knowledge of the kinetic energy, T, of the mechanical system as a function of the coordinates and their derivatives, and also of the potential energy, V, as a function of coordinates and possibly the time. From the functional form of T and V it then permits the deduction of the coordinates as functions of the time.

The principle postulates that the integral

$$\int_{t_1}^{t_2} (T - V) dt$$

shall have a stationary value.   The integrand, $T - V$, is called the *Lagrangian function*.   We shall consider only *conservative* mechanical systems, that is, systems for which V is a function of the coordinates only.

Let us first treat the motion of a simple mass point in three dimensions, using rectangular coordinates for its description.   Then

$$T = \tfrac{1}{2}m(x_t^2 + y_t^2 + z_t^2)$$

and,

$$V = V(x,y,z)$$

so that

$$I = \tfrac{1}{2}m(x_t^2 + y_t^2 + z_t^2) - V$$

Eqs. (10) are then seen to be Newton's laws of motion:

$$\frac{d}{dt}(mx_t) = -\frac{\partial V}{\partial x}, \quad \frac{d}{dt}(my_t) = -\frac{\partial V}{\partial y}, \quad \frac{d}{dt}(mz_t) = -\frac{\partial V}{\partial z}$$

An advantage of Hamilton's principle becomes apparent when the problem is such that another system of coordinates is more natural for its solution.   In that case Newton's laws require the transformation of the force components to the new coordinates, which is sometimes inconvenient, while the scalars T and V are more easily transformed.   Thus consider the motion of a particle in a central field of force, that is, $V = V(r)$.   Using polar coordinates we have

$$T = \frac{m}{2}(r_t^2 + r^2\varphi_t^2), \quad V = V(r)$$

Dependent variables are r and φ.   Hence Euler's equations read

$$\frac{d}{dt}(mr_t) - mr\varphi_t^2 = -\frac{\partial V}{\partial r}$$

$$\frac{d}{dt}(mr^2\varphi_t) = 0$$

The first of these is the well known radial equation of the problem of planetary motion $(-\partial V/\partial r = \text{const.}/r^2)$: the term $mr\varphi_t^2$ represents the centripetal force, which appears automatically in this theory.   The second equation is Kepler's second law for it states that $r^2(d\varphi/dt) = \text{const.}$ Its meaning is obvious when it is remembered that the area swept out by the radius vector is $\frac{1}{2}r^2(d\varphi/dt)$.

Turning now to the consideration of more complicated physical systems containing more than one mass point, we first introduce general coordinates, $q_1, q_2, q_3, \cdots q_n$, where n is the number of degrees of freedom. V will be a function of the q's, but it will not depend on the q_t.   The kinetic energy, T, however, will be a function of the q's as well as the q_t's (except when Cartesian coordinates are used).

Hamilton's principle then states that

$$\int_{t_1}^{t_2} \delta[T(q_1 q_2 \cdots q_n, q_{1t} q_{2t} \cdots q_{nt}) - V(q_1 \cdots q_n)]dt = 0$$

Eqs. (10) become[1]

$$\frac{\partial T}{\partial q_i} - \frac{d}{dt}\frac{\partial T}{\partial \dot{q}_i} = \frac{\partial V}{\partial q_i}, \quad i = 1, 2, \cdots n \qquad (6\text{--}11)$$

These are the famous Lagrangian equations of motion, first derived by Lagrange (without the use of the calculus of variations).

To illustrate their applicability and also the use of generalized coordinates we discuss one further example taken from the field of electricity. If q is the charge and $i = q_t$ the current in a simple circuit which has capacitance C and self-inductance L, its total energy at any instant may be shown to be

$$\tfrac{1}{2}Lq_t^2 + \tfrac{1}{2}\frac{q^2}{C}$$

It is clear from the foregoing remarks that the first of these two terms may be regarded as kinetic energy T, the second as potential energy V provided q is chosen as a generalized coordinate.   The intuitive meaning

[1] Here $\dot{q}_i$ has been written for dq_i/dt.

of T and V here becomes lost, as it does in many problems of advanced dynamics. Lagrange's equation for the present case takes the form

$$L\frac{d^2q}{dt^2} + \frac{q}{C} = 0$$

and this will be recognized as the differential equation describing the natural oscillations of an electrical circuit having no resistance. More complicated examples of the application of Lagrange's equations to electrical and indeed even thermal phenomena are available.[2]

Problem. For a simple harmonic oscillator, $V = \frac{1}{2}kx^2$. Use Hamilton's principle to obtain its equation of motion.

6.4. Several Independent Variables.—Next, it is necessary to extend the simple theory so as to permit the integrand to contain several *independent* variables. The problem then is to find a function $u(x,y,z)$ such that

$$\int_{x_1}^{x_2}\int_{y_1}^{y_2}\int_{z_1}^{z_2} I(x,y,z,u,u_x,u_y,u_z)dxdydz$$

is stationary. Here we are treating x, y, z as independent variables, u as the one dependent variable, and we define again: $u_x = du/dx$, etc. As before, we require

$$\int\int\int \delta I dxdydz = 0 \qquad (6\text{--}12)$$

Here δu represents the increment incurred in the passage from the extremal u to some neighboring function U, x, y, and z being held fixed. Hence $\delta x = \delta y = \delta z = 0$. Therefore

$$\delta I = \frac{\partial I}{\partial u}\delta u + \frac{\partial I}{\partial u_x}\delta u_x + \frac{\partial I}{\partial u_y}\delta u_y + \frac{\partial I}{\partial u_z}\delta u_z$$

In evaluating an integral like $\int\int\int \dfrac{\partial I}{\partial u_x}\delta u_x dxdydz$ we first perform the integration with respect to x, obtaining

$$\int_{x_1}^{x_2}\frac{\partial I}{\partial u_x}\frac{d}{dx}\delta u dx = 0 - \int_{x_1}^{x_2}\frac{\partial}{\partial x}\left(\frac{\partial I}{\partial u_x}\right)\delta u dx$$

and (12) reads

$$\int\int\int\left(\frac{\partial I}{\partial u} - \frac{\partial}{\partial x}\frac{\partial I}{\partial u_x} - \frac{\partial}{\partial y}\frac{\partial I}{\partial u_y} - \frac{\partial}{\partial z}\frac{\partial I}{\partial u_z}\right)\delta u dxdydz$$

[2] See Thomson, J. J., "Applications of Dynamics to Physics and Chemistry," Macmillan Co., 1888. A less extensive account may be found in Lindsay and Margenau, "Foundations of Physics," John Wiley and Sons, 1936, pp. 188–212.

The corresponding Euler equation is therefore

$$\frac{\partial I}{\partial u} - \frac{\partial}{\partial x}\frac{\partial I}{\partial u_x} - \frac{\partial}{\partial y}\frac{\partial I}{\partial u_y} - \frac{\partial}{\partial z}\frac{\partial I}{\partial u_z} = 0 \qquad (6\text{-}13)$$

If, in addition to u, there are other dependent variables v, w, etc., eq. (13) is augmented by other equations in which u is replaced by v, w, etc.

Examples.

a. Let us find the function $u(x,y,z)$ which has a minimum average value of the square of its gradient in a certain region of space. Although this requirement seems artificial at first sight, it is nevertheless of considerable significance in electrostatic and quantum-mechanical problems. If

$$\int\int\int (\nabla u)^2 dx dy dz$$

is to be stationary, $I = u_x^2 + u_y^2 + u_z^2$ (cf. Chapter 4 for the definition of the operator ∇), and (13) becomes

$$u_{xx} + u_{yy} + u_{zz} = \nabla^2 u = 0$$

This is Laplace's equation which must be satisfied, for instance, by the electric potential in free space. (Cf. Chapter 7.)

b. *Vibrating String.* Let a string of length l be under tension F. When it executes small vibrations, it suffers the displacement $u(x)$ at right angles to its length, which will be taken along x. For any distortion, l changes to l', and

$$l' = \int_0^l \sqrt{1 + u_x^2}\, dx$$

If the distortion is small, the integrand may be expanded to read $1 + \frac{1}{2}u_x^2$, so that

$$l' = l + \frac{1}{2}\int_0^l u_x^2 dx$$

The potential energy of the entire string will then be

$$V = Fl' - Fl = \frac{1}{2}F\int_0^l u_x^2 dx$$

provided the tension F is not changed by the small displacements $u(x)$. The kinetic energy is, clearly,

$$T = \frac{1}{2}m\int_0^l u_t^2 dx$$

if m represents the mass of the string per unit length, considered constant. Hamilton's principle now states:

$$\int_{t_1}^{t_2} \int_0^l (\tfrac{1}{2}mu_t^2 - \tfrac{1}{2}Fu_x^2)dxdt$$

shall be stationary. The two variables, x and t, are here to be regarded as the independent ones. The Euler equation (13) for this case is the wave equation:

$$u_{tt} = \frac{F}{m} u_{xx}$$

6.5. Accessory Conditions; Lagrangian Multipliers.—Problems sometimes arise in which it is necessary to make an integral stationary while at the same time one or more integrals involving the same variables are to be kept constant. A typical example, discussed below, is that of finding the closed plane curve of given perimeter and maximum area. This example, being one of the earliest to engage mathematical interest, has given this class of problems the name " isoperimetric."

In general, the presence of accessory conditions can be dealt with by means of " Lagrange's method of undetermined multipliers," as follows. We wish to find the stationary value of

$$\int I d\tau$$

provided that

$$\int I_1 d\tau = c_1, \quad \int I_2 d\tau = c_2, \cdots \int I_n d\tau = c_n \qquad (6\text{--}14)$$

All I's contain the same variables; the limits are fixed and identical in all integrations, and the integrations may be multiple; in the latter case $d\tau$ stands for a product of differentials. The c's are understood to be constants.

We introduce a set of n constant parameters, $\lambda_1, \lambda_2, \cdots \lambda_n$, the values of which are not at once specified. It is clear that, if $\int I d\tau$ is stationary,

$$\int K d\tau$$

where $K = I + \lambda_1 I_1 + \lambda_2 I_2 + \cdots + \lambda_n I_n$, is also stationary whatever the values of the λ's, because of (14). We are thus confronted with a problem similar to the foregoing, the minimization (or maximization) of a single integral, but with a modified integrand: I must be replaced by

$K = I + \sum_i \lambda_i I_i$. If now the same steps are pursued in evaluating $\int \delta K d\tau$

as were outlined in sec. 6.1, we arrive at the equivalent of eq. (6) in which I is now replaced by K. But the passage from (6) to (7) is now obstructed because δy is no longer an *arbitrary* function: the variations must be in accord with the relations (14). One may say that δy has lost n degrees of freedom. But here the unspecified character of the λ's comes to our rescue. They are precisely n in number and can be so adjusted that the parentheses vanish.[3] Hence the transition from eq. (6) to (7) is permitted in this case as well. The extremals must satisfy Euler's equation

$$\frac{\partial K}{\partial y} - \frac{d}{dx}\frac{\partial K}{\partial y_x} = 0 \qquad (6\text{--}15)$$

or its equivalent (7a). If there are several dependent and independent variables, eqs. (10) and (13) take the place of (15).

In solving Euler's equation the λ's which are now presumably fixed but unknown appear as constants in the extremals. They may be eliminated formally by means of conditions (14), but their meaning can usually be recognized more directly at some stage of the solution.

Examples.

a. To find the plane curve of fixed perimeter and maximum area. We seek that $r(\varphi)$ which maximizes

$$A = \tfrac{1}{2}\int_0^{2\pi} r^2 d\varphi$$

and has a fixed

$$S = \int_0^{2\pi} (r^2 + r_\varphi^2)^{1/2} d\varphi$$

Here

$$K = \tfrac{1}{2}r^2 + \lambda(r^2 + r_\varphi^2)^{1/2}$$

so that (15) reads:

$$r + \lambda r(r^2 + r_\varphi^2)^{-1/2} - \frac{d}{d\varphi}[\lambda r_\varphi(r^2 + r_\varphi^2)^{-1/2}] = 0$$

This leads to

$$\frac{rr_{\varphi\varphi} - 2r_\varphi^2 - r^2}{(r^2 + r_\varphi^2)^{3/2}} = \frac{1}{\lambda}$$

The left of this equation will be recognized as the curvature, $1/\rho$, of the curve. This is to be constant, hence the curve is a circle with radius $\rho = \lambda$.

[3] For a more detailed discussion of Lagrange's method of undetermined multipliers see Page, L., " Introduction to Theoretical Physics," Third Edition, D. Van Nostrand Co., New York, 1952.

b To prove that the sphere is the solid figure of revolution which, for a given surface area, has maximum volume. The area is

$$A = 2\pi \int yds = 2\pi \int_0^a y(1 + y_x^2)^{1/2}dx$$

the volume:

$$V = \pi \int_0^a y^2 dx$$

Therefore

$$K = y^2 + \lambda y(1 + y_x^2)^{1/2}$$

since we are here permitted to drop constant factors. As K does not contain x explicitly, it is convenient to use (7a) instead of (7) or (15):

$$\frac{\partial K}{\partial x} - \frac{d}{dx}\left(K - y_x \frac{\partial K}{\partial y_x}\right) = 0$$

whence

$$K - y_x \frac{\partial K}{\partial y_x} = y^2 + \lambda y(1 + y_x^2)^{1/2} - \lambda y y_x^2(1 + y_x^2)^{-1/2} = c$$

But clearly, $y = 0$ at $x = 0$ and at $x = a$, which can only be true if $c = 0$. Hence

$$y^2 + \lambda y(1 + y_x^2)^{-1/2} = 0$$

or

$$y = -\lambda(1 + y_x^2)^{-1/2}$$

Solving this for y_x we obtain

$$\frac{dy}{dx} = \frac{1}{y}\sqrt{\lambda^2 - y^2}$$

which on integration leads to

$$-\sqrt{\lambda^2 - y^2} = x - x_0$$

or,

$$(x - x_0)^2 + y^2 = \lambda^2$$

We note that the figure is a sphere with center on the X-axis at x_0 and of radius λ.

It is possible to work this problem without the use of Lagrangian multipliers by means of an ingenious method due to Euler. He uses in place of the independent variable x a new one, ξ, which measures essentially the area of revolution formed by the arc $y(x)$ between $x = 0$ and the variable point $x = x$:

$$\xi = \int_0^x y(1 + y_x^2)^{1/2}dx$$

In terms of this variable,

$$dx = \left[\left(\frac{d\xi}{y} \right)^2 - dy^2 \right]^{1/2}$$

so that

$$V = \pi \int^b y(1 - y_\xi^2 y^2)^{1/2} d\xi \qquad (6\text{-}16)$$

Here b represents the value of ξ when $x = a$, that is, the given area divided by 2π. By keeping b fixed the accessory condition $A = $ const. is automatically satisfied. This method, while very elegant, cannot be applied generally.

The stationarity condition for (16), if written in the form (7b), yields

$$y(1 - y^2 y_\xi^2)^{1/2} + y^3 y_\xi^2 (1 - y^2 y_\xi^2)^{-1/2} = y(1 - y^2 y_\xi^2)^{-1/2} = c \qquad (6\text{-}17)$$

whence

$$y_\xi = \frac{dy}{d\xi} = \frac{1}{y} \sqrt{1 - \frac{y^2}{c^2}}$$

After integration,

$$d - \frac{\xi}{c^2} = \left[1 - \left(\frac{y}{c} \right)^2 \right]^{1/2}$$

The new constant d must be 1 if the curves are to pass through $\xi = y = 0$. To obtain the result in terms of x and y, we substitute for y_ξ in (17) the value obtained by solving

$$y_x = y(1 + y_x^2)^{1/2} y_\xi$$

Eq. (17) then reads

$$y(1 + y_x^2)^{1/2} = c$$

and this is precisely the equation solved above with $-c = \lambda$.

c. *Wave equation.* In sec. 6.4 we have seen that Laplace's equation is the necessary condition that the average of the square of the gradient of a function shall have a stationary value. If the same quantity is to be made stationary, but with the additional requirement that $\int u^2 dx dy dz$ shall have a fixed value, another interesting equation results. In that case

$$I = \left(\frac{\partial u}{\partial x} \right)^2 + \left(\frac{\partial u}{\partial y} \right)^2 + \left(\frac{\partial u}{\partial z} \right)^2 ; \quad I_1 = u^2$$

The integral to be minimized is therefore $\int (I + \lambda I_1) dxdydz$. Euler's equation [in the form (13)] then reads

$$\frac{\partial^2 u}{\partial x^2} + \frac{\partial^2 u}{\partial y^2} + \frac{\partial^2 u}{\partial z^2} - \lambda u = 0$$

which is a special form of the wave equation, namely, that describing sinusoidal waves of a single frequency. (Cf. Chapter 7.) Such a wave may therefore be characterized as a disturbance in which the displacement u has a fixed mean square value and at the same time a minimum square gradient.

6.6. Schrödinger Equation.—The fundamental equation of quantum mechanics (sec. 11.9) can be derived from a variation principle, as will now be shown. We define an operator, known as the *Hamiltonian operator*, as follows:

$$H \equiv -k\nabla^2 + V(x,y,z)$$

The physical meaning of k is seen from the relation $k = h^2/8\pi^2 m$ where h is Planck's constant and m the mass of the particle whose motion is considered; V is its potential energy. We now seek a function ψ, possibly complex, which satisfies the following two conditions:

$$\int \int \int \psi^*(H\psi) dxdydz \qquad (6\text{-}18a)$$

shall be stationary;

$$\int \int \int \psi^*\psi dxdydz = 1 \qquad (6\text{-}18b)$$

The integrations are taken over fixed domains of x, y, and z. It will be supposed, furthermore, that the permissible functions ψ and ψ^* either vanish sufficiently strongly at the boundaries of the volume of integration, or take on the same values and derivatives at corresponding points on opposite boundaries.

When this is true, the following transformation may be made:

$$\int \psi^* \frac{\partial^2 \psi}{\partial x^2} dx = \left[\psi^* \frac{\partial \psi}{\partial x} \right]_{x_1}^{x_2} - \int \frac{\partial \psi^*}{\partial x} \cdot \frac{\partial \psi}{\partial x} dx$$

The integrated part vanishes. As a consequence

$$\int \int \int \psi^* \nabla^2 \psi dxdydz = - \int \int \int (\nabla \psi^*) \cdot \nabla \psi dxdydz$$

and condition (18a) may be modified to read:

$$\int\int\int \delta[k(\nabla\psi^*)\cdot(\nabla\psi) + V(x,y,z)\psi^*\psi]dxdydz = 0$$

The function K which appears in Euler's equation [(15) but generalized in accordance with (13) to take care of the fact that there are now three independent variables] is

$$K = k(\psi_x^*\psi_x + \psi_y^*\psi_y + \psi_z^*\psi_z) + V\psi^*\psi - \lambda\psi^*\psi$$

Euler's equations are (ψ^* and ψ are both dependent variables !)

$$\left. \begin{array}{l} \dfrac{\partial K}{\partial \psi} - \dfrac{\partial}{\partial x}\dfrac{\partial K}{\partial \psi_x} - \dfrac{\partial}{\partial y}\dfrac{\partial K}{\partial \psi_y} - \dfrac{\partial}{\partial z}\dfrac{\partial K}{\partial \psi_z} = 0 \\[3mm] \dfrac{\partial K}{\partial \psi^*} - \dfrac{\partial}{\partial x}\dfrac{\partial K}{\partial \psi_x^*} - \dfrac{\partial}{\partial y}\dfrac{\partial K}{\partial \psi_y^*} - \dfrac{\partial}{\partial z}\dfrac{\partial K}{\partial \psi^*} = 0 \end{array} \right\}$$

They reduce to

$$-k(\psi_{xx} + \psi_{yy} + \psi_{zz}) + V\psi = \lambda\psi \tag{6-19}$$

and a similar equation for ψ^*. To identify the constant λ, we note that eq. (19) may be written

$$H\psi = \lambda\psi$$

If we multiply this equation by ψ^* and integrate over x, y, z the left side becomes the stationary integral (18a), which will be denoted by E. The right is λ in view of (18b). Hence $\lambda = E$. With this substitution for λ, eq. (19) is Schrödinger's equation.

This result is worth summarizing. Schrödinger's equation serves the purpose of selecting the extremals ψ which make $\int\int\int \psi^*(H\psi)dxdydz$ stationary, provided $\int\int\int \psi^*\psi dxdydz$ is held constant. If the latter constant is unity, then, $\int\int\int \psi^*(H\psi)dxdydz$ is the energy which appears in the Schrödinger problem. Further inspection shows the energy to be a minimum rather than a maximum in most cases of physical interest. Upon these results is based one of the most powerful methods of obtaining approximate solutions of eq. (19). (Cf. sec. 11.18.)

6.7. Concluding Remarks.—In concluding this chapter, we note a few possible generalizations of the theory given here. In the first place, one may remove the restriction that $\delta y = 0$ at the limits of integration. This means, with reference to Fig. 1, that the curves $y(x)$ and $Y(x)$ do not have the same termini.   The integrated term which appears in the partial integration leading to eq. (6) will then no longer vanish, and there arise three conditions in place of eq. (7):

$$\frac{\partial I}{\partial y} - \frac{d}{dx}\frac{\partial I}{\partial y_x} = 0; \quad \left[\frac{\partial I}{\partial y_x}\right]_{x_1} = 0; \quad \left[\frac{\partial I}{\partial y_x}\right]_{x_2} = 0$$

The second and third of these then serve to fix the arbitrary constants in the solution of Euler's equation.

A further generalization is needed when the limits x_1 and x_2 themselves are no longer fixed. Whenever this happens, introduction of a new parameter, in terms of which both x and y may be expressed, reduces the problem to the forms here discussed.[4] The Principle of Least Action involves a variation problem with variable limits. Since Hamilton's principle is in general more powerful the former, in spite of its historical interest, will here be omitted.

When the integrand I involves higher derivatives than the first, no great complications arise. The Euler equation then contains additional terms.[5] The point where our simple treatment has been most deficient is in its omission of all considerations establishing the actual existence of maximizing and minimizing curves. It will be recalled that Euler's equations are merely necessary conditions. They furnish no assurance whatever that the curves sought are indeed present among the extremals. For these more mathematical questions we refer the reader to the treatises by Bolza, Bliss, and Kneser.

REFERENCES

Other good general texts on the calculus of variations are:
Bliss, G. A., " Calculus of Variations," Open Court, La Salle, 1925.
Bolza, O., " Vorlesungen über Variationsrechnung," Teubner, Leipzig, 1909.
Forsyth, A. R., " Calculus of Variations," Cambridge University Press, 1927.
For modern applications, see
Lanczos, C., " The Variational Principles of Dynamics," University of Toronto Press, 1949.
Morse, P. M., and Feshbach, H., " Methods of Theoretical Physics," McGraw-Hill Book Co., 1953.
Wentzel, G., " Quantum Theory of Fields," Interscience Publishers, Inc., New York, 1949.

[4] See Byerly, W. E., " Introduction to the Calculus of Variations," Harvard University Press, 1917.

[5] See Kneser, A., " Lehrbuch der Variationsrechnung," Vieweg & Sohn, Braunschweig, 1925.

CHAPTER 7

PARTIAL DIFFERENTIAL EQUATIONS OF CLASSICAL PHYSICS

7.1. General Considerations.—The general theory of partial differential equations is well beyond the scope of this book and will not be developed in a systematic way.[1] Attention will here be limited to a small number of partial differential equations which are of frequent occurrence, almost all of which may be resolved by a powerful method known as the separation of variables. Before we proceed to consider specific examples, however, a few remarks about the meaning and variety of the solutions are in order.

The simplest type of an *ordinary* differential equation, that of the first order, has a general solution which contains one arbitrary constant; geometrically it may be interpreted as a set of plane curves labeled by different values of the arbitrary constant. In particular, if the equation is linear, there is but one curve passing through a given point, and this is uniquely specified when the value of y for some value of x is prescribed.

The simplest type of partial differential equation is one with two independent variables (x and y), and the dependent variable (z), which is linear and of the first order. Its solutions represent, geometrically, a set of surfaces constructed over the X–Y plane. The question may be asked: Is one such surface uniquely determined by requiring that it include a given point? If this were true, the manifold of solutions of the differential equation, $z(x,y)$, would reduce to a single surface when it is specified that the solution shall contain that point.

This, however, is not the case. For consider the simple equation $\partial z/\partial x + \partial z/\partial y = 0$. It is clear that any function of the form $z = \varphi(x - y)$ will satisfy it. This function is not uniquely determined by fixing one point of it. The origin, for instance, is contained in all surfaces $z = c(x - y)$, and yet every different value of c defines a different surface.

Neither does a prescribed *curve* fix a surface. For let it be required that the solution z of the partial equation above shall pass through the line $x = y$ in the X–Y plane. This is certainly accomplished by taking $z = (x - y)^n$, but there is an infinite number of such surfaces depending on the parameter n. It is clear from these elementary considerations that

---

[1] A more complete discussion is found in the references at the end of this chapter.

in dealing with the solutions of partial differential equations we are confronted with a variety of functions which far transcends the degree of complexity encountered in connection with ordinary differential equations. In fact one must not be surprised to find that the complete geometric specification of a solution of a partial equation even of the simplest type usually requires the fixation of an infinite number of parameters.

7.2. Laplace's Equation.—An equation which arises in almost all branches of analysis is Laplace's:

$$\nabla^2 V = 0 \qquad\qquad (7\text{-}1)$$

Its intuitive meaning was discussed in the chapter on the calculus of variation (sec. 6.4), where eq. (6–1) was shown to be equivalent to the postulate that V shall have the least mean gradient. The function V satisfying (1) may be said to be the " smoothest " of all functions. This is obvious when Laplace's equation is solved in one dimension, for then it simply reads: $d^2V/dx^2 = 0$ and has as its solutions all straight lines.

To indicate briefly the range of application of eq. (1) we state three instances in which it occurs:

a. A fundamental theorem of function theory states:

Let $z = x + iy$; then the function $f(z)$ takes the form

$$f(z) = u(x,y) + iv(x,y)$$

wherein u, v, x, y are all real, if and only if the functions u and v satisfy:

$$\nabla^2 u = 0, \quad \nabla^2 v = 0$$

b. In sec. 4.12 it was shown that the velocity $\mathbf{v}$ of an indestructible fluid, as a function of space coordinates and the time, must be a solution of the equation of continuity, which reads

$$\frac{\partial \rho}{\partial t} + \nabla \cdot (\rho \mathbf{v}) = 0$$

If the fluid is incompressible, its density ρ is constant, and the equation reads

$$\nabla \cdot \mathbf{v} = 0$$

If, furthermore, the motion is irrotational, the velocity vector is the gradient of a scalar function V, known as the velocity potential: $\mathbf{v} = -\nabla V$, and the equation of continuity thus becomes equivalent to Laplace's: $\nabla^2 V = 0$.

c. The electrostatic potential in a region of space not occupied by charges satisfies Laplace's equation.

Before discussing a partial differential equation of this general form one must realize, of course, that its solutions for different numbers of dimensions (independent variables) are quite different; moreover, that the form of the solution even for the same number of dimensions will be different in different systems of coordinates.

7.3. Laplace's Equation in Two Dimensions.—a. *Rectangular Coordinates.* The equation reads:

$$\frac{\partial^2 V}{\partial x^2} + \frac{\partial^2 V}{\partial y^2} = 0 \tag{7-2}$$

A method, not of universal applicability but suitable for this particular problem, involves the transformation to a new set of independent variables:

$$\xi = x + iy, \quad \eta = x - iy$$

In terms of these,

$$\frac{\partial^2}{\partial x^2} = \frac{\partial^2}{\partial \xi^2} + 2\frac{\partial^2}{\partial \xi \partial \eta} + \frac{\partial^2}{\partial \eta^2}; \quad \frac{\partial^2}{\partial y^2} = -\frac{\partial^2}{\partial \xi^2} + 2\frac{\partial^2}{\partial \xi \partial \eta} - \frac{\partial^2}{\partial \eta^2}$$

so that

$$\nabla^2 V = 4\frac{\partial^2 V}{\partial \xi \partial \eta} = 0$$

Clearly, this equation admits both $V = f(\xi)$ and $V = f(\eta)$ as solutions, hence

$$V = f_1(\xi) + f_2(\eta) = f_1(x + iy) + f_2(x - iy)$$

where f_1 and f_2 are any two functions which are twice differentiable. The reader will hardly fail to see the connection between this result and the statement above concerning the functions of a complex variable.

For many problems another form of solution, obtainable by the method of separation of variables, is more satisfactory. Let us make the assumption, justifiable by its success, that V may be written in the form

$$V = X(x) \cdot Y(y) \tag{7-3}$$

where X and Y are functions of only one independent variable, x and y, respectively. When (3) is substituted in (2) there results, after division by V,

$$\frac{X''}{X} + \frac{Y''}{Y} = 0 \tag{7-4}$$

an equation in which primes denote differentiation of a function with respect to its own variable. If (4) is to have a solution at all, then each term on the left must separately be equal to a constant; for a change in x would not alter the value of Y''/Y, and a change in y would not affect

X''/X. One may therefore conclude:

$$\frac{X''}{X} = k^2, \quad \frac{Y''}{Y} = -k^2 \tag{7-5}$$

where the constant parameter k^2, written in this form for convenience, may have any value, real or complex. These are two *ordinary* equations which may easily be solved by the methods of Chapter 2. Eq. (5) leads at once to

$$X = c_1 e^{\pm kx}, \quad Y = c_2 e^{\pm iky}$$

Hence a solution of (2), characterized by a given value of the parameter k, will be

$$V_k = c_k e^{\pm k(x \pm iy)} \tag{7-6}$$

Since (2) is a linear equation, a sum of expressions like (6) is also a solution. Hence a more general solution is

$$V = \sum_k c_k e^{\pm k(x \pm iy)}$$

or even

$$V = \int c(k) e^{\pm k(x \pm iy)} dk \tag{7-6a}$$

For the value $k = 0$ the result is of a more special form. Eq. (5) then leads to

$$X = a_1 x + a_2, \quad Y = b_1 y + b_2$$

so that

$$V = axy + cx + dy + e \tag{7-7}$$

Which of the solutions, (6), (6a), or (7), is to be chosen depends entirely on the nature of the problem at hand. (Cf. examples.)

b. *Polar Coordinates.* Laplace's equation reads:

$$\frac{\partial^2 V}{\partial \rho^2} + \frac{1}{\rho} \frac{\partial V}{\partial \rho} + \frac{1}{\rho^2} \frac{\partial^2 V}{\partial \varphi^2} = 0 \tag{7-8}$$

Using again the method of separation of variables, we put

$$V = P(\rho) \Phi(\varphi)$$

When this is substituted into (8) there results, after multiplication by ρ^2/V,

$$\rho^2 \frac{P''}{P} + \rho \frac{P'}{P} + \frac{\Phi''}{\Phi} = 0 \tag{7-9}$$

Here the first two terms are independent of φ, the third is independent of ρ. Hence we may write

$$\rho^2 \frac{P''}{P} + \rho \frac{P'}{P} = k^2; \qquad \frac{\Phi''}{\Phi} = -k^2$$

The solution of the first equation is at once seen to be $P = \rho^{\pm k}$, that of the second, $\Phi = e^{\pm ik\varphi}$. Hence

$$V_k = c_k \rho^{\pm k} e^{\pm ik\varphi} \tag{7-10}$$

or, more generally,

$$V = \sum_k c_k \rho^{\pm k} e^{\pm ik\varphi} \tag{7-10a}$$

For $k = 0$, (9) becomes

$$P'' + \frac{P'}{\rho} = 0, \quad \Phi'' = 0$$

When integrated once, the first of these yields $P' = c\rho^{-1}$; after another integration $P = a_1 \ln \rho + a_2$. On the other hand, $\Phi'' = 0$ leads to $\Phi = b_1\varphi + b_2$. Hence a particular solution is

$$V = (a_1 \ln \rho + a_2)(b_1\varphi + b_2) \tag{7-11}$$

Again, further information must be available before a special one of these results can be selected as a suitable solution of a given problem.

7.4. Laplace's Equation in Three Dimensions.[2]—a. *Rectangular Coordinates.* An application of the method of separation of variables to

$$\frac{\partial^2 V}{\partial x^2} + \frac{\partial^2 V}{\partial y^2} + \frac{\partial^2 V}{\partial z^2} = 0$$

follows precisely along the lines of sec. 7.3a. We put $V = X(x)Y(y)Z(z)$ and obtain

$$\frac{X''}{X} + \frac{Y''}{Y} + \frac{Z''}{Z} = 0$$

Each of these terms must separately equal a constant, and the sum of these constants (which we write as k_1^2, k_2^2, k_3^2) must vanish. Thus

$$V_{k_1 k_2 k_3} = e^{k_1 x + k_2 y + k_3 z}, \quad k_1^2 + k_2^2 + k_3^2 = 0 \tag{7-12}$$

If k_1, k_2, or k_3 is zero, the corresponding factor in (12) must be replaced by $a_1 x + a_2$, etc. A more general solution would be

$$V = \sum_{k_1 k_2 k_3} c_{k_1 k_2 k_3} e^{k_1 x + k_2 y + k_3 z} \tag{7-12a}$$

[2] A solution of Laplace's equation in three dimensions is often called an " harmonic " function.

In this connection it is sometimes convenient to regard k_1, k_2, k_3 formally as the components of a vector **k**. Eq. (12) may then be written

$$V_{\mathbf{k}} = c(\mathbf{k})e^{\mathbf{k}\cdot\mathbf{r}}, \quad |k| = 0$$

b. *Cylindrical Coordinates.* In accordance with the results of Chapter 5,

$$\nabla^2 V = \frac{\partial^2 V}{\partial \rho^2} + \frac{1}{\rho}\frac{\partial V}{\partial \rho} + \frac{\partial^2 V}{\partial z^2} + \frac{1}{\rho^2}\frac{\partial^2 V}{\partial \varphi^2} = 0$$

Put

$$V = \mathrm{P}(\rho)Z(z)\Phi(\varphi)$$

substitute, and divide by V. The result is

$$\frac{\mathrm{P}''}{\mathrm{P}} + \frac{1}{\rho}\frac{\mathrm{P}'}{\mathrm{P}} + \frac{1}{\rho^2}\frac{\Phi''}{\Phi} + \frac{Z''}{Z} = 0$$

Clearly, the last term on the left must be constant; let us put it equal to $-k^2$. Then

$$Z = c_1 e^{\pm ikz} \tag{7-13}$$

The remaining equation,

$$\frac{\mathrm{P}''}{\mathrm{P}} + \frac{1}{\rho}\frac{\mathrm{P}'}{\mathrm{P}} + \frac{1}{\rho^2}\frac{\Phi''}{\Phi} - k^2 = 0$$

when multiplied by ρ^2, separates again into two equations:

$$\frac{\Phi''}{\Phi} = -l^2, \quad \rho^2 \mathrm{P}'' + \rho \mathrm{P}' - (k^2\rho^2 + l^2)\mathrm{P} = 0$$

The first has the solution

$$\Phi = c_2 e^{\pm il\varphi}$$

the second turns into *Bessel's* differential equation (2-57) when the substitution $ik\rho = x$ is made, for it then reads:

$$x^2 \frac{d^2\mathrm{P}}{dx^2} + x\frac{d\mathrm{P}}{dx} + (x^2 - l^2)\mathrm{P} = 0$$

The solution of this equation was discussed in sec. 2.14. It will here be denoted by Z_l. Collecting these results we have

$$V_{kl} = c_{kl}Z_l(ik\rho)e^{\pm i(kz \pm l\varphi)} \tag{7-14}$$

When $l = 0$, $\Phi = a_1\varphi + a_2$; hence we obtain as another solution of lesser generality than (14) the expression:

$$V_{k0} = c_{k0}Z_0(ik\rho)e^{\pm ikz}(a_1\varphi + a_2) \tag{7-14a}$$

When $k = 0$, $Z = (b_1 z + b_2)$ instead of the function (13). The equation for P takes the form

$$\rho^2 P'' + \rho P' - l^2 P = 0$$

which was already encountered in sec. 3b. It has the solution $\rho^{\pm l}$. Hence

$$V_{0l} = \rho^{\pm l}(b_1 z + b_2)e^{\pm il\varphi} \tag{7-14b}$$

Finally, when both l and k are zero, the solution may be seen to take the form

$$V_{00} = (a_1 \ln \rho + a_2)(b_1 z + b_2)(c_1 \varphi + c_2) \tag{7-14c}$$

The most general function satisfying Laplace's equation is a superposition of solutions (14)–(14c).

c. *Polar (Spherical) Coordinates.* As was shown in the chapter on coordinate systems (sec. 5.4), the equation $\nabla^2 V = 0$, when transformed to polar coordinates, reads:

$$\frac{1}{r^2}\frac{\partial}{\partial r}\left(r^2 \frac{\partial V}{\partial r}\right) + \frac{1}{r^2 \sin\theta}\frac{\partial}{\partial \theta}\left(\sin\theta \frac{\partial V}{\partial \theta}\right) + \frac{1}{r^2 \sin^2\theta}\frac{\partial^2 V}{\partial \varphi^2} = 0 \tag{7-15}$$

Multiplication by $r^2 \sin^2\theta$ will isolate the term $\partial^2 V/\partial\varphi^2$ as the only one depending on φ from the remainder of the equation. If, therefore, we put it equal to $-m^2$ so that

$$\Phi = ce^{\pm im\varphi} \tag{7-16}$$

(V being written as $R(r) \cdot \Theta(\theta) \cdot \Phi(\varphi)$), then eq. (15) takes the form

$$\frac{\sin^2\theta}{R}\frac{d}{dr}\left(r^2 R'\right) + \frac{\sin\theta}{\Theta}\frac{d}{d\theta}(\sin\theta\Theta') - m^2 = 0$$

When this is divided through by $\sin^2\theta$ the terms involving r are cleanly separated from those involving θ. Hence we obtain

$$\frac{1}{\Theta}\frac{1}{\sin\theta}\frac{d}{d\theta}(\sin\theta\Theta') - \frac{m^2}{\sin^2\theta} + c = 0 \tag{7-17}$$

$$\frac{1}{R}\frac{d}{dr}(r^2 R') - c = 0 \tag{7-18}$$

where c denotes the same constant in both equations. It will prove convenient to write this constant in the form $c = l(l + 1)$. Let us now make the substitution $\cos\theta = x$ in eq. (17), obtaining (after multiplication by Θ)

$$(1 - x^2)\frac{d^2\Theta}{dx^2} - 2x\frac{d\Theta}{dx} + \left[l(l + 1) - \frac{m^2}{1 - x^2}\right]\Theta = 0 \tag{7-19}$$

This, however, is none other than the differential equation (cf. eq. 2–41) for associated spherical harmonics discussed in sec. 2.11. Special solutions were studied in sec. 3.6. They were written in the form

$$\Theta = P_l^m(x)$$

It must here be noted that these functions do not represent the *general* solution of eq. (19), but a particular one having the property of being finite for all values of x between -1 and $+1$, including these limits. In most physical problems this is a condition naturally to be imposed on the solution of Laplace's equation; there are cases, however, in which a more general solution of (19) must be chosen. It was also found in sec. 2.11 that the constant l must, for the sake of finiteness, be a positive integer. We shall restrict the present consideration to problems in which these conditions hold, and assume

$$\Theta = P_l^m(\cos\theta) \tag{7–20}$$

This expression has no meaning unless m, also, is a positive integer. Again, the nature of most physical problems imposes this requirement. For if V represents the distribution of any physical quantity in space, it must obviously be periodic in φ and have a period of 2π, since otherwise $V(\varphi)$ and $V(2\pi + \varphi)$ would have different values although φ and $2\pi + \varphi$ denote the same angle. But the function (16) does not possess this periodicity unless m is an integer.

The function R is now easily obtained by solving (18) which reads on expansion:

$$r^2 R'' + 2rR' - l(l+1)R = 0$$

Its solution is obviously of the form $R = r^\alpha$, and on substitution we find

$$\alpha(\alpha - 1) + 2\alpha - l(l+1) = 0$$

so that α is either l or $-(l+1)$. Hence

$$R = a_1 r^l + a_2 r^{-l-1} \tag{7–21}$$

In view of (16), (20) and (21) we conclude that a solution of Laplace's equation in polar coordinates has the form

$$V_{lm} = (a_1 r^l + a_2 r^{-l-1})P_l^m(\cos\theta)e^{\pm im\varphi} \tag{7–22}$$

and the general solution will be a superposition of any number of such functions.

Other systems of coordinates in which the equation $\nabla^2 V = 0$ can be solved by the method of separation of variables are listed in Chapter 5. It is felt, however, that the foregoing special cases illustrate the procedure.

EXAMPLES OF SOLUTIONS OF LAPLACE'S EQUATION

7.5. Sphere Moving through an Incompressible Fluid without Vortex Formation.—Since the motion of the liquid is irrotational, its velocity, **v**, at every point is the gradient of a scalar potential, V, which satisfies Laplace's equation. Thus

$$\mathbf{v} = -\nabla V, \quad \text{and} \quad \nabla^2 V = 0$$

Which of all the solutions derived above is to be chosen, depends entirely on the boundary conditions of the problem. These are, clearly:

a. The radial velocity of the fluid at the surface of the sphere of radius r_0 shall be equal to the velocity of the sphere times the cosine of the angle which r makes with the direction of motion of the sphere. Taking the latter as the polar axis, we have

$$-\frac{\partial V}{\partial r}\bigg|_{r=r_0} = v_0 \cos \theta \tag{a}$$

b. The distant portions of the liquid are not affected by the motion of the sphere. Hence

$$\frac{\partial V}{\partial r}\bigg|_{r\to\infty} = 0 \tag{b}$$

The form of these conditions at once prescribes the use of polar coordinates. The solution is, therefore, of the form (22). To satisfy (a) we must put the *angular* part of this expression equal to $\cos\theta$; there is no dependence on φ at all. The only possible value of m which produces freedom from φ is zero, and of all the functions $P_l^0 (\cos\theta)$, only $P_1^0 (\cos\theta)$ is equal to $\cos\theta$. Hence $l = 1$. Condition (a) now states:

$$-\frac{\partial}{\partial r}(a_1 r + a_2 r^{-2})\bigg|_{r=r_0} \cos\theta = v_0 \cos\theta$$

whence

$$-a_1 + 2a_2 r_0^{-3} = v_0$$

But condition (b) cannot be satisfied unless $a_1 = 0$. Therefore $2a_2 r_0^{-3} = v_0$. Eq. (22) has thus been reduced to

$$V = \frac{v_0 r_0^3}{2r^2} \cos\theta$$

This represents the velocity potential for the case in question.

7.6. Simple Electrostatic Potentials.—As a matter of illustration, we consider the simplest electrostatic potentials from the point of view of Laplace's equation: that due to a charged conducting sphere, a uniformly charged cylinder (wire) of infinite length, and a uniformly charged infinite plane.

a. The boundary condition in the first case is obviously $V(r_0,\theta,\varphi) = V_0$, a constant, provided we write r_0 for the radius of the sphere. Since spherical polar coordinates are used in describing this condition, the general solution of Laplace's equation must be taken in the form (22). The condition also prescribes that V shall be independent of θ and φ, for otherwise $V(r_0,\theta,\varphi)$ could not be constant. Hence m and l must both be zero. We conclude, therefore, that $V = a_1 + a_2/r$, and since $a_1 + a_2/r_0 = V_0$, we find on eliminating a_2:

$$V = a_1\left(1 - \frac{r_0}{r}\right) + \frac{V_0 r_0}{r}$$

If we require in addition that V be zero at $r = \infty$ (which would be true if the potential were produced entirely by the charged sphere) the constant $a_1 = 0$.

b. The boundary condition in the case of the cylinder reads $V(\rho_0,z,\varphi) = V_0$, ρ_0 denoting the radius of the cylinder. Solution (14) is now relevant; but the observation that V is independent of φ and z leads at once to (14c) with $b_1 = c_1 = 0$. Hence we have $V = a_1 \ln \rho + a_2$. On eliminating a_2 by means of the boundary condition we find

$$V = V_0 + a_1 \ln \frac{\rho}{\rho_0}$$

The constant a_1 can be determined only when further facts, e.g., the charge density on the cylinder, are known. (In fact, $a_1 = -2\lambda$ where λ is linear charge density.)

c. In the case of the charged plane we require $V(x,y,0) = V_0$, supposing $z = 0$ to define the plane. This leads at once to a solution of the form (12), but with $k_1 = k_2 = 0$; for otherwise V would depend on x and y. Since then k_3 must also vanish,

$$V = (a_1 x + a_2)(b_1 y + b_2)(c_1 z + c_2)$$

Again, to satisfy the boundary condition, $a_1 = b_1 = 0$, so that

$$V = c_1 z + V_0$$

The constant c_1 can be eliminated when the charge density on the plane is known. ($c_1 = -4\pi\sigma$ if σ is surface density of charge.)

All these results could have been obtained much more simply by applying Gauss' law of electrostatics; our purpose here was to exhibit them as solutions of Laplace's equation.

Problem. To find the potential produced when a conducting sphere is placed in an originally uniform field of strength E_0, extending along the Z-axis. Use as boundary conditions:

$$V = 0 \text{ at } r = r_0 \quad \text{(radius of sphere)}$$
$$V = -E_0z = -E_0r \cos \theta \text{ at } r \to \infty$$

Ans.
$$V = -E_0r \cos \theta \left[1 - \left(\frac{r_0}{r} \right)^3 \right]$$

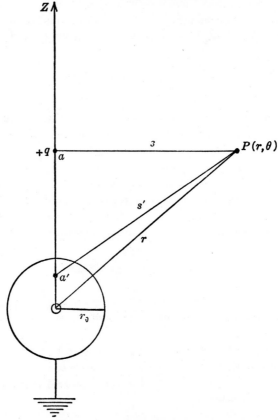

FIG. 7–1

7.7. Conducting Sphere in the Field of a Point Charge.—We wish to find the potential at P due to a point charge $+q$ situated at $z = a$ on the Z-axis when a conducting earthed sphere, distorting the field, is placed with its center at O(cf. Fig. 1). Clearly,

$$V(r,\theta) = \frac{q}{s} + U$$

if U is the potential due to the induced charge on the sphere. Like the first term q/s, U must be a solution of Laplace's equation and may conveniently be written in the form (22). Here, however, it becomes necessary to retain full generality and use a superposition of harmonic functions:

$$U = \sum_{l,m} (a_{l,m}r^l + b_{l,m}r^{-l-1})P_l^m (\cos \theta)e^{\pm im\varphi}$$

From the symmetry of the physical distribution about the Z-axis it is clear that U cannot depend on φ; hence $m = 0$. Also, since U must vanish at $r = \infty$, every $a_{l,m} = 0$. Hence

$$U = \sum_l b_l r^{-l-1} P_l (\cos \theta) \tag{7-23}$$

The coefficients b_l are to be determined by the condition that V shall be zero on the surface of the sphere:

$$\frac{q}{s}\bigg|_{\text{on sphere}} + \sum_l b_l r_0^{-l-1} P_l \cos (\theta) = 0$$

The first term on the left can be expanded by means of a theorem proved in the discussion of the Legendre polynomials (eq. 3-24 et seq.)

$$\frac{1}{s} = \frac{1}{a} \sum_{l=0}^{\infty} \left(\frac{r}{a}\right)^l P_l (\cos \theta) \quad \text{if} \quad a > r \tag{7-24a}$$

Hence the foregoing condition becomes:

$$\sum_l \left[\frac{q}{a}\left(\frac{r_0}{a}\right)^l + b_l r_0^{-l-1}\right] P_l (\cos \theta) = 0$$

But this is satisfied only if the coefficient of every P_l is zero, so that

$$b_l = -qa^{-l-1}r_0^{2l+1}$$

On substituting this back into (23) we find

$$U = -\frac{q}{r}\frac{r_0}{a} \sum_l \left(\frac{r_0^2}{ar}\right)^l P_l (\cos \theta) \tag{7-25}$$

a result which permits a very simple and interesting interpretation. Consider a point, such as a' (cf. Fig. 1) on the Z-axis. If $r > a'$, the expansion of $1/s'$ may be seen to be (see derivation leading to eq. 3-27)

$$\frac{1}{s'} = \frac{1}{r} \sum_l \left(\frac{a'}{r}\right)^l P_l (\cos \theta); \quad r > a' \tag{7-24b}$$

in contrast to (24a). But (25) is of the same form as this; indeed it becomes

$$U = -q\frac{r_0}{a}\frac{1}{s'}$$

if we put $a' = r_0^2/a$. Our final result may now be written

$$V = \frac{q}{s} - \frac{q'}{s'} \tag{7-26}$$

provided q' is identified with $(r_0/a)q$. In words: when a conducting (earthed) sphere is placed near a point charge $+q$ it changes the potential in the same manner as would a point charge of opposite sign and magnitude $q' = (r_0/a)q$, placed at the point $a' = r_0^2/a$. The charge q' is said to be the *image* of q.

The same reasoning holds when an earthed plane is placed near a charge. For, suppose we put $a = r_0 + A$, $a' = r_0 - A'$ and let r_0 go to infinity. From $a'a = r_0^2$ we then get $A = A'$, and r_0/a approaches 1. It is seen that the effect of the plane can also be expressed by means of an image charge which, in this case, has the same magnitude as the real charge and is located at its mirror image.

Problem. Find the potential of an electric dipole, and of an axial electric quadrupole. $A \begin{Bmatrix} \text{dipole} \\ \text{quadrupole} \end{Bmatrix}$ may be defined as a distribution of charge whose potential, while vanishing at infinity, is proportional to $\begin{Bmatrix} P_1(\cos\theta) \\ P_2(\cos\theta) \end{Bmatrix}$.

Ans. $c_1 \dfrac{\cos\theta}{r^2}$, $\dfrac{c_2}{r^3}(3\cos^2\theta - 1)$.

7.8. The Wave Equation.—To give a concise definition of a wave in physical descriptive terms is not an easy matter; mathematically it is defined as the condition of a physical quantity, U, which satisfies the differential equation

$$v^2\nabla^2 U - \frac{\partial^2 U}{\partial t^2} = 0 \tag{7-27}$$

For a reason which will soon be evident, v is called the *phase velocity* of the wave. In general, v may be a function of space coordinates (wave traveling in a non-homogeneous medium). When this is true, eq. (27) has an enormous variety of solutions, some of which would hardly conform to the more intuitive conception commonly attached to the word wave. This general case is of special interest in quantum or wave mechanics and in certain branches of optics and will be dealt with in Chapter 11.

In the present section, v will be considered constant, that is, independent of space and time. Before examining eq. (27) by the method of separation of variables, we discuss a form of solution which is interesting from a physical point of view. For it happens that this equation can be solved by the introduction of a single independent variable

$$\xi = \alpha x + \beta y + \gamma z + vt$$

α, β, γ being constants, provided ∇^2 is written in its Cartesian form. On substituting this, (27) takes the form

$$[v^2(\alpha^2 + \beta^2 + \gamma^2) - v^2]\frac{d^2U}{d\xi^2} = 0$$

which is clearly satisfied if we put

$$\alpha^2 + \beta^2 + \gamma^2 = 1 \qquad (7\text{-}28)$$

Subject to this condition, the substitution

$$\eta = \alpha x + \beta y + \gamma z - vt$$

will also lead to a solution $U(\eta)$. The functional form of U is left entirely arbitrary aside from the requirement that it must permit of two differentiations. We conclude, therefore, that

$$U = f_1(\xi) + f_2(\eta) \qquad (7\text{-}29)$$

is a general solution of the wave equation (with constant v).

Relation (28), however, allows the interpretation of α, β, and γ as direction cosines,[3] that is, as components of a *unit* vector, $\boldsymbol{\sigma}$. Eq. (29) then takes the form

$$U = f_1(\boldsymbol{\sigma} \cdot \mathbf{r} + vt) + f_2(\boldsymbol{\sigma} \cdot \mathbf{r} - vt) \qquad (7\text{-}30)$$

Now constant values of $f_1(\boldsymbol{\sigma} \cdot \mathbf{r} + vt)$ are defined by $\boldsymbol{\sigma} \cdot \mathbf{r} = -vt$; they lie on a plane traveling along $-\boldsymbol{\sigma}$ with a velocity v. Constant values of $f_2(\boldsymbol{\sigma} \cdot \mathbf{r} - vt)$ are given by $\boldsymbol{\sigma} \cdot \mathbf{r} = vt$; they lie on a plane traveling along $+\boldsymbol{\sigma}$ with velocity v. The representation (30) therefore describes two *plane* waves traveling in opposite directions with the same speed.

A solution of equal simplicity may be obtained when (27) is written in polar coordinates provided we assume that U is a function of the radius vector and t alone. (The solution here derived is therefore far from general.) In that case, ∇^2 reduces to $\partial^2/\partial r^2 + (2/r)(\partial/\partial r)$, and the equation reads

$$\frac{v^2}{r}\frac{\partial^2(rU)}{\partial r^2} - \frac{\partial^2 U}{\partial t^2} = 0$$

The substitution $\xi = r + vt$, $rU = P$ converts it into $v^2(d^2P/d\xi^2) - v^2(d^2P/d\xi^2) = 0$; hence $P = f_1(r + vt)$. A similar result would have been achieved by choosing $\eta = r - vt$ in place of ξ. Hence

$$P = f_1(r + vt) + f_2(r - vt)$$

or

$$U = \frac{1}{r}[f_1(r + vt) + f_2(r - vt)] \qquad (7\text{-}31)$$

---

[3] This interpretation destroys generality; α, β, γ need not be real if only they satisfy (28).

This solution represents two *spherical* waves, one traveling in toward the origin, the other out from the origin. The factor $1/r$, without which U would not be a solution of (27) and therefore not a wave, accounts for the attenuation of a spherical wave as it moves out from its source.

By suitable choices of f_1 and f_2 a great variety of wave complexes can be formed, of which *standing* waves, defined by the condition $U(r,t) = F(r) \cdot G(t)$ where F and G represent new functions, are perhaps the simplest.

Problem. Show that, if f_1 and f_2 are both sine functions, written in the customary form $\sin (2\pi/\lambda)(r\pm vt)$, U represents a standing wave.

We now turn to a more detailed analysis of the wave equation, based on the method of separation of variables. On assuming that

$$U = ST$$

where S is a function of space coordinates and T a function of t only, (27) is changed to the form

$$v^2 \frac{\nabla^2 S}{S} = \frac{\ddot{T}}{T}$$

the dots denoting time derivatives. Each side of this equation must equal the same constant which, for convenience, we shall call $-\omega^2$. No supposition concerning the reality of ω is here implied, although ω will turn out to be real in the more interesting practical applications. The equation

$$\ddot{T} + \omega^2 T = 0$$

has the general solution

$$T_{\omega} = c_1 e^{i\omega t} + c_2 e^{-i\omega t} \tag{7-32}$$

The constant ω, clearly, has the meaning of an " angular " frequency.

Now the space part of the wave function is defined by the equation

$$\nabla^2 S + \frac{\omega^2}{v^2} S = 0$$

The constant ω/v will henceforth be denoted by k; in terms of the *wave length* λ, which is related to ω and v by the well known formula

$$\frac{\omega}{2\pi} \lambda = v$$

$k = 2\pi/\lambda$. It signifies the number of waves of given ω per 2π units of length and is called the *wave number*. The equation

$$\nabla^2 S + k^2 S = 0 \tag{7-33}$$

is the basis of the entire theory of vibrations and will be referred to as the *space form* of the *wave equation.* The remainder of the present section will be devoted to its study.

7.9. One Dimension.—Eq. (33) reduces to the simple form

$$\frac{d^2 S}{dx^2} + k^2 S = 0$$

which has the solution

$$S_k = ae^{ikx} + be^{-ikx}$$

One such solution is obtained for every value of k. For $k = 0$, $S_0 = ax + b$. It should be noted that

$$S = \sum_k S_k$$

is *not* a solution of (33), but that

$$U = \sum_k S_k T_k$$

is a solution of (27). (We are writing T_k in place of T_ω because k is fixed when ω is chosen.) Similar caution is required in all subsequent considerations.

7.10. Two Dimensions.—a. *Rectangular Coordinates.* The work goes as in sec. 7.3. In place of eq. 4 we now have

$$\frac{X''}{X} + \frac{Y''}{Y} + k^2 = 0$$

Separation is achieved by putting

$$\frac{X''}{X} = -k_1^2, \quad \frac{Y''}{Y} = -k_2^2$$

and requiring that

$$k_1^2 + k_2^2 = k^2$$

Hence

$$S_{k_1 k_2} = XY = c_{k_1 k_2} e^{\pm i(k_1 x \pm k_2 y)} \tag{7-34}$$

b. *Polar Coordinates.* In place of (9) there results

$$\rho^2 \frac{P''}{P} + \rho \frac{P'}{P} + \frac{\Phi''}{\Phi} + \rho^2 k^2 = 0$$

On equating Φ''/Φ to $-m^2$, the radial equation becomes

$$\rho^2 P'' + \rho P' + (k^2 \rho^2 - m^2) P = 0$$

It is identical with Bessel's (eq. 2-57) when the independent variable is taken to be $k\rho$. Hence

$$S_{k,m} = Z_m(k\rho) e^{\pm im\varphi}$$

or, more generally,

$$S_k = \sum_m a_m Z_m(k\rho)e^{\pm im\varphi} \tag{7-35}$$

7.11. Three Dimensions.—a. *Rectangular Coordinates.* Immediate generalization of eq. (34) shows that

$$S_{k_1 k_2 k_3} = c_{k_1 k_2 k_3} e^{\pm i(k_1 x \pm k_2 y \pm k_3 z)} \tag{7-36}$$

provided that $k_1^2 + k_2^2 + k_3^2 = k^2$. If k_1, k_2, k_3 are taken to be real (an assumption destroying the generality of the solution) they may be regarded as the components of a vector $\mathbf{k}$, and (36) may be written

$$S(\mathbf{k}) = c(\mathbf{k})e^{i\mathbf{k}\cdot\mathbf{r}} \tag{7-37}$$

When this result is combined with (32) one sees that a solution of the wave equation (27) has the form

$$U = \sum_{\mathbf{k}} c(\mathbf{k})e^{i(\mathbf{k}\cdot\mathbf{r} \pm kvt)}$$

or

$$U = \int c(\mathbf{k})e^{i(\mathbf{k}\cdot\mathbf{r} \pm kvt)}d\mathbf{k} \tag{7-38}$$

The notation[4] used here, which is rather common in modern physics, is to be understood as follows: A function of a vector, such as $c(\mathbf{k})$, is simply to be regarded as a function of the three real variables k_1, k_2, and k_3; $d\mathbf{k}$ is an abbreviation for the product of three differentials: $dk_1 dk_2 dk_3$. Summations and integrations over $\mathbf{k}$ are therefore threefold.

Eq. 38 is a very useful form of the solution of the wave equation. Physically, it corresponds to the construction of a general wave by superposition of plane sinusoidal waves. It also permits initial conditions to be included in the calculation rather easily. For suppose that we know the form of the disturbance at $t = 0$, $U_0(x,y,z)$. The $c(\mathbf{k})$ are then given at once by the Fourier analysis of this function, viz.:

$$U_0(x,y,z) = \int c(\mathbf{k})e^{i\mathbf{k}\cdot\mathbf{r}}d\mathbf{k}$$

and (38) represents the wave at any other time.

Problem a. Show that, in general,

$$U(x,y,z,t) = (2\pi)^{-3} \int\int\int\int\int\int_{-\infty}^{\infty} U_0(x'y'z')e^{i[\mathbf{k}\cdot(\mathbf{r}-\mathbf{r}')\pm kvt]}\,dx'dy'dz'dk_1 dk_2 dk_3$$

---

[4] This notation is indeed ambiguous. In vector analysis, $d\mathbf{k}$ is the element of a vector, and hence itself a vector. Here it means the element of volume in k-space, which is *not* a vector. But the convenience of the present notation is so great that we shall occasionally employ it when confusion is not likely to arise.

If U_0 is concentrated at the origin, that is, if U_0 is the limit of a function which tends to ∞ at the origin, but in such a way that $\iiint U_0(x,y,z)dxdydz = 1$, then

$$U(x,y,z,t) = (2\pi)^{-3} \iiint e^{\,i\,(\mathbf{k}\cdot\mathbf{r} - \, kvt)}dk_1dk_2dk_3$$

Problem b. Show that

(α) U decreases continually with time at $\mathbf{r} = 0$.
(β) U is zero wherever $|\,r\,| > vt$.

Note the physical significance of these results.

   b. *Cylindrical Coordinates.* The substitutions in sec. 7.4b lead to the ordinary differential equations

$$Z'' = -\kappa^2 Z$$

$$\Phi'' = -l^2 \Phi$$

$$\rho^2 P'' + \rho P' - [(\kappa^2 - k^2)\rho^2 + l^2]P = 0$$

The last equation has the solution

$$P = Z_l(\sqrt{k^2 - \kappa^2}\rho)$$

Consequently

$$S_{k,\kappa,l} = ce^{\pm i(\kappa z \pm l\varphi)}Z_l(\sqrt{k^2 - \kappa^2}\rho) \tag{7–39}$$

   If this function is to be single-valued in φ, l must be an *integer*. Constructing a solution of the wave equation wherein the space function has the form (39) we thus obtain

$$U = \sum_{k\kappa l} S_{k,\kappa,l}e^{\pm ikvt}$$

But it is usually more satisfactory to indicate the nature of the summations (l is integral, k and κ may vary continuously) more explicitly. If, furthermore, we limit $\sqrt{k^2 - \kappa^2}$ to real, positive values (thus again destroying generality) and call this quantity μ, the following useful representation is obtained:

$$U = \int e^{\pm ikvt}dk \sum_{l=-\infty}^{\infty} e^{il\varphi} \int_0^{\infty} g_l(k,\mu)e^{\pm i\sqrt{k^2-\mu^2}z}\, J_l(\mu\rho)\mu d\mu \tag{7–40}$$

where we have written $c = g\mu$ for convenience later, and J_l is the Bessel function defined in sec. 2.14.

   The type of problem in which eq. (40) is used is this. Suppose that at $t = 0$, the disturbance is confined to the plane $z = 0$ where it has the form $U_0(\rho,\varphi)$. Also, let the wave be monochromatic ($k = $ const.; so that integration over dk is absent). **Then**

$$U_0(\rho,\varphi) = \sum_{l=-\infty}^{\infty} e^{il\varphi} \int_0^{\infty} g_l(\mu) J_l(\mu\rho)\mu d\mu \qquad (7\text{--}41)$$

and from this relation all coefficients $g_l(\mu)$ can be determined. For if we multiply both sides of (41) by $e^{-il'\varphi}$ and integrate over φ from 0 to 2π, we obtain

$$\int_0^{\infty} g_{l'}(\mu) J_{l'}(\mu\rho)\mu d\mu = \frac{1}{2\pi} \int_0^{2\pi} U_0(\rho,\varphi)e^{-il'\varphi}\, d\varphi \equiv U_{l'}(\rho)$$

This, however, is nothing other than a Fourier-Bessel transformation[5] of $U_{l'}(\rho)$, and it follows that

$$g_l(\mu) = \int_0^{\infty} U_l(\rho) J_l(\mu\rho)\rho d\rho$$

Problem. Show that the diffraction pattern due to a plane monochromatic wave passing through a circular aperture of radius a is given by

$$U(\rho,z) = \text{const.} \int_0^{\infty} J_0(\mu\rho) J_1(\mu a) e^{i\sqrt{k^2-\mu^2}z}\, d\mu$$

c. *Spherical (Polar) Coordinates.* The equation for S is similar to (15), except that the term $+k^2 S$ is also present on the left. The substitution

$$S = R(r) \cdot \Theta(\theta) \cdot \Phi(\varphi)$$

now leads to the three equations

$$\Phi'' = -m^2\Phi \qquad (7\text{--}42a)$$

$$\frac{1}{\sin\theta} \frac{d}{d\theta}(\sin\theta\Theta') - \frac{m^2}{\sin^2\theta}\Theta + l(l+1)\Theta = 0 \qquad (7\text{--}42b)$$

$$\frac{1}{r^2} \frac{d}{dr}(r^2 R') + \left[k^2 - \frac{l(l+1)}{r^2}\right]R = 0 \qquad (7\text{--}42c)$$

The second of these is the equation for associated Legendre functions (l and m are integers again: m to insure single-valuedness in φ, l in order that the solution Θ be a polynomial, i.e., that it should not diverge for $\cos\theta = \pm 1$). The third equation may be transformed as follows. Put $R = P/r$, and change the independent variable to $t = kr$. Eq. 42c then takes the form

$$\frac{d^2P}{dt^2} + \left[1 - \frac{l(l+1)}{t^2}\right]P = 0$$

[5] For further details, see sec. 8.3.

Again, put $P = \sqrt{t}\, Q$, so that the last equation reads

$$\frac{d^2Q}{dt^2} + \frac{1}{t}\frac{dQ}{dt} + \left[1 - \frac{(l+\frac{1}{2})^2}{t^2}\right]Q = 0$$

This is at once recognized as Bessel's equation (2–57); hence

$$Q = Z_{l+1/2}(t)$$

so that

$$R = cr^{-1/2}Z_{l+1/2}(kr)$$

For the space part of the wave function, we thus find

$$S_k = \sum_{m,l} c_{k,l,m}P_l^m(\cos\theta)e^{im\varphi}r^{-1/2}Z_{l+1/2}(kr) \qquad (7\text{--}43)$$

A sum of the form

$$\sum_{m=-l}^{l} c_m P_l^m(\cos\theta)e^{im\varphi}$$

with arbitrary coefficients c_m is often called a *spherical harmonic* and denoted by the symbol $Y_l(\theta,\varphi)$. In using this symbol one must remember that the function which it represents is not unique, but contains $2l+1$ arbitrary constants. With this abbreviation, then,

$$S_k = \sum_{l=0}^{\infty} c_{k,l}Y_l(\theta,\varphi)r^{-1/2}Z_{l+1/2}(kr)$$

and the wave function is

$$U = \sum_k S_k e^{\pm ikvt} = \int dk \sum_{l=0}^{\infty} c_{k,l}Y_l(\theta,\varphi)r^{-1/2}Z_{l+1/2}(kr)e^{\pm ikvt} \qquad (7\text{--}44)$$

7.12. Examples of Solutions of the Wave Equation.—The local pressure P in a gas traversed by a sound wave, satisfies the wave equation.

a. The simplest type of a wave is that emitted by a "breathing" sphere, i.e., a sphere performing volume oscillations without distortion. It is characterized by the two boundary conditions:

(α) $$P_{r=r_0} = \text{const.}\; e^{-i\omega t},$$

(β) $$P_{r\to\infty} = f(r,\theta)e^{i(kr-\omega t)}$$

Condition (α) states that at the surface of the sphere (of radius r_0) all points shall be in phase; condition (β) implies that at infinity the wave shall be an outgoing one. We limit ourselves to *monochromatic* waves (pure tones), so that there is only *one* value of k or ω. Clearly, spherical polar coordinates must here be used. Considering then eq. (44), we must first omit the integration over k. Since in accordance with condition (α)

there must, at $r = r_0$, be no functional dependence either on φ or on θ, both l and m are zero. Hence (44) reduces to

$$P = Cr^{-1/2}Z_{1/2}(kr)e^{-i\omega t}$$

But the general Bessel function

$$Z_{1/2}(x) = a_1 J_{1/2}(x) + a_2 J_{-1/2}(x)$$

as was shown in Chapter 3.   Inserting these, we have

$$P = C\left(a_1 \frac{\sin kr}{kr} - a_2 \frac{\cos kr}{kr}\right)e^{-i\omega t}$$

In order to satisfy condition (β) we put $a_1 = i$, $a_2 = -1$, obtaining

$$P = \frac{C}{kr} e^{i(kr-\omega t)}$$

as our final result.

b. When the sphere of the preceding example vibrates, not with spherical symmetry, but in such a way that condition (α) reads

(α) $$P_{r=r_0} = \text{const. } \cos\theta e^{-i\omega t}$$

it is said to emit *dipole* waves.   Condition (β) remains unchanged.   Of all the functions composing $Y_l(\theta,\varphi)$, only P_1^0 ($\cos\theta$) is a cosine function. Therefore l must be 1.   Hence (44) now reduces to

$$P = Cr^{-1/2}Z_{3/2}(kr) \cos\theta e^{-i\omega t}$$

But

$$r^{-1/2}Z_{3/2}(kr) = r^{-1/2}[a_1 J_{3/2}(kr) + a_2 J_{-3/2}(kr)]$$

and this is proportional to

$$a_1\left[\frac{\sin kr}{(kr)^2} - \frac{\cos kr}{kr}\right] + a_2\left[-\frac{\sin kr}{kr} - \frac{\cos kr}{(kr)^2}\right]$$

If this expression is to satisfy condition (β), it is necessary to choose

$$a_1 = -1, \quad a_2 = -i$$

so that

$$r^{-1/2}Z_{3/2}(kr) \propto \left[\frac{1}{kr} + \frac{i}{(kr)^2}\right]e^{ikr}$$

and

$$P = C\left[\frac{1}{kr} + \frac{i}{(kr)^2}\right]\cos\theta e^{i(kr-\omega t)}$$

The constant C may be complex.   If it is written $C = C_1 + iC_2$, the real part of P, which alone is of interest, will be

$$\mathcal{R}P = \left(\frac{C_1}{kr} - \frac{C_2}{k^2r^2}\right) \cos\theta \cos(kr - \omega t) - \left(\frac{C_1}{k^2r^2} + \frac{C_2}{kr}\right) \cos\theta \sin(kr - \omega t)$$

For small values of r,

$$\mathcal{R}P \doteq -\frac{\cos\theta}{k^2r^2}[C_2 \cos(kr - \omega t) + C_1 \sin(kr - \omega t)],$$

for large r,

$$\mathcal{R}P \doteq \frac{\cos\theta}{kr}[C_1 \cos(kr - \omega t) - C_2 \sin(kr - \omega t)]$$

If C_1 is zero, the disturbance is of the form $\cos(kr - \omega t)$ near the surface of the sphere, but of the sine form at infinity. If $C_2 = 0$, the reverse is true. There occurs, therefore, a curious change of phase as the wave moves outward.

7.13. Equation of Heat Conduction and Diffusion.—The temperature U in a homogeneous medium, in which $A(x,y,z)$ calories of heat are generated (by some unspecified agency) per unit of volume surrounding the point (x,y,z) per second, and which has density ρ, specific heat s, and thermal conductivity κ, satisfies the partial differential equation

$$\frac{\partial U}{\partial t} = \frac{\kappa}{\rho s} \nabla^2 U + \frac{A}{\rho s} \tag{7-45}$$

Various simplifying conditions may arise: In the first place, attention may be confined to "steady states," that is, to temperature distributions which do not change with time. Such states will always occur in physical and chemical problems after heat conduction has taken place for a sufficiently long time. In that case, $\partial U/\partial t$ is zero, and the equation reads

$$\nabla^2 U = -\frac{A}{\kappa} \tag{7-46}$$

It is of the form of *Poisson's* equation which will be discussed in sec. 7.17. If, in addition, it is assumed that no heat is generated anywhere within the body, $A = 0$ and (46) becomes identical with Laplace's equation which we have already studied.

Of greater interest is the situation in which, to be sure, A is taken to be zero, but consideration is given to non-steady states. The temperature is then subject to the equation

$$\frac{\kappa}{\rho s} \nabla^2 U - \frac{\partial U}{\partial t} = 0 \tag{7-47}$$

which is very similar to the wave equation. (This equation is derived by vector methods in sec. 4.18.)

In the kinetic theory, one meets the equation of diffusion which regulates the flow of fluid matter within another material medium. It states in its basic form:

$$\frac{\partial U}{\partial t} = \nabla \cdot (D\nabla U) \tag{7-48}$$

U represents the concentration of fluid matter, D its coefficient of diffusion. Strictly speaking, D is a function of U and hence of (x,y,z). But for small concentrations D is found to be very nearly constant. For that case, then, (48) may be written

$$D\nabla^2 U - \frac{\partial U}{\partial t} = 0 \tag{7-49}$$

All parameters appearing in (49) as well as in (47) are positive, hence both of these equations will be written in the form

$$a^2\nabla^2 U - \frac{\partial U}{\partial t} = 0 \tag{7-50}$$

and we remember that, for heat conduction, U = temperature and $a^2 = \kappa/\rho s$, while for diffusion, U = concentration and $a^2 = D$. The remainder of this section is devoted to the solutions of eq. (50).

Separation may at once be achieved by putting $U = S(x,y,z) \cdot T(t)$, and it is found that $a^2\nabla^2 S/S = \dot{T}/T$. On equating the right-hand side to $-a^2k^2$, k being an arbitrary constant, it is seen that

$$T_k = \text{const. } e^{-a^2k^2t} \tag{7-51}$$

while S must satisfy

$$\nabla^2 S + k^2 S = 0 \tag{7-52}$$

an equation identical with the space form of the wave equation, (33). If, therefore, we combine the solutions of (33), discussed in the preceding section, with T_k in the form (51), we have an answer to the problems of heat conduction and diffusion.

7.14. Example: Linear Flow of Heat.—Suppose that heat flows in a linear filament placed along the X-axis. The solution of (52) is then

$$S_k = c_k e^{ikx} + d_k e^{-ikx}$$

and this may be taken as

$$S_k = c_k e^{ikx}$$

if we assign both positive and negative values to k. The general solution reads:

$$U = \sum_k S_k T_k = \int_{-\infty}^{\infty} c(k)e^{ikx - a^2k^2t}dk \tag{7-53}$$

Every choice for $c(k)$ will satisfy eq. (47), but the proper selection is to be made in accordance with initial conditions. Let us suppose, then, that $U = U_0(x)$ at $t = 0$. Eq. (53) now states:

$$U_0(x) = \int_{-\infty}^{\infty} c(k)e^{ikx}dk$$

and $c(k)$ may be obtained from this by means of a Fourier transformation. In view of eq. (8–13)$'$

$$c(k) = \frac{1}{2\pi} \int_{-\infty}^{\infty} U_0(x')e^{-ikx'}dx'$$

so that (53) becomes

$$U(x,t) = \frac{1}{2\pi} \int_{-\infty}^{\infty} \int_{-\infty}^{\infty} U_0(x')e^{ik(x-x')-a^2k^2t}dx'dk$$

The integration with respect to k can be performed:

$$\int_{-\infty}^{\infty} e^{-a^2tk^2+i(x-x')k}dk = \sqrt{\frac{\pi}{a^2t}}\, e^{-(x-x')^2/4a^2t}$$

whence

$$U(x,t) = \frac{1}{2a\sqrt{\pi t}} \int_{-\infty}^{\infty} U_0(x')e^{-(x-x')^2/4a^2t}dx' \qquad (7\text{-}54)$$

Problems.

   a. Prove that (54) reduces to $U_0(x)$ for $t = 0$.

   b. Show that, if $U_0(x)$ is a step function such that

$$U_0 = \begin{cases} 1 \text{ if } & |x| < 1 \\ 0 \text{ if } & |x| > 1 \end{cases}$$

$$U = \tfrac{1}{2}\left\{\phi\left(\frac{1-x}{2a\sqrt{t}}\right) + \phi\left(\frac{1+x}{2a\sqrt{t}}\right)\right\}$$

where $\phi(x)$ is the error integral $\dfrac{2}{\sqrt{\pi}} \displaystyle\int_{0}^{x} e^{-\xi^2}d\xi$.

   c. Show that, if

$$U_0 = \begin{cases} 1 \text{ for } & x > 0 \\ 0 \text{ for } & x \lessgtr 0, \end{cases} \quad \text{then} \quad U = \tfrac{1}{2}\left\{1 + \phi\left(\frac{x}{2a\sqrt{t}}\right)\right\}$$

Interpret the last two problems from the point of view of diffusion.

   d. Suppose U_0 is a " function " which is everywhere zero except at $x = 0$, where it tends to ∞ in such a way that $\displaystyle\int U_0(x)dx = 1$. (Such a " function " was introduced by Dirac and is commonly known to physicists as a δ-function. Strictly speaking it is no function at all.) Then, clearly,

$$U = \frac{1}{2a\sqrt{\pi t}}e^{-x^2/4a^2t}$$

Discuss the temperature at any point x, and show in particular that it will rise to a maximum at $t = x^2/2a^2$. This fact affords a simple experimental determination of a, and hence of D and the thermal quantities.

7.15. Two-Dimensional Flow of Heat.—In polar coordinates, S is given by eq. (35)   Hence

$$U = \int_{-\infty}^{\infty} c(k)dk (\sum_m a_m Z_m(k\rho)e^{im\varphi})e^{-a^2k^2t} \qquad (7\text{-}55)$$

If, as we shall suppose, the temperature distribution at $t = 0$ is radially symmetrical, so that U does not depend on φ, the only value permitted to m is zero.   Also, since Z_0 is an even function, the integration in (55) may be taken from 0 to ∞ without error.   For Z_0 we shall take the J_0-function, because it will at once be seen that most temperature distributions can be expressed in terms of J_0 alone.   Thus

$$U = \int_0^{\infty} c(k)J_0(k\rho)e^{-a^2k^2t}dk \qquad (7\text{-}56)$$

Let us write

$$c(k) = k\, g(k)$$

and suppose that $U = U_0(\rho)$ at $t = 0$.   It is then easy to determine $g(k)$ formally and hence $U(\rho,t)$.   For in accordance with (56)

$$U_0(\rho) = \int_0^{\infty} g(k)J_0(k\rho)kdk$$

in other words, $g(k)$ is the Fourier-Bessel transform of $U_0(\rho)$.   (Cf. Sec. 8.3.) Hence

$$g(k) = \int_0^{\infty} U_0(\rho')J_0(k\rho')\rho'd\rho'$$

When this is put back into (56) the final form of $U(\rho,t)$ is obtained:

$$U(\rho,t) = \int\int_0^{\infty} U_0(\rho')J_0(k\rho')J_0(k\rho)e^{-a^2k^2t}k\rho'dkd\rho' \qquad (7\text{-}57)$$

Problem.   Show that, if $U_0(\rho)$ is concentrated at $\rho = 0$, and $\int_0^{\infty} U_0(\rho)\rho d\rho = 1$,

$$U(\rho,t) = \frac{1}{2a^2t}\, e^{-\rho^2/4a^2t}$$

Compare this with problem (d) of Sec. 7.14.   Interpret above as a diffusion problem.

7.16. Heat Flow in Three Dimensions.—In rectangular coordinates, U is given as a generalization of eq. (53) (cf. eq. (36) for the form of S):

$$U = \int\int\int_{-\infty}^{\infty} c(k_1 k_2 k_3)e^{i(k_1 x + k_2 y + k_3 z) - a^2(k_1^2 + k_2^2 + k_3^2)t}dk_1 dk_2 dk_3$$

or, with the use of the vector notation previously explained (cf. footnote on p. 227)

$$U = \iiint c(\mathbf{k}) e^{i\mathbf{k}\cdot\mathbf{r} - a^2 k^2 t} d\mathbf{k} \tag{7-58}$$

We now repeat essentially the procedure leading from (53) to (54), but using three variables instead of one.

$$U_0(x,y,z) = \iiint c(\mathbf{k}) e^{i\mathbf{k}\cdot\mathbf{r}} d\mathbf{k}$$

hence $c(\mathbf{k})$ is the Fourier transform of U_0:

$$c(\mathbf{k}) = \frac{1}{8\pi^3} \iiint_{-\infty}^{\infty} U_0(x',y',z') e^{-i\mathbf{k}\cdot\mathbf{r}} d\mathbf{r}'$$

whence

$$U(x,y,z,t) = \frac{1}{8\pi^3} \iiint \iiint_{-\infty}^{\infty} U_0(x',y',z') e^{i\mathbf{k}\cdot(\mathbf{r}-\mathbf{r}') - a^2 k^2 t} d\mathbf{r}' d\mathbf{k}$$

$$= (2a\sqrt{\pi t})^{-3} \iiint U_0(x',y',z') e^{-(\mathbf{r}-\mathbf{r}')^2/4a^2 t} d\mathbf{r}' \tag{7-59}$$

If U_0 is a function of x' alone, the integration over y' and z' may be performed, and the result is identical with (54), as it should be. Of greatest practical importance is the case where U_0 is a function of r' alone. The volume element $d\mathbf{r}'$ may then be written in polar form: $r'^2 dr' \sin\theta' d\theta' d\varphi'$, and the integration over θ' and φ' can be performed. It is to be observed in this connection that

$$(\mathbf{r} - \mathbf{r}')^2 = r^2 + r'^2 - 2rr' \cos\theta$$

One then finds

$$U(r,t) = (2ar\sqrt{\pi t})^{-1} \int_0^{\infty} U_0(r') [e^{-[(r-r')/2a\sqrt{t}]^2} - e^{-[(r+r')/2a\sqrt{t}]^2}] r' dr'$$

Problem. Show that, if

$$U_0 = \begin{cases} 1 & \text{for } r \leq 1 \\ 0 & \text{for } r > 1 \end{cases}$$

$$U = \tfrac{1}{2}[\phi(\xi_+) + \phi(\xi_-)] + \frac{a}{r}\sqrt{\frac{t}{\pi}}\,(e^{-\xi_+^2} - e^{-\xi_-^2})$$

where

$$\xi_\pm = \frac{1 \pm r}{2a\sqrt{t}}; \quad \phi(x) = \frac{2}{\sqrt{\pi}}\int_0^x e^{-u^2} du$$

7.17. Poisson's Equation.—All partial differential equations treated thus far in the present chapter are linear and homogenous in the dependent variables (cf. footnote in sec. 2.5). It is only for this type of equation that the method of separation of variables may work.   The variety of linear and *inhomogeneous* equations of importance in scientific analysis is also great, but there exists for their solution no method nearly so powerful as the separation of variables.

An equation like (33), the space form of the wave equation, would become inhomogeneous if the right-hand side were not zero but some function $f(x,y,z)$.   One remarkable feature of an inhomogeneous equation, which will here only be mentioned, is that it may not possess solutions for every value of k even though the homogeneous equation, with the same boundary condition, has solutions.   The inhomogeneity selects, as it were, special values of the parameter k for which solutions are possible.   This phenomenon, which is the rule for inhomogeneous equations, may also occur for homogeneous ones if the boundary or initial conditions of the problem are sufficiently stringent.   It will be discussed under the heading " characteristic values " or " eigenvalues " in Chapter 8.

An inhomogeneous equation which is rather common is *Poisson's;* it will here be chosen to illustrate a process of solution.   Its general form is:

$$\nabla^2 \Phi = f(x,y,z) \tag{7-60}$$

One encounters it (1) in electrostatics, where Φ is the ordinary potential and f represents a constant times the distribution of charge,[6] $\rho(x,y,z)$, the constant depending on the units chosen; (2) in the theory of heat flow, where eq. (45) takes the form (60) when $\partial U/\partial t = 0$, as shown in eq. (46).

To solve (60) we first recall Green's theorem (see sec. 4.19) which states that, for any two functions of space coordinates, u and v which are finite, continuous and have continuous first and second derivatives,

$$\int_\tau (u\nabla^2 v - v\nabla^2 u)d\tau = \int_\sigma (u\nabla v - v\nabla u) \cdot d\sigma \tag{7-61}$$

Here τ represents a certain closed volume and σ its surface; $d\sigma$ is taken positive in the direction outward from the volume.   In our problem we are given the function $f(x,y,z)$ and we wish to find $\Phi(x'y'z')$ for a fixed point of observation $(x'y'z')$.   In the following it is necessary to distinguish between this fixed point, which will be denoted by primes, and the variable point (xyz) over which integrations are to be performed.

It will prove convenient to consider, in connection with theorem (61), a volume τ such as that depicted in Fig. 2.   It is bounded by the outer surface σ_2 and the inner surface σ_1, a spherical cavity of radius s_0 about the fixed point P'.   The function u will be specified to be

$$u = \frac{1}{|\mathbf{r} - \mathbf{r'}|} \equiv \frac{1}{s}$$

---

[6] If $\rho = 0$ the equation reduces to Laplace's, as pointed out in sec. 7.2.

it satisfies Laplace's equation $\nabla^2 u = 0$, as may readily be verified. Then eq. (61) reads:

$$\int_{\tau} \frac{\nabla^2 v}{s} \, d\tau = \int_{\sigma} \left[\frac{\Delta v}{s} - v\nabla \left(\frac{1}{s} \right) \right] \cdot d\boldsymbol{\sigma} \qquad (7\text{--}62)$$

If now we interpret v as Φ, we may replace $\nabla^2 v$ by f in accordance with (60). The right-hand side of (62) consists of two integrations, one over $d\boldsymbol{\sigma}_1$ and the other over $d\boldsymbol{\sigma}_2$.

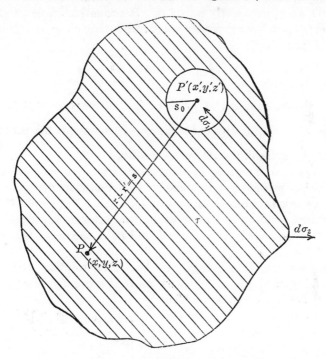

FIG. 7–2

other over $d\boldsymbol{\sigma}_2$. Consider first that over $d\boldsymbol{\sigma}_1$. Clearly, $\nabla \Phi \cdot d\boldsymbol{\sigma}_1$ approaches $-\partial\Phi/\partial r|_{P'} d\sigma_1$ as s_0 tends to zero, the minus sign coming from the fact that $d\boldsymbol{\sigma}_1$ is *inward* with respect to the cavity. Hence

$$\int_{\sigma_1} \frac{\nabla \Phi \cdot d\boldsymbol{\sigma}_1}{s} \to \frac{\partial\Phi}{\partial r}\Big|_{P'} \cdot \frac{4\pi s_0^2}{s_0} \to 0 \quad \text{as} \quad s_0 \to 0$$

provided Φ has a finite derivative. The second integral on the right of (62), when taken over σ_1 becomes, in the limit as $s_0 \to 0$,

$$-\int_{\sigma_1} \Phi\nabla \left(\frac{1}{s} \right) \cdot d\boldsymbol{\sigma} \to -\Phi\Big|_{P'} \left(-\frac{1}{s_0^2} \right) (-4\pi s_0^2) = -4\pi\Phi\Big|_{P'}$$

Hence, if it is assumed that $s_0 \to 0$, eq. (62) reduces to

$$\int_\tau \frac{f}{s}\, d\tau = -4\pi \Phi(x',y',z') + \int_{\sigma_2}\left[\frac{\nabla\Phi}{s} - \Phi\nabla\left(\frac{1}{s}\right)\right]\cdot d\boldsymbol{\sigma} \qquad (7\text{--}63)$$

Here the remaining integral over σ_2 on the right has a rather simple meaning. It is a solution of Laplace's equation in the form: $\nabla'^2\Phi(x',y',z') = 0$, for the only quantities which depend on the primed coordinates are $1/s$ and $\nabla(1/s)$, and these clearly satisfy it. Hence if this whole integral were subtracted from Φ, the remainder would still satisfy eq. (60). It is indeed easily seen that

$$\int_{\sigma_2}\left[\frac{\nabla\Phi}{s} - \Phi\nabla\left(\frac{1}{s}\right)\right]\cdot d\boldsymbol{\sigma}$$

represents the contribution to Φ coming from those parts of $f(x,y,z)$ which lie outside of τ. In the electrostatic case, $\displaystyle\int_{\sigma_2}$ represents the potential due to the charge outside of the volume τ considered.

The integral over σ_2 may be eliminated in another way. Suppose we allow τ to become infinite and impose on Φ the boundary condition that, at infinity, it vanish at least as strongly as $1/r$. Then $\nabla\Phi/s$ and $\Phi\nabla(1/s)$ are both of order $1/r^3$ at ∞, and after the surface integration, which amounts to multiplication by r^2, the result will still be of the order $1/r$ and hence vanish.

Of interest, therefore, is chiefly the particular solution which remains when the integral over σ_2 in (63) is omitted; it is usually referred to as the solution of Poisson's equation. Thus

$$\Phi(x',y',z') = -\frac{1}{4\pi}\int_\tau \frac{f(x,y,z)}{|\mathbf{r}-\mathbf{r}'|}\, dx\,dy\,dz \qquad (7\text{--}64)$$

Problem. Show that, when $f(x,y,z)$ is different from zero only within a finite volume τ_0 such that $\displaystyle\int_{\tau_0} f(x,y,z)d\tau = q$, then for any point $(x'y'z')$ far removed from τ_0,

$$\Phi(x',y',z') = -\frac{1}{4\pi}\frac{q}{r'}$$

the origin being chosen inside τ_0. Interpret this result in electrostatics.

REFERENCES

Frank, P. and von Mises, R., " Die Differential und Integralgleichungen der Mechanik and Physik," Vieweg, Brunswick, 1935. (This is the eighth edition of the famous old Riemann-Weber, " Partielle Differentialgleichungen der mathematischen Physik," edited by the first-named authors.)

245

Sommerfeld, A., " Partial Differential Equations in Physics," Academic Press, New York, 1949.

Bateman, H., " Partial Differential Equations of Mathematical Physics," Dover Publications, New York, 1944.

Tamarkin, J. D. and Feller, W., " Partial Differential Equations," Brown University, Providence, 1941.

Webster, A. G., " Partial Differential Equations of Mathematical Physics," Stechert, New York, 1933.

Coulson, C. A., " Waves," Oliver and Boyd, Edinburgh, 1941.

CHAPTER 8

EIGENVALUES AND EIGENFUNCTIONS

8.1. Simple Examples of Eigenvalue Problems.—It frequently happens in mathematical analysis that a given equation, or a set of equations, yields solutions which are in general uninteresting or trivial, except when a certain parameter appearing in the equations is given a definite value. Such circumstances give rise to eigenvalues[1] and eigenfunctions.[1] Their occurrence is so common that it often goes unrecognized. For illustration, let us take a very simple (and useless) example.

Suppose one wishes to solve the two simultaneous equations

$$(1 - \lambda)x + 2y = 0, \quad 2x + (1 - \lambda)y = 0$$

To be sure, they always possess solutions; but they are almost always $x = 0, y = 0$. Only for two values of the parameter λ will this not be true: for $\lambda = 3$ the solution is $x = y$; for $\lambda = -1$ it is $x = -y$. (The numerical values of x or y are of course never fixed by the linear homogeneous equations above.) The two values of λ for which the equations possess *nonvanishing* solutions are said to be *eigenvalues;* the two corresponding solutions are called *eigenfunctions*.

Eigenvalues are not always denumerable and discrete, as in the foregoing example. To show this, we choose an even more trivial illustration. The equation

$$x^2 = \lambda$$

always possesses a solution. If, however, we wish a *real* solution, λ is at once limited to the domain of positive numbers. Hence we may properly say that $x^2 = \lambda$ is an equation leading to eigenvalues: $\lambda \geqq 0$ and corresponding eigenfunctions $x = \sqrt{\lambda}$.

In both examples eigenvalues were called into being by the imposition of special conditions: in the first that the solutions shall not vanish everywhere; in the second that the solution shall be real. This is generally true; eigenvalues are always produced by special requirements placed upon the solutions of equations. In the most interesting cases of physics and chemistry, these equations are differential or integral equations, and the con-

---

[1] The terms *eigen*value and *eigen*function, because of their brevity, appear to be rapidly displacing their classical synonyms: characteristic value and characteristic function, at least in the physical and chemical literature.

ditions are boundary conditions. We now turn to some cases of greater scientific interest.

8.2. Vibrating String; Fourier Analysis.—In classical physics, many eigenvalue problems occur in connection with vibrating systems. The simplest of these is the problem of a vibrating string. Consider the string to extend along the X-axis, to be fastened with its left end at the origin and its right end at $x = l$. From elementary physics we recall that if its mass per unit length is m and its tension T, the speed of waves along the string is given by $v = \sqrt{T/m}$. The wave equation, discussed in sec. 7.8, will then read

$$v^2 \frac{\partial^2 U}{\partial x^2} - \frac{\partial^2 U}{\partial t^2} = 0 \qquad (8\text{-}1)$$

U is the vertical displacement of the points along x.

We restrict our attention for the moment to types of vibration having a single frequency ν, or angular frequency $\omega = 2\pi\nu$, so that $U = S(x)e^{i\omega t}$ or, if we care to limit the analysis to real functions, $S(x)\begin{pmatrix}\sin \omega t \\ \text{or} \\ \cos \omega t\end{pmatrix}$. The function S will then satisfy the ordinary differential equation

$$\frac{d^2 S}{dx^2} + k^2 S = 0 \qquad (8\text{-}2)$$

where, in conformity with the usage of Chapter 7, the abbreviation

$$k = \frac{\omega}{v} = \frac{2\pi}{\lambda} \qquad (8\text{-}3)$$

has been used. Here λ stands again for the wave length of the disturbance produced. The general solution of (2) is, clearly,

$$S = A \sin (kx + \delta) \qquad (8\text{-}4)$$

where A and δ are arbitrary constants. Every solution of the form (4) is perfectly acceptable as far as the differential equation is concerned, but it does not describe the behavior of the string. Solution (4) permits the ends of the string to vibrate, whereas the physical condition requires them to be fixed. It is therefore necessary to impose the following boundary conditions upon the solutions (4):

(a) $\qquad\qquad\qquad\qquad S(0) = 0$
(b) $\qquad\qquad\qquad\qquad S(l) \ = 0$

Both can of course be satisfied by putting $A = 0$, but this would lead to the unwanted solution $U = 0$ everywhere. Hence there is only the second arbitrary constant, δ, left for adjustment. It must be taken to be zero in order to satisfy condition (a). (Choice of π, 2π, etc., leads to the same final result.) But the function $S = A \sin kx$ will not obey (b). Thus the problem can be solved only if we are willing to tamper with k: we are led to eigenvalues. If $\sin kl$ is to be zero, k must be 0, or $\pi/l, 2\pi/l \cdots n\pi/l$. The value 0, however, is excluded for the same reason that $A = 0$ was rejected. To each eigenvalue of $k = n\pi/l$ (n integral), there corresponds an eigenfunction $S_n = A_n \sin n\pi x/l$. These eigenfunctions are of course undetermined with respect to the constant multipliers, A_n, which may be chosen at will, and which may be different for every n.

Since k is related to λ by eq. (3), there is thus generated a corresponding set of eigenvalues for λ, namely $\lambda = 2l/n$, n integral. This is the well known equation for the wave lengths of standing waves supported by a vibrating string. In the simplest mode of vibration, corresponding to the fundamental frequency, $\lambda = 2l$, the string has nodes only at the end points. For the first harmonic, $\lambda = l$, there is in addition a node at the center of the string, and so on. In general, the number of nodes is $n + 1$.

The eigenfunctions under consideration have two important properties which, as we shall see in sec. 8.5 et seq., are common to a large class of eigenfunctions arising in connection with different problems. They are (1) *orthogonality*, (2) *completeness*. To explain the meaning of these terms, let us arrange the eigenvalues of eq. (2) in a definite order, $k_n = n\pi/l$, $n = 1, 2, 3 \cdots$; and write again $S_n = A_n \sin n\pi x/l$. Orthogonality means:

$$\int_0^l S_n(x) S_m(x) dx = c_n \delta_{nm} \tag{8-5}$$

The word comes originally from vector analysis (cf. Chapter 4) where two vectors, $\mathbf{A}$ and $\mathbf{B}$, are said to be orthogonal if $\mathbf{A} \cdot \mathbf{B} \equiv A_x B_x + A_y B_y + A_z B_z = 0$. Similarly, vectors in N dimensions having components A_i, B_i ($i = 1, 2, \cdots N$) are said to be orthogonal when $\sum_{i=1}^{N} A_i B_i = 0$. If now we imagine a vector space of an infinite number of dimensions, in which the components A_i and B_i become continuously distributed and everywhere dense, i is no longer a denumerable index but a continuous variable (x) and the scalar product $\sum_i A_i B_i$ turns into $\int A(x)B(x)dx$. If it is zero, the functions A and B are said to be orthogonal, and this is the sense in which the word is used above.

The idea of orthogonality is indefinite unless reference is made to a specific range of integration, which in the present case is from 0 to l. Here the validity of eq. (5) is at once verified on substitution of the S-functions,

and the constant c_n is seen to be

$$A_n^2 \int_0^l \sin^2 \frac{n\pi}{l} \, x dx = A_n^2 \cdot \frac{l}{n\pi} \int_0^{n\pi} \sin^2 u du = \frac{l}{2} A_n^2$$

For many purposes it is convenient to have c_n equal to unity. This can always be achieved by a suitable choice of A_n. In the present case, every $c_n = 1$ if $A_n = \sqrt{2/l}$. If, therefore, we write $S_n = \sqrt{2/l} \sin n\pi x/l$, the orthogonality relation (5) reads

$$\int_0^l S_n(x) S_m(x) dx = \delta_{nm} \qquad (8\text{--}5')$$

When the constants A_n are thus chosen the eigenfunctions are said to be *normalized;* functions satisfying (5′) will henceforth be termed *orthonormal.* It is clear that a set of functions having the property of orthogonality (expressed by eq. 5) can always be made ortho-normal by a proper choice of multiplicative constants.

A simple modification in the idea of orthogonality is to be made when *complex* functions are considered. For these, condition (5) must be replaced by

$$\int S_n^*(x) S_m(x) dx = c_n \delta_{nm} \qquad (8\text{--}5^*)$$

where S^* represents the complex conjugate of S. This definition will be used in later work.

We turn to the second property, that of *completeness.* A set of functions is said to be complete if an arbitrary function, $f(x)$, satisfying the same boundary conditions as the functions of the set, can be expanded as follows:

$$f(x) = \sum_{n=1}^{\infty} a_n S_n(x) \qquad (8\text{--}6)$$

the a_n being constant coefficients.

In the present instance, eq. (6) is equivalent to the theorem of *Fourier* which states, in its simplest form, that a function $f(x)$ which vanishes both at $x = 0$ and at $x = \pi$ (and has but a finite number of finite discontinuities) may always be written[2]

$$f(x) = \sum_{n=1}^{\infty} a_n \sin nx \qquad (8\text{--}7a)$$

[2] See, for instance, Byerly, W. E., "Fourier Series and Spherical Harmonics," Ginn and Co., 1893, p. 38. The formulas 7a, b are special cases of eq. 42, developed later in this chapter. Note also the more precise definition of completeness given in sec. 8.8.

the coefficients being given by

$$a_n = \frac{2}{\pi} \int_0^\pi f(\xi) \sin n\xi d\xi \qquad (8\text{--}7b)$$

Eqs. (7) may be modified by using, in place of x and ξ, the variables $\pi x/l$ and $\pi\xi/l$. This has the effect of changing the range of x from $(0,\pi)$ to $(0,l)$, and the results are

$$f(x) = \sum_{n=1}^\infty a_n \sin\left(\frac{n\pi}{l} x\right) \qquad (8\text{--}8a)$$

$$a_n = \frac{2}{l} \int_0^l f(\xi) \sin\left(\frac{n\pi}{l} \xi\right) d\xi \qquad (8\text{--}8b)$$

If S is taken in its *normalized* form, these equations read simply

$$f(x) = \sum_{n=1}^\infty a_n S_n(x), \quad a_n = \int_0^l f(\xi) S_n(\xi) d\xi$$

The fact of completeness has an important bearing on the problem of the vibrating string which we originally set out to solve. While it is true that only a particular S_n, for which k assumes a specific eigenvalue, is a solution of eq. (2) [the series (6) would not be a solution of (2)!], the value of k is not prescribed by eq. (1). Hence eq. (1) is satisfied by

$$U = \sum_n c_n S_n(x) \cos \omega_n t, \quad \omega_n = vk_n$$

with arbitrary coefficients c_n. This, then, is the most general solution of the string problem. It reduces to a series like (6) for $t = 0$, a series which can be chosen to represent any function $f(x)$ which vanishes at the end points. Hence it is seen that *any* initial configuration of the string will yield a solution of eq. (1), that is, a (standing) wave.

Fourier analysis is so useful a tool in applied mathematics that it seems well here to digress for a moment and summarize its essential features beyond the needs of the present problem. Details may be found in the book by Byerly already mentioned. The general theory, including proofs for the statements here made, will be found in Secs. 8.5–8.8. A function $f(x)$ defined between $x = 0$ and $x = \pi$ may also be expanded as a cosine series:

$$f(x) = \tfrac{1}{2}b_0 + \sum_{n=1}^\infty b_n \cos nx \qquad (8\text{--}9a)$$

where

$$b_n = \frac{2}{\pi} \int_0^\pi f(\xi) \cos n\xi d\xi \qquad (8\text{--}9b)$$

except that the series may not yield the same values as $f(x)$ at discontinuities and at the end points. Otherwise the developments (7) and (9) are equivalent. There is, however, an interesting difference in the values of the two series when they are extrapolated to the range $(-\pi,0)$. Here the series (7) changes sign in such a way that $f(-x) = -f(x)$, while series (9) yields $f(-x) = f(x)$, as is evident from the fact that $\sin x$ is an odd, $\cos x$ an even function. Thus, if it is desired to expand a function between $-\pi$ and $+\pi$, series (7) can be used only if the function is odd, series (9) when it is even. Now any function can be represented as the sum of an even and an odd one. Hence, if an arbitrary function $f(x)$ is to be developed between $-\pi$ and π, both cosine and sine series must be used. It is evident, therefore, that in this more general case

$$f(x) = \sum_{n=1}^{\infty} a_n \sin nx + \tfrac{1}{2}b_0 + \sum_{n=1}^{\infty} b_n \cos nx \qquad (8\text{--}10a)$$

where

$$a_n = \frac{1}{\pi}\int_{-\pi}^{\pi} f(\xi) \sin n\xi d\xi; \quad b_n = \frac{1}{\pi}\int_{-\pi}^{\pi} f(\xi) \cos n\xi d\xi \qquad (8\text{--}10b)$$

The coefficients in front of the integrals are most easily checked as follows: Multiply (10a) by $\sin mx$ and integrate over x between $-\pi$ and π. Because of the mutual orthogonality of the functions $\sin nx$, $\sin mx$, $\cos nx$, for $n \neq m$ the relations (10b) are at once apparent.

If $f(x)$ is defined, not in the range $(-\pi,\pi)$, but in $(-l,l)$, a simple change of variable from x, ξ to $(\pi/l)x$, $(\pi/l)\xi$ in eq. (10) will produce the required modification. The result is

$$\left. \begin{array}{l} f(x) = \sum_{n} a_n \sin \dfrac{n\pi}{l} x + \tfrac{1}{2}b_0 + \sum_{n} b_n \cos \dfrac{n\pi}{l} x \\[2mm] a_n = \dfrac{1}{l}\int_{-l}^{l} f(\xi) \sin \dfrac{n\pi}{l} \xi d\xi, \quad b_n = \dfrac{1}{l}\int_{-l}^{l} f(\xi) \cos \dfrac{n\pi}{l} \xi d\xi \end{array} \right\} \qquad (8\text{--}11)$$

This may be expressed more simply in complex form. For if the sine and cosine functions are written in their exponential form, the reader will verify without difficulty[3] that

$$f(x) = \sum_{-\infty}^{\infty} c_n e^{in\pi x/l}, \quad c_n = \frac{1}{2l}\int_{-l}^{l} f(\xi)e^{-in\pi\xi/l}d\xi \qquad (8\text{--}12)$$

The coefficients c_n in this expansion are complex.

When the series $f(x)$ as given by (12) is extrapolated beyond the range $(-l,l)$ the function $f(x)$ is repeated periodically in every interval between

---

[3] Note that $e^{in\pi x/l}$ and $e^{im\pi x/l}$ are orthogonal functions in the sense of eq. 5*, the range of integration being $(-l,l)$.

$(2n + 1)l$ and $(2n + 3)l$.   Hence formula (12) permits representation of periodic functions only.   One may wonder, therefore, whether it is possible to perform a Fourier analysis of a non-periodic function, defined in the range of the entire real axis.   Highly technical considerations for which the reader is referred to more specific treatises[4] affirm this possibility, provided the function, $f(x)$, to be expanded is piecewise continuous and such that the integral $\int_{-\infty}^{\infty} |f(x)|\, dx$ exists.   In that case

$$f(x) = \int_{-\infty}^{\infty} c(k)e^{ikx}dk$$

$$c(k) = \frac{1}{2\pi} \int_{-\infty}^{\infty} f(\xi)e^{-ik\xi}d\xi \qquad (8\text{--}13')$$

These equations may be written more symmetrically by putting $c(k) = (1/\sqrt{2\pi})g(k)$.   They then become

$$\left. \begin{aligned} f(x) &= \frac{1}{\sqrt{2\pi}} \int_{-\infty}^{\infty} g(k)e^{ikx}dk \\ g(k) &= \frac{1}{\sqrt{2\pi}} \int_{-\infty}^{\infty} f(\xi)e^{-ik\xi}d\xi \end{aligned} \right\} \qquad (8\text{--}13)$$

Two functions f and g related by eq. (13) are called a pair of *Fourier transforms;* i.e., g is the Fourier transform of f and vice versa.   Such pairs are of great importance in the analysis of electrical impulses[5] and in quantum mechanics, where they effect the transformation from coordinate to momentum space.

Problems.

 a. Show that the Fourier transform of $f(x) = e^{-x^2/2}$ is $g(k) = e^{-k^2/2}$.   (This fact is occasionally expressed by saying: the error function $e^{-x^2/2}$ is its own Fourier transform.)

 b. Show that the F.T. of the step function $f(x) = \sqrt{2\pi}/2l$ if $|x| < l$ and vanishing if $|x| > l$ is $g(k) = \sin kl/kl$.   Note: as l approaches zero, $f(x)$ becomes ∞ at $x = 0$.   It is then called a " unit impulse " function, or a δ-function.   Its transform $g(k) = 1$.

[4] E.g., Titchmarsh, E. C., " Introduction to the Theory of Fourier Integrals," Oxford University Press, 1937.

 [5] For further considerations see v. Kármán, T., and Biot, M., " Mathematical Methods in Engineering," McGraw-Hill Book Co., 1940.   An extensive list of Fourier transforms has been compiled by Campbell, G. A., and Foster, R. M., "Fourier Integrals for Practical Applications," *Bell Tel. Syst. Tech. Pub.* Monograph B–584, 1931.   See also Magnus, W., and Oberhettinger, F., " Formulas and Theorems for the Special Functions of Mathematical Physics," Chelsea Publishing Co., New York, 1949.

c. Show that the F.T. of $f(x) = \begin{cases} \sqrt{\dfrac{\pi}{2}} \cos k_0 x & \text{if } |x| < l \\ 0 & \text{if } |x| > l \end{cases}$ is $\dfrac{\sin[(k_0 - k)l]}{k_0 - k}$

The *Fourier Integral Theorem* may be deduced immediately from (13). On putting $g(k)$ into the integral for $f(x)$, there results

$$f(x) = \frac{1}{2\pi} \int \int_{-\infty}^{\infty} f(\xi) e^{ik(x-\xi)} dk d\xi \qquad (8\text{--}14)$$

When $f(x)$ is real, the imaginary part of $e^{ik(x-\xi)}$ may clearly be neglected, and the Fourier integral theorem takes the more customary form:

$$f(x) = \frac{1}{2\pi} \int_{-\infty}^{\infty} dk \int_{-\infty}^{\infty} f(\xi) \cos k(x - \xi) d\xi \qquad (8\text{--}14')$$

Finally one may derive from (14) a result sometimes called the Dirichlet integral. On performing the integration not between infinite limits but between $-A$ and A, and then passing to the limit $A \to \infty$ we find

$$f(x) = \frac{1}{\pi} \lim_{A \to \infty} \int_{-\infty}^{\infty} f(\xi) \frac{\sin[A(x - \xi)]}{x - \xi} d\xi \qquad (8\text{--}15)$$

As a special form of (15) we note:

$$f(0) = \frac{1}{\pi} \lim_{A \to \infty} \int_{-\infty}^{\infty} f(x) \frac{\sin Ax}{x} dx$$

The expression $\dfrac{1}{\pi} \lim\limits_{A \to \infty} \dfrac{\sin[A(x - \xi)]}{x - \xi}$ or $\dfrac{1}{2\pi} \int_{-\infty}^{\infty} e^{i(x-\xi)t} dt$ is called the Dirac δ-function and denoted by $\delta(x,\xi)$. Eq. (15) may therefore be written

$$f(x) = \int_{-\infty}^{\infty} f(\xi) \delta(x,\xi) d\xi \qquad (8\text{--}15')$$

All the foregoing results can be generalized[6] to permit expansion of functions of several variables, provided they satisfy the condition

$$\int |f(x,y,z \cdots)| \, dx \, dy \, dz \cdots$$

exists. For instance, in place of (12) we have

$$\left.\begin{array}{l} f(x,y) = \displaystyle\sum_{m,n = -\infty}^{\infty} c_{m,n} e^{i(\pi/l)(mx+ny)} \\[2em] c_{m,n} = \dfrac{1}{4l^2} \displaystyle\int \int_{-l}^{l} f(\xi,\eta) e^{-i(\pi/l)(m\xi+n\eta)} d\xi d\eta \end{array}\right\} \qquad (8\text{--}16)$$

---

[6] See Courant, R., "Vorlesungen über Differential- und Integralrechnung," Vol. 1, Second Edition, p. 373.

and in place of (13),

$$f(x,y) = \frac{1}{2\pi} \int \int_{-\infty}^{\infty} g(k_1 k_2) e^{i(k_1 x + k_2 y)} dk_1 dk_2$$

$$g(k_1 k_2) = \frac{1}{2\pi} \int \int_{-\infty}^{\infty} f(\xi,\eta) e^{-i(k_1\xi + k_2\eta)} d\xi d\eta$$

$$(8\text{-}17)$$

8.3. Vibrating Circular Membrane; Fourier-Bessel Transforms.—The mathematical description of the vibrating membrane also leads to an interesting eigenvalue problem. The wave equation, when written in polar coordinates, was shown in sec. 7.10 (cf. eq. 7–35) to have the solution

$$U = S \cdot T, \quad S_{k,m} = Z_m(k\rho)e^{\pm im\varphi} \qquad (8\text{-}18)$$

The fact, pointed out before, that m must here be an integer to insure the function to be physically meaningful ($e^{\pm im\varphi}$ must be the same as $e^{\pm im(\varphi + 2\pi)}$ because φ and $\varphi + 2\pi$ denote the same angle in the problem of the membrane) may also be expressed by saying: the eigenvalues of m in the differential equation $\Phi'' = -m^2\Phi$ are all integers. Note that the corresponding eigenfunctions, $e^{im\varphi}$, are orthogonal and form a complete set, the range being $(0,2\pi)$. But we wish here to discuss another, less simple eigenvalue problem.

Consider modes of vibration of the membrane which have circular symmetry. This limits m to the value zero, and (18) becomes

$$S_k = Z_0(k\rho) \qquad (8\text{-}19)$$

We now impose the boundary condition: $U = 0$ at all times at the periphery of the membrane, corresponding to the physical condition of having the edge fixed. If the radius of the membrane is a, this means

$$Z_0(ka) = 0 \qquad (8\text{-}20)$$

The function Z_0 is a linear combination of the Bessel functions J_0 and N_0, a Bessel function of the second kind which is linearly independent of J_0 (sometimes called a *Neumann* function). But the latter may be shown to be infinite at $\rho = 0$ and must therefore be excluded. The Z_0 in (19) and (20) must therefore be interpreted as J_0. To satisfy (20) the parameter k must be so adjusted as to make ka a root of J_0, and since J_0 has an infinite number of roots,[7] the eigenvalues of k will form an infinite set $k_i = x_i/a$, where x_i is the i-th root of $J_0(x)$. The corresponding eigenfunctions are $J_0(k_i\rho)$.

[7] The values of the roots of $J_0(x)$ are listed in books on Bessel functions. See also Jahnke, E., and Emde, F., " Tables of Functions," Dover Publications, New York, 1943.

Are these functions orthogonal? It is not difficult to show that $\int_0^a J_0(k_1\rho)J_0(k_2\rho)d\rho$ is different from zero (an inspection of the graph of the integrand will convince the reader). Thus it seems that eq. (5) fails in this example. But we have overlooked an important feature: the element of area of the circular membrane is not $d\rho$, but $2\pi\rho d\rho$. And now it will be found that

$$\int_0^a J_0(k_m\rho)J_0(k_n\rho)\rho d\rho = c_n\delta_{m,n} \qquad (8\text{--}21)$$

As the present problem shows, specification of a *range* of integration is not sufficient in defining orthogonality of functions; it is also necessary to state the *weighting* factor associated with each differential range of the coordinate. In the problem of the vibrating string, the weighting factor $w(x)$ happened to be unity; here it is $w(\rho) = \rho$. In the next example it will be seen to be ρ^2. The same w which appears in the orthogonality relation will also occur in the integrals defining expansion coefficients (cf. eq. 42).

To prove eq. (21) for $m \neq n$ we use the last of the formulas in sec. 3.9, according to which the left-hand side has the value 0 because both $J_0(k_1 a)$ and $J_0(k_2 a)$ vanish. According to another formula in this list,

$$\int_0^a [J_0(k_n\rho)]^2\rho d\rho = -\frac{a^2}{2}J_{-1}(k_n a)J_1(k_n a)$$

But in view of eq. (3–69), $J_{-1} = -J_1$, so that the constant c_n in (21) has the value $(a^2/2)[J_1(k_n a)]^2$.

The question of the completeness of the functions $J_0(k_n\rho)$, i.e., the possibility of the expansion

$$f(\rho) = \sum_{n=1}^{\infty} a_n J_0(k_n\rho) \qquad (8\text{--}22)$$

will be investigated in sec. 8. We shall here anticipate completeness provided, of course, that $f(\rho)$ vanishes also at $\rho = a$. Granting this, the coefficients a_n may be computed in the manner already illustrated in connection with Fourier series:

Multiply both sides of (22) by $J_0(k_m\rho)\rho d\rho$ and integrate. The result is, again in view of (21),

$$\int_0^a f(\rho)J_0(k_m\rho)\rho d\rho = a_m \cdot \frac{a^2}{2}[J_1(k_m a)]^2$$

If we use the *normalized* function $S_n = (\sqrt{2}/a)[J_1(k_n a)]^{-1}J_0(k_n\rho)$, the expansion reads

$$f(\rho) = \sum_1^{\infty} a_n S_n(\rho)$$

and the coefficients are

$$a_n = \int_0^a f(\rho)S_n(\rho)\rho d\rho$$

The problem of the circular membrane has been simplified by our assumption of circular symmetry. One may wonder what happens if types of vibrations are permitted in which the displacement is a function of both ρ and φ, for these certainly occur. It is then necessary to use the function $S_{k,m}$ defined in eq. (18). These may easily be seen to be orthogonal with respect to both indices, i.e.,

$$\int_0^a \rho d\rho \int_0^{2\pi} J_{m_1}(k_{n_1}\rho)e^{-im_1\varphi}J_{m_2}(k_{n_2}\rho)e^{im_2\varphi}d\varphi = c\delta_{m_1 m_2}\delta_{n_1 n_2}$$

Moreover it is possible to expand

$$f(\rho,\varphi) = \sum_{n,m} a_{nm}J_m(k_n\rho)e^{im\varphi}$$

The details of this development may be left as an exercise to the interested reader; they are worked out fully in some works on sound.[8]

The condition $f(a) = 0$, upon which the expansion (22) was based, may be removed; the range of integration must then be extended from 0 to ∞. Now it is clear that, as $a \rightarrow \infty$, the values k_n move closer and closer together. In the limit they will, in fact, form a continuum. When the passage to this limit is performed, eq. (22) becomes what is known as a *Fourier-Bessel integral*,[9] an equation which is useful in the theory of radiation. While the transition to the limit is difficult, the result may be obtained quite simply by a method used by Stratton,[10] which will here be given.

Suppose $f(x,y)$ can be expanded according to eq. (17). In these equations, we transform the variables of integrations to polar form:

$$x = \rho\cos\varphi, \quad y = \rho\sin\varphi; \quad k_1 = k\cos\alpha, \quad k_2 = k\sin\alpha$$

They then read

$$f(\rho,\varphi) = \frac{1}{2\pi}\int_0^\infty kdk\int_0^{2\pi} g(k,\alpha)e^{ik\rho\cos(\varphi-\alpha)}d\alpha \qquad (8\text{--}23a)$$

$$g(k,\alpha) = \frac{1}{2\pi}\int_0^\infty \rho d\rho\int_0^{2\pi} f(\rho,\varphi)e^{-ik\rho\cos(\varphi-\alpha)}d\varphi \qquad (8\text{--}23b)$$

[8] See particularly Morse, P. M., " Vibration and Sound," McGraw-Hill Book Co., 1936, p. 153 et seq.

[9] We are here following a terminology which seems to be gaining ground, although we have been unable to discover its origin. It appears that relations of the form (24) were first discovered by *Hankel*.

[10] Stratton, J. A., " Electromagnetic Theory," McGraw-Hill Book Co., 1941.

We now take for $f(\rho,\varphi)$ the special function $f(\rho)e^{im\varphi}$. The integration over φ appearing in (23b) may then be performed with the use of eq. (3–72a), according to which

$$\int_0^{2\pi} e^{i[m\varphi - k\rho\cos(\varphi-\alpha)]}d\varphi = \int_0^{2\pi} e^{i[m\varphi - k\rho\sin(\varphi-\alpha+\pi/2)]}d\varphi$$

$$= e^{im(\alpha-\pi/2)} \cdot \int_0^{2\pi} e^{i[m\theta - k\rho\sin\theta]}\,d\theta = 2e^{im(\alpha-\pi/2)}\int_0^{\pi} \cos\,[m\theta - k\rho\sin\theta]d\theta$$

$$= 2\pi e^{im(\alpha-\pi/2)} \cdot J_m(k\rho)$$

Thus

$$g(k,\alpha) = \int_0^{\infty} f(\rho)J_m(k\rho)\rho d\rho \cdot e^{im(\alpha-\pi/2)} \equiv g(k) \cdot e^{im(\alpha-\pi/2)} \qquad (8\text{–}24\text{b})$$

On putting this answer into (23a) we find

$$f(\rho)e^{im\varphi} = \frac{1}{2\pi}\int_0^{\infty} g(k)kdk \int_0^{2\pi} e^{i[m(\alpha-\pi/2)+k\rho\cos(\varphi-\alpha)]}d\alpha$$

$$= \frac{1}{2\pi}\int_0^{\infty} g(k)kdk \cdot 2\pi e^{im\varphi}J_m(k\rho) \qquad (8\text{–}24\text{a})$$

These results may be expressed in the symmetrical form

$$f(\rho) = \int_0^{\infty} g(k)J_m(k\rho)kdk \qquad (8\text{–}24\text{a})$$

$$g(k) = \int_0^{\infty} f(\rho)J_m(k\rho)\rho d\rho \qquad (8\text{–}24\text{b})$$

The functions f and g satisfying relations (24a, b) are said to be a pair of *Fourier-Bessel* transforms. It is to be noted that the expansion (24a) of the function $f(\rho)$ holds for every value of the integer m. Eq. (22), therefore, is a special case of a Fourier-Bessel expansion.

Problem a. Show, using the formulas of sec. 3.9, that the Fourier-Bessel transform of $f(\rho) = \rho^r$, with respect to J_n, is

$$g(k) = \frac{2^{r+1}\,\Gamma\left(\dfrac{n+r}{2}+1\right)k^{-r-2}}{\Gamma\left(\dfrac{n-r}{2}\right)}$$

Problem b. Verify the identity

$$f(\xi) = \int_0^{\infty}\int_0^{\infty} f(\rho)J_m(k\rho)J_m(k\xi)\rho k d\rho dk$$

8.4. Vibrating Sphere with Fixed Surface.—The problem of a sphere vibrating with a node at its surface is of little interest in acoustics, for if there is never any displacement at the surface, the sphere cannot radiate. However, the same problem, interpreted quantum-mechanically, describes the motion of a particle within a spherical cavity and has as such enjoyed some attention in nuclear physics. For the sake of simplicity, we shall here maintain the acoustic interpretation.

The solution, S, of the space part of the wave equation was shown in eq. (7–43) et seq. to be of the form

$$S_k = \sum_{l=0}^{\infty} c_{kl} Y_l(\theta,\varphi) r^{-1/2} Z_{l+1/2}(kr) \tag{8-25}$$

As usual, k determines the frequency of the vibration: $\nu = kv/2\pi$, v being the velocity of the waves inside the spherical medium. Eigenvalues in k, and hence in the frequency "spectrum," are induced by the boundary condition

$$S_k = 0 \quad \text{at} \quad r = a, \quad \text{the radius of the sphere}$$

According to (25), this is satisfied only if $Z_{l+1/2}(ka) = 0$. Thus, for every integer l, there exists an infinite sequence of k_i such that $k_i a$ is a root of $Z_{l+1/2}$. But $Z_{l+1/2}$ is a linear combination of $J_{l+1/2}$ and $J_{-l-1/2}$, of which only the former can be retained because $r^{-1/2} J_{-l-1/2}$ is always infinite at $r = 0$ and does not, therefore, represent a possible mode of vibration. Hence it is

$$J_{l+1/2}(ka) = 0$$

which determines the eigenvalues of k.

When $l = 0$ the situation is very simple indeed, for $J_{1/2}(x) = \sqrt{2/\pi x}\sin x$. Thus the k's are determined by $\sin(ka) = 0$, which means that for this case

$$k_{0,n} = \frac{n\pi}{a}, \quad n \text{ an integer}$$

The frequency spectrum is much the same as for the vibrating string. Let us now see what is the physical meaning of the condition $l = 0$. A glance at eq. (25) shows that $Y_l(\theta,\varphi)$ is a constant, and this means there are no radial nodes. The sphere vibrates in spherical symmetry.

In addition to these eigenvalues, which have a linear distribution, there are the other sets given by $J_{l+1/2}(k_{l,n}a) = 0$. These are irregularly distributed and interspersed between the k_{0n}.

The orthogonality of the S_k (eq. 25) is at once evident. Orthogonality with respect to the index l arises from a property of the spherical harmonics proved in sec. 3.53. But even for the same l and different k the functions retain their orthogonality. The *weighting factor* in this case is r^2 because the volume element contains this factor. Thus

$$\int_0^a S_{k_1} S_{k_2} r^2 dr \propto \int_0^a J_{l+1/2}(k_1 r) J_{l+1/2}(k_2 r) r dr$$

an expression which vanishes unless $k_1 = k_2$ as is seen from the last formula of sec. 3.9.

By more special considerations it may also be shown that the set of functions is complete in the sense that any $f(r,\theta,\varphi)$ which vanishes at $r = a$ and is piecewise continuous can be expanded in the form $\sum_l Y_l \sum_n c_{l,k_n} r^{-1/2} J_{l+1/2}(k_{l,n} r)$. For the special case $l = 0$ this expansion reduces to a Fourier series.

Problem. Compute the lowest 12 eigenfrequencies of the vibrating sphere.

8.5. Laplace and Related Transformations.—A Laplace transformation of the function $F(t)$ is

$$\int_0^\infty F(t)e^{-st}dt = f(s) \qquad (8\text{--}26)$$

The function $f(s)$ is the *Laplace transform* of $F(t)$. Symbolically, we write

$$f(s) = \mathfrak{L}\{F\} \qquad (8\text{--}27)$$

If, in eq. (26), we put $t = -\ln z$, we get

$$f(s) = -\int_1^0 \phi(z)z^{s-1}dz = \int_0^1 \phi(z)z^{s-1}dz$$

This is called a *Mellin transformation.*

If $s = -ix$, eq. (26) reads

$$f(-ix) = \int_0^\infty F(t)e^{ixt}dt$$

It represents a *Fourier transformation* of the function $f(-ix)$. In eq. (8–13′) we have encountered a formula very similar to this, except that there the transformation was " two-sided," or bilateral, i.e., the integral was extended from $-\infty$ to $+\infty$, and the function was called $f(x)$ instead of $f(-ix)$. In this section we limit our study to one-sided Laplace transformations, which are the ones usually encountered in practice.

We shall now derive a formula expressing $F(t)$ in terms of $f(s)$, i.e., a formula which represents the inversion of eq. (26).

By Cauchy's integral theorem, eq. (3–1),

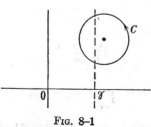

FIG. 8–1

$$f(s) = \frac{1}{2\pi i}\int_C \frac{f(z)dz}{z - s}$$

Suppose $f(z)$ is analytic to the right of the line $x = \gamma$ (see Fig. 1). We can then distort the contour C and integrate from $\gamma + i\infty$ to $\gamma - i\infty$, thence to the right to $\infty - i\infty$, up to $\infty + i\infty$ and back to $\gamma + i\infty$. Only the part from $\gamma + i\infty$ to $\gamma - i\infty$ contributes to the integral. Hence,

$$f(s) = \frac{1}{2\pi i} \int_{\gamma+i\infty}^{\gamma-i\infty} \frac{f(z)dz}{z-s} = \frac{1}{2\pi i} \int_{\gamma-i\infty}^{\gamma+i\infty} \frac{f(z)dz}{s-z}$$

Clearly, γ must be smaller than the real part of s, in symbols: $\gamma < R(s)$. To the last equation we apply the inverse operator $\mathcal{L}^{-1}$, understanding that $\mathcal{L}^{-1}f(s) = F(t)$:

$$\mathcal{L}^{-1}f(s) = \frac{1}{2\pi i} \int_{\gamma-i\infty}^{\gamma+i\infty} f(z)dz\,\mathcal{L}^{-1}\left(\frac{1}{s-z}\right) \tag{8-28}$$

But we now show that $\mathcal{L}^{-1}\left(\dfrac{1}{s-z}\right) = e^{zt}$. We have

$$\mathcal{L}(e^{zt}) = \int_0^\infty e^{-st}e^{zt}dt$$

$$= \frac{1}{s-z} \quad \text{if} \quad R(s) > R(z)$$

This condition is satisfied in eq. (28). Hence,

$$F(t) = \frac{1}{2\pi i} \int_{\gamma-i\infty}^{\gamma+i\infty} f(z)e^{zt}dz = \frac{1}{2\pi i} \int_{\gamma-i\infty}^{\gamma+i\infty} f(s)e^{st}ds \tag{8-29}$$

Here γ is any real number such that, to the right of $R(z) = \gamma$ (a vertical line through γ), the function $f(z)$ is analytic. Equations (26) and (29) represent a Laplace transformation of $F(t)$ and its inverse. $F(t)$ and $f(s)$ are said to be a pair of Laplace transforms.

When the function $F(t)$ is changed to some other function, $f(s)$ will likewise undergo a change. It is useful to study such correlated changes.

Suppose $\qquad\qquad f(s) = \mathcal{L}[F(t)]$

Now " operate " on the function $F(t)$ with some operator P, converting it into $PF(t)$. The transform of this function may be called $pf(s)$, so that

$$pf(s) = \mathcal{L}[PF(t)]$$

We want to know what operator p corresponds to P.

(1) Let P be a *linear substitution*:

$$PF(t) = F(at - b), \quad a \geq 0, \quad b > 0$$

$$\mathcal{L}[F(at - b)] = \int_0^\infty F(at - b)e^{-st}dt = \frac{1}{a}e^{-bs/a}\int_{-b}^\infty F(\tau)e^{-s\tau/a}d\tau$$

Insofar as the Laplace transformation is concerned, only the behavior of

$F(t)$ for positive t and zero is important.   We shall now define F to be zero for $t < 0$.   Thus,

if F vanishes for negative arguments,

$$\mathcal{L}[F(at - b)] = \frac{e^{-bs/a}}{a} \int_0^\infty F(\tau)e^{-s\tau/a}d\tau,$$

or

$$\mathcal{L}[F(at - b)] = \frac{e^{-bs/a}}{a} f\left(\frac{s}{a}\right) \qquad (8\text{--}30)$$

The operator p which corresponds to our linear substitution is: multiplication by $\dfrac{e^{-bs}}{a}$ and substitution of $\dfrac{s}{a}$ for s.

(2) *Integration.*   Let $P = \displaystyle\int_0^t dt$.

We wish to find $\displaystyle\int_0^\infty dt e^{-st} \int_0^t F(\tau)d\tau = \int_0^\infty dt e^{-st}\phi(t)$ where $\dfrac{d\phi}{dt} = F(t)$.
Integrate by parts, obtaining

$$-\frac{1}{s}\phi e^{-st}\Big|_0^\infty + \frac{1}{s}\int_0^\infty F(t)e^{-st}dt$$

The integrated part vanishes; hence

$$\mathcal{L}\left[\int_0^t F(\tau)d\tau\right] = \frac{1}{s}f(s)$$

To integration, there corresponds *division by s*.   Also, for iterated integration, one finds by repeated application of this formula

$$\mathcal{L}\left[\left(\int_0^t d\tau\right)^n F(\tau)\right] = s^{-n}\mathcal{L}[F] \qquad (8\text{--}31)$$

(3) *Differentiation.*

$$\mathcal{L}[F'] = \int_0^\infty e^{-st}\frac{dF}{dt}\,dt = e^{-st}F\Big|_0^\infty + s\int_0^\infty e^{-st}F\,dt$$

$$= s\mathcal{L}[F] - F(0) = sf(s) - F(0) \qquad (8\text{--}32)$$

provided $F(0)$ is the value of F at $t = 0$.

(4) *Convolution.*   If F_1 and F_2 are both functions of t, the integral

$$\int_0^t F_1(\tau)F_2(t - \tau)d\tau \qquad (8\text{--}33)$$

is often denoted by $F_1 * F_2$ and called the *convolution* or *Faltung* of F_1 and F_2.   It is of frequent occurrence in physical problems.   Suppose, for instance, that an error, ϵ, is the linear result of two individual errors,

$\epsilon = \epsilon_1 + \epsilon_2$, and that we know the distributions, or probabilities of ϵ_1 and ϵ_2. These are $w_1(\epsilon_1)$ and $w_2(\epsilon_2)$. The distribution of ϵ is then clearly

$$w(\epsilon) = \int\int_{\epsilon_1+\epsilon_2=\epsilon} w_1(\epsilon_1)w_2(\epsilon_2)d\epsilon_1 d\epsilon_2 = \int w_1(\epsilon_1)w_2(\epsilon - \epsilon_1)d\epsilon_1 = w_1 * w_2$$

The German word " Faltung " means folding; it arises from the following simple fact. If a line of length t be folded back in the middle, as in Fig. 8–2, the points adjacent to each other on the two segments are those which lie, respectively, at distances τ and $t - \tau$ from the origin 0. These,

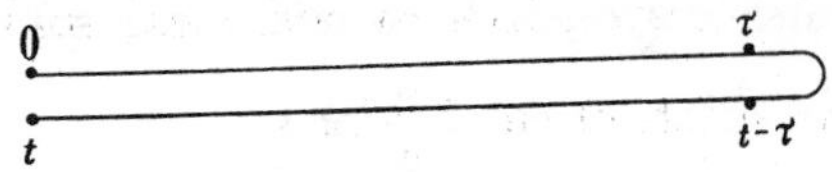

FIG. 8–2

however, are the arguments of the functions F_1 and F_2 that occur in the convolution integral. One final comment regarding this integral: If F_2 is defined to be zero for negative arguments, the upper limit, instead of being t, can be taken to be ∞.

The Laplace transform of a convolution is very easy to compute.

$$\mathfrak{L}[F_1 * F_2] = \int_0^\infty dt e^{-st}\int_0^t F_1(\tau)F_2(t - \tau)d\tau$$

$$= \int_{-\tau}^\infty dr' \int_0^t d\tau e^{-s(\tau+\tau')}F_1(\tau)F_2(\tau')$$

$$= \int_0^\infty F_2(\tau')e^{-s\tau'}d\tau' \int_0^\infty F_1(\tau)e^{-s\tau}d\tau$$

if $F_2(t) = 0$ for $t < 0$.

Hence,

$$\mathfrak{L}[F_1 * F_2] = f_1 f_2 \tag{8–34}$$

The transform of the convolution is the product of the transforms of F_1 and F_2. Note also that convolution is commutative,

$$F_1 * F_2 = F_2 * F_1$$

and associative,

$$(F_1 * F_2) * F_3 = F_1 * (F_2 * F_3)$$

(5) *Multiplication.* To find

$$\mathfrak{L}[F_1(t) \cdot F_2(t)]$$

we consider the triple integral

$$\frac{1}{2\pi}\int_{-\infty}^\infty dz \int_0^\infty e^{-(s-iz)t_1}F_1(t_1)dt_1 \int_0^\infty e^{-izt_2}F_2(t_2)dt_2$$

which is, by definition, $\dfrac{1}{2\pi}\displaystyle\int_{-\infty}^{\infty} dz f_1(s-iz)f_2(iz)$. On integrating over z,

there results $\dfrac{1}{2\pi}\displaystyle\int_{-\infty}^{\infty} e^{iz(t_1-t_2)}dz = \delta(t_1,t_2)$, the δ-function defined in (8–15$'$).
Hence the integral becomes

$$\int_0^\infty e^{-st_1}F_1(t_1)dt_1 \int_0^\infty F_2(t_2)\delta(t_1,t_2)dt_2 = \int_0^\infty e^{-st_1}F_1(t_1)F_2(t_1)dt_1$$

Therefore

$$\mathfrak{L}(F_1F_2) = \frac{1}{2\pi}\int_{-\infty}^{\infty} dz f_1(s-iz)f_2(iz)$$

If the variable is changed from iz to z,

$$\mathfrak{L}[F_1F_2] = \frac{1}{2\pi i}\int_{-i\infty}^{i\infty} dz f_1(s-z)f_2(z) \tag{8–35}$$

The transform of a product is a convolution along the imaginary axis.

For Fourier transformations, we have

$$\mathfrak{F}[F_1] = \int_{-\infty}^{\infty} e^{-ist}F_1(t)dt, \quad \mathfrak{F}[F_2] = \int_{-\infty}^{\infty} e^{-ist}F_2 dt$$

$$\frac{1}{2\pi}\int_{-\infty}^{\infty} dz f_1(s-z)f_2(z) = \frac{1}{2\pi}\int_{-\infty}^{\infty} dz \int_{-\infty}^{\infty} e^{-i(s-z)t_1}F_1(t_1)\int_{-\infty}^{\infty} e^{-izt_2}F_2(t_2)dt_2$$

The integration over z yields $\delta(t_1,t_2)$; hence

$$\frac{1}{2\pi}\int_{-\infty}^{\infty} dz f_1(s-z)f_2(z) = \int_{-\infty}^{\infty} e^{-ist}F_1(t)F_2(t)dt = \mathfrak{F}[F_1F_2]$$

The Fourier transform of a product is $\dfrac{1}{2\pi}$ times the ordinary convolution:

If $\mathfrak{F}(F_1) = f_1$, $\mathfrak{F}(F_2) = f_2$, then

$$\mathfrak{F}(F_1F_2) = \frac{1}{2\pi}\int_{-\infty}^{\infty} dz f_1(s-z)f_2(z) = \frac{1}{2\pi}f_1 * f_2 \tag{8–36}$$

8.6. Use of Transforms in Solving Differential Equations.—A. Consider
the differential equation

$$Y'' + k^2 Y = 0 \tag{8–37}$$

where the primes denote differentiation with respect to t.

Multiply by $\displaystyle\int e^{-st}dt$ to obtain

$$\mathfrak{L}(Y'') + k^2\mathfrak{L}(Y) = 0$$

Now by eq. (32),

$$\mathfrak{L}(Y'') = s\mathfrak{L}(Y') - Y'(0)$$

$$\mathfrak{L}(Y') = s\mathfrak{L}(Y) - Y(0)$$

hence,

$$\mathfrak{L}(Y'') = s^2\mathfrak{L}(Y) - Y'(0) - sY(0)$$

If we write $y(s)$ for $\mathfrak{L}(Y)$, eq. (37) becomes

$$s^2 y - Y'(0) - sY(0) + k^2 y = 0$$

Note how nicely the initial conditions on Y and on Y' introduce themselves into the calculation!

On solving, we obtain

$$y = \frac{Y'(0) + sY(0)}{k^2 + s^2} = \frac{a_1}{s + ik} + \frac{a_2}{s - ik}$$

where

$$a_1 = \left[\frac{1}{2}Y(0) + \frac{iY'(0)}{k}\right], \quad a_2 = \frac{1}{2}\left[Y(0) - \frac{iY'(0)}{k}\right]$$

By eq. (29)

$$Y(t) = \frac{1}{2\pi i}\int_{\gamma - i\infty}^{\gamma + i\infty} y(s)e^{st}ds$$

Now

$$\frac{1}{2\pi i}\int_{\gamma - i\infty}^{\gamma + i\infty} \frac{e^{st}ds}{s + ik} = \frac{1}{2\pi i}\int_{\gamma - i\infty}^{\gamma + i\infty} \frac{e^{(s'-ik)t}}{s'}ds' \quad \text{(on putting } s' = s + ik)$$

$$= e^{-ikt}\frac{1}{2\pi i}\int_{\gamma - i\infty}^{\gamma + i\infty} \frac{e^{s't}}{s'}ds'.$$

This integral can be evaluated by the method of residues (see sec. 3.2). Since γ must be positive in order that the singularity at $s' = 0$ be avoided, we integrate along the square drawn in Fig. 3. The extension of the path

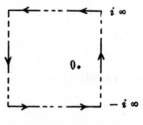

Fɪɢ. 8–3

to close the contour is harmless, for the added parts contribute nothing to the integral. The residue of $\dfrac{e^{s't}}{s'}$ within the contour lies at the origin and equals 1. Hence,

$$\frac{1}{2\pi i} \int_{\gamma-i\infty}^{\gamma+i\infty} \frac{e^{st}ds}{s+ik} = e^{-ikt}$$

The other part of Y, coming from $\dfrac{1}{s-ik}$, yields e^{ikt}. Thus

$$Y = a_1 e^{-ikt} + a_2 e^{ikt} \tag{8-38}$$

The reader can easily verify that this is a solution, indeed the solution which satisfies the initial conditions.

B. Consider the inhomogeneous equation

$$Y'' + k^2 Y = F(t) \tag{8-39}$$

This leads to $\qquad s^2 y - Y'(0) - sY(0) + k^2 y = f(s)$

if f is the transform of F.

Hence,

$$y = \frac{f(s) + Y'(0) + sY(0)}{k^2 + s^2}$$

We thus obtain, in addition to the solution of the homogeneous equation (38), a solution Y_p whose transform is $\dfrac{f(s)}{k^2 + s^2}$. This can often be found in tables. Otherwise we proceed as follows: Let

$$f(s) = f_1, \quad \frac{1}{k^2 + s^2} = f_2$$

Then, by theorem (34), Y is the convolution of $F_1(t)$ and $F_2(t)$. F_1 is the inhomogeneity in the differential equation (39).

We have

$$\frac{1}{k^2 + s^2} = \frac{i}{2k}\left[\frac{1}{s+ik} - \frac{1}{s-ik}\right].$$

Hence

$$F_2(t) = \frac{i}{2k} \cdot \frac{1}{2\pi i} \int_{-i\infty}^{i\infty}\left[\frac{e^{st}}{s+ik} - \frac{e^{st}}{s-ik}\right] ds = \frac{i}{2k}[e^{-ikt} - e^{ikt}] = \frac{\sin kt}{k}$$

$$Y_p = F_1 * F_2 = \int_0^t F_1(\tau)\frac{\sin k(t-\tau)}{k}\, d\tau$$

Problem. Solve the equation, $Y'' + 2bY + w^2 y = f_0 \sin \alpha t$, by this method and compare the result with example a of sec. 2.8.

Table 1 presents a list of Laplace pairs. Such tables have to be used with care, for it is not always easy to state explicitly the conditions under which the integrals converge.

TABLE 1

$F(t)$	$f(s)$
1	$\dfrac{1}{s}$
$t^z,\ (Rz > -1)$	$\dfrac{\Gamma(z+1)}{s^{z+1}}$
e^{at}	$\dfrac{1}{s-a}$
$t^b e^{at}$	$\dfrac{\Gamma(b+1)}{(s-a)^{b+1}}$
$\cos \omega t$	$\dfrac{s}{s^2+\omega^2}$
$\sin \omega t$	$\dfrac{\omega}{s^2+\omega^2}$
$\cosh \omega t$	$\dfrac{s}{s^2-\omega^2}$
$\sinh \omega t$	$\dfrac{\omega}{s^2-\omega^2}$
$\delta(t,\tau)$	$e^{-s\tau}$
$F(t) = \begin{cases} 0 \text{ if } 0 \leqslant t \leqslant \tau \\ 1 \text{ if } t \geqslant \tau \end{cases}$	$\dfrac{e^{-s\tau}}{s}$
$(1 - e^{-t})t^{-1}$	$\ln\left(1 + \dfrac{1}{s}\right)$
$\cos(x\sqrt{t})/\pi\sqrt{t}\ (x \text{ real})$	$e^{\dfrac{-x^2}{4s}} \Big/ \sqrt{\pi s}$
$\sin(x\sqrt{t})/\pi\ (x \text{ real})$	$\dfrac{x}{2\sqrt{\pi}\,s^{3/2}} e^{-x^2/4s}$
$J_o(t)$	$(1 + s^2)^{-1/2}$
$J_n(t);\ R(n) > -1$	$(1 + s^2)^{-1/2}[(1 + s^2)^{1/2} - s]^n$
$L_n(t)$ (See sec. 3.11)	$s^{-n-1}(s - 1)^n$

More extensive tables of Laplace transforms may be found in G. Doetsch, *Laplace-Transformation*, Dover Publishing Co., 1943.

See also:

Carslaw, H. S., and Jaeger, J. C., " Conduction of Heat in Solids," Clarendon Press, Oxford, 1948.

Murnaghan, F. D., " Introduction to Applied Mathematics," John Wiley and Sons, New York, 1948.

Jeffreys, H., "Operational Methods in Mathematical Physics," Cambridge University Press, 1927.

Widder, D., "The Laplace Transform," Princeton University Press, 1941.

Magnus, W. and Oberhettinger, F., "Formulas and Theorems for the Special Functions of Mathematical Physics," Chelsea Publishing Co., New York, 1949. This contains tables of Fourier, Laplace, Hankel, Mellin, and Gauss transforms. For Fourier transforms see:

Carslaw, H. S., and Jaeger, J. C., "Operational Methods in Applied Mathematics," Oxford Press, 1941.

Churchill, R. V., "Fourier Series and Boundary Value Problems," McGraw-Hill Book Co., New York, 1941.

Sneddon, I. N., "Fourier Transforms," McGraw-Hill Book Co., New York, 1951.

Titchmarsh, E. C., "Introduction to the Theory of Fourier Integrals," Oxford Press, 1937.

Wiener, N., "The Fourier Integral and Certain of Its Applications," Cambridge University Press, 1933.

8.7. Sturm-Liouville Theory.—Deeper insight into the nature of eigenvalue problems which arise in connection with second order differential equations is obtained from a study of a theory at once simple and beautiful, the theory of the Sturm-Liouville equation. Nearly every eigenvalue problem encountered in physics and chemistry leads to an equation of the general form

$$L(u) + \lambda wu = 0 \qquad (8\text{--}40)$$

where the *differential operator* L is defined by

$$L(u) \equiv (pu')' - qu \qquad (8\text{--}41)$$

The quantities p, q, and w are understood to be functions of the independent variable x, and we shall suppose that w, which will soon be recognized as the former weighting function, satisfies

$$w(x) \geq 0$$

in the entire range of the variable x. This range is different in different problems, but it will be assumed to be *finite* and to extend from a to b. Finally, λ is a constant; it will turn out to be the eigenvalue parameter.

An operator[11] of the form (41) is said to be *self-adjoint*. The necessary and sufficient condition for the general second order differential operator

$$D(u) = fu'' + gu' + hu$$

(in which f, g, and h are functions of x) to be self-adjoint is simply that $g = f'$. Eq. (40), however, is not a very special one. Every second order

[11] For a general definition of an operator and its adjoint the reader is referred to Courant-Hilbert, "Methoden der Mathematischen Physik," Vol. II, Second Edition, p. 434, or Frank, P. and v. Mises, R., "Differentialgleichungen der Physik," Vol. I, p. 780.

TABLE 2

Name of equation	F	$p(x)$	$q(x)$	λ	$w(x)$	Number of eq. in Chapter 2
Legendre	1	$1-x^2$	0	$l(l+1)$	1	28
Associated Legendre	1	$1-x^2$	$\dfrac{m^2}{1-x^2}$	$l(l+1)$	1	41
Gauss	$-x^{\gamma-1}(1-x)^{\alpha+\beta-\gamma}$	$x^{\gamma}(1-x)^{\alpha+\beta-\gamma+1}$	0	$-\alpha\beta$	$x^{\gamma-1}(1-x)^{\alpha+\beta-\gamma}$	45
Bessel	*	ξ	$\dfrac{n^2}{\xi}$	a^2	ξ	57
Hermite polynomial	e^{-x^2}	e^{-x^2}	0	2α	e^{-x^2}	62
Hermite orthogonal functions	1	1	x^2-1	2α	1	66
Tschebyscheff	$(1-\xi^2)^{-1/2}$	$(1-\xi^2)^{1/2}$	0	n^2	$(1-\xi^2)^{-1/2}$	52
Jacobi	$-x^{q-1}(1-x)^{p-q}$	$x^{q}(1-x)^{p-q+1}$	0	$n(n+p)$	$x^{q-1}(1-x)^{p-q}$	56
Laguerre polynomial	e^{-x}	xe^{-x}	0	α	e^{-x}	67
Associated Laguerre polynomial	$x^k e^{-x}$	$x^{k+1}e^{-x}$	0	$\alpha-k$	$x^k e^{-x}$	70
Associated Laguerre function	x	x^2	$k^2-1+2(k-1)x+x^2$	n	x	71
Mathieu	1	1	$-16b\cos 2x$	a	1	73

* Of Sturm-Liouville type when written in the form

$$\xi \frac{d^2y}{d\xi^2} + \frac{dy}{d\xi} - \frac{n^2}{\xi}y + a^2\xi y = 0, \quad \text{with } \xi = \frac{x}{a}$$

differential operator $D(u)$ can be made self-adjoint; it need only be multiplied from the left by $\exp \int \dfrac{g - f'}{f} \, dx$. Thus *all* differential equations encountered in Chapter 2 may be written in self-adjoint form, and the theory we are presenting applies to them all. In Table 2 we list the factor F by which the equation named on the left, written in the customary form in which it appears in Chapter 2, must be multiplied in order to be self-adjoint, and also the quantities p, q, and w in (40).

The function u is subject to boundary conditions. In the examples of the preceding sections these were of different types: in the problem of the string every u had to vanish at both end points, in the other problem it was to be finite at $r = 0$ but zero at $r = a$. Examination of these and many other examples (see Chapter 11) will show that the boundary condition in most problems of interest may be expressed in the uniform way

$$ puu' \Big|_a = puu' \Big|_b = 0 $$

for usually either p or u or u' vanishes at the end points of the range. But it is equally satisfactory to state these conditions in a somewhat milder form: Let u and v be any permissible solutions of eq. (40); we then require

$$ vpu' \Big|_a = vpu' \Big|_b \qquad (8\text{--}42) $$

On the basis of this condition it is possible to establish the important theorem:

$$ \int_a^b vL(u)dx = \int_a^b uL(v)dx \qquad (8\text{--}43) $$

The proof is straightforward:

$$ \int_a^b vL(u)dx = \int v(pu')'dx - \int vqudx = vpu' \Big|_a^b - \int v'pu'dx - \int vqudx $$

The first term on the right vanishes because of (42); the second may be transformed by another partial integration into $-v'pu \Big|_a^b + \int u(pv')'dx$, of which the first vanishes also. But the remaining integral, $\int [u(pv')' - uqv]dx$, is nothing other than $\int uL(v)dx$. The result (43) is often expressed by saying that the operator L is *Hermitian* with respect to functions satisfying condition (42). The importance of Hermitian operators will be more evident in the next chapters.

8.8. Variational Aspects of the Eigenvalue Problem.[12]—Before proceeding further, the reader is advised to review the main points of Chapter 6. It will be shown that the Sturm-Liouville equation (40) is the *Euler* condition which the function u must satisfy in order that (1) the integral[13]

$$\int (pu'^2 + qu^2)dx \equiv \Lambda(u) \tag{8–44}$$

take on a stationary value, (2) the function u be normalized:

$$\int wu^2 dx = 1 \tag{8–45}$$

The proof is simple. In the notation of sec. 6.5 we have, on writing $\lambda_1 = -\lambda$ for convenience,

$$K = I - \lambda I_1 = pu'^2 + qu^2 - \lambda wu^2$$

and the Euler equation (6–15) is

$$\frac{\partial K}{\partial u} - \frac{d}{dx}\frac{\partial K}{\partial u'} = 0 \tag{8–46}$$

This is clearly identical with (40). The eigenvalue λ here plays the role of a Lagrangian multiplier. We have thus seen that the process of solving the Sturm-Liouville equation is tantamount to a search for those functions $u(x)$ which maximize or minimize Λ, subject to condition (45). This condition is important, for the integral Λ has usually only a single stationary value; but when eq. (45) is imposed Λ has numerous values each of which is stationary for a given neighborhood of functions $u(x)$, although of course only one of them is an absolute minimum or maximum.

Example. Let us see whether the procedure here outlined will actually lead to a simple type of function defined by a Sturm-Liouville equation, say the Legendre polynomial. We start by assuming

$$u = a + bx + cx^2$$

with a, b, and c unknown. From Table 1 we see that $p = 1 - x^2$; hence,

$$\Lambda = \int_{-1}^{1} (1 - x^2)(b^2 + 4bcx + 4c^2x^2)dx = \tfrac{4}{3}b^2 + \tfrac{16}{15}c^2$$

We require that

$$\int u^2 dx = 2a^2 + \tfrac{2}{3}(b^2 + 2ac) + \tfrac{2}{5}c^2 = 1$$

---

[12] The development in this and the following sections leans heavily on Courant-Hilbert, " Methoden der Mathematischen Physik," Vol. I, Second Edition.

[13] Henceforth in this chapter limits of integration will not be indicated when the range is from a to b.

Thus it is necessary to minimize

$$\tfrac{4}{3}b^2 + \tfrac{16}{15}c^2 - \lambda[2a^2 + \tfrac{2}{3}(b^2 + 2ac) + \tfrac{2}{5}c^2]$$

by choice of a, b, and c. On putting the partial derivatives with respect to a, b, and c equal to zero and finally rewriting the normalization condition, four equations are obtained for the determination of the quantities a, b, c, and λ:

$$\lambda\left(a + \frac{c}{3}\right) = 0 \tag{1}$$

$$b(2 - \lambda) = 0 \tag{2}$$

$$c(8 - 3\lambda) - 5a\lambda = 0 \tag{3}$$

$$a^2 + \tfrac{1}{3}(b^2 + 2ac) + \frac{c^2}{5} = \tfrac{1}{2} \tag{4}$$

Suppose we put $c = 0$. Then, according to (1) and (3), $a\lambda = 0$, while (4) yields a relation between a and b. Hence we can put either $a = 0$ or $\lambda = 0$. In the latter instance, i.e., if

$$\lambda = 0$$

we get from (2), $b = 0$, and from (4), $a = \sqrt{\tfrac{1}{2}}$. In the former instance, namely $a = 0$, (2) yields

$$\lambda = 2$$

and (4) gives $b = \sqrt{\tfrac{3}{2}}$.

Now instead of assuming $c = 0$, let us take $b = 0$. Consistency then requires that neither a nor c nor λ can be zero. Hence we find from (1), $c = -3a$, and from (3) and (4),

$$\lambda = 6$$

and $a = \sqrt{\tfrac{5}{8}}$. We have thus determined altogether three solutions, corresponding to three possible values of λ:

λ	u
0	$\sqrt{\tfrac{1}{2}}$
2	$\sqrt{\tfrac{3}{2}}x$
6	$\sqrt{\tfrac{5}{8}}(1 - 3x^2)$

The reader will notice that the λ's are the first three values of $l(l + 1)$, and the u's the first three normalized Legendre polynomials. We now return to eq. (40) in its general form.

Let us assume for definiteness that the extremals of Λ are *minima;* the argument to be presented is equally valid when they are maxima. Also, let $u_1(x)$ be that function which produces the *lowest* minimum of Λ while satisfying (45), and let λ_1 be the eigenvalue corresponding to it. We now seek a function $u_2(x)$ which will also produce a minimum of Λ and satisfy (45), but which, in addition, shall be orthogonal to u_1:

$$\int w u_1 u_2 dx = 0 \qquad (8\text{–}47)$$

The Euler equation for u_2 is more complicated than that for u_1 since u_2 must satisfy *two* accessory conditions and u_1 only one. In fact,

$$K = p u_2'^2 + q u_2^2 - \lambda_2 w u_2^2 - \mu w u_1 u_2$$

μ being a new Lagrangian multiplier. Hence eq. (46) becomes

$$2 q u_2 - 2 \lambda_2 w u_2 - \mu w u_1 - 2(p u_2')' = 0$$

and this is identical with

$$L(u_2) + \lambda_2 w u_2 + \tfrac{1}{2} \mu w u_1 = 0$$

To determine the value of μ we multiply this equation by u_1 and integrate, making use of relation (43) which, of course, we require u_1 and u_2 to obey. The result is

$$\int u_2 L(u_1) dx + \lambda_2 \int w u_1 u_2 dx + \tfrac{1}{2}\mu \int w u_1^2 dx = 0 \qquad (8\text{–}48)$$

Here the first term is $-\lambda_1 \int w u_1 u_2 dx$ because u_1 satisfies (40), and this equals zero because of (47). For the latter reason, the second term of (48) also vanishes. But the integral appearing in the third term is certainly finite. Hence, we conclude that the multiplier $\mu = 0$; we might as well not have required relation (47): u_2 satisfies the same equation as u_1, but for a different eigenvalue λ_2. Moreover, it is *automatically* orthogonal to u_1.

This process may be continued. Suppose we seek a function u_3 which will minimize Λ, subject to the three conditions

$$\int w u_3^2 dx = 1, \quad \int w u_1 u_3 dx = \int w u_2 u_3 dx = 0$$

The minimum thus obtained will lie at least as high as that produced by u_2, for the choice of functions has been further restricted. The quantity K appearing in Euler's equation now contains three undetermined parameters, λ_3, μ, and ν. The last two of these may be shown to vanish by a method similar to that above. By further extension of this process we are led to this result: If we desire a set of functions which (1) minimize Λ, (2) are normalized, (3) are mutually orthogonal, they are found as solutions of eq. (40).

Conversely, it is easy to show that all solutions of (40) belonging to different eigenvalues are orthogonal. To do this, one need only multiply two specific forms of (40):

$$L(u_i) + \lambda_i w u_i = 0, \quad L(u_j) + \lambda_j w u_j = 0$$

by u_j and u_i respectively, integrate each equation and subtract. When (43) is used, the result is simply

$$(\lambda_i - \lambda_j) \int w u_i u_j dx = 0 \tag{8-49}$$

Hence, either $\lambda_i = \lambda_j$, or u_i and u_j are orthogonal.

The case in which $\lambda_i = \lambda_j$, where two (or more) eigenfunctions belong to the same eigenvalue, is not of very great interest under the simple conditions we are here considering (*real* eigenfunctions, *one* independent variable). It is very much more important in the more general eigenvalue problems of Chapter 11. As to terminology, whenever several eigenfunctions, i.e., *linearly independent* eigenfunctions, are associated with one eigenvalue, that eigenvalue is said to be *degenerate*.

It may seem strange to find eq. (49) predicting orthogonality only for non-degenerate cases, while the variational argument of the preceding paragraphs implies no restriction of this sort. Harmony is restored when we realize that a set of linearly independent solutions of eq. (40) can always be combined in such a way as to form an equally numerous, equivalent set of orthogonal solutions. (Cf., for instance, the method of Schmidt,[14] sec. 10.8). Hence we may, if we like, speak of the orthogonality of *all* solutions of eq. (40), assuming tacitly that the process of orthogonalization has been carried out on all sets of functions belonging to a degenerate eigenvalue.

One further point is to be made in connection with the variational property of the solutions of the Sturm-Liouville equation. We have seen that the u_i minimize the integral Λ. What are the stationary values of Λ thus produced? Let us compute them.

$$\Lambda(u_i) = \int (p u_i'^2 + q u_i^2) dx = u_i p u_i' \Big|_a^b - \int [u_i (p u_i')' - u_i q u_i] dx$$

$$= - \int u_i L(u_i) dx = \lambda_i \int w u_i^2 dx = \lambda_i \tag{8-51}$$

The simple and interesting answer is, then, that the stationary values of Λ are the eigenvalues λ_i.

---

[14] The use of this method for functions instead of vectors is illustrated in Lindsay and Margenau, " Foundations of Physics," John Wiley and Sons, p. 425.

Example. Degeneracy arises when, in the vibrating string problem expressed by eq. (2), one replaces the ordinary boundary conditions (a) and (b), sec. 2, by one requiring only *periodicity:*

$$S(a) = S(b) \tag{8–50}$$

The eigenvalue parameter, λ, in this equation is k^2. Moreover, it is to be noted that the periodicity condition (50) conforms to our general requirement (42). The solution satisfying (50) is easily seen to be

$$S = A \sin\left(\delta + \frac{2\pi n}{l} x\right)$$

where $l = b - a$, and δ is arbitrary, the quantity k taking on the values $2\pi n/l$. But n may be a positive or a negative integer. Hence to the same value of k^2, namely $4\pi^2 n^2/l^2$, there correspond the *two* functions

$$S_1 = A_1 \sin\left(\delta + \frac{2\pi n}{l} x\right), \quad S_2 = A_2 \sin\left(\delta - \frac{2\pi n}{l} x\right)$$

Except when δ is an integral multiple of π, as it must be when the ordinary boundary condition is imposed, S_1 and S_2 are linearly independent. Yet they are not orthogonal (except in the special case when $\delta = \pi/4$). It is easily seen, however, that if we put

$$\Sigma_1 = \sqrt{\frac{2}{l}} \sin\left(\delta + \frac{2\pi n}{l} x\right)$$

$$\Sigma_2 = \sqrt{\frac{2}{l(1 - s^2)}}\left[s \sin\left(\delta + \frac{2\pi n}{l} x\right) - \sin\left(\delta - \frac{2\pi n}{l} x\right)\right]$$

wherein $s = \sin^2 \delta - \cos^2 \delta$, we have a pair of functions, satisfying the differential equation for the same k^2, which are both orthogonal and normal.

Problem. The integral $\Lambda\,(u)$ for the differential equation $u'' + k^2 u = 0$ is $\int_0^l (u')^2 dx$. Assume for u any normalized polynomial containing the factors x and $x - l$, and show that Λ computed for this u is greater than the lowest eigenvalue π^2/l^2.

8.9. Distribution of High Eigenvalues.—Preceding considerations indicate no uniform law according to which the eigenvalues of any differential equation are arranged; regularity does, however, prevail for the "high" eigenvalues, as will now be shown. Let all λ's be arranged in numerical order, so that λ_0 is the lowest. Although no proof of the existence of an infinite number of eigenvalues has here been given, their variational meaning strongly suggests[15] and the examples confirm this expectation. We shall now prove the theorem:

$$\lim_{n \to \infty} \lambda_n = \text{const. } n^2 \tag{8-52}$$

Under the substitutions

$$z = (pw)^{1/4}u, \quad t = \int_a^x \left(\frac{w}{p}\right)^{1/2} dx$$

the Sturm-Liouville equation takes the form

$$\frac{d^2z}{dt^2} - f(t)z + \lambda z = 0 \tag{8-53}$$

Detailed consideration which may be left to the reader shows that the function $f(t)$ is bounded.

Now consider, in place of (53), the differential equation

$$\frac{d^2z}{dt^2} + \lambda'z = 0 \tag{8-53'}$$

Its eigenvalues are minima of $\int_0^\tau \left(\frac{dz}{dt}\right)^2 dt$, where τ is the value of t at $x = b$, namely $\tau = \int_a^b \left(\frac{w}{p}\right)^{1/2} dx$. On the other hand, the eigenvalues of (53) are the minima of

$$\int_0^\tau \left[\left(\frac{dz}{dt}\right)^2 + fz^2\right] dt$$

Thus

$$\lambda \equiv \text{minimum of } \int_0^\tau \left[\left(\frac{dz}{dt}\right)^2 + fz^2\right] dt$$

$$\lambda' \equiv \text{minimum of } \int_0^\tau \left(\frac{dz}{dt}\right)^2 dt$$

Assume that z' is the specific function which produces the minimum λ', whereas z produces λ. If we compute λ using z', we shall obtain a value for the integral that is greater than its minimum, λ. Hence,

$$\lambda \leq \int_0^\tau \left[\left(\frac{dz'}{dt}\right)^2 + fz'^2\right] dt = \lambda' + \int_0^\tau fz'^2 dt$$

But since f is bounded and z' is normalized, $\int fz^2 dt$ has some finite value

---

[15] Suppose u_n produces the minimum λ_n. Of the function u_{n+1} we require that it be orthogonal not only to the $n - 1$ functions with respect to which u_n has this property, but also to u_n itself. Hence the class of functions from which u_{n+1} must be chosen is more restricted, and the minimum produced by u_{n+1} cannot lie below λ_n. Now there is an infinite number of functions orthogonal to the set $u_0, u_1 \cdots u_n$, and it is hard to believe that they will all produce the same eigenvalue λ_n.

F', so that

$$\lambda \leq \lambda' + F'$$

If we proceed in the reverse manner and use z in computing λ', we obtain

$$\lambda' \leq \int \left(\frac{dz}{dt}\right)^2 dt = \lambda - \int fz^2 d\tau = \lambda - F$$

where F is again finite. Upon combining the last two inequalities we find

$$\lambda' + F \leq \lambda \leq \lambda' + F'$$

which means that λ can differ from λ' by only a finite amount.   If the λ'-values tend to ∞, the λ's also do.

But the eigenvalues of (53′) are well known.   They depend, of course, on the boundary conditions for z, and hence for u.   If u vanishes at both a and b, so that z vanishes at 0 and τ, the eigenvalues are

$$\lambda_n' = \frac{n^2\pi^2}{\tau^2}$$

In case only periodicity of z is required (see example of preceding section), the eigenvalues are

$$\lambda_n' = \frac{4n^2\pi^2}{\tau^2}$$

In any case,

$$\lambda_n' = \text{const. } n^2$$

Since the " high " eigenvalues λ approach the " high " values of λ', theorem (52) is established.   It is to be observed that our result in this particular form is conditional upon the assumption of a *finite* τ, which is usually equivalent to a finite range of x.   Several of the equations listed in Table 2 are ordinarily treated for infinite ranges of the independent variable; for these, theorem (52) is not valid because τ becomes ∞.   Hermite's equation is a case in point: its eigenvalues are proportional to n rather than n^2 even asymptotically.   But here, as well as in all other cases, it is still true that

$$\lambda_n \to \infty \quad \text{as} \quad n \to \infty \tag{8–54}$$

It is interesting to note that the solutions of eq. (53′) are asymptotically (for large λ) equal to those of (53).   Thus

$$\lim_{n \to \infty} z_n = A_n \sin \frac{n\pi}{\tau} t$$

provided the boundary condition is: $z(0) = z(\tau) = 0$.   In terms of u this reads

$$\lim_{n \to \infty} u_n = A_n (pw)^{-1/4} \sin \left\{ n\pi \left[\int_a^x \left(\frac{w}{p}\right)^{1/2} dx \right] \left[\int_a^b \left(\frac{w}{p}\right)^{1/2} dx \right]^{-1} \right\}$$

8.10. Completeness of Eigenfunctions.—In sec. 2 there appeared a qualitative, although crude, definition of completeness. We now wish to give that definition greater precision and to prove it under the conditions outlined in sec. 8.7. A system of functions u_1, u_2, $\cdots$ is complete if it is possible to " approximate in the mean " any function $f(x)$, satisfying the same boundary conditions as the u's, by means of a series $\sum_1^n a_i u_i$, that is, if

$$\lim_{n \to \infty} \int \left(f - \sum_1^n a_i u_i \right)^2 w \, dx = 0 \tag{8-55}$$

We are here concerned with functions u which are solutions of eq. (40); hence we know them to be orthogonal. This permits at once the determination of the coefficients a_i. If, for any given, finite n, we wish to make the quantity

$$N = \int (f - \sum_1^n a_i u_i)^2 w \, dx$$

as small as possible, then

$$\frac{\partial N}{\partial a_j} = 0 \quad \text{for} \quad j = 1, 2, \cdots, n$$

The differentiations may be carried out under the integral sign, so that

$$\int \left(f - \sum_1^n a_i u_i \right) u_j w \, dx = 0$$

whence

$$a_j = \int f u_j w \, dx \tag{8-56}$$

This, then, is the best choice of coefficients with which we may hope to satisfy (55).

Now introduce the following abbreviations

$$\Delta_n = f - \sum_1^n a_i u_i, \quad c_n = \left[\int \Delta_n^2 w \, dx \right]^{1/2}$$

We shall show that the function Δ_n/c_n has the following properties: (1) it is normalized, (2) it is orthogonal to every u_i up to and including u_n. The first property is obvious; the second is easily seen as follows:

$$\int \left(\frac{\Delta_n}{c_n} \right) u_i w \, dx = \frac{1}{c_n} \left\{ \int f u_i w \, dx - \sum_{j=1}^n a_j \int u_j u_i w \, dx \right\}$$

$$= \begin{cases} \dfrac{1}{c_n} (a_i - a_i) & \text{if } i \le n \\ \dfrac{1}{c_n} (a_i - 0) & \text{if } i > n \end{cases}$$

But if Δ_n/c_n has these two properties, it satisfies all the conditions which, in the variational procedure, we imposed upon u_{n+1}, *except that of minimizing* Λ. Hence it is clear that

$$\Lambda\left(\frac{\Delta_n}{c_n}\right) \geq \Lambda(u_{n+1})$$

and this means:

$$\frac{1}{c_n^2}\,\Lambda(\Delta_n) \geq \lambda_{n+1} \tag{8-57}$$

The remainder of our argument consists in proving that $\Lambda(\Delta_n)$ is finite. If the reader will accept this fact,[16] which is almost obvious from the meaning of Δ_n, the last inequality leads at once to (55); for as n approaches infinity, the right-hand side tends to infinity in view of (54), hence

$$\lim_{n \to \infty} c_n^2 = 0$$

This is the same as (55).

[16] For the more exacting reader, we here indicate the proof. The integral $\Lambda(\Delta_n)$ may be transformed in accordance with the first three steps of (51) into

$$\Lambda(\Delta_n) = -\int \Delta_n L(\Delta_n) dx$$

But

$$\int \Delta_n L(\Delta_n) dx = \int (f - \textstyle\sum a_i u_i) L(f - \textstyle\sum a_i u_i) dx = \int f L(f) dx + \sum_i a_i^2 \lambda_i$$

because

$$L(u_i) = -\lambda_i w u_i, \quad \text{and} \quad \int u_i L(f) dx = \int f L(u_i) dx = -a_i \lambda_i$$

Hence,

$$\Lambda(\Delta_n) = \Lambda(f) - \sum_{i=1}^{n} a_i^2 \lambda_i$$

The existence of $\Lambda(f)$ must be assumed, for otherwise an expansion of f in terms of the u's may be impossible. Moreover, f and therefore the approximating function $\varphi_n = \sum_{i=1}^{n} a_i u_i$ must possess integrable squares. Let us suppose that

$$\int \varphi_n^2 w\, dx = \sum_{i=1}^{n} a_i^2 = M_n$$

If we add zero in the form $\sum_1^n a_i^2 \lambda_1 - M_n \lambda_1$, where λ_1 is the lowest of all eigenvalues, to the last expression for $\Lambda(\Delta_n)$ we obtain

$$\Lambda(\Delta_n) = \Lambda(f) - \sum_1^n a_i^2(\lambda_i - \lambda_1) - M_n \lambda_1$$

The difference $\Lambda(f) - M_n \lambda_1$ is certainly finite for all n. Let us call it A. The summation on the right consists of positive terms only. Inequality (57) may therefore certainly be written

$$\frac{A}{c_n^2} \geq \lambda_{n+1}$$

This forces c_n to become zero for large n since λ_{n+1} tends to ∞.

8.11. Further Comments and Generalizations.—In the last section we have shown, not that

$$f = \sum_{i=1}^{\infty} a_i u_i \qquad (8\text{–}58)$$

but rather that the series on the right approximates f " in the mean " in accordance with eq. (55). To put the difference more concretely: Eq. (55) may be true and yet (58) may not hold *for all points of the range* $(a \le x \le b)$. It is clear that if (58) is true almost everywhere but fails at a finite set of points, the contribution of these points to the integral in (55) would be nil and that equation would be true. To prove (58) in addition to (55) would involve the establishment of absolute and uniform convergence of the series $\sum_{1}^{\infty} a_i u_i$. For the solutions of eq. (40) with boundary conditions of the type here chosen this can indeed be done,[17] and the reader need not be excessively concerned over the difference between " completeness " (expressed by eq. 55) and the possibility of expansion of an arbitrary function (indicated by 58).

The preceding theory has always involved the assumption of a finite range, b–a, of the independent variable. This is clearly a serious limitation, for it excludes the usual solutions of a number of the equations listed in Table 2. To develop a rigorous account of the situation arising when the range is extended to infinity is not easy, but what happens qualitatively under such conditions can be readily seen.

Consider again the vibrating string with eigenvalues $k^2 = n^2\pi^2/l^2$. As l tends to infinity, these eigenvalues move closer together until in the limit they form a continuum. The eigenfunctions are still of the form $A \sin (kx + \delta)$, but they refuse to be normalized in the former sense; for clearly the integral $\int \sin^2 kx\, dx$, when taken over an infinite range, diverges. Also, since the eigenvalues are no longer discrete, our definition of orthogonality loses its sense. However, completeness is still guaranteed since what was originally a Fourier series will now become a Fourier integral (cf. eq. 14). The difficulty concerning orthogonality and normalization can, however, be avoided by introducing " eigendifferentials " instead of eigenfunctions.[18]

The situation brought about by an extension of the range may be even more complicated than this. We shall see in Chapter 11 that the differential equation describing the hydrogen atom (eq. 11–55), which is closely

---

[17] See Courant-Hilbert, p. 370.

[18] See, for instance, Kemble, E. C., " The Fundamental Principles of Quantum Mechanics," McGraw-Hill Book Co., 1937, p. 162 et seq.

related to Laguerre's, admits, because of its infinite range, both a discrete and a continuous set of eigenvalues (" spectrum "). This phenomenon is of very frequent occurrence. On the other hand, eigenvalues associated with a Sturm-Liouville problem of infinite range are not necessarily continuous, as the example of the simple harmonic oscillator (cf. Chapter 11) or Hermite's differential equation (eq. 2–62) clearly shows.

No mention has thus far been made of the possibility that the solutions of the Sturm-Liouville equation may possess singularities in the range $a \leq x \leq b$. Troubles of this sort might have been circumvented by postulating that the function p appearing in eq. (41) be always of one sign and never zero, as is sometimes done in treatments of the eigenvalue problem. This, however, would have excluded some interesting cases from Table 2, notably Legendre's equation which has (non-essential) singular points at $x = \pm 1$, and Hermite's equation which has an essential singularity at ∞. Suffice it to say here that these matters, although of considerable fundamental interest, occasion no modification of the conclusions here derived. Attention is given to them in Kemble's book (*loc. cit.*).

The solutions of eq. (40) have been assumed to be *real* functions throughout this section. If the functions p, q, and w are real, this entails no loss in generality. Suppose that a complex function $u = X + iY$ were admitted as solution of the differential equation[19]; this would merely imply that both X and Y are real solutions belonging to the same eigenvalue. Thus, whenever complex solutions arise and are compatible with the boundary conditions, we may at once conclude that the corresponding eigenvalue is degenerate. (In the complex scheme, both u and $u^* = X - iY$ are linearly independent solutions.) If now we require as normalizing condition

$$\int u^* u w dx = 1$$

we are merely postulating that, in place of the usual normalization

$$\left(\int X^2 w dx = \int Y^2 w dx = 1 \right),$$

$$\int X^2 w dx + \int Y^2 w dx = 1$$

shall hold. In other words, we are operating, in the complex scheme, with linear combinations of the real functions, and with a different normalization. Orthogonality, if defined by eq. (5*) instead of (5), reverts to its ordinary meaning, for

$$\int u_1^* u_2 w dx = \int [X_1 X_2 + Y_1 Y_2 + iX_1 Y_2 - iY_1 X_2] w dx = 0$$

[19] We may, for instance, write the solution of eq. (2) in the form $S = Ae^{ikx}$.

is an immediate consequence of the fact that eigenfunctions belonging to different eigenvalues λ_1 and λ_2 are orthogonal. Furthermore, if we require u^* and u to be orthogonal,

$$\int u^2 wdx = \int (X^2 - Y^2 + 2iXY)wdx = \int X^2 wdx - \int Y^2 wdx = 0$$

provided X and Y have been chosen orthogonal. Thus both X and Y are normalized to $\frac{1}{2}$ when the complex formalism is used. In view of these simple facts the validity of the completeness proof remains intact for complex f and complex u; only formal changes are necessary. The a_i become complex, and completeness is defined by the relation $\lim_{n \to \infty} \int \Delta_n^* \Delta_n wdx = 0$. Complete revision of the theory is necessary when the coefficients p, q, w are permitted to be complex.

Finally, it is appropriate to remark that our development has been restricted to one dimension. The Sturm-Liouville theory can be generalized without great difficulty to certain partial differential equations with much the same results. For this generalization we refer the reader to Courant-Hilbert.

Eigenvalue problems arise in the most diverse fields of physics and chemistry. Many of them are treated in:

Morse, P. M., and Feshbach, H., " Methods of Theoretical Physics," McGraw-Hill, New York, 1953.

Jeffreys, H., and Jeffreys, B. S., " Methods of Mathematical Physics," Cambridge University Press, Second Edition, 1950.

See also the bibliography on quantum mechanics.

CHAPTER 9

MECHANICS OF MOLECULES

9.1. Introduction.—As an illustration of the mathematical methods used in mechanics, we discuss in this chapter an important physical and chemical problem, namely, the motion of a molecule containing n atoms. We limit ourselves to this single topic for several reasons: its complexity requires us to describe most of the mathematics used in mechanics; the same methods may be extended to other problems, for example, the motions of particles within the atomic nucleus or the motions of a macroscopic body such as an aeroplane[1]; and finally because the structure of the polyatomic molecule and its spectra are matters of considerable interest to many chemists and physicists. This chapter will also present an opportunity for dealing with the purely mathematical question of how to describe the configuration of a rigid body (Euler's angles, etc.), a matter which is of great generality and must be included in a survey of mathematical methods used in science. Many adequate accounts of classical mechanics[2] exist so that we do no more here than recall briefly some of the principles of that subject before proceeding to the special problem in which we are interested.

9.2. General Principles of Classical Mechanics.—A *free particle* is one whose motion is completely unrestricted. It is said to have three *degrees of freedom*, for its position is uniquely determined at any instant by three independent coordinates. Consider a system containing n such particles, where the instantaneous position of the i-th particle of mass m_i is specified by the vector $\mathbf{r}_i$. If $\mathbf{F}_i$ is the vector resultant of all the forces acting upon the particle then the motion of the system is described by *Newton's equations* which may be written in the form

$$m_i \frac{d^2\mathbf{r}_i}{dt^2} = m_i \ddot{\mathbf{r}}_i = \mathbf{F}_i; \quad (i = 1, 2, \cdots, n) \tag{9-1}$$

In many cases, the particles composing the system are not free but restricted. For example, a member of the system may be allowed to

---

[1] See Frazer, R. A., Duncan, W. J., and Collar, A. R., "Elementary Matrices," Cambridge University Press, 1938.

[2] Whittaker, E. T., "Analytical Dynamics of Particles and Rigid Bodies," Third Edition, Cambridge University Press, 1927; Corbin, H. C., and Stehle, P., "Classical Mechanics," John Wiley and Sons, Inc., New York, 1950; Goldstein, Herbert, "Classical Mechanics," Addison-Wesley Press, Inc., Cambridge, Mass., 1951.

move only on a surface, so that its degrees of freedom become two.   Under such circumstances the equation of the surface is called the *constraint*.   In a similar way if the particle is required to move along a line, there is only one degree of freedom and the two equations which define the line are the constraints.   If the sum of the degrees of freedom of all the particles is $k < 3n$, then the system may be regarded as a collection of free particles subjected to $3n - k$ independent constraints so that only k coordinates are needed to describe the motion of the system.   These new coordinates q_1, q_2, $\cdots$, q_k are related to the Cartesian coordinates of the particles (cf. eqs. 5-1 and 5-2); they are called the *generalized coordinates of Lagrange*.

If, for convenience, we let the Cartesian components of r_1 be x_1, x_2, x_3; the components of r_2 be x_4, x_5, x_6 and so on (remembering also that $m_1 = m_2 = m_3$; $m_4 = m_5 = m_6$; etc.), then the *kinetic energy* T of the system is given by

$$2T = \sum_{i=1}^{3n} m_i \dot{x}_i^2 = \sum_{r=1}^{k} \sum_{s=1}^{k} A_{rs} \dot{q}_r \dot{q}_s \tag{9-2}$$

where

$$\dot{x}_i = \frac{\partial x_i}{\partial t}\,; \quad A_{rs} = \sum_{i=1}^{3n} m_i \frac{\partial x_i}{\partial q_r} \frac{\partial x_i}{\partial q_s} \tag{9-3}$$

Since the components of *momentum* in Cartesian coordinates are

$$p_i = m_i \dot{x}_i = \frac{\partial T}{\partial \dot{x}_i}$$

we define, by analogy, the *generalized momenta* as

$$p_r(q_1 q_2 \cdots ; \dot{q}_1 \dot{q}_2 \cdots) = \frac{\partial T}{\partial \dot{q}_r} = \sum_{s=1}^{k} A_{rs} \dot{q}_s \tag{9-4}$$

In many physical problems, the system is conservative, that is, a potential function $V(q_1, q_2, \cdots, q_k)$ exists such that

$$Q_i = -\frac{\partial V}{\partial q_i}\,; \quad (i = 1, 2, \cdots, k) \tag{9-5}$$

Then, as was shown in sec. 6.3 (cf. eq. 6-11), *Lagrange's equations of motion* are

$$\frac{d}{dt}\left(\frac{\partial T}{\partial \dot{q}_i}\right) - \frac{\partial T}{\partial q_i} = Q_i; \quad (i = 1, 2, \cdots, k) \tag{9-6}$$

This is a set of k differential equations of second order with q_1, q_2, $\cdots$, q_k as dependent variables and t as independent variable.

If we introduce the *Lagrangian function*

$$L(q_i, \dot{q}_i) = T(q_i, \dot{q}_i) - V(q_i) \tag{9-7}$$

eq. (5) becomes

$$\frac{d}{dt}\left(\frac{\partial L}{\partial \dot{q}_i}\right) - \frac{\partial L}{\partial q_i} = 0; \quad (i = 1, 2, \cdots, k) \tag{9-8}$$

The solution of Lagrange's equations in either form (5) or (8) will result in an expression for each generalized coordinate q_i as a function of time and $2k$ constants of integration. The latter must be determined from the initial conditions of the n particles of the system.

It is often of advantage to transform (5) or (8) to a set of $2k$ first order differential equations. From (4), (8), and the definition $L = T - V$, we have

$$p_i = \frac{\partial L}{\partial \dot{q}_i}; \quad \dot{p}_i = \frac{\partial L}{\partial q_i} \tag{9-9}$$

We now define the *Hamiltonian function*

$$H = \sum_{i=1}^{k} p_i \dot{q}_i - L \tag{9-10}$$

Its total differential is

$$dH = \sum_{i=1}^{k} p_i d\dot{q}_i + \sum_{i=1}^{k} \dot{q}_i dp_i - \sum_{i=1}^{k} \frac{\partial L}{\partial q_i} dq_i - \sum_{i=1}^{k} \frac{\partial L}{\partial \dot{q}_i} d\dot{q}_i$$

But by using (9), the first and last terms cancel, giving

$$dH = \sum_{i=1}^{k} \dot{q}_i dp_i - \sum_{i=1}^{k} \frac{\partial L}{\partial q_i} dq_i \tag{9-11}$$

This equation depends only on dp_i and dq_i but not on $d\dot{q}_i$, hence H is a function of q and p alone and we may write

$$dH = \sum_{i=1}^{k} \frac{\partial H}{\partial p_i} dp_i + \sum_{i=1}^{k} \frac{\partial H}{\partial q_i} dq_i \tag{9-12}$$

Comparison of (11) with (12) shows us that

$$\frac{\partial H}{\partial p_i} = \dot{q}_i; \quad \frac{\partial H}{\partial q_i} = -\frac{\partial L}{\partial q_i} = -\dot{p}_i; \quad (i = 1, 2, \cdots, k) \tag{9-13}$$

The resulting first order differential equations (13), $2k$ in number, are *Hamilton's canonical equations of motion*; p_i and q_i are said to be *canonically conjugate variables*.

Problem. Show that $2T = \sum p_i \dot{q}_i$ and $H = T + V$.

9.3. The Rigid Body in Classical Mechanics.—As a crude first approximation to the motion of a molecule we consider a *rigid body* which is defined as a system of n particles bound together by interior forces in such a way that the distance between the i-th and j-th particles is constant and unaffected by any external force to which the system is subjected. Suppose x_i, y_i, z_i are the Cartesian coordinates of the i-th particle, then the distance between the i-th and j-th particle is

$$r_{ij} = \sqrt{(x_i - x_j)^2 + (y_i - y_j)^2 + (z_i - z_j)^2} = \text{constant} \quad (9\text{-}14)$$
$$(i, j = 1, 2, \cdots, n)$$

It is readily shown that the most general displacement of a body of this sort may be obtained in a variety of ways by a combination of *translation* and *rotation* about an axis fixed in the body. The proof of this fact, known as *Chasles's theorem*, may be found, for example, in Whittaker, loc. cit. The choice of a reference point, that is, the origin of the vector which locates the fixed axis, is entirely arbitrary. For a given displacement, this point may be chosen in such a way that the translation is parallel to the axis of rotation. With this choice of reference point, each displacement can be effected in one and only one way, the resulting motion being similar to the displacement of a nut on a threaded screw. It is thus only necessary to consider translation and rotation in order to study the most general motion of a rigid body. It should be remembered, however, that the axis of rotation may be continually changing its direction, hence we usually refer to an instantaneous axis of rotation.

9.4. Velocity, Angular Momentum, and Kinetic Energy.—Suppose a rigid body is rotating about an axis with a constant *angular velocity* $\boldsymbol{\omega}$; then the *linear velocity* of any point P in the body is given by

$$\mathbf{v} = \boldsymbol{\omega} \times \mathbf{r} \quad (9\text{-}15)$$

where r is a radius vector drawn to P from a fixed point O on the axis of rotation (see eq. 4–16). If the point P has a mass m, its *momentum* is

$$m\mathbf{v} = m(\boldsymbol{\omega} \times \mathbf{r}) \quad (9\text{-}16)$$

and its *moment of momentum* or *angular momentum* (see sec. 4.5) about the point O is

$$\mathbf{M} = \mathbf{r} \times m\mathbf{v} = m[\mathbf{r} \times (\boldsymbol{\omega} \times \mathbf{r})] \quad (9\text{-}17)$$

Suppose the fixed point O about which the body is rotating is taken as the axis of a Cartesian coordinate system $OXYZ$, the components of $\boldsymbol{\omega}$ are ω_x, ω_y, ω_z and the components of r are x, y, z. Then in accordance with eq. (4–13), the components of v are:

$$v_x = z\omega_y - y\omega_z$$
$$v_y = x\omega_z - z\omega_x \quad (9\text{-}18)$$
$$v_z = y\omega_x - x\omega_y$$

and the components of **M** are:

$$M_x = m(yv_z - zv_y)$$
$$M_y = m(zv_x - xv_z) \qquad (9\text{--}19)$$
$$M_z = m(xv_y - yv_x)$$

On combining (18) and (19) there results

$$M_x = A\omega_x - F\omega_y - E\omega_z$$
$$M_y = B\omega_y - D\omega_z - F\omega_x \qquad (9\text{--}20)$$
$$M_z = C\omega_z - E\omega_x - D\omega_y$$

where A, B, C are *moments of inertia* and D, E, F are *products of inertia:*

$$A = m(y^2 + z^2); \quad D = myz$$
$$B = m(z^2 + x^2); \quad E = mzx \qquad (9\text{--}21)$$
$$C = m(x^2 + y^2); \quad F = mxy$$

The kinetic energy T of the particle at P is given by

$$2T = m\mathbf{v} \cdot (\boldsymbol{\omega} \times \mathbf{r}) = m[\mathbf{v}\boldsymbol{\omega}\mathbf{r}] = m[\boldsymbol{\omega}\mathbf{r}\mathbf{v}]$$
$$= m\boldsymbol{\omega} \cdot (\mathbf{r} \times \mathbf{v}) = \boldsymbol{\omega} \cdot \mathbf{M} \qquad (9\text{--}22)$$

where we have used eqs. (4–17), (4–18), and (9–17). Thus, in view of (20) we find

$$2T = A\omega_x^2 + B\omega_y^2 + C\omega_z^2 - 2D\omega_y\omega_z - 2E\omega_z\omega_x - 2F\omega_x\omega_y \quad (9\text{--}23)$$

9.5. The Eulerian Angles.—We digress here to give explicit relations useful for locating a point P in a rigid body. Six parameters are needed. Three of them will specify a fixed reference point in the body, which is not necessarily at the origin of the coordinate system as in the preceding discussion. Two more parameters are required to define the position of a line fixed in the body and passing through the fixed point, while the sixth parameter defines a rotation of the body about this line.

Suppose we attach a rigid framework $O'X'Y'Z'$ to the body and denote the position of its origin relative to a coordinate system $OXYZ$ fixed in space by x_0, y_0, z_0. We will also suppose that we know the nine direction cosines a_{ij} of $O'X'Y'Z'$ relative to $OXYZ$. The point P may then be located in either coordinate system at will for we have the relations (see sec. 4.1)

$$x = x_0 + a_{11}x' + a_{12}y' + a_{13}z'$$
$$y = y_0 + a_{21}x' + a_{22}y' + a_{23}z'$$
$$z = z_0 + a_{31}x' + a_{32}y' + a_{33}z'$$

where x,y,z refer to $OXYZ$ and x',y',z' refer to $OX'Y'Z'$. Let us choose x_0,y_0,z_0 as three of the parameters required. The nine direction cosines which remain, and which we know are not linearly independent, may then be combined in a variety of ways in order to obtain the three additional independent parameters needed. Some useful combinations are the *Euler-Rodrigues parameters*, the *Cayley-Klein parameters*,[3] and the *Eulerian angles*. The latter are suitable for the present purpose and will now be described.

Unfortunately, the Euler angles have been defined in several different ways in the scientific literature and great confusion occurs when one attempts to compare the results of various writers. The one which we adopt in the following is that favored by the majority of more than fifty references[4] which have been consulted. A possible advantage of it lies in the fact that our angles α and β become the polar angles, ϕ and θ, respectively, in spherical polar coordinates. Moreover, a rotation about the OX-axis toward the OY-axis, as is required in the second step of our procedure, seems to be a natural operation. This step does, however, introduce additional imaginary factors into the Cayley-Klein parameters and the representations of the three-dimensional group (see sec. 15.15). Further complications also result when one compares the wave-functions of the asymmetric top in quantum mechanics with those of its limiting case, the symmetric top. These latter objections are removed if the second rotation is made about the OY-axis, instead of the OX-axis, as is done by Whittaker (loc. cit.) and by Wigner.[5] Note, however, that Wigner has used a left-handed coordinate system.

Let us return to the problem of describing the Eulerian angles, which we show according to our definition in Fig. 1. Perhaps a clearer conception of the relations involved may be obtained from the cross-section diagrams of Fig. 2, which give the planes XOY, ZOZ', and $X'OY'$ and which show, in parentheses, the axes perpendicular to the plane of the page. It will be seen that the axis OK, called the line of nodes, is the intersection of the XOY and $X'OY'$ planes. The axis OL is perpendicular to OK in the XOY plane, and OM is perpendicular to OK in the $X'OY'$ plane. Study of Fig. 2 will show that $OXYZ$ may be superimposed on $OX'Y'Z'$ by the following rotations, provided that they are performed in the order given

[3] The *Cayley-Klein parameters* are related to the *Pauli spin matrices* used in quantum mechanics, as will be shown in sec. 15.15.

[4] It agrees with that chosen by Goldstein (loc. cit.), who has also commented on the conflicting definitions of the Euler angles. In his notation, our symbols are: $\alpha = \phi$, $\beta = \theta$, $\gamma = \psi$. It should be noted that our present equations differ from those in the first edition of this book, since we inadvertently used a left-handed coordinate system there.

[5] Wigner, E., "Gruppentheorie und ihre Anwendung auf die Quantenmechanik der Atomspektren," Friedr. Vieweg und Sohn, Braunschweig, 1931.

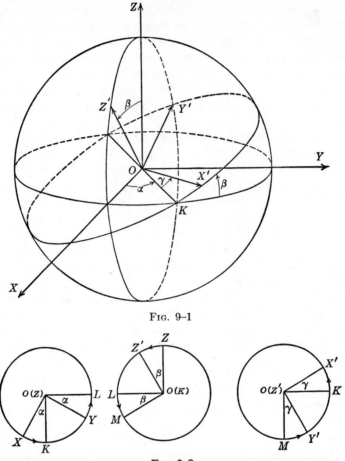

Fig. 9–1

Fig. 9–2

and always in a counterclockwise direction: (1) rotate about OZ by the angle α; (2) rotate through β about OK (OK and OZ are now identical because of the first rotation) which will bring OZ into coincidence with OZ'; (3) rotate about OZ' by γ which then brings OX to OX' and OY to OY'.

Relations between $OXYZ$ and $OX'Y'Z'$ may be found most simply by matrix methods (see Chapter 10). Suppose a vector $\mathbf{x}$ in the space-fixed system becomes $\mathbf{x}'$ in the body-fixed system; then the matrix which connects the two vectors is $R(\alpha,\beta,\gamma)$, where

$$\mathbf{x}' = R(\alpha,\beta,\gamma)\mathbf{x}$$

and

$$R(\alpha,\beta,\gamma) = R_z(\gamma)\,R_x(\beta)\,R_z(\alpha)$$

The first and last of these matrices are like the matrix R^+ of sec. **10.17**, with γ or α in place of ϕ. The remaining matrix is similar in form but rearranged to represent a rotation about the OX-axis.

When the matrix product is evaluated, the result is that given in Table 1. It should be interpreted in a manner similar to that of Table 4–1.

TABLE 1

	OX	OY	OZ
OX'	$\cos\alpha\cos\gamma$ $-\sin\alpha\cos\beta\sin\gamma$	$\sin\alpha\cos\gamma$ $+\cos\alpha\cos\beta\sin\gamma$	$\sin\beta\sin\gamma$
OY'	$-\cos\alpha\sin\gamma$ $-\sin\alpha\cos\beta\cos\gamma$	$-\sin\alpha\sin\gamma$ $+\cos\alpha\cos\beta\cos\gamma$	$\sin\beta\cos\gamma$
OZ'	$\sin\alpha\sin\beta$	$-\cos\alpha\sin\beta$	$\cos\beta$

In order to obtain the angular velocity in terms of the Euler angles, it is convenient to use the body-fixed system, $OX'Y'Z'$, with components $\dot\alpha$, $\dot\beta$, and $\dot\gamma$ along OZ, OK, and OZ', respectively. Since $\dot\alpha$ is parallel to the space-fixed axis, OZ, its components are given by the last column of Table 1. The components of $\dot\beta$, which is parallel to OK, may be found from the first column of the matrix $R_z(\gamma)$. Finally, since $\dot\gamma$ is parallel to OZ', its only component is $\dot\gamma$. Collecting these results, we have

$$\omega_x' = \sin\beta\sin\gamma\,\dot\alpha + \cos\gamma\,\dot\beta$$
$$\omega_y' = \sin\beta\cos\gamma\,\dot\alpha - \sin\gamma\,\dot\beta \qquad (9\text{--}24)$$
$$\omega_z' = \qquad\cos\beta\,\dot\alpha \qquad\quad + \dot\gamma$$

for the three components of angular velocity along OX', OY', and OZ'. In terms of the Eulerian angles, the kinetic energy of a rotating symmetric top $(A = B)$, which we shall need later, is seen from eq. (23) to be

$$T = \tfrac{1}{2}[A\dot\beta^2 + A\dot\alpha^2\sin^2\beta + C(\dot\gamma + \dot\alpha\cos\beta)^2] \qquad (9\text{--}25)$$

provided we choose OX', OY', and OZ' to coincide with the principal axes of inertia of the top, for then the products of inertia D, E, and F all vanish.

9.6. Absolute and Relative Velocity.—We now return to a more general consideration of the motion of a rigid body. Suppose a point P in it is located, relative to $OXYZ$ by the vector $\mathbf{r}_0$ and relative to $O'X'Y'Z'$ by the vector $\mathbf{r}$. Let the instantaneous position of the origin of $O'X'Y'Z'$ be measured relative to $OXYZ$ by $\mathbf{r}'$, where the prime here and in the remainder of this chapter never means differentiation. Then the absolute position of P is given by

$$\mathbf{r}_0 = \mathbf{r}' + \mathbf{r} \qquad (9\text{--}26)$$

and its absolute velocity by

$$v_0 = v' + v'' \tag{9-27}$$

where $v' = dr'/dt$ measures the velocity of the origin of $O'X'Y'Z'$ relative to $OXYZ$ and v'' is the velocity of the point in the moving system. Now suppose that the latter system is rotating with constant angular velocity of ω radians per second; then the point P has a linear velocity $\omega \times r$ in addition to its translational velocity v relative to $O'X'Y'Z'$. Its components are $v_x = dr_x/dt = \dot{r}_x$; $v_y = \dot{r}_y$; $v_z = \dot{r}_z$. Thus,

$$v_0 = v' + \omega \times r + v \tag{9-28}$$

It is important to have a clear understanding of the separate terms in (28). The absolute velocity of the point is v_0; v is the apparent velocity of P measured by an observer in the system $O'X'Y'Z'$ who does not know that his coordinate axes are rotating, while $\omega \times r$ is the absolute velocity which the terminus of r must have in order to maintain its position in the moving body. The last velocity is often called the *velocity of following*. If the point P is rigidly attached to the moving system, $v = 0$; if the moving system and the fixed system have coincident origins, $v' = 0$.

9.7. Motion of a Molecule.—In a molecule, we may consider the electrons and nuclei as bound together in a rigid framework which moves through space in translational motion and which rotates around its center of gravity. Both of these types of motion are included in the equations already given. One further motion is needed, however, for the nuclei execute oscillations around an equilibrium position. In order to allow for this vibrational motion, let r_i be the instantaneous position vector of the i-th particle and a_i, ρ_i be the equilibrium and displacement vectors, respectively, so that

$$r_i = a_i + \rho_i \tag{9-29}$$

while

$$r_{0i} = r' + r_i \tag{9-30}$$

is the instantaneous position of the point relative to $OXYZ$ as shown in Fig. 3

Then from (28)

$$v_{0i} = v' + (\omega \times r_i) + v_i \tag{9-31}$$

and

$$2T = \sum m_i v_{0i}^2 = v'^2 \sum m_i + \sum m_i v_i^2 + \sum m_i (\omega \times r_i) \cdot (\omega \times r_i)$$
$$+ 2v' \cdot \sum m_i v_i + 2 \sum m_i r_i \cdot (v' \times \omega) + 2\omega \cdot \sum (m_i r_i \times v_i) \tag{9-32}$$

The reason for writing the last two terms of (32) as given comes from eq. (4–18) since

$$\mathbf{v}' \cdot (\boldsymbol{\omega} \times \mathbf{r}) = \mathbf{r} \cdot (\mathbf{v}' \times \boldsymbol{\omega}); \quad \mathbf{v} \cdot (\boldsymbol{\omega} \times \mathbf{r}) = \boldsymbol{\omega} \cdot (\mathbf{r} \times \mathbf{v})$$

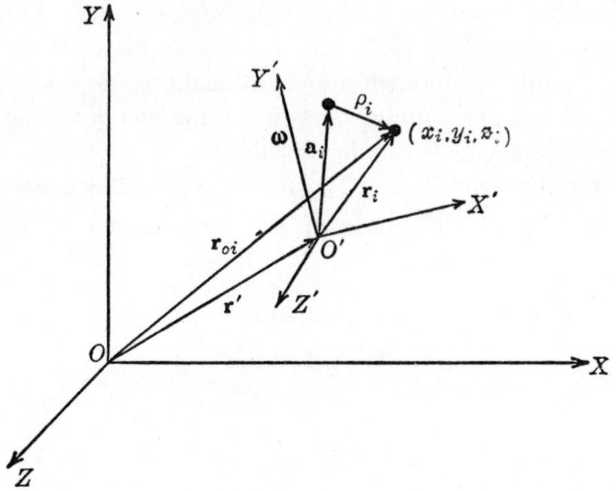

FIG. 9–3

Six further relations are needed to define the rotating coordinate system. These[6] may conveniently be taken as

$$\sum m_i \mathbf{v}_i = 0 \tag{9-33}$$

$$\sum m_i \mathbf{a}_i \times \mathbf{v}_i = 0 \tag{9-34}$$

The first three of these equations locate the origin of $O'X'Y'Z'$ at the center of gravity of the system, for that point is given by

$$\bar{\mathbf{r}} = \frac{\sum m_i \mathbf{r}_i}{\sum m_i}$$

and if $\bar{\mathbf{r}} = 0$, $\sum m_i \mathbf{r}_i = 0$ and $\sum m_i \mathbf{v}_i = 0$. The second condition, eq. (34), states that there is no angular momentum relative to $O'X'Y'Z'$, when all particles occupy their equilibrium positions, i.e., when every $\mathbf{r}_i = \mathbf{a}_i$.

Using (29), (33), and (34), eq. (32) becomes

$$2T = \mathbf{v}'^2 \sum m_i + \sum m_i \mathbf{v}_i^2 + \sum m_i (\boldsymbol{\omega} \times \mathbf{r}_i) \cdot (\boldsymbol{\omega} \times \mathbf{r}_i) + 2\boldsymbol{\omega} \cdot \sum (m_i \boldsymbol{\rho}_i \times \mathbf{v}_i)$$
$$= 2(T_t + T_v + T_r + T_{\text{int}}) \tag{9-35}$$

Inspection of (35) shows that the kinetic energy is a sum of four terms which may be interpreted in order as due to the *translational motion* of the molecule as a whole through space (T_t); the *vibrational motion* of the nuclei

[6] See Eckart, *Phys. Rev.* **47**, 552 (1935); Sayvetz, *J. Chem. Phys.* **7**, 383 (1939).

about an equilibrium position (T_v); the *rotation* of the molecule as a rigid body about its center of gravity (T_r); *interaction* between vibration and rotation (T_{int}).

9.8. The Kinetic Energy of a Molecule.—It is necessary to obtain (35) in explicit form before further calculations can be made. As shown previously T_r becomes equal to (23), but it must be remembered that A, B, $\cdots$, F are *instantaneous* moments and products of inertia relative to the moving axes. They are not constants but functions of the position of the atoms and they change as the molecule vibrates.

In discussing the terms T_v and T_{int}, it is convenient to use normal coordinates (see sec. 10.17). Suppose ρ_i has components $\xi_i/\sqrt{m_i}$, $\eta_i/\sqrt{m_i}$, $\zeta_i/\sqrt{m_i}$ where,

$$\xi_i = \sum l_{ik}Q_k$$
$$\eta_i = \sum m_{ik}Q_k \qquad\qquad (9\text{-}36)$$
$$\zeta_i = \sum n_{ik}Q_k$$

and l_{ik}, m_{ik}, n_{ik} are constant coefficients such that

$$\sum_k l_{ki}l_{kj} = \tfrac{1}{3}\delta_{ij}; \quad \sum_k m_{ki}m_{kj} = \tfrac{1}{3}\delta_{ij}; \quad \sum_k n_{ki}n_{kj} = \tfrac{1}{3}\delta_{ij}$$

Then,

$$\sum m_i \mathbf{v}_i^2 = \sum(\dot{\xi}_i^2 + \dot{\eta}_i^2 + \dot{\zeta}_i^2) = \sum \dot{Q}_k^2 \qquad (9\text{-}37)$$

Moreover,

$$\sum m_i(\rho_i \times \mathbf{v}_i)_x = \sum(\eta_i \dot{\zeta}_i - \zeta_i \dot{\eta}_i) = \sum X_k \dot{Q}_k$$
$$\sum m_i(\rho_i \times \mathbf{v}_i)_y = \sum(\zeta_i \dot{\xi}_i - \xi_i \dot{\zeta}_i) = \sum Y_k \dot{Q}_k \qquad (9\text{-}38)$$
$$\sum m_i(\rho_i \times \mathbf{v}_i)_z = \sum(\xi_i \dot{\eta}_i - \eta_i \dot{\xi}_i) = \sum Z_k \dot{Q}_k$$

where,

$$X_k = \sum_{i,l}(n_{ik}m_{il} - m_{ik}n_{il})Q_l$$
$$Y_k = \sum_{i,l}(l_{ik}n_{il} - n_{ik}l_{il})Q_l \qquad (9\text{-}39)$$
$$Z_k = \sum_{i,l}(m_{ik}l_{il} - l_{ik}m_{il})Q_l$$

Collecting terms, (35) finally appears as

$$2T_t = \mathbf{v}'^2 \sum m_i$$
$$2T_v = \sum \dot{Q}_k^2$$
$$2T_r = A\omega_x^2 + B\omega_y^2 + C\omega_z^2 - 2D\omega_x\omega_y - 2E\omega_z\omega_y - 2F\omega_x\omega_z$$
$$2T_{\text{int}} = 2\omega_x \sum X_k \dot{Q}_k + 2\omega_y \sum Y_k \dot{Q}_k + 2\omega_z \sum Z_k \dot{Q}_k \qquad (9\text{-}40)$$

9.9. The Hamiltonian Form of the Kinetic Energy.—In order to obtain the Hamiltonian form of (40), we must change from angular velocities to angular momenta. From (17) or (22) we see that the components of total angular momentum are

$$P_x = \frac{\partial T}{\partial \omega_x} = A\omega_x - D\omega_y - F\omega_z + \sum X_k \dot{Q}_k$$

$$P_y = \frac{\partial T}{\partial \omega_y} = -D\omega_x + B\omega_y - E\omega_z + \sum Y_k \dot{Q}_k \qquad (9\text{-}41)$$

$$P_z = \frac{\partial T}{\partial \omega_z} = -F\omega_x - E\omega_y + C\omega_z + \sum Z_k \dot{Q}_k$$

Similarly, the momenta conjugate to Q_k are

$$p_k = \frac{\partial T}{\partial \dot{Q}_k} = \dot{Q}_k + X_k \omega_x + Y_k \omega_y + Z_k \omega_z \qquad (9\text{-}42)$$

Solving this equation for $\dot{Q}_k$ and substituting in (41) gives

$$P_x = A\omega_x - D\omega_y - F\omega_z + \sum X_k(p_k - X_k\omega_x - Y_k\omega_y - Z_k\omega_z) \qquad (9\text{-}43)$$

with similar expressions for P_y and P_z. The following abbreviations may be used to simplify the final results.

$$A' = A - \sum X_k^2; \quad D' = D + \sum X_k Y_k$$
$$B' = B - \sum Y_k^2; \quad E' = E + \sum Y_k Z_k \qquad (9\text{-}44)$$
$$C' = C - \sum Z_k^2; \quad F' = F + \sum Z_k X_k$$

In terms of them, we may write

$$P_x = A'\omega_x - D'\omega_y - F'\omega_z + \sum X_k p_k$$
$$P_y = -D'\omega_x + B'\omega_y - E'\omega_z + \sum Y_k p_k \qquad (9\text{-}43\text{a})$$
$$P_z = -F'\omega_x - E'\omega_y + C'\omega_z + \sum Z_k p_k$$

If we also write

$$p_x = \sum X_k p_k; \quad p_y = \sum Y_k p_k; \quad p_z = \sum Z_k p_k \qquad (9\text{-}45)$$

(43a) may be further simplified to read

$$P_x = p_x + A'\omega_x - D'\omega_y - F'\omega_z$$
$$P_y = p_y - D'\omega_x + B'\omega_y - E'\omega_z \qquad (9\text{-}46)$$
$$P_z = p_z - F'\omega_x - E'\omega_y + C'\omega_z$$

The quantities p_x, p_y, p_z arise from vibration alone as may be seen from their definition, eq. (45); they are called components of *internal angular momentum*.

Adding together all the terms of (40) and using (41), (42) and (45), we obtain

$$2T = 2T_t + (P_x - p_x)\omega_x + (P_y - p_y)\omega_y$$
$$+ (P_z - p_z)\omega_z + \sum p_k^2 \tag{9-47}$$

Finally, we find by solving (46) for the ω's that

$$\omega_i = \sum_j \mu_{ij}\mathcal{P}_j; \quad (i, j = x, y, z)$$
$$\mu_{ij} = \mu_{ji}; \quad \mathcal{P}_j = (P_j - p_j) \tag{9-48}$$

With the use of these variables, eq. (47) takes the more elegant form

$$2T = 2T_t + \sum \mu_{ij}\mathcal{P}_i\mathcal{P}_j + \sum p_k^2 \tag{9-49}$$

Explicitly the μ's are:

$$\mu_{xx} = \frac{B'C' - E'^2}{\Delta}; \quad \mu_{yy} = \frac{A'C' - F'^2}{\Delta}$$

$$\mu_{zz} = \frac{A'B' - D'^2}{\Delta}; \quad \mu_{xy} = \frac{C'D' + E'F'}{\Delta}$$

$$\mu_{xz} = \frac{D'E' + B'F'}{\Delta}; \quad \mu_{yz} = \frac{A'E' + D'F'}{\Delta}$$

$$\Delta = \begin{vmatrix} A' & -D' & -F' \\ -D' & B' & -E' \\ -F' & -E' & C' \end{vmatrix} \tag{9-50}$$

9.10. The Vibrational Energy of a Molecule.[7]—The first term in (47), the translational energy, is of little interest in physical problems. We shall have no more to say about it. The only other term of that equation which can be treated further by classical mechanics is the last one, corresponding to the vibrational energy of the molecule. We first consider the potential energy of the system due to the vibration of the particles. It will be some function of the mutual positions of the nuclei and it is most convenient to specify these in terms of the mass-adjusted components of a displacement vector. We formerly took these as $\xi_i/\sqrt{m_i}$, $\eta_i/\sqrt{m_i}$, $\zeta_i/\sqrt{m_i}$ (see eq. 36), $3n$ in number. Following convention we now use $q_1, q_2, \cdots$, q_{3n} for the same coordinates. If the system is placed originally in the equilibrium configuration (all $q_i = 0$) and if the particles have very small

---

[7] This section as well as secs. 9.11 and 9.12 makes use of some of the results of Chapter 10. It should be omitted or postponed by readers not familiar with orthogonal transformations. The authors suggest that the reader, rather than endeavor to understand normal coordinates by "elementary considerations," acquaint himself with the more powerful methods of the next chapter.

initial velocities, we assume that they will never depart to any large distance from that configuration, nor will they ever acquire large velocities. Under these conditions, we may develop the potential energy V by Taylor's theorem in terms of ascending powers of the q_i.

$$V(q_1, q_2, \cdots, q_{3n}) = V_0 + \sum_i \left(\frac{\partial V}{\partial q_i}\right) q_i + \tfrac{1}{2}\sum_{i,j} \left(\frac{\partial^2 V}{\partial q_i \partial q_j}\right) q_i q_j + \cdots \tag{9-51}$$

The constant term V_0 which is independent of the q_i can be omitted since it has no effect on the equations of motion of the system. The term linear in the q_i must also vanish since $\partial V/\partial q_i = 0$ is the condition for equilibrium. Finally if we omit all terms beyond the third, we obtain as an approximation to the vibrational potential energy

$$2V = \sum_{i,j} b_{ij} q_i q_j \tag{9-52}$$

where $b_{ij} = (\partial^2 V/\partial q_i \partial q_j)$. From (37) we have, in terms of the coordinates q_i

$$2T = \sum \dot{q}_i^2 \tag{9-53}$$

where T is now written for the former T_v.

If we now subject both T and V to an orthogonal transformation (see sec. 10.17), we obtain

$$2T = \sum \dot{Q}_k^2; \quad 2V = \sum \lambda_k Q_k^2 \tag{9-54}$$

where the normal coordinates Q_k are related to the q's by

$$q_i = \sum \alpha_{ik} Q_k \tag{9-55}$$

The constants λ_k are the $3n$ *eigenvalues* found from the characteristic equation

$$\left| \lambda \delta_{ij} - b_{ij} \right| = 0 \tag{9-56}$$

and α_{ik} is the matrix formed from the eigenvectors.

Knowing T and V we may obtain the motion of the molecule by solving Lagrange's equations (8). They appear as

$$\frac{d}{dt}\left(\frac{\partial T}{\partial \dot{Q}_k}\right) + \frac{\partial V}{\partial Q_k} = 0; \quad (k = 1, 2, \cdots, 3n)$$

or

$$\ddot{Q}_k = -\lambda_k Q_k \tag{9-57}$$

Three different possibilities arise: (a) $\lambda_k > 0$; (b) $\lambda_k = 0$; (c) $\lambda_k < 0$.

a. $\lambda_k > 0$. The solutions of (57) are

$$Q_k = A_k \cos(\sqrt{\lambda_k}\, t + \delta_k); \quad (k = 1, 2, \cdots, 3n) \tag{9-58}$$

This is the equation of simple harmonic motion with two constants of integration; A_k is the *amplitude* and δ_k, the *phase constant*. Eq. (55) now reads

$$q_i = \sum_k \alpha_{ik} A_k \cos (\sqrt{\lambda_k} t + \delta_k) \qquad (9\text{-}59)$$

If all of the A_k are zero except one, say A_1, then all of the nuclei are acting as simple harmonic oscillators with a frequency of

$$\nu_1 = \frac{\sqrt{\lambda_1}}{2\pi}$$

about their equilibrium position. Each nucleus has the same phase constant and reaches its equilibrium position at the same time. The amplitudes will vary because of the factor α_{ik}. Such a motion is called a *normal mode of vibration*. Actually the situation is much more complex, for many of the A_k will be different from zero. Thus the motion of the nuclei consists of a superposition of all the normal modes of vibration, each with its own frequency $\sqrt{\lambda_k}/2\pi$ and amplitude.

It frequently happens that some of the λ_k will be equal to each other in pairs or threes. This phenomenon, called double or triple degeneracy,[8] means that two or three equivalent motions of the molecule have the same frequency and differ only with respect to their orientation in space. The phase factors and amplitudes must be evaluated from the initial positions and velocities of the n nuclei. We show in the next section how the normal modes and coordinates may be determined for a specific example.

b. $\lambda_k = 0$. The solution of (57) is

$$Q_k = A_k t + \delta_k$$

hence the resulting motion is not a vibration. The nuclei will not oscillate about the equilibrium position but will continually move away. Since the whole treatment of the problem is based upon small oscillations from the equilibrium position, we are no longer justified in this case in omitting higher terms in the potential energy, and the method fails. Actually, it will be found that six of the λ_k vanish in the molecular problem (five if the equilibrium arrangement of the nuclei is linear). Three of these zero frequencies may be associated with translation of the molecule along three mutually perpendicular axes and the remaining three with rotation about the same axes. When it is desired, the zero frequencies may be removed from the problem before solving (56). This is done by reducing the number of coordinates from $3n$ to $3n - 6$, the equations of conservation

---

[8] Wu, Ta-You, " Vibrational Spectra and Structure of Polyatomic Molecules," Second Edition, Edwards Brothers, Inc., Ann Arbor, 1946; Mathieu, Jean-Paul, " Spectres de Vibration et Symétrie des Molécules et des Cristaux," Hermann et Cie, Paris, 1945; Herzberg, G., " Infrared and Raman Spectra of Polyatomic Molecules," D. Van Nostrand Co., New York, 1945.

of linear and angular momentum (eqs. 33 and 34) being used for that purpose.

c. $\lambda_k < 0$. The solution becomes imaginary and again does not correspond to a vibration. This case never occurs if the potential energy is a positive definite quadratic form (see sec. 10.12) which is always true in the molecular problem.

9.11. Vibrations of a Linear Triatomic Molecule.—As an example of the preceding theory, we consider a linear symmetrical triatomic molecule XY_2 such as carbon dioxide. Let the central atom X have a mass m_2 and the two end particles have mass m_1. Let the equilibrium positions be x_1^0 and x_3^0 for Y and x_2^0 for X. In order to simplify the problem, we arbitrarily assume that the only motion which the nuclei can make is along the line adjoining them, hence the displaced positions are $x_i = x_i^0 + \delta x_i$. If we now take the potential energy[9] as proportional to the square of the relative displacements of the particles, in accordance with eq. (51) we have

$$2V = k\{(\delta x_1 - \delta x_2)^2 + (\delta x_2 - \delta x_3)^2\} \tag{9-60}$$

and

$$2T = m_1(\delta \dot{x}_1^2 + \delta \dot{x}_3^2) + m_2 \delta \dot{x}_2^2$$

In terms of mass adjusted coordinates $q_i = \sqrt{m_i} \delta x_i$

$$2T = \sum \dot{q}_i^2$$

$$2V = k\left\{\left[\frac{q_1}{\sqrt{m_1}} - \frac{q_2}{\sqrt{m_2}}\right]^2 + \left[\frac{q_2}{\sqrt{m_2}} - \frac{q_3}{\sqrt{m_1}}\right]^2\right\} \tag{9-61}$$

Comparison with (52), shows us that

$$b_{11} = k/m_1; \quad b_{12} = b_{21} = -k/\sqrt{m_1 m_2}; \quad b_{13} = b_{31} = 0$$

$$b_{22} = 2k/m_2; \quad b_{23} = b_{32} = -k/\sqrt{m_1 m_2}; \quad b_{33} = k/m_1$$

When these values are substituted in (56) and the determinantal equation is solved we obtain

$$\lambda_1 = k/m_1; \quad \lambda_2 = k\mu; \quad \lambda_3 = 0$$

$$\mu = (2m_1 + m_2)/m_1 m_2 \tag{9-62}$$

In order to find the coefficients α_{ik} of eq. (55) which relate the q_i to the normal coordinates Q_i it is necessary to find the transformation which reduces T and V simultaneously to a sum of squares (see sec. 10.17). According to sec. 10.15 the matrix effecting this transformation has as its columns the eigenvectors of the matrix $[b_{ij}]$, and these eigenvectors are the

---

[9] See Herzberg, loc. cit., for remarks concerning the choice of the potential energy expression in special cases.

solutions (x_1, x_2, x_3) of the equations

$$\sum b_{ij} x_j = \lambda x_i$$

corresponding to the three eigenvalues λ_i already found. Simple computation yields for these eigenvectors

$$[-x_3,\, 0,\, x_3] \qquad\qquad \lambda = \frac{k}{m_1}$$

$$\left[x_3,\, -2\sqrt{\frac{m_1}{m_2}}\, x_3,\, x_3 \right] \qquad \lambda = k\mu$$

$$\left[x_3,\, \sqrt{\frac{m_2}{m_1}}\, x_3,\, x_3 \right] \qquad \lambda = 0$$

They are already orthogonal; when x_3 is also fixed by normalization, i.e., by equating the sum of the squares of the components of each vector to unity, they may be compounded to give

$$\alpha_{ik} = \begin{bmatrix} -1/\sqrt{2} & 1/\sqrt{2\mu m_1} & 1/\sqrt{\mu m_2} \\ 0 & -2/\sqrt{2\mu m_2} & 1/\sqrt{\mu m_1} \\ 1/\sqrt{2} & 1/\sqrt{2\mu m_1} & 1/\sqrt{\mu m_2} \end{bmatrix} \tag{9-63}$$

We can now find the normal modes of vibration from (59). Taking $\lambda_k = \lambda_1$, we see that the two end atoms move in opposite directions while the central atom is stationary. The other normal modes are found in the

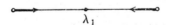

$$\lambda_1$$

$$\lambda_2 \qquad\qquad\qquad\qquad \lambda_3$$

Fig. 9-4

same way. They are shown in Fig. 4. It will be observed that for the zero frequency[10] λ_3, the motion is translational, since $x_1 = m_1^{-1/2} q_1 = m_1^{-1/2} \alpha_{13}(A_3 t + \delta_3) = (2m_1 + m_2)^{-1/2}(A_3 t + \delta_3)$, and x_2, x_3 also equal this expression.

[10] This frequency could have been removed from the problem by applying the condition $m_1(\delta \ddot{x}_1 + \delta \ddot{x}_3) + m_2 \delta \ddot{x}_2 = 0$ to (60).

The treatment of this molecule is not complete because of the artificial assumption that the motion is only along the line of nuclei. For a complete treatment the reader is referred to Wu, Mathieu, or Herzberg (loc. cit.).

9.12. Quantum Mechanical Hamiltonian.—Lack of space forbids the transcription of the results thus far obtained into the quantum mechanical language of Chapter 11. To provide a general view, however, we shall append here a few comments indicating the line of attack to be taken on the problem of the polyatomic molecule from the quantum point of view. The material of this section is not needed in other parts of this book. The expression for the classical kinetic energy found in (49) contains momenta $p_a (a = x, y, z)$ defined in eq. (45) and p_k defined in eq. (42). Both of these are conjugate to the normal coordinates Q_k. On the other hand the momenta P_a of (43) are not conjugate to Q_k. In order to obtain a suitable expression for use in quantum mechanical calculations, all of the coordinates and momenta must be conjugate to each other. It is true that the P_a which are functions of the angular velocities could be written in terms of some set of coordinates such as the Eulerian angles and then the Eulerian angles α, β, γ, the normal coordinates and the conjugate momenta p_α, p_β, p_γ, p_a and p_k would be appropriate. The coordinates used in (49) may be retained, however, as shown by several authors. The correct quantum mechanical Hamiltonian[11] is

$$H = \tfrac{1}{2}\mu^{1/2}\sum_{a,b}(P_a - p_a)\mu_{ab}\mu^{-1/2}(P_b - p_b)$$
$$+ \tfrac{1}{2}\mu^{1/2}\sum_k p_k \mu^{-1/2} p_k + V \tag{9-64}$$

where a, b denote x, y, z and μ is the determinant of μ_{ab} (cf. eq. 50). This expression may be simplified by noting that P_a commutes with p_k and that the μ_{ab} are functions only of the Q_k. We thus obtain

$$H = \tfrac{1}{2}\sum_{a,b}\mu_{ab}P_a P_b - \sum h_a P_a + \tfrac{1}{2}\sum \mu^{1/2} p_a \mu_{ab}\mu^{-1/2} p_b$$
$$+ \tfrac{1}{2}\sum \mu^{1/2} p_k \mu^{-1/2} p_k + V \tag{9-65}$$

where

$$h_a = \tfrac{1}{2}\sum_b 2\mu_{ab} p_b + p_b \mu_{ab} + \mu_{ab}\mu^{1/2}(p_b \mu^{-1/2}) \tag{9-66}$$

and p_b does not commute with the μ's.

For the sake of greater generality, we no longer need confine ourselves to the potential energy expression previously used but write the most general function consistent with the symmetry of the molecule

$$V = V_0 + V_1 + V_2 + \cdots$$

[11] See Wilson, E. B. and Howard, J. B., *J. Chem. Phys.* **4**, 260 (1936) or Dennison, D. M., and Darling, B. T., *Phys. Rev.* **57**, 128 (1940).

The first term, V_0, is identical with that given in (52) or (54); V_1 is homogeneous in the third powers of the normal coordinates and their cross-products; V_2 is of the fourth power, etc. When the Hamiltonian is expanded, it is found that it can be divided into terms of different orders as follows:

$$H = H_0 + H_1 + H_2 + \cdots \tag{9-67}$$

The explicit form of H_0 is

$$2H_0 = \left\{\frac{P_x^2}{A_0} + \frac{P_y^2}{B_0} + \frac{P_z^2}{C_0}\right\} + \Sigma p_k^2 + V_0 \tag{9-68}$$

where A_0, B_0, C_0 are the equilibrium moments of inertia. It is seen that this represents the sum of the Hamiltonians of a rigid rotator and a harmonic oscillator; hence this part of H may be treated exactly by the methods of quantum mechanics as outlined in Chapter 11.

Even to this order of approximation the details are tedious, for the vibrational part of the Hamiltonian for an n-atomic molecule involves the solution of a secular determinant like (56) with $(3n - 6)$ rows and columns. Utilization of molecular symmetry, however, makes it possible to factor this determinant.[12] Moreover, if the potential energy term, V_0, is expressed in coordinates parallel or perpendicular to chemical bonds (so-called valence-bond coordinates), then vector and matrix methods, developed by Wilson and others,[13] prove to be powerful tools for even quite complex molecules.

Still further difficulties arise if higher terms in the Hamiltonian are included but these are important if interactions between the rotational and vibrational energies are considered. Such interactions are often detectable experimentally and higher order rotational effects are also observed, especially in the microwave region.[14] A suitable perturbation technique for such cases, which involves contact transformations, has been developed by Nielsen[15] for the general n-atomic molecule and has been applied to many special molecules.

[12] See Chapter 15, or Herzberg, loc. cit.

[13] Wilson, E. B., Jr., Decius, J. C., and Cross, P. C., " Molecular Vibrations. The Theory of Infrared and Raman Vibrational Spectra," McGraw-Hill Book Co., Inc., New York, 1955.

[14] Gordy, W., Smith, W. V., and Trambarulo, R. F., " Microwave Spectroscopy," John Wiley and Sons, New York, 1953.

[15] Nielsen, H. H., *Revs. Mod. Phys.* **23**, 90 (1951).

CHAPTER 10

MATRICES AND MATRIX ALGEBRA

In ordinary arithmetic, attention is focused upon *single* numbers. These numbers may be combined by various operations, such as addition, subtraction, multiplication and so on, to yield new numbers. In many branches of algebra, the student is forced to confer interest, not upon single numbers, but on *collections* of numbers (or functions). These collections can be simple sequences like a_1, a_2, $\cdots$, a_n, in which the order of the individuals may, or may not be of importance. A vector is an example of this kind. When such a sequence is written down, no understanding prevails that the numbers are to be combined in a certain way; it is the collection itself which matters. Meaning is imparted to the collection by specifying how it is to be combined with other collections.

Besides simple sequences, collections of two-dimensional character are often objects of interest in mathematics, and recently in physics and chemistry. They may have a great variety of forms; they may be *triangular*, as

$$
\begin{array}{ccc}
 & a_1 & \\
 b_1 & & b_2 \\
 c_1 & c_2 & c_3
\end{array}
$$

or *rectangular*, as

$$
\begin{array}{ccccc}
a_1 & a_2 & a_3 & \cdots & a_n \\
b_1 & b_2 & b_3 & \cdots & b_n \\
\multicolumn{5}{c}{\cdots\cdots\cdots\cdots\cdots} \\
e_1 & e_2 & e_3 & \cdots & e_n
\end{array}
$$

or *quadratic*, as

$$
\begin{array}{cccc}
a_1 & a_2 & a_3 & a_4 \\
b_1 & b_2 & b_3 & b_4 \\
c_1 & c_2 & c_3 & c_4 \\
d_1 & d_2 & d_3 & d_4
\end{array}
$$

Of these, the rectangular and quadratic ones are of greatest value. Without further specification, they are simply *arrays*, devoid of meaning. But when rules are laid down, stating how they may be combined to form new arrays, they become objects of mathematical importance, such as

301

determinants and *matrices*. This is usually indicated by enclosing the array in bars or brackets of different form, bars being frequently used for determinants, brackets for matrices. It is also convenient to use a single letter for the individuals of a collection, and to distinguish the individuals of a simple or linear collection by single subscripts, those of a two-dimensional collection by two subscripts.

10.1. Arrays.—A collection of real or complex quantities is called an *array* if it can be displayed in an orderly table of *rows* and *columns*. The individual members of the array are its *elements*. Each is equipped with a pair of *indices*, the first one referring to the row and the second one to the column in which the element is located. For example, the element A_{pq} will appear in the p-th row and the q-th column. If the number of rows n equals the number of columns, the array is said to be *square* (or *quadratic*) and of order n; if there are n rows and m columns $(n \neq m)$, the array is *rectangular* and of order $(n \times m)$.

10.2. Determinants.—The most familiar type of array is the *determinant*,[1] which always has an equal number of rows and columns. It will be written in one of the forms:

$$\det A = \begin{vmatrix} A_{11} & A_{12} & A_{13} & \cdots & A_{1n} \\ A_{21} & A_{22} & A_{23} & \cdots & A_{2n} \\ A_{31} & A_{32} & A_{33} & \cdots & A_{3n} \\ \cdots & \cdots & \cdots & \cdots & \cdots \\ A_{n1} & A_{n2} & A_{n3} & \cdots & A_{nn} \end{vmatrix}$$

The *value* of the determinant is obtained by the following procedure. First, a total of $n!$ products is formed by taking one element from each row and column. Each product is then arranged so that the first subscripts of the elements are in their natural order $1, 2, \cdots, n$. When this has been done, it will be found that the products may be separated into *even* and *odd classes* each containing $n!/2$ terms, as follows. In the even class, an even number of interchanges of the elements is required to bring the second subscripts into their natural order while in the odd class, an odd number of interchanges is needed. For example, $A_{12}A_{23}A_{31}$ is in the even class while $A_{12}A_{21}A_{33}$ is in the odd class. If a plus sign is affixed to the even products and a minus sign to the odd ones, the algebraic sum of the $n!$ terms, by definition, is the value of the determinant. We may thus write

$$| A | = \sum (-1)^h A_{1r_1} A_{2r_2} \cdots A_{nr_n} \tag{10–1}$$

---

[1] References will be found at the end of this chapter. The most complete accounts of determinants are those of Muir, T., "Theory of Determinants in the Historical Order of Development," 4 vols., 1906–1923, and "Contributions to the History of Determinants, 1900–1920", Blackie and Son, Ltd., London, 1930.

where the summation is made over all permutations of $r_1\ r_2, \cdots, r_n$, and h is the number of interchanges required to restore the natural order.

The following properties are direct consequences[2] of this definition. In each statement, the word row may be replaced by the word column and the reverse.

1. The value of a determinant vanishes, $\mid A \mid\ = 0$, when:

   a. All elements of a row are zero.

   b. All elements of one row are identical with, or multiples of, the corresponding elements of another row.

2. The value of a determinant is unchanged, if:

   a. Rows and columns are interchanged.

   b. A linear combination of any number of rows is added to any one row; i.e., if A_{ij} is replaced by $\sum_{k=1}^{n} c_k A_{kj}, j = 1, 2, \cdots, n$, provided the c_k are fixed numbers.

3. The value of a determinant changes sign if two rows are interchanged.

4. If each element in any one row appears as the sum (or difference) of two or more quantities, the determinant may be written as a sum (or difference) of two or more determinants of the same order. Thus if the order is two

$$
\begin{vmatrix} A_{11} \pm B_{11} & A_{12} \pm B_{12} \\ A_{21} & A_{22} \end{vmatrix} = \begin{vmatrix} A_{11} & A_{12} \\ A_{21} & A_{22} \end{vmatrix} \pm \begin{vmatrix} B_{11} & B_{12} \\ A_{21} & A_{22} \end{vmatrix}
$$

5. If all elements of a row are multiplied by a constant factor, the value of the determinant is multiplied by the same factor.

10.3. Minors and Cofactors.—The *complementary minor* of an element A_{pq} is the determinant obtained by striking out the row and column in which A_{pq} appears. The *cofactor* of A_{pq} is $(-1)^{p+q}$ times its complementary minor. It will be indicated by A^{pq}. It follows from eq. (1) that

$$
\mid A \mid\ = \sum_{i=1}^{n} A_{ik}A^{ik} = \sum_{i=1}^{n} A_{ki}A^{ki}; \quad (k = 1, 2, \cdots, n) \qquad (10\text{-}2)
$$

However,

$$
\sum_{i=1}^{n} A_{ik}A^{ij} = \sum_{i=1}^{n} A_{ki}A^{ji} = 0; \quad j \neq k \qquad (10\text{-}3)
$$

for comparison with (2) shows that these equations are the expansion of a determinant whose k-th and j-th columns are identical with the k-th column of $\mid A \mid$, and according to property 1-b of sec. 10.2, if two columns are identical, $\mid A \mid\ = 0$. Eq. (2), called the *Laplace development*, is commonly

---

[2] Details of the proofs may be found in texts on determinants (see references at end of chapter).

used for numerical evaluation of determinants, but if their order is larger than three or four, the number of terms and the labor involved is so great that other procedures are to be preferred. We describe one in sec. 13.27.

10.4. Multiplication and Differentiation of Determinants.—If $|A|$ and $|B|$ are determinants of order n, the *product* $|C|$

$$|A||B| = |C|$$

is a determinant of the same order. Its elements are given by one of the four equivalent (though not equal!) expressions

$$C_{ij} = \sum_{k=1}^{n} A_{ik}B_{kj} \text{ or } \sum_{k=1}^{n} A_{ik}B_{jk} \text{ or } \sum_{k=1}^{n} A_{ki}B_{kj} \text{ or } \sum_{k=1}^{n} A_{ki}B_{jk} \qquad (10\text{--}4)$$

The proof for determinants of order two follows. Using the first form of (4) we obtain

$$|C| = \begin{vmatrix} A_{11}B_{11} + A_{12}B_{21} & A_{11}B_{12} + A_{12}B_{22} \\ A_{21}B_{11} + A_{22}B_{21} & A_{21}B_{12} + A_{22}B_{22} \end{vmatrix}$$

but according to property 4 of sec. 10.2, the product may also be written

$$|C| = \begin{vmatrix} A_{11}B_{11} & A_{11}B_{12} \\ A_{21}B_{11} & A_{21}B_{12} \end{vmatrix} + \begin{vmatrix} A_{11}B_{11} & A_{11}B_{12} \\ A_{22}B_{21} & A_{22}B_{22} \end{vmatrix} +$$

$$\begin{vmatrix} A_{12}B_{21} & A_{12}B_{22} \\ A_{21}B_{11} & A_{21}B_{12} \end{vmatrix} + \begin{vmatrix} A_{12}B_{21} & A_{12}B_{22} \\ A_{22}B_{21} & A_{22}B_{22} \end{vmatrix}$$

The first and last terms of this sum vanish, for if the constant factor $A_{11}A_{21}$ is removed from the first determinant its first row is identical with its second row. Removal of the constant term $A_{12}A_{22}$ from the last determinant leaves it with two identical columns. Constant factors may also be removed from the remaining determinants but they do not vanish. The result is

$$|C| = A_{11}A_{22} \begin{vmatrix} B_{11} & B_{12} \\ B_{21} & B_{22} \end{vmatrix} + A_{12}A_{21} \begin{vmatrix} B_{21} & B_{22} \\ B_{11} & B_{12} \end{vmatrix}$$

Referring to property 3 of sec. 10.2 we see that this becomes

$$|C| = (A_{11}A_{22} - A_{12}A_{21}) \begin{vmatrix} B_{11} & B_{12} \\ B_{21} & B_{22} \end{vmatrix}$$

Finally we note that $(A_{11}A_{22} - A_{12}A_{21})$ is just the Laplace development of $|A|$ so that we have shown the equivalence of the determinant $|C|$ with the product $|A||B|$. The proof with the other forms of (4) is similar. The method is also clearly applicable to determinants of higher order.

From (2), the partial derivative of a determinant with respect to an element equals the cofactor:

$$\frac{\partial \, | \, A \, |}{\partial A_{ik}} = A^{ik}$$

10.5. Preliminary Remarks on Matrices.—If two or more arrays may be combined in a certain way described in sec. 10.6, they are called matrices.[3] We indicate them by

$$A = [A_{ij}] = \begin{bmatrix} A_{11} & A_{12} & A_{13} & \cdots & A_{1m} \\ A_{21} & A_{22} & A_{23} & \cdots & A_{2m} \\ A_{31} & A_{32} & A_{33} & \cdots & A_{3m} \\ \cdots & \cdots & \cdots & \cdots & \cdots \\ A_{n1} & A_{n2} & A_{n3} & \cdots & A_{nm} \end{bmatrix}$$

Unlike determinants, matrices may be square or rectangular. Matrices of infinite order[4] will not be discussed here. When a matrix contains only one row or column, it is called a vector. For a *row vector*, we will write

$$[\mathbf{x}] = [x_1, x_2, x_3, \cdots, x_n] \tag{10-5a}$$

in order to save space, we write a *column vector* as

$$\{\mathbf{x}\} = \{x_1, x_2, x_3, \cdots, x_n\} \tag{10-5b}$$

although its matrix form would be

$$\begin{bmatrix} x_1 \\ x_2 \\ x_3 \\ \cdot \\ \cdot \\ \cdot \\ x_n \end{bmatrix}$$

A small letter $\mathbf{u}$, $\mathbf{v}$, $\cdots$, $\mathbf{z}$ written without brace or bracket *always* means a column vector. Matrices with two or more rows or columns will be indicated by capital letters.

The elements of a square matrix A may be written and evaluated as a determinant. If $| \, A \, | = 0$, the corresponding matrix A is called *singular*. Since determinants do not exist for rectangular (non-quadratic) arrays, all rectangular matrices, by definition, are singular. Suppose we formed determinants of all possible orders by taking successively 1, 2, $\cdots$, n rows and columns of A. If at least one determinant of order r does not vanish

---

[3] For treatises on matrix theory, see references at end of this chapter.

[4] For their properties, see Wintner, A., " Spektraltheorie der unendlichen Matrizen," S. Hirzel, Leipzig, 1929.

and all determinants of order greater than r do vanish, A is said to be of *rank* r. Thus if A is singular and of order n, $r < n$; if non-singular, $r = n$.

Problem. Show that the rank of the following matrix is two:

$$\begin{bmatrix} 1 & 1 & 1 & 1 \\ 2 & 2 & 3 & -1 \\ 0 & 0 & 1 & -3 \\ 3 & 3 & 5 & -3 \end{bmatrix}$$

10.6. Combination of Matrices.—Two matrices A and B are equal if and only if they are identical. If $A = B$, then $A_{pq} = B_{pq}$ for every p and q.

The addition or subtraction of two matrices of order n gives a new matrix of the same order according to the following rule. If $A \pm B = C$, then $C_{pq} = A_{pq} \pm B_{pq}$. Addition and subtraction are both commutative and associative.

$$A \pm B = \pm B + A; \quad (A \pm B) \pm C = A + (\pm B \pm C)$$

Multiplication of a matrix by a scalar quantity α is defined by

$$\alpha A = \alpha[A_{ij}] = [\alpha A_{ij}] = A\alpha$$

Two matrices A and B may be multiplied together in the order AB only when the number of columns in A equals the number of rows in B. Under this condition, the matrices are said to be *conformable*. If A is of order $(n \times h)$, B of order $(h \times m)$, the product C is of order $(n \times m)$. Its elements are given by

$$C_{pq} = \sum_{s=1}^{h} A_{ps}B_{sq}; \quad (p = 1, 2, \cdots, n; \quad q = 1, 2, \cdots, m)$$

$$AB = [C_{ij}] = C$$

(10–6)

This rule for multiplying matrices is not as arbitrary as it might seem; it is suggested by the properties of *linear transformations* and the reason for defining it in this way will be given in sec. 10.10. We note at this point, however, that the law of matrix multiplication is identical with the first form of eq. (4) which defines the multiplication of determinants. Hence $det\ (AB) = (det\ A) \cdot (det\ B)$ if A, B are square, but $det\ (A + B) \neq det\ A + det\ B$. In general, $AB \neq BA$, but when the order of multiplication is of no importance, so that $AB = BA$, the two matrices are said to *commute* or to be *permutable*. The ordinary laws regarding distribution and association apply.

$$A(B + C)F = ABF + ACF; \quad (AB)C = A(BC) = ABC$$

Provided A, B, $\mathbf{x}$ and $\mathbf{y}$ are properly conformable

$$A\{\mathbf{x}\} = \{\mathbf{y}\}; \quad [\mathbf{x}]A = [\mathbf{y}]$$
$$[\mathbf{x}]\{\mathbf{y}\} = \text{a scalar}; \quad \{\mathbf{x}\}[\mathbf{y}] = B \tag{10-7}$$

In the last case, B is a square matrix which has the same number of rows as $\{\mathbf{x}\}$ (or columns as $[\mathbf{y}]$). Its rows (or columns) are proportional to each other.

A given matrix may be divided into smaller matrices, the result being a partitioned matrix. For example, a square matrix of order three may be divided into four submatrices as shown.

$$A = \begin{bmatrix} A_{11} & A_{12} & A_{13} \\ A_{21} & A_{22} & A_{23} \\ A_{31} & A_{32} & A_{33} \end{bmatrix} = \begin{bmatrix} a_{11} & a_{12} \\ a_{21} & a_{22} \end{bmatrix}$$

where

$$a_{11} = \begin{bmatrix} A_{11} & A_{12} \\ A_{21} & A_{22} \end{bmatrix}, \quad a_{12} = \begin{bmatrix} A_{13} \\ A_{23} \end{bmatrix}$$
$$a_{21} = [\, A_{31} \quad A_{32} \,], \quad a_{22} = A_{33}$$

If B is a similar matrix and is similarly partitioned, then each submatrix a_{ij} and b_{ij} may be treated as a single element so that

$$AB = C = \begin{bmatrix} a_{11}b_{11} + a_{12}b_{21} & a_{11}b_{12} + a_{12}b_{22} \\ a_{21}b_{11} + a_{22}b_{21} & a_{21}b_{12} + a_{22}b_{22} \end{bmatrix}$$

Finally, the elements of C are completely evaluated by the usual rules for matrix multiplication and addition.

If $A = [A_{kj}]$ is a square matrix of order m and $B = [B_{pq}]$ is a square matrix of order n, then the *direct product*

$$A \times B = [A_{kj}B_{pq}]$$

is a square matrix of order mn. The index pairs (k,p) and (j,q) refer to the row and column, respectively. A suitable convention for arranging the rows and columns consists in taking these pairs in such a way that (j,q) precedes (j',q') if $j < j'$, $q < q'$ or if $j = j'$, $q < q'$ (dictionary order). If A, C are of order m and B, F of order n then

$$(A \times B)(C \times F) = AC \times BF$$

is a matrix of order mn. The direct product of matrices has of course nothing to do with the cross product of vectors, for which the same symbol, $\times$, is used.

Problem a. Prove eq. (7).
Problem b. Prove that $(A \times B)(C \times F) = AC \times BF$.

10.7. Special Matrices.—When all the elements of a matrix are zero, the matrix is called *null* and indicated by O. For any matrix A,

$$O + A = A; \quad OA = AO = O$$

It should not be inferred, however, that the vanishing of a matrix product implies that either or both of the matrices multiplied together are the null matrix (cf. Problem, sec. 10.7).

The *unit matrix* E has unity for elements along the " main " diagonal.[5] All other elements are zero. The matrix elements are conveniently symbolized by the *Kronecker delta* (cf. sec. 3.4)

$$\delta_{pq} = \begin{cases} 0; & p \neq q \\ 1; & p = q \end{cases}$$

For every matrix,

$$EA = AE = A$$

If all matrix elements vanish except diagonal ones, the matrix is called *diagonal*. The general element of a diagonal matrix is thus of the form $D_i\delta_{ij}$. All diagonal matrices commute with each other, for if D and D' are diagonal

$$(DD')_{ik} = \sum_{j=1}^{n} D_i\delta_{ij}D_i'\delta_{jk} = D_iD_i'\delta_{ik} = (D'D)_{ik}$$

If a matrix A commutes with a diagonal one, D, the elements A_{ij} will all vanish, except those for which the diagonal matrix has equal elements, $D_i = D_j$. The proof is as follows. Assume that $AD = DA$ and, of course, $D_{ij} = D_i\delta_{ij}$. Then

$$\sum_k A_{ik}D_k\delta_{kj} = \sum_k D_i\delta_{ik}A_{kj}$$

and

$$A_{ij}D_j = D_iA_{ij}; \quad A_{ij}(D_i - D_j) = 0$$

Hence, either $D_i = D_j$ or $A_{ij} = 0$.

If all of the diagonal elements of D are different, A must be truly diagonal with all different elements, say $A_1, A_2, \cdots, A_n$. It is sometimes convenient to write such a matrix in the form

$$A = diag\ (A_1, A_2, \cdots, A_n)$$

If some of the diagonal elements of D are repeated so that its form is

$$D = diag\ (D_1, D_1, D_2, D_2, D_2, \cdots)$$

then the commuting matrix A will have the form

$$A = diag\ (a_1, a_2, \cdots)$$

where the square matrices a_i are arranged in symmetric positions about the main diagonal and the other elements of A are zero. The forms of the submatrices will be

$$a_1 = \begin{bmatrix} A_{11} & A_{12} \\ A_{21} & A_{22} \end{bmatrix} \quad a_2 = \begin{bmatrix} A_{33} & A_{34} & A_{35} \\ A_{43} & A_{44} & A_{45} \\ A_{53} & A_{54} & A_{55} \end{bmatrix}$$

[5] The main or principal diagonal is that running from the upper left to the lower right of the array.

The sum of the diagonal elements of a square matrix is called the *trace* (German " Spur ").

$$\text{Tr } A = \sum_{i=1}^{n} A_{ii}$$

The trace of the product of two or more matrices is independent of the order of multiplication. The proof is simple.

$$\text{Tr } AB = \sum_{i}(AB)_{ii} = \sum_{i}\sum_{j}A_{ij}B_{ji} = \text{Tr } BA$$

If $A \times B = C$, $\text{Tr } C = \text{Tr } A \cdot \text{Tr } B$.

The *transposed matrix* to A, indicated by $\tilde{A} = [A_{ji}]$ is formed from A by interchanging rows and columns. If A and B of (6) are transposed, $\tilde{A}$ becomes of order $(h \times n)$ and $\tilde{B}$ of order $(m \times h)$. They may be multiplied together only in the order $\tilde{B}\tilde{A}$ and the product $\tilde{C}$ is of order $(m \times n)$. Thus when a matrix product is transposed, the sequence of the matrices forming the product must be reversed. This holds true for any number of factors

$$F = ABCD \cdots X; \quad \tilde{F} = \tilde{X} \cdots \tilde{D}\tilde{C}\tilde{B}\tilde{A}$$

The matrix $\hat{A} = [A^{ji}]$ is the *adjoint matrix*.[6] Note that the adjoint is formed by first finding the cofactor A^{pq} of the element A_{pq} in $|\,A\,|$ and then *transposing the resulting matrix*. From the properties of determinants, it follows that

$$A\hat{A} = \hat{A}A = |\,A\,|\,E \tag{10-8}$$

hence if A is singular

$$A\hat{A} = \hat{A}A = O \tag{10-9}$$

However, the adjoint matrix exists even when A is singular.

When A is a non-singular square matrix, we may divide $\hat{A}$ by $|\,A\,|$ to obtain a matrix A^{-1} which is the *reciprocal* of A. Only square matrices have reciprocals.

$$A^{-1} = \frac{\hat{A}}{|\,A\,|}\,; \quad AA^{-1} = A^{-1}A = E \tag{10-10}$$

Suppose the matrices of (6) are square and non-singular. Multiply both sides of the equation by $B^{-1}A^{-1}$ and then by C^{-1} in the order shown:

$$B^{-1}A^{-1}ABC^{-1} = B^{-1}A^{-1}CC^{-1}$$

Thus $C^{-1} = B^{-1}A^{-1}$. Reciprocation of a matrix product requires reversal of the order of the factors as in the case of the transposed matrix product. The rule holds for any number of factors.

If the elements of A are complex numbers, the *complex conjugate* of A is

---

[6] This name seems to be in agreement with the usual mathematical convention. Writers on quantum mechanics frequently call that matrix adjoint which we later call associate. The reader should take care not to be confused by this situation.

defined as $A^* = [A_{ij}^*]$. Unlike the preceding case, if $F = ABC \cdots X$, $F^* = A^*B^*C^* \cdots X^*$.

The matrix formed by taking the complex conjugate of all the elements and then transposing the matrix is called the *associate matrix*,[7] $A^\dagger = (\widetilde{A^*}) = (\widetilde{A})^*$. If $F = ABC$; $F^\dagger = C^\dagger B^\dagger A^\dagger$.

At this point we have defined four important operations on a matrix A. These result in $-A$, $\widetilde{A}$, A^{-1} and A^*. It is important to note that each of these operations has the reflexive property, so that when the operation is performed twice, the original matrix is reproduced:

$$-(-A) = A; \quad (\widetilde{\widetilde{A}}) = A; \quad (A^{-1})^{-1} = A; \quad (A^*)^* = A$$

By combining these operations in all possible ways, the following 16 matrices may be derived from A: $\pm A$, $\pm \widetilde{A}$, $\pm A^{-1}$, $\pm A^*$, $\pm(\widetilde{A})^{-1}$, $\pm(A^*)^{-1}$, $\pm A^\dagger$, $\pm(A^\dagger)^{-1}$. In certain cases, A may be identical with some other member of this set. Such matrices have been given special names. We shall have occasion to discuss the properties of most of them later, but for convenience we list them now in Table 1. We will have no need of the types: $A = A^{-1}$ (*involutary*) and $A = (A^*)^{-1}$.

TABLE 1

Relation	Name of A	Matrix Elements
$A = \widetilde{A}$	symmetric	$A_{pq} = A_{qp}$
$A = -\widetilde{A}$	skew symmetric	$A_{pp} = 0$; $A_{pq} = -A_{qp}$
$A = \widetilde{A}^{-1}$	orthogonal	cf. eq. (42)
$A = A^*$	real	$A_{pq} = A_{pq}^*$
$A = -A^*$	pure imaginary	$A_{pq} = iB_{pq}$; B_{pq} real
$A = A^\dagger$	Hermitian	$A_{pq} = A_{qp}^*$
$A = -A^\dagger$	skew Hermitian	$A_{pp} = 0$; $A_{pq} = -A_{qp}^*$
$A = (A^\dagger)^{-1}$	unitary	cf. eq. (50)

Note that a *real symmetric matrix* is a special case of an *Hermitian matrix*. Suppose $H = A + iB$ is Hermitian with both A and B real; then $H^\dagger = \widetilde{A} - i\widetilde{B}$; but by definition $H = H^\dagger$. Thus the real part is symmetric and the imaginary part *skew symmetric*; in other words, a real Hermitian matrix is also symmetric. Similarly, a *real orthogonal matrix* is *unitary*, for if $U = A + iB$ is unitary then by definition $U = (U^\dagger)^{-1}$, $U^\dagger U = E$ and $(\widetilde{A} - i\widetilde{B})(A + iB) = E$. If $B = \widetilde{B} = O$, then $A\widetilde{A} = E$ which defines the orthogonal matrix. However, a complex symmetric matrix is not Hermitian nor is a complex orthogonal matrix unitary.

Problem. Show that $AB = O$ but $BA \neq O$ where

$$A = \begin{bmatrix} -6 & -4 & -2 \\ -9 & -6 & -3 \\ 3 & 2 & 1 \end{bmatrix}; \quad B = \begin{bmatrix} 0 & 1 & -2 \\ -1 & 0 & 3 \\ 2 & -3 & 0 \end{bmatrix}$$

[7] This is the matrix called adjoint by writers on quantum mechanics. It is also called the *Hermitian conjugate*.

10.8. Real Linear Vector Space.—Let us consider a space of two dimensions, that is, a plane in ordinary three-dimensional space. A vector in this space, as we have shown in sec. 4.1, is completely described by its two components or by the coordinates of its origin and terminus. It is also described by the matrix **x** of one column, or its transposed, the row vector $[\mathbf{x}] = \tilde{\mathbf{x}}$, the two real numbers which are its components being the two matrix elements. After we have chosen one vector it is possible to find another vector **y** in the same plane which is not a multiple of **x**. In fact, **y** is completely independent of **x**. But no matter how we draw a third vector **z**, it may always be represented as

$$ a\mathbf{x} + b\mathbf{y} = \mathbf{z} $$

where a and b are numbers. There is nothing unique about **x** and **y**, the point being that two and only two vectors are linearly independent in two dimensions and a third vector is linearly dependent on the other two. The situation may further be characterized as follows. If two vectors are linearly independent, no relation

$$ a\mathbf{x} + b\mathbf{y} = 0 $$

can exist unless $a = b = 0$, for as we have seen a linear combination of two vectors gives a new vector. For the purposes of this chapter, we shall need more than two or three dimensions, hence we shall speak of a space of n dimensions, where n is an integer. When n is greater than three, it is, of course, impossible to visualize the situation, but the geometric concepts of ordinary space will be used wherever convenient. Thus an n-dimensional coordinate system will consist of n mutually perpendicular axes, a point will require n coordinates for its location and a vector will be described by means of its n components or by the coordinates of its origin and terminus.

Suppose the components of a vector in such a space are real numbers $x_1, x_2, \cdots, x_n$, then we may write the vector **x** as a matrix of either a single row or a single column as in (5a) or (5b).

The *scalar product* of two vectors[8] is a scalar

$$ \tilde{\mathbf{x}}\mathbf{y} = x_1 y_1 + x_2 y_2 + \cdots + x_n y_n \qquad (10\text{–}11) $$

The square of the length of a vector is defined as in sec. 4.1

$$ l^2 = \tilde{\mathbf{x}}\mathbf{x} = \mathbf{x}^2 = x_1^2 + x_2^2 + \cdots + x_n^2 \qquad (10\text{–}12) $$

The *vector product*, usually denoted by $\mathbf{y} \times \mathbf{x}$, is more difficult to formulate by matrix methods. To obtain it, we first construct from **y** the skew-symmetric matrix

$$ Y = \begin{bmatrix} 0 & -y_3 & y_2 \\ y_3 & 0 & -y_1 \\ -y_2 & y_1 & 0 \end{bmatrix} $$

---

[8] For definiteness, we suppose that **x** and **y** are both column vectors. Note that $\tilde{\mathbf{x}}\mathbf{y}$ is the equivalent, in matrix notation, of $\mathbf{x} \cdot \mathbf{y}$ in vector notation.

In terms of this,

$$\mathbf{y} \times \mathbf{x} = Y\mathbf{x}$$

The *vectors* $\mathbf{u}_1$, $\mathbf{u}_2$, $\cdots$, $\mathbf{u}_n$ are *linearly independent* if there exists no set of scalar quantities c_1, c_2, $\cdots$, c_n not all zero such that

$$c_1\mathbf{u}_1 + c_2\mathbf{u}_2 + \cdots + c_n\mathbf{u}_n = 0 \tag{10-13}$$

The simplest way of testing vectors for linear independence is to evaluate the Gram determinant (see sec. 3.13)

$$| \Gamma | = \begin{vmatrix} \tilde{\mathbf{u}}_1\mathbf{u}_1 & \tilde{\mathbf{u}}_1\mathbf{u}_2 & \cdots & \tilde{\mathbf{u}}_1\mathbf{u}_n \\ \tilde{\mathbf{u}}_2\mathbf{u}_1 & \tilde{\mathbf{u}}_2\mathbf{u}_2 & \cdots & \tilde{\mathbf{u}}_2\mathbf{u}_n \\ \cdots & \cdots & \cdots & \cdots \\ \tilde{\mathbf{u}}_n\mathbf{u}_1 & \tilde{\mathbf{u}}_n\mathbf{u}_2 & \cdots & \tilde{\mathbf{u}}_n\mathbf{u}_n \end{vmatrix}$$

If $| \Gamma |$ vanishes, the vectors are linearly dependent; if $| \Gamma | > 0$, linearly independent.

When n linearly independent vectors have been chosen they form an n-dimensional coordinate system or *basis*, being equivalent to a set of n coordinate axes. Any other vector $\mathbf{v}$ may then be expressed as a linear combination of the chosen vectors $\mathbf{u}_1$, $\mathbf{u}_2$, $\cdots$, $\mathbf{u}_n$, the linear combination being unique. It should be emphasized that there is nothing unique about the choice of the basis, for any n linearly independent vectors are suitable for that purpose although the most convenient choice, in general, is a set of unit vectors. The latter are defined by the relations[9]

$$\mathbf{e}_1 = \{1,0,0,0,\cdots,0\}$$
$$\mathbf{e}_2 = \{0,1,0,0,\cdots,0\}$$
$$\mathbf{e}_3 = \{0,0,1,0,\cdots,0\}$$
$$\cdots\cdots\cdots\cdots\cdots$$
$$\mathbf{e}_n = \{0,0,0,0,\cdots,1\}$$

or similarly as row vectors. Clearly they are of unit length and mutually perpendicular, for

$$\tilde{\mathbf{e}}_i\mathbf{e}_j = \delta_{ij} \tag{10-14}$$

In terms of the unit vectors, any vector $\mathbf{x}$ may be written

$$\mathbf{x} = x_1\mathbf{e}_1 + x_2\mathbf{e}_2 + \cdots + x_n\mathbf{e}_n \tag{10-15}$$

If the origin of $\mathbf{x}$ is taken as coincident with the origin of the basis formed by the $\mathbf{e}_i$, the components of $\mathbf{x}$ are the coordinates of the terminus of $\mathbf{x}$.

It is often necessary to use a particular set of linearly independent vectors as a basis, constructing from them a set whose members are mutually perpendicular and of unit length. This procedure, known as *Schmidt's orthogonalization method*, is effected in the following way. Suppose the n given vectors are $\mathbf{u}_1$, $\mathbf{u}_2$, $\cdots$, $\mathbf{u}_n$. Select any one of them, say $\mathbf{u}_1$, and let $\mathbf{v}_1 = \mathbf{u}_1$, $\mathbf{e}_1 = \mathbf{v}_1/l_1$, where l_1 is the length of $\mathbf{v}_1$. Now

---

[9] Notice that the subscripts on the vectors $\mathbf{e}_i$ do not designate components.

take $\mathbf{v}_2 = \mathbf{u}_2 - c_{21}\mathbf{e}_1$, choosing c_{21} so that $\tilde{\mathbf{e}}_1\mathbf{v}_2 = \tilde{\mathbf{e}}_1\mathbf{u}_2 - c_{21}\tilde{\mathbf{e}}_1\mathbf{e}_1 = 0$ which requires $c_{21} = \tilde{\mathbf{e}}_1\mathbf{u}_2$ or $\mathbf{v}_2 = \mathbf{u}_2 - (\tilde{\mathbf{e}}_1\mathbf{u}_2)\mathbf{e}_1$. If we put $\mathbf{e}_2 = \mathbf{v}_2/l_2$, where l_2 is the length of $\mathbf{v}_2$, we shall have $\tilde{\mathbf{e}}_1\mathbf{e}_2 = \delta_{12}$. Next let $\mathbf{v}_3 = \mathbf{u}_3 - c_{31}\mathbf{e}_1 - c_{32}\mathbf{e}_2$, determining the constants so that $\tilde{\mathbf{e}}_1\mathbf{v}_3 = \tilde{\mathbf{e}}_1\mathbf{u}_3 - c_{31} = 0$ and $\tilde{\mathbf{e}}_2\mathbf{v}_3 = \tilde{\mathbf{e}}_2\mathbf{u}_3 - c_{32} = 0$, which means that $c_{31} = \tilde{\mathbf{e}}_1\mathbf{u}_3$ and $c_{32} = \tilde{\mathbf{e}}_2\mathbf{u}_3$. Finally let $\mathbf{e}_3 = \mathbf{v}_3/l_3$. Continuing in this way, we may construct the complete set of n unit vectors with

$$\mathbf{e}_{n+1} = \frac{\mathbf{v}_{n+1}}{l_{n+1}}\,; \quad \mathbf{v}_{n+1} = \mathbf{u}_{n+1} - \sum_{k=1}^{n} (\tilde{\mathbf{e}}_k\mathbf{u}_{n+1})\mathbf{e}_k$$

Problem a. Consider the columns of the matrix of Problem, sec. 10.5, as the components of four vectors. Test them for linear dependence.

Problem b. Prove that $Y^3 = -\,[y]\{y\}Y$. This relation is known as the "Cayley identity."

10.9. Linear Equations.—Matrix methods are useful in solving and discussing *linear equations* of the form

$$\begin{aligned}
A_{11}x_1 + A_{12}x_2 + \cdots + A_{1n}x_n &= y_1 \\
A_{21}x_1 + A_{22}x_2 + \cdots + A_{2n}x_n &= y_2 \\
&\cdots\cdots\cdots\cdots\cdots \\
A_{n1}x_1 + A_{n2}x_2 + \cdots + A_{nn}x_n &= y_n
\end{aligned} \qquad (10\text{–}16)$$

which are *inhomogeneous*.[10] They may also be written as

$$A\mathbf{x} = \mathbf{y}$$

The corresponding *homogeneous* equation is

$$A\mathbf{x} = O$$

The matrix A and the vector $\mathbf{y}$ are to be considered as known while the n components of $\mathbf{x}$ are unknown. The questions of chief interest concern the number of possible solutions and the method of finding them. Several cases arise depending on the rank of A, but for our purposes we consider [11] only three possibilities.

a. $\mathbf{y} \neq 0;\ |\,A\,| \neq 0$. According to (10), A^{-1} exists, hence

$$\mathbf{x} = A^{-1}\mathbf{y} = \frac{\hat{A}\mathbf{y}}{|\,A\,|}$$

is the unique solution. From the definition of $\hat{A}$ and the rule for matrix multiplication, it also follows that

$$x_i = \frac{1}{|\,A\,|}\,(y_1 A^{1i} + y_2 A^{2i} + \cdots + y_n A^{ni})$$

which is commonly known as *Cramer's rule*.[12]

---

[10] See sec. 2.5 for the meaning of the term homogeneous.

[11] The others are discussed by Bôcher, M., "Introduction to Higher Algebra," Macmillan Co., New York, 1907.

[12] In actual calculations, it is usually simpler to solve (16) by direct elimination (cf. sec. 13.26).

b. $\mathbf{y} = 0$; $|A| \neq 0$. The only solutions are the trivial ones $x_1 = x_2 = \cdots = x_n = 0$.

c. $\mathbf{y} = 0$; $|A| = 0$; $A^{ik} \neq 0$ for at least one value of i and k. If we knew the value of one of the unknowns we could find the values of the remaining $(n - 1)$ unknowns, since we could then form from the original set $(n - 1)$ inhomogeneous equations with non-vanishing determinant. In other words, we are confronted in case (c) with n unknowns but only $(n - 1)$ equations. However, we note that the k-th row of our set of n equations is

$$A_{k1}x_1 + A_{k2}x_2 + \cdots + A_{kn}x_n = 0 \qquad (10\text{--}17)$$

so that if we take

$$x_i = cA^{ji} \qquad (10\text{--}18)$$

where c is any constant, it follows from (3) that (17) is satisfied. Even if $j = k$, we still have a solution, for (17) is then identical with (2) but $|A| = 0$. We thus have an infinite number of solutions of the homogeneous equation when $|A| = 0$ as j may take any value from 1 to n and c is completely arbitrary. Of course, some of the solutions (18) may be worthless, since several of the cofactors may vanish; but it will be found that there are always enough non-vanishing ones so that the ratio of all the unknowns is determined.[13] The fact that the set of homogeneous equations $A\mathbf{x} = O$ possesses non-trivial solutions only when $|A| = 0$ is of great importance in many problems and will often be used in the next chapter.

Problem a. Sometimes chemical analysis must be done in an indirect way. Solve the following problem by means of determinants. A mixture of sodium chloride, sodium bromide, and sodium iodide weighed 0.5000 gram. Upon the addition of silver nitrate, the mixed silver halides weighed 1.0369 g. The iodine in the mixture was precipitated as palladous iodide, which weighed 0.3006 g. Find the composition of the original mixture.

Ans. NaCl, 50%; NaBr, 25%; NaI, 25%.

Problem b. Solve the set of linear equations which result from application of Kirchhoff's laws to the network known as a Wheatstone bridge, out of balance, and obtain the current through the galvanometer. Label resistances R_1, R_2, R_3, R_4, going clockwise around the " diamond." Let galvanometer resistance be R_g.

$$\textit{Ans.} \quad i_g = \frac{(R_2R_4 - R_1R_3)E}{R_1R_2R_3 + R_2R_3R_4 + R_3R_4R_1 + R_4R_1R_2 + R_g(R_1 + R_2)(R_3 + R_4)}$$

where E is the external electromotive force.

10.10. Linear Transformations.—Consider a set of linear, inhomogeneous equations, similar to (16), relating m quantities $\mathbf{x}$ and h quantities $\mathbf{x}'$, which can be written in the form

$$x'_s = \sum_q B_{sq}x_q; \quad \begin{array}{l} s = 1, 2, \cdots, h; \\ q = 1, 2, \cdots, m \end{array} \qquad (10\text{--}19)$$

---

[13] The proof is given by Bôcher, p. 4.

These equations define a *linear transformation*, which may be interpreted in two different ways, as we shall see later.

Furthermore, let n quantities $\mathbf{x}''$ be related to $\mathbf{x}'$:

$$x_p'' = \sum_s A_{ps}x_s'; \qquad p = 1, 2, \cdots, n$$

Then, on combining these equations, we find

$$x_p'' = \sum_{s,q} A_{ps}B_{sq}x_q$$

The same equations could also be written as a single sum

$$x_p'' = \sum_q C_{pq}x_q \qquad (10\text{--}20)$$

provided that we agree to take C_{pq} as in eq. (6) which defines the law of matrix multiplication.

The importance of linear transformations thus indicates that this definition of matrix multiplication would be a useful one. It could, of course, be defined in other ways, and one possibility is

$$(AB)_{ik} = A_{ij}B_{kj}$$

where, for convenience, the summation sign is omitted and summation over the repeated index, as in tensor analysis, is required. The consequence of this definition is interesting, for if we multiply by a third matrix C, we find

$$(AB \cdot C)_{ik} = (AB)_{ij}C_{kj} = A_{is}B_{js}C_{kj}$$

but

$$(A \cdot BC)_{ik} = A_{is}(BC)_{ks} = A_{is}B_{kj}C_{sj}$$

and the matrix elements are not the same in the two cases. Therefore, the associative law of multiplication would not hold.

A linear transformation is frequently interpreted as a rotation of coordinate axes. In sec. 4.1, we have shown how direction cosines may be used to relate the components of a vector in two different coordinate systems. Let us now consider these relations in matrix notation. Suppose that two systems $OXYZ$ and $OX'Y'Z'$, or bases with unit vectors $\mathbf{e}$ and $\mathbf{e}'$, coincide originally and that $OX'Y'Z'$ is then rotated in the positive (counterclockwise) direction through the angle ϕ about the OZ-axis. A vector $\mathbf{x}'$ in the new system is related to the same vector $\mathbf{x}$ in the original system by an equation similar to (19), which in matrix form would read

$$\mathbf{x}' = B(\phi)\mathbf{x} \qquad (10\text{--}19a)$$

Rotate the system again, this time through the angle θ, to obtain $OX''Y''Z''$ and the vector becomes $\mathbf{x}''$, where $\mathbf{x}'' = A(\theta)\mathbf{x}'$. We could then write

$$\mathbf{x}'' = A\mathbf{x}' = AB\mathbf{x} = C\mathbf{x} \qquad (10\text{--}20a)$$

from which we see that $C = AB$ is a matrix which transforms $\mathbf{x}$ directly to $\mathbf{x}''$, without passing through the intermediate system in which it was called $\mathbf{x}'$. We also see from (20a) that, if physical significance is to be attached to the matrices A and B, the order of performing the operations involved must be preserved, for in general $AB \neq BA$ although the matrices do commute, in the case now being discussed. As suggested by this example, we always understand that the order [14] is from *right* to *left*, B first and then A in the case of eq. (20a).

Provided C is not singular, we may find C^{-1}, hence

$$C^{-1}\mathbf{x}'' = C^{-1}C\mathbf{x} = \mathbf{x} \tag{10-21}$$

We may thus use (20) or (21) to determine the components of the same vector in either of two different coordinate systems.

Sometimes one wishes to think of linear transformations in another way. Suppose that there is only *one* coordinate system and that a vector $\mathbf{x}$ is rotated through the angle ϕ in the positive direction to give a new vector $\mathbf{y}$ in the same coordinate system. If the reader will think about the situation for a moment (or draw a simple figure, if necessary) he will be convinced that the operation is equivalent to rotation of the coordinate system through the angle ϕ in the *negative* (clockwise) direction. The matrix elements will therefore not be identical with those used when we assume two different coordinate systems. They are, however, closely related as will be shown in sec. 10.17.

10.11. Equivalent Matrices.—Let P and Q be non-singular matrices. Then A and B are said to be *equivalent* when

$$B = PAQ \tag{10-22}$$

Equivalent matrices have many properties in common as the subsequent discussion will show; their importance is due to the fact that it is often possible by means of a linear transformation like (22) to find an equivalent matrix which has simpler properties than the original one. When the equivalent matrix is in its simplest form, usually diagonal, it is said to be canonical.[15] The problem of finding an equivalent matrix of canonical form is analogous to that of finding a suitable coordinate system in ordinary scalar algebra (cf. Chapter 5).

Several special cases of equivalent matrices are possible, depending on the nature of the matrices P and Q effecting the transformation.

[14] The opposite convention is often used and endless confusion may result if this fact is overlooked when the equations of various writers are compared. The elements of the matrix products, of course, must always be evaluated from *left* to *right*, as required by eq. (6).

[15] Canonical matrices are exhaustively discussed by Turnbull, H. W., and Aitken, A. C., "The Theory of Canonical Matrices," Blackie and Son, London, 1932.

a. If $PQ = E$, then

$$B = Q^{-1}AQ \qquad (10\text{-}23)$$

The transformation is called *collineatory* or a *similarity transformation* (*Ähnlichkeitstransformation*). The two matrices A and B are said to be the *transforms* of each other.

b. If $P = \tilde{Q}$, the transformation is called *congruent*:

$$B = \tilde{Q}AQ \qquad (10\text{-}24)$$

c. If $P = Q^{\dagger}$, then

$$B = Q^{\dagger}AQ \qquad (10\text{-}25)$$

the transformation is *conjunctive*. If the matrices are all real, this becomes identical with (24).

d. If $PQ = E$, $P = \tilde{Q}$ (i.e., Q is orthogonal) and all the matrix elements are real, then

$$B = \tilde{Q}AQ = Q^{-1}AQ \qquad (10\text{-}26)$$

represents a *real orthogonal transformation*. It is both collineatory and congruent.

e. If the matrix elements are complex and $PQ = E$, $P = Q^{\dagger} = Q^{-1}$ (i.e., Q is unitary), then

$$B = Q^{\dagger}AQ = Q^{-1}AQ \qquad (10\text{-}27)$$

is called a *unitary* transformation. It is collineatory and conjunctive.

10.12. Bilinear and Quadratic Forms.—A homogeneous polynomial of the second degree in $2n$ variables $x_1, x_2, \cdots, x_n; y_1, y_2, \cdots, y_n$ is called a *bilinear form*. It may be abbreviated as

$$A(\mathbf{x,y}) = \tilde{\mathbf{x}}A\mathbf{y} = \sum_{i,j}^{n} A_{ij}x_i y_j \qquad (10\text{-}28a)$$

where $A = [A_{ij}]$. If both $\mathbf{x}$ and $\mathbf{y}$ undergo non-singular transformations

$$\mathbf{x} = P\mathbf{x'}; \quad \mathbf{y} = Q\mathbf{y'}$$

then

$$A(\mathbf{x,y}) = \tilde{\mathbf{x}}'\tilde{P}AQ\mathbf{y'} = \tilde{\mathbf{x}}'B\mathbf{y'} = A'(\mathbf{x',y'}) \qquad (10\text{-}29)$$

If $P = Q^{-1}$, so that $\mathbf{x} = Q^{-1}\mathbf{x'}$; $\mathbf{y} = Q\mathbf{y'}$ then $\mathbf{x}$ and $\mathbf{y}$ are called *contragradient* variables since they undergo opposite transformations.

As a special case of a bilinear form suppose $\mathbf{x} = \mathbf{y}$. Then the coefficient of $x_i x_j$ ($i \neq j$) in (29) is $(A_{ij} + A_{ji})$, and the matrix A becomes symmetric if we write $(A_{ij} + A_{ji})/2$ for every A_{ij} and A_{ji}. Eq. (28a) may then be written

$$A(\mathbf{x,x}) = \sum_{i,j}^{n} A_{ij}x_i x_j = \tilde{\mathbf{x}}A\mathbf{x}; \quad A = \tilde{A} \qquad (10\text{-}28b)$$

Such a function is a *quadratic form*; if it is positive for all real values of the variables, it is called a *positive definite* quadratic form; if it is positive or zero, it is called *semi-definite*.

10.13. Similarity Transformations.—Suppose the same vector is called $\mathbf{x}$ when referred to the basis $\mathbf{e}$ and $\mathbf{x}'$ when referred to another basis $\mathbf{e}'$. Let another vector be $\mathbf{y}$ or $\mathbf{y}'$, where

$$\mathbf{x} = Q\mathbf{x}'; \quad \mathbf{y} = Q\mathbf{y}' \tag{10-30}$$

Such variables which undergo the same transformation are said to be *cogredient*. Now consider the transformation

$$\mathbf{x} = A\mathbf{y} \tag{10-31}$$

which changes $\mathbf{y}$ into $\mathbf{x}$ in the basis $\mathbf{e}$. Then,

$$Q\mathbf{x}' = A\mathbf{y} = AQ\mathbf{y}'$$

or, if Q is non-singular

$$\mathbf{x}' = Q^{-1}AQ\mathbf{y}' = B\mathbf{y}' \tag{10-32}$$

Hence, (32) is a transformation which changes $\mathbf{y}'$ into $\mathbf{x}'$ in the basis $\mathbf{e}'$ while A, the transform of B, performs the corresponding transformation from $\mathbf{y}$ to $\mathbf{x}$ in the basis $\mathbf{e}$. This is the reason for the name similarity transformation.

An alternative interpretation of the transform may be given. Let $\mathbf{x}$, $\mathbf{x}'$, $\mathbf{y}$, $\mathbf{y}'$ be four different vectors all in the same basis. Then (30) changes $\mathbf{x}'$ into $\mathbf{x}$ and $\mathbf{y}'$ into $\mathbf{y}$ while (31) changes $\mathbf{y}$ into $\mathbf{x}$. The single transformation that changes $\mathbf{y}'$ directly into $\mathbf{x}'$ is (32), since $\mathbf{x}' = Q^{-1}\mathbf{x} = Q^{-1}A = Q^{-1}AQ\mathbf{y}'$. Here as in sec. 10.10, the form of the matrix equations is similar for different vectors in the same basis or the same vectors in different reference frames. The matrix elements, however, will not be identical in the two cases.

10.14. The Characteristic Equation of a Matrix.—If λ is a scalar parameter, A is a square matrix of order n and E the unit matrix of the same order, the matrix

$$K = [\lambda E - A] \tag{10-33}$$

is called the *characteristic matrix* of A. The equation

$$K(\lambda) = |K| = |\lambda E - A| = 0$$

or its equivalent

$$K(\lambda) = \lambda^n + a_1\lambda^{n-1} + a_2\lambda^{n-2} + \cdots + a_n = 0 \tag{10-34}$$

where the a_s are functions of the elements of A, is the *characteristic equation* of A. The n roots of $K(\lambda)$, $\lambda_1, \lambda_2, \lambda_3, \cdots, \lambda_n$, not necessarily all different, are the *characteristic* (or *latent*) *roots*. On writing (34) in the form

$$(\lambda - \lambda_1)(\lambda - \lambda_2)(\lambda - \lambda_3) \cdots (\lambda - \lambda_n) = 0$$

and comparing coefficients of the different powers of λ it will be seen that

$$\lambda_1 + \lambda_2 + \cdots + \lambda_n = -a_1$$
$$\lambda_1\lambda_2 + \lambda_1\lambda_3 + \cdots + \lambda_{n-1}\lambda_n = a_2$$
$$\lambda_1\lambda_2\lambda_3 + \cdots + \lambda_{n-2}\lambda_{n-1}\lambda_n = -a_3$$
$$\cdots\cdots\cdots\cdots\cdots\cdots\cdots\cdots\cdots\cdots\cdots\cdots\cdots\cdots$$
$$\lambda_1\lambda_2\lambda_3 \cdots \lambda_n = (-1)^n a_n$$

If $B = Q^{-1}AQ$, then $[\lambda E - B] = [\lambda E - Q^{-1}AQ] = Q^{-1}[\lambda E - A]Q$. Moreover,

$$| \lambda E - B | = | Q^{-1} ||\lambda E - A || Q | = |\lambda E - A | \qquad (10\text{–}35)$$

Hence two matrices related by a similarity transformation have the same characteristic roots.

We leave the proof of the following statements to the reader.

$$\mathrm{Tr}\ Q^{-1}AQ = \mathrm{Tr}\ A$$
$$| Q^{-1}AQ | = | A |$$

If $C = A \times B$, the characteristic roots of C are the products, taken in pairs, of the roots of A and B.

Problem.   Prove the statements of the preceding paragraph.

10.15. Reduction of a Matrix to Diagonal Form.—Consider the linear transformation

$$Ax = \lambda x \qquad (10\text{–}36)$$

The only effect of the matrix A on the vector $\mathbf{x}$ is to multiply it by the constant scalar factor λ. Rewriting (36) in the form

$$[\lambda E - A]\mathbf{x} = K\mathbf{x} = O \qquad (10\text{–}37)$$

we see that, except for the trivial case where all the components of $\mathbf{x}$ are zero, $| K |$ must vanish (cf. sec. 10.9b, c). Hence, as shown in the previous section, λ can only take the values $\lambda_1, \lambda_2, \cdots, \lambda_n$ where λ_i is one of the characteristic roots of $K(\lambda)$. These quantities are the *eigenvalues* of the matrix A; the accompanying sets of vectors $\mathbf{x}$ are the *eigenvectors*. Eq. (36) is the matrix form of an eigenvalue equation, other examples of which were discussed in Chapter 8.

Now suppose B is a diagonal matrix; then the roots of its characteristic equation are identical with its diagonal elements. If A is not a diagonal matrix but is related to B by a similarity transformation, $B = Q^{-1}AQ$, then it follows from (35) that its characteristic roots and equation are the same as those of B. The problem of reducing A to diagonal form by means of a similarity transformation is thus closely related to the problem of finding its eigenvalues. We shall now show how such a reduction may be made.

The eigenvalues themselves must first be obtained [16] by solving (34).

[16] Numerical methods of finding them are discussed in sec. 13.28.

Having found the λ_i, we wish to determine a matrix X such that

$$X^{-1}AX = \Lambda = [\lambda_i \delta_{ij}] \tag{10-38}$$

We distinguish the following two cases.

a. *The Eigenvalues are all Different.* Let us consider the case in which all eigenvalues of A are different. Select one, say λ_k, and form the n linear equations

$$A\mathbf{x} = \lambda_k \mathbf{x} \tag{10-39}$$

They are homogeneous, but as shown in sec. 10.9, we may solve them for the ratio of the components of the eigenvector $\mathbf{x}_k$. Remembering that each component contains an arbitrary constant, we write them as a column vector

$$\mathbf{x}_k = \{x_{1k}, x_{2k}, \cdots, x_{nk}\}$$

The remaining eigenvectors are determined in the same way using each eigenvalue in turn. Finally we form a matrix X whose columns are the eigenvectors of A. This matrix clearly satisfies the equation

$$AX = X[\lambda_i \delta_{ij}]$$

When this result is multiplied by X^{-1}, eq. (38) is obtained. We have thus shown that the matrix X which diagonalizes A, may be found by compounding the eigenvectors of A into a matrix. The reduction to diagonal form here described is unique except for the order in which the eigenvalues occur along the diagonal.

Although not required by the method, the eigenvectors may be orthogonalized and normalized by the Schmidt process, which fixes the undetermined constant appearing in the solution of (39). We return to this question in sec. 10.17.

b. *The Eigenvalues are Not all Different.* When two or more of the eigenvalues of A are equal to each other, reduction to true diagonal form is not always possible. Suppose λ_1 is an eigenvalue of A repeated r_1 times. Proceeding as before, we find an eigenvector $\mathbf{x}_1$ so that

$$A\mathbf{x}_1 = \lambda_1 \mathbf{x}_1$$

Then if $\mathbf{x}_1$ is the first column of a square matrix X, the first column of AX will be $\lambda_1 \mathbf{x}_1$ and the first column of $X^{-1}AX$ will consist of λ_1 followed by $(n-1)$ zeros. Call this matrix B:

$$B = X^{-1}AX = \left[\begin{array}{c|c} \lambda_1 & B_j \\ \hline 0 & B_{ij} \end{array} \right]; \quad i, j = 2, 3, \cdots, n$$

Here B_j is a row matrix with $(n-1)$ elements and B_{ij} is square of order $(n-1)$. Since B is the transform of A, it also has the eigenvalue λ_1 repeated r_1 times, but B_{ij} contains that eigenvalue only $(r_1 - 1)$ times. This matrix is subjected to the same procedure as A: we find an eigenvector and form a new matrix Y whose first column is that eigenvector. Note however that Y has only $(n-1)$ rows and columns and

$$Y^{-1}B_{ij}Y = \left[\begin{array}{c|c} \lambda_1 & C_k \\ \hline 0 & C_{kl} \end{array}\right]; \quad k, l = 3, 4, \cdots, n$$

so that the matrix

$$\left[\begin{array}{c|c} 1 & 0 \\ \hline 0 & Y \end{array}\right]$$

will transform B into the form

$$\left[\begin{array}{c|c} \lambda_1 & B_jY \\ \hline 0 & \lambda_1 & C_k \\ \hline 0 & 0 & C_{kl} \end{array}\right]$$

Continued applications of similar transformations will eventually result in a single matrix Z such that

$$Z^{-1}AZ = \left[\begin{array}{c|c} A_1 & F \\ \hline 0 & G \end{array}\right] \tag{10--40}$$

where

$$A_1 = \begin{bmatrix} \lambda_1 & H_{12} & H_{13} & \cdots & H_{1r_1} \\ 0 & \lambda_1 & H_{23} & \cdots & H_{2r_1} \\ 0 & 0 & \lambda_1 & \cdots & H_{3r_1} \\ 0 & 0 & \cdot & & \\ & & & \cdot & \\ & & & & \lambda_1 \end{bmatrix} \tag{10--41}$$

The matrix F is rectangular with r_1 rows and $(n-r_1)$ columns while G is square and of order $(n-r_1)$. Now if Z_1 is a rectangular matrix composed from the first r_1 columns of Z, then we may remove the unwanted matrix F from (40), for

$$Z_1^{-1}AZ_1 = A_1$$

The next step is to treat the matrix G in a similar way until it is reduced to the form of A_1, with its eigenvalue λ_2 along the diagonal. We then continue with each remaining matrix until every eigenvalue has been used. Finally if we join together all of the rectangular matrices Z_i to form a square matrix W, we will have

$$W^{-1}AW = \text{diag }(A_1, A_2, \cdots, A_r)$$

where r is the number of distinct eigenvalues of A. Note that we are using the notation of sec. 10.7 to denote a diagonal matrix, but in this case the diagonal elements A_i are really matrices themselves. Each is of the triangular form of (41).

In the general case, it is possible to make a further transformation so that the eigenvalues occur along the diagonal of A_i, while unity appears in each position immediately above the eigenvalue and zero elsewhere.[17] In the special cases where A is symmetric, Hermitian, or unitary, the non-diagonal elements of the triangular matrices may be completely removed so that the final form is truly diagonal. We consider these cases in secs. 10.17, 10.19, 10.20.

Problem. Reduce to diagonal form

$$\begin{bmatrix} 8 & -8 & -2 \\ 4 & -3 & -2 \\ 3 & -4 & 1 \end{bmatrix} \cdot \quad Ans. \quad \lambda_1 = 1; \; \lambda_2 = 2; \; \lambda_3 = 3.$$

10.16. Congruent Transformations.—When a change of variable, $\mathbf{x} = Q\mathbf{y}$ is applied to a quadratic form, (28b) becomes

$$A(\mathbf{x},\mathbf{x}) = \tilde{\mathbf{x}}A\mathbf{x} = \tilde{\mathbf{y}}\tilde{Q}AQ\mathbf{y} = \tilde{\mathbf{y}}B\mathbf{y}$$

Thus the transformation of the matrix A which corresponds to this change of variable is congruent. Its importance is due to the fact that by its use a quadratic form may be reduced to a sum of squared terms, as will now be shown. Provided B is diagonal, $\tilde{\mathbf{y}}B\mathbf{y}$ will be a sum of squares. Hence our problem is that of diagonalizing the symmetric matrix A by means of a congruent transformation. Suppose A_{11} in A is not equal to zero. Then A may be written as

$$A = \begin{bmatrix} A_{11} & \mathbf{V} \\ \tilde{\mathbf{V}} & A' \end{bmatrix}$$

where A' is the matrix obtained from A by striking out the first row and first column and $\mathbf{V} = [A_{12}, A_{13}, \cdots, A_{1n}]$. Now let

$$Q_1 = \begin{bmatrix} 1 & -\mathbf{V}/A_{11} \\ 0 & E_{n-1} \end{bmatrix}$$

where E_{n-1} is the unit matrix of order $(n-1)$. Then

$$\tilde{Q}_1 A Q_1 = \begin{bmatrix} A_{11} & 0 \\ 0 & A'' \end{bmatrix}$$

and A'' is a matrix of order $(n-1)$ whose elements are

$$B_{ij} = A_{i+1,j+1} - \frac{A_{1,i+1}A_{1,j+1}}{A_{11}}$$

and for which the last row and column are designated $(n-1)$—not n. The matrix A'' may be treated in the same way and the process continued until A is completely reduced to diagonal form:

$$\tilde{Q}AQ = \text{diag} \, (\alpha_1, \alpha_2, \cdots, \alpha_n)$$

---

[17] The proof requires the theory of elementary divisors; see Turnbull and Aitken, loc. cit.

The final result is

$$\tilde{x}Ax = \tilde{\xi}D\xi = \alpha_1\xi_1^2 + \alpha_2\xi_2^2 + \cdots + \alpha_n\xi_n^2$$

with

$$D = \tilde{Q}AQ = [\alpha_i\delta_{ij}]$$

$$Q = Q_1Q_2Q_3\cdots Q_n$$

and

$$x = Q\xi$$

The matrix Q will have the form

$$Q = \begin{bmatrix} 1 & -A_{12}/A_{11} & -A_{13}/A_{11} & \cdots & -A_{1n}/A_{11} \\ 0 & 1 & 0 & \cdots & 0 \\ 0 & 0 & 1 & \cdots & 0 \\ \cdot & \cdots & \cdots & \cdots & \cdots \\ 0 & 0 & 0 & \cdots & 1 \end{bmatrix} \times$$

$$\begin{bmatrix} 1 & 0 & 0 & \cdots & 0 \\ 0 & 1 & -B_{12}/B_{11} & \cdots & -B_{1,n-1}/B_{11} \\ 0 & 0 & 1 & \cdots & 0 \\ \cdot & \cdots & \cdots & \cdots & \cdots \\ 0 & 0 & 0 & \cdots & 1 \end{bmatrix} \cdots \begin{bmatrix} 1 & 0 & 0 & \cdots & 0 \\ 0 & 1 & 0 & \cdots & 0 \\ 0 & 0 & 1 & \cdots & 0 \\ \cdot & \cdots & \cdots & \cdots & \cdots \\ 0 & 0 & 0 & \cdots & 1 \end{bmatrix} =$$

$$\begin{bmatrix} 1 & Q_{12} & Q_{13} & \cdots & Q_{1n} \\ 0 & 1 & Q_{23} & \cdots & Q_{2n} \\ 0 & 0 & 1 & \cdots & Q_{3n} \\ \cdot & \cdots & \cdots & \cdots & \cdots \\ 0 & 0 & 0 & \cdots & 1 \end{bmatrix}$$

The determinants, Δ_m, formed from A by omitting all but the first m rows and columns are called the *discriminants* of the quadratic form. Moreover, as the reader may show (Problem b),

$$\alpha_1 = \Delta_1; \quad \alpha_2 = \Delta_2/\Delta_1; \quad \alpha_3 = \Delta_3/\Delta_2; \quad \cdots; \quad \alpha_n = \Delta_n/\Delta_{n-1}$$

If it is so desired, a further linear transformation $\eta_i = \sqrt{\alpha_i}\xi_i$ will reduce $\tilde{x}Ax$ to the form

$$\tilde{\eta}E\eta = \eta_1^2 + \eta_2^2 + \cdots + \eta_n^2$$

Assuming that no element A_{ii} is zero, we note that instead of starting with A_{11} at the first step, as we have done here, any of the remaining A_{ii}, $(n - 1)$ in number might have been chosen. At the second step, there are $(n - 2)$ choices available and so on. Thus the final forms of Q and D are not unique. When some of the A_{ii} are zero or when some of the discriminants vanish, modifications [18] are required in the method.

[18] These cases are discussed by Bôcher, loc. cit.

Problem a. Show that for $n = 3$, the elements of Q may be taken as $Q_{12} = -A_{12}/A_{11}$; $Q_{13} = A^{13}/A^{33}$; $Q_{23} = A^{23}/A^{33}$.

Problem b. Verify the relation given in the preceding paragraph between the diagonal element α_i and the discriminant.

Problem c. Reduce the following expression to a sum of squares: $2x_1^2 + 7x_2^2 + 3x_3^2 + 4x_1x_2 + 8x_1x_3 - 2x_2x_3$. For one answer, $\alpha_1 = 2$; $\alpha_2 = 5$; $\alpha_3 = -10$.

10.17 Orthogonal Transformations.—In this section we limit our discussion to *real* orthogonal matrices, since we shall have no need for those containing complex elements. By definition, if R is orthogonal, $R = \tilde{R}^{-1}$ hence $R\tilde{R} = \tilde{R}R = E$. One of their most important properties arises from the fact that transformation by them leaves the length of a vector unchanged. Suppose $\mathbf{x}$ and $\mathbf{y}$ are related by an orthogonal transformation

$$\mathbf{x} = R\mathbf{y}; \quad \tilde{\mathbf{x}} = \tilde{\mathbf{y}}\tilde{R}$$

then

$$\tilde{\mathbf{x}}\mathbf{x} = \tilde{\mathbf{y}}\tilde{R}R\mathbf{y} = \tilde{\mathbf{y}}\mathbf{y}$$

Our assertion is proved since $\tilde{\mathbf{x}}\mathbf{x}$ is the square of the length of $\mathbf{x}$ and $\tilde{\mathbf{y}}\mathbf{y}$ is the square of the length of $\mathbf{y}$. On expanding $R\tilde{R} = E$ we find that

$$\sum_{s=1}^{n} R_{ps}R_{qs} = \sum_{s=1}^{n} R_{sp}R_{sq} = \delta_{pq} \tag{10-42}$$

These relations are the necessary and sufficient conditions that a matrix be orthogonal.

From the definition $R\tilde{R} = E$, we also see that $|R| \times |\tilde{R}| = |R|^2 = 1$, hence

$$|R| = \pm 1$$

Let us consider two matrices

$$R^+ = \begin{bmatrix} \cos\phi & \sin\phi & 0 \\ -\sin\phi & \cos\phi & 0 \\ 0 & 0 & 1 \end{bmatrix}; \quad R^- = \begin{bmatrix} \cos\phi & \sin\phi & 0 \\ -\sin\phi & \cos\phi & 0 \\ 0 & 0 & -1 \end{bmatrix}$$

which are easily shown to be orthogonal, the first having the determinant $+1$, the second -1. If we refer to eqs. (4-2) and (4-3), we see that they are both contained in eq. (42) and that the matrix R^+ represents a rotation of the coordinate system about OZ through the angle ϕ in the positive direction. Similarly, R^- is the matrix for the same rotation, followed by a reflection in the XY-plane. These two cases are called *proper* and *improper* rotations. If $\phi = 0$ in the latter case, the operation is a simple reflection; if $\phi = \pi$, it is called an *inversion*, the matrix R^- becomes diagonal, with -1 for its elements, and the result is equivalent to a change in sign of the three components of a vector (x,y,z).

Matrices similar to R^+ are to be used in eqs. (10-20a) and (10-21), if we interpret the orthogonal transformation as a rotation of coordinate axes. However, if we prefer to rotate the vector $\mathbf{x}$ and obtain a new vector $\mathbf{y}$, as we also discussed in sec. 10.10, we must change the sign of the angle in R^+. But $\sin(-\phi) = -\sin\phi$, $\cos(-\phi) = \cos\phi$ and the following results are obtained since the matrices are orthogonal:

$$R(\phi) = \tilde{R}^{-1}(\phi) = R^{-1}(-\phi) = \tilde{R}(-\phi)$$

The matrix relations are thus

$$\mathbf{y} = R(-\phi)\mathbf{x} = R^{-1}(\phi)\mathbf{x} = \tilde{R}(\phi)\mathbf{x}$$

and

$$\mathbf{x} = R(\phi)\mathbf{y}$$

It follows that successive rotations of the same vector in one coordinate system to give new vectors $\mathbf{y}$ and $\mathbf{z}$ must be written in the form

$$\mathbf{y} = B(-\phi)\mathbf{x}; \quad \mathbf{z} = A(-\theta)\mathbf{y}; \quad \mathbf{z} = A(-\theta)B(-\phi)\mathbf{x}$$

or, if we prefer to retain the matrix elements shown in R^+, we must change the order of the matrix product to read

$$\mathbf{x} = R(\phi)R(\theta)\mathbf{x}$$

The fact that an orthogonal transformation is both congruent and collineatory makes it useful for the following reason: It has been seen that the congruent transformation may be used to reduce a quadratic form $A(\mathbf{x},\mathbf{x})$ to a sum of squares, but the reduction is by no means unique. On the other hand, suppose the quadratic form has been reduced to a sum of squares by a congruent transformation and the elements of the transforming matrix are real. They can then be orthogonalized and normalized according to (42) and the resulting matrix R is both congruent and collineatory (hence orthogonal). In symbols,

$$\mathbf{x} = R\mathbf{y}; \quad A(\mathbf{x},\mathbf{x}) = \tilde{\mathbf{x}}A\mathbf{x} = \tilde{\mathbf{y}}R^{-1}AR\mathbf{y} = \tilde{\mathbf{y}}\Lambda\mathbf{y}$$

where Λ is diagonal with the eigenvalues of A for elements. It will be remembered that when a matrix is reduced to diagonal form by a similarity transformation, the eigenvectors which form the columns of the transforming matrix X are not completely determined because the equations to be solved for the components of the eigenvectors are homogeneous. This arbitrariness now disappears, for we must fix the ratio of the components of (42) so that the transforming matrix is orthogonal and $R\tilde{R} = E$.

We are now in a position to prove a statement made in sec. 10.15, namely that if a matrix A is symmetric and has multiple eigenvalues it may still be reduced to true diagonal form by an orthogonal transformation. Suppose A undergoes a congruent transformation by the matrix Q. Then the new matrix $\tilde{Q}AQ$ is symmetric if A is symmetric, for $\overline{(\tilde{Q}AQ)} = \tilde{Q}AQ$.

It thus follows that an orthogonal transformation will leave the symmetry of A unchanged. This only can be true if the off-diagonal elements of the triangular matrices of (41) are zero, but then A is diagonal.

Orthogonal transformations are often called *principal axis transformations* since they are used in the problem of reducing a conic to principal axes and in finding the principal axes of a rotating body, or in reducing kinetic and potential energy expressions to sums of squared terms. The eigenvectors are frequently called *normal coordinates* in these cases.[19]

A similar procedure serves to reduce[20] simultaneously two quadratic forms to a sum of squares. Suppose the two forms are $A(\mathbf{x,x}) = \tilde{\mathbf{x}}A\mathbf{x}$ and $B(\mathbf{x,x}) = \tilde{\mathbf{x}}B\mathbf{x}$. First reduce $A(\mathbf{x,x})$ to a sum of squares by a congruent transformation, $\mathbf{x} = Q\mathbf{y}$, which will give

$$\tilde{\mathbf{x}}A\mathbf{x} = \tilde{\mathbf{y}}\tilde{Q}AQ\mathbf{y} = \tilde{\mathbf{y}}D\mathbf{y} = \tilde{\mathbf{y}}[\alpha_i\delta_{ij}]\mathbf{y}$$

The same transformation applied to B will give

$$\tilde{\mathbf{x}}B\mathbf{x} = \tilde{\mathbf{y}}\tilde{Q}BQ\mathbf{y} = \tilde{\mathbf{y}}C\mathbf{y}$$

but C is not diagonal. Now make the substitution $\eta_i = \sqrt{\alpha_i}y_i$ which results in

$$\tilde{\mathbf{y}}D\mathbf{y} = \tilde{\boldsymbol{\eta}}E\boldsymbol{\eta}; \quad \tilde{\mathbf{y}}C\mathbf{y} = \tilde{\boldsymbol{\eta}}C'\boldsymbol{\eta}$$

where the α_i have been absorbed into C to give C'. Finally, an orthogonal transformation, $\boldsymbol{\eta} = R\boldsymbol{\xi}$ will reduce C' to diagonal form, yielding

$$\tilde{\mathbf{y}}D\mathbf{y} = \tilde{\boldsymbol{\xi}}\tilde{R}ER\boldsymbol{\xi} = \tilde{\boldsymbol{\xi}}E\boldsymbol{\xi} = \xi_1^2 + \xi_2^2 + \cdots + \xi_n^2$$

$$\tilde{\mathbf{y}}C\mathbf{y} = \tilde{\boldsymbol{\xi}}\tilde{R}C'R\boldsymbol{\xi} = \tilde{\boldsymbol{\xi}}\Lambda\boldsymbol{\xi} = \lambda_1\xi_1^2 + \lambda_2\xi_2^2 + \cdots + \lambda_n\xi_n^2$$

Even when the two quadratic forms are not functions of the same variables, the transformation may often be made. For example, in the mechanical problem of small oscillations where it is required to find normal coordinates for the kinetic and potential energies, the two quadratic forms appear as $T = \tilde{\mathbf{v}}A\mathbf{v}$ and $V = \tilde{\mathbf{x}}B\mathbf{x}$ where $\mathbf{v} = d\mathbf{x}/dt$, T being positive definite. The reduction causes no difficulty since the cogredient variables $\mathbf{x}$ and $\mathbf{v}$ both undergo the same transformation.[21]

We show in eq. (53) that for a unitary matrix, $\lambda_i\lambda_i^* = 1$ for every i. Since a real orthogonal matrix is also unitary, it follows that the only possible eigenvalues for a real orthogonal matrix are ± 1 or $e^{\pm i\phi}$. In the latter

---

[19] See Chapter 9; for a fuller discussion, see Whittaker, E. T., " A Treatise on the Analytical Dynamics of Particles and Rigid Bodies," Third Edition, Cambridge Press, 1927.

[20] The reduction is not always possible but it can be made if one of the forms is positive definite, as is the case in most physical problems.

[21] An example of this case was discussed in sec. 9.11; see also Whittaker, loc. cit., Chapter VII.

case, ϕ must be real and the exponentials occur in pairs with opposite signs. If k is the number of eigenvalues equal to -1, the determinant of the matrix equals $(-1)^k$.

In the previous discussion of this section, we have shown that when the eigenvalues are real it is possible to reduce matrices to diagonal form by means of an orthogonal transformation. Now suppose that R, the matrix to be reduced, is itself orthogonal and of order n, and that the n eigenvalues are $+1$ occurring j_1 times, -1 occurring j_2 times $(j_1 + j_2 = j \leq n)$ and $e^{\pm i\phi_k}$. Since the latter must appear in pairs there are an even number of them, i.e., $n - j = 2m$ and $k = 1, 2, \cdots, m$. Some simplification of the final diagonal matrix may be made by noting that if $\phi_k = 0$ or π, $e^{\pm i\phi_k}$ equals ± 1. Thus we may write an *even* number of the eigenvalues ± 1 as exponentials and obtain the following special cases for Λ, the diagonalized form of R.

n even, $n = 2k$

$$|R| = +1; \quad \Lambda = \text{diag } (e^{i\phi_1}, e^{i\phi_2}, \cdots, e^{i\phi_k}, e^{-i\phi_1}, \cdots, e^{-i\phi_k})$$

$$|R| = -1; \quad \Lambda = \text{diag } (1, -1, e^{i\phi_1}, \cdots, e^{i\phi_{k-1}}, e^{-i\phi_1}, \cdots, e^{-i\phi_{k-1}})$$

n odd, $n = 2k + 1$

$$|R| = +1; \quad \Lambda = \text{diag } (1, e^{i\phi_1}, \cdots, e^{i\phi_k}, e^{-i\phi_1}, \cdots, e^{-i\phi_k})$$

$$|R| = -1; \quad \Lambda = \text{diag } (-1, e^{i\phi_1}, \cdots, e^{i\phi_k}, e^{-i\phi_1}, \cdots, e^{-i\phi_k})$$

$$(10\text{--}43)$$

Now consider the form of the matrix X which diagonalizes R:

$$X^{-1}RX = \Lambda$$

Its r-th column is an eigenvector $\mathbf{x}_r$ of R and its eigenvalue will be assumed to be $e^{i\phi_r}$. But according to (36)

$$R\mathbf{x}_r = e^{i\phi_r}\mathbf{x}_r \qquad (10\text{--}44)$$

hence $\mathbf{x}_r$ will in general be complex and of the form

$$\mathbf{x}_r = \mathbf{x}'_r + i\mathbf{x}''_r$$

where $\mathbf{x}'_r$ and $\mathbf{x}''_r$ are real. In a similar manner, it follows that the eigenvector and column of X corresponding to $e^{-i\phi_r}$ is $\mathbf{x}'_r - i\mathbf{x}''_r$. We conclude that usually the transforming matrix X will have some complex elements. Recalling the fact that in the previous case, where X was orthogonal, the transformation was both collineatory and congruent we see that the necessary modification here is that the transformation be collineatory and conjunctive. The transforming matrix, then, is unitary, hence we can only diagonalize an orthogonal matrix by means of a unitary matrix.

Let us see what would happen if we transform a real orthogonal matrix

R by another real orthogonal matrix S.

$$S^{-1}RS = Z$$

We write (44) as

$$R(\mathbf{x}_r' + i\mathbf{x}_r'') = e^{i\phi_r}(\mathbf{x}_r' + i\mathbf{x}_r'') \qquad (10\text{--}45)$$

for the particular eigenvalue $e^{i\phi_r}$. Since $e^{i\phi_r} = \cos\,\phi_r + i \sin\,\phi_r$ we may equate the real and imaginary parts of (45) to get

$$R\mathbf{x}_r' = \mathbf{x}_r' \cos\,\phi_r - \mathbf{x}_r'' \sin\,\phi_r$$
$$R\mathbf{x}_r'' = \mathbf{x}_r' \sin\,\phi_r + \mathbf{x}_r'' \cos\,\phi_r$$

with a similar expression for the column of X that comes from $e^{-i\phi_r}$. If we replace the complex eigenvector $\mathbf{x}_r = \mathbf{x}_r' + i\mathbf{x}_r''$ by $\mathbf{x}_r'$ and $\mathbf{x}_r' - i\mathbf{x}_r''$ by $\mathbf{x}_r''$ the resulting matrix contains only real elements and may be made diagonal by requiring that (42) be fulfilled. Let us call this matrix S. Transformation by it will give the following forms for Z:

n *even*, $n = 2k$

$$|\,R\,| = +1; \quad Z = \text{diag}\,(C_1, C_2, \cdots, C_k)$$
$$|\,R\,| = -1; \quad Z = \text{diag}\,(1, -1, C_1, C_2, \cdots, C_{k-1})$$

n *odd*, $n = 2k + 1$

$$|\,R\,| = +1; \quad Z = \text{diag}\,(1, C_1, C_2, \cdots, C_k)$$
$$|\,R\,| = -1; \quad Z = \text{diag}\,(-1, C_1, C_2, \cdots, C_k)$$

where
$$C_k = \begin{bmatrix} \cos\,\phi_k & \sin\,\phi_k \\ -\sin\,\phi_k & \cos\,\phi_k \end{bmatrix}$$

It is worth while to point out that the only other possible two-dimensional real orthogonal matrix is of the type

$$\begin{bmatrix} \cos\,\phi & \sin\,\phi \\ \sin\,\phi & -\cos\,\phi \end{bmatrix}$$

Its eigenvalues are ± 1, hence such matrices cannot occur in the reduced form of R as we have already included all real eigenvalues in the preceding expressions for Z.

Problem. Prove that R^+ and R^- are orthogonal matrices. Reduce each to diagonal form.

10.18. Hermitian Vector Space.—Since many of the matrices occurring in physical problems[22] contain complex elements, it is necessary to

---

[22] For the use of matrix theory in quantum mechanics, see Chapter 11. For further discussion, see Born, M., and Jordan, P., " Elementare Quantenmechanik," J. Springer, Berlin, 1930; Wigner, E., " Gruppentheorie und ihre Anwendung auf die Quantenmechanik der Atomspektren," Vieweg, Braunschweig, 1931,

amplify the vector concept presented in sec. 10.8. We write in place of (11), the *Hermitian scalar product*

$$\mathbf{x}^\dagger\mathbf{y} = x_1^*y_1 + x_2^*y_2 + \cdots + x_n^*y_n \qquad (10\text{--}46)$$

The square of the absolute length of a vector is then real,

$$\mathbf{x}^\dagger\mathbf{x} = x_1^*x_1 + x_2^*x_2 + \cdots + x_n^*x_n$$

If $\mathbf{x}^\dagger\mathbf{y} = \mathbf{y}^\dagger\mathbf{x} = 0$, the two vectors are orthogonal or mutually perpendicular. If $\mathbf{x}^\dagger\mathbf{x} = 1$, the vector is a unit vector or normalized. For a scalar α,

$$\mathbf{x}^\dagger\alpha\mathbf{y} = \alpha\mathbf{x}^\dagger\mathbf{y}; \quad (\alpha\mathbf{x})^\dagger\mathbf{y} = \alpha^*\mathbf{x}^\dagger\mathbf{y}$$

The Hermitian scalar product is associative

$$\mathbf{x}^\dagger(\mathbf{y} + \mathbf{z}) = \mathbf{x}^\dagger\mathbf{y} + \mathbf{x}^\dagger\mathbf{z}$$

If A is any matrix,

$$\mathbf{x}^\dagger A\mathbf{y} = (A^\dagger\mathbf{x})^\dagger\mathbf{y}; \quad (A\mathbf{x})^\dagger\mathbf{y} = \mathbf{x}^\dagger(A^\dagger\mathbf{y}) \qquad (10\text{--}47)$$

10.19. Hermitian Matrices.—If the variables in the bilinear form (28) are complex conjugate to each other and if its matrix is Hermitian, the form is called *Hermitian*. Thus,

$$H(\mathbf{x},\mathbf{x}) = \sum_{i,j}H_{ij}x_i^*x_j = \mathbf{x}^\dagger H\mathbf{x}; \quad H_{ij} = H_{ji}^* \qquad (10\text{--}48)$$

In spite of the fact that the elements of (48) are complex, the form itself is real.

The eigenvalues of an Hermitian matrix are also all real. Suppose λ_i is an eigenvalue corresponding to an eigenvector $\mathbf{x}$, then

$$H\mathbf{x} = \lambda_i\mathbf{x}; \quad \mathbf{x}^\dagger H\mathbf{x} = \lambda_i\mathbf{x}^\dagger\mathbf{x}$$

Since $\mathbf{x}^\dagger H\mathbf{x}$ and $\mathbf{x}^\dagger\mathbf{x}$ are both real, it follows that λ_i is real.

An Hermitian matrix H remains Hermitian when transformed by either an orthogonal or a unitary matrix. To prove this statement for a real orthogonal matrix R, suppose H_1 is known to be Hermitian and $R^{-1}H_1R = H_2$. Then, since $R = \tilde{R}^{-1}$, we have $\tilde{H}_2 = \tilde{R}\tilde{H}_1\tilde{R}^{-1} = \tilde{R}\tilde{H}_1R$ and $H_2^\dagger = R^\dagger H_1^\dagger R^*$. But $H_1^\dagger = H_1$ and R is assumed to be real, so $R^\dagger = \tilde{R}$, $R^* = R$. Thus $H_2^\dagger = \tilde{R}H_1^\dagger R = R^{-1}H_1R = H_2$. The proof for a unitary matrix is similar.

As we have previously stated, a real symmetric matrix is a special case of the Hermitian matrix. Thus, except for slight modifications, the reduction of Hermitian matrices to diagonal form is similar to the procedure used for real matrices. For example, an Hermitian form may be converted to a sum of squares in many ways by a conjunctive transformation. On the other hand, we saw in sec. 10.15 that a matrix could be converted to diagonal form, with its eigenvalues on the diagonal, by means of a collineatory transformation. If the matrix is Hermitian, we may require that the transformation be both collineatory and conjunctive, hence unitary, and the diagonal form is then unique. The same argument which was used for a real symmetric matrix shows us that even if the eigenvalues are not all different, the true diagonal form may be obtained since transformation by a unitary matrix leaves the symmetry of an Hermitian matrix unchanged.

The necessary condition that two Hermitian forms be simultaneously reducible to a sum of squares is that they commute.[23] Suppose that $H(\mathbf{x},\mathbf{x}) = \mathbf{x}^\dagger H \mathbf{x}$ and $K(\mathbf{x},\mathbf{x}) = \mathbf{x}^\dagger K \mathbf{x}$ are given and that both H and K are Hermitian or unitary.[24] Let S be a unitary matrix that reduces H and K simultaneously to diagonal forms, H' and K':

$$H' = S^{-1}HS; \quad K' = S^{-1}KS$$

Clearly H' and K' commute since they are both diagonal, hence we may write

$$H'K' = S^{-1}HSS^{-1}KS = S^{-1}HKS$$

$$K'H' = S^{-1}KSS^{-1}HS = S^{-1}KHS$$

or

$$S^{-1}HKS = S^{-1}KHS$$

since $K'H' = H'K'$. It thus follows that $HK = KH$.

Problem. Prove that an Hermitian matrix remains Hermitian after transformation by a unitary matrix.

10.20. Unitary Matrices.—If we indicate a unitary matrix by U, then from its definition

$$U = (U^\dagger)^{-1}$$

hence

$$U^\dagger = U^{-1}; \quad UU^\dagger = U^\dagger U = E \qquad (10\text{-}49)$$

Suppose the elements in a single column of U are given by $\mathbf{U}_j$, then the Hermitian scalar product of two columns

$$\mathbf{U}_j^\dagger \mathbf{U}_k = \delta_{jk}$$

A similar relation may be found between the rows. Hence the rows and columns of a unitary matrix of order n form a set of n mutually perpendicu-

---

[23] The sufficiency of this condition is proved by Weyl, H., " The Theory of Groups and Quantum Mechanics," Methuen, London, 1931.

[24] If these matrices were not Hermitian or unitary, neither of them could be reduced to diagonal form (unless all eigenvalues were distinct, a case which is not very interesting).

lar unit vectors in Hermitian space.   This may be seen at once by writing
(49) explicitly:

$$\sum_s U_{ps} U_{qs}^* = \sum_s U_{sp}^* U_{sq} = \delta_{pq} \tag{10-50}$$

These equations are analogous to (42) for orthogonal matrices.

If x and y are any two vectors,

$$(Ux)^\dagger Uy = x^\dagger (U^\dagger Uy) = x^\dagger y$$

hence a transformation by a unitary matrix leaves a bilinear or quadratic
form invariant.   In particular, if $x = Uy$, then

$$x^\dagger x = (Uy)^\dagger Uy = y^\dagger y$$

This is the analogue of the fact that an orthogonal matrix in real vector
space leaves the length of a vector unchanged.   In fact, the unitary matrix
in Hermitian vector space is the generalization of the orthogonal matrix
for real vector space.

The product of two unitary matrices U and V is also unitary:

$$(UV)^\dagger = V^\dagger U^\dagger = V^{-1} U^{-1} = (UV)^{-1} \tag{10-51}$$

The reciprocal of a unitary matrix is unitary:

$$(U^{-1})^\dagger = (U^\dagger)^\dagger = U = (U^{-1})^{-1} \tag{10-52}$$

The eigenvalues of a unitary matrix may be real or complex but of
absolute value 1.   Suppose λ_i is an eigenvalue of U, then

$$Ux = \lambda_i x; \quad (Ux)^\dagger Ux = x^\dagger x = \lambda_i \lambda_i^* x^\dagger x \tag{10-53}$$

Since $x^\dagger x$ is real and does not vanish, $\lambda_i \lambda_i^* = 1$.

A unitary matrix may be transformed into diagonal form by another
unitary matrix V, the diagonal elements being the eigenvalues of U.   The
procedure is similar to that for similarity and orthogonal transformations.
The eigenvectors must be normalized to satisfy $U^\dagger U = E$.   The result is

$$V^{-1} UV = V^\dagger UV = \Lambda = \text{diag} (\lambda_1, \lambda_2, \cdots, \lambda_n)$$

10.21. Summary on Diagonalization of Matrices.—The matter of
diagonalizing matrices is so useful in practice that a final and simple
statement regarding conditions for the feasibility of this reduction seems
in order.

A matrix may be diagonalized (a) if all its eigenvalues are distinct (for
procedure, see sec. 10.15a), (b) if it is Hermitian or symmetric (see sec.
10.16 and 10.19), (c) if it is unitary (see sec. 10.20).   In cases (b) and (c)
a unitary matrix can always be found to effect the transformation while in
(a) a more general type of transforming matrix will be needed.

REFERENCES

Aitken, A. C., " Determinants and Matrices," Interscience Publishers, Inc., New York, 1948.

Frazer, R. A., Duncan, W. J., and Collar, A. R., " Elementary Matrices," Cambridge University Press, New York, 1938.

Kowalewski, G., " Determinantentheorie einschliesslich der Fredholmschen Determinanten," Third Edition, Chelsea Publishing Co., New York, 1942.

MacDuffee, C. C., " The Theory of Matrices," Second Edition, Chelsea Publishing Co., New York.

Muir, T., " Theory of Determinants," London, 1906–1923.

Perlis, S., " Theory of Matrices," Addison-Wesley Press, Inc., Cambridge, 1952.

Schreier, O. and Sperner, E., " Introduction to Modern Algebra and Matrix Theory," Chelsea Publishing Co., New York, 1950.

Wade, T. L., " Algebra of Vectors and Matrices," Addison-Wesley Press, Inc., Cambridge, 1951.

CHAPTER 11

QUANTUM MECHANICS

11.1. In conformity with the scope of this book, the emphasis of the present chapter is on the mathematics of quantum mechanics, the physical ideas entering the discussion only in a secondary way. Limitation of space further demands that only the important, and this happily implies the more elementary, portions of the wide field be presented. Complete exclusion of physical ideas would, however, leave its subject matter so poorly joined and so incomprehensible to the student who has no prior knowledge of quantum mechanics that the value of an entirely formal treatment appears questionable. It is also true that no part of applied mathematics exacts from its student a more radical change from his customary habits of thought, a greater tolerance for new methods of inquiry, than does this latest branch. In order to provide the proper attitude of mind, we preface the later mathematical developments by a few qualitative remarks whose relevance to the present book is but auxiliary.

The central notion of classical mechanics is the mass point, or particle. Classical theory therefore presupposes, tacitly, that a physical system can in principle be recognized as a particle, or a set of particles. Until the advent of quantum physics this dogma has never been questioned; in fact scientific philosophers have frequently inflated it to the dimensions of a universal proposition claiming that all physical systems are composed of particles. The method of physical description in best accord with this fundamental attitude is clearly this: To correlate instantaneous positions of a given particle with instants of time, assuming motion to be continuous in space and time. Thus, if a particle moves along the X-axis, the complete description of its motion would appear in the form $x = f(t)$.

Now it is conceivable that such a correlation becomes impossible, and the question then arises whether this fundamental mode of description should be abandoned in such circumstances. The answer which has often been given and which the modern physicist emphatically rejects is the flatly negative one, the answer alleging that classical description is intrinsically evident and that the relation $x = f(t)$ has meaning even when the functional relation cannot be established. On the other hand, one would not like to discard this successful description lightly, for instance because of certain practical and accidental difficulties in the procedure of measuring x

as a function of t. The criterion which has ultimately produced clarity is this: A method of description must be abandoned when it becomes impossible, not because of experimental difficulty, but because its use contradicts known laws of science. Classical description *has* become impossible for the latter reason, as the following simple example will show.

Imagine an oscillating mass point, e.g., the bob of a pendulum. As long as the eye can follow the bob, correlations between x and t can certainly be made. But suppose the mass point is made to increase its frequency of vibration. The eye will soon be unable to perceive instantaneous positions, but the camera can still establish them. When the camera fails, oscillographic methods may be available, and after that, ingenious devices perhaps not yet invented may serve. But ultimately, a barrier of an essential kind will be encountered. Let us assume that the bob oscillates 10^{10} times per second. It is a fact of atomic physics that visible light requires about 10^{-8} seconds to be emitted (or reflected). Thus if it were used as the medium of report, the light-emitting mass would have to remain in a given position for approximately that length of time. In the present instance, however, the bob executes 100 vibrations within this period. A similar argument can finally be used to invalidate every other means for establishing the classical correspondence. The latter has to be ultimately abandoned because its use contradicts the laws of optics.

What, then, can be done? Perhaps the example suggests an answer. While a snapshot can in principle no longer be taken of the rapidly oscillating bob, a time exposure would reveal *some* features of its dynamical behavior. It would give essentially a correlation between the time the bob spends within a given interval dx and the location of that interval, in other words between x and the probability $w\,dx$ of encountering it in dx. This leads to a less pretentious description of the physical system called a mass point, of the form $w = p(x)$, and this description is characteristic of quantum mechanics. It is to be noted that $p(x)$ can be inferred from the classical relation $x = f(t)$, but not $f(t)$ from $w = p(x)$.

Quantum mechanics provides the means for deducing probability relations of the type described, and it does so in a logically consistent fashion. But before turning to this central issue, let us see what has become of the concept: particle. Our time exposure has left it very ill defined. Indeed if the system called a mass point were invisibly small or never sufficiently stationary to permit the classical description, the customary properties of particles would never be exhibited. By the criterion of essential observability, the concept would lose its physical significance. From a misunderstanding of this situation there has arisen a claim that quantum mechanics leads to a dualism, to the monstrous conception that ultimate entities of physics like electrons are both particles and waves: the correct statement is that they are *neither* particles *nor* waves, but more abstract entities for the description of which quantum mechanics gives most simple and successful rules. The question as to the particle or wave nature of an electron must be put in the same class as that regarding its color — or, to use a lighter metaphor due to the philosopher Dingle, as the question concerning the color of an elephant's egg if an elephant laid eggs.

Despite this fundamental situation we shall place no ban upon the use of the terms particle, wave, etc.; we shall even adhere to universal practice in calling the electron one of the elementary particles of nature; we do this only, of course, as a concession to usage. But whenever a paradox arises, the reader should endeavor to resolve it by recalling that the " classical language " when applied to atomic entities is in fact metaphoric.

AXIOMATIC FOUNDATION

11.2. Definitions.—For the sake of brevity all historical considerations are omitted here. Nor will any attempt be made to " deduce " quantum mechanics either from classical physics or from outstanding experimental facts, for in a strict logical sense this cannot be done. We shall, however, present the framework of the theory with utmost economy of thought and space, committing the reader to the tacit understanding that all experimental consequences of the theory outlined have been verified as far as they could hitherto be tested.

On a *physical system*, by which is meant any object of interest to physics or chemistry, numerous observations or measurements can be made. The quantities so observed or measured, such as size, energy, position and momentum, are called *observables*. It is well to think of these observables without ascribing to them the intuitive qualities they possess in classical mechanics. Position, or energy, is not so much *possessed by* a system as it is characteristic of a certain measuring process which can be carried out upon it. The measurement of an observable upon a system yields a number.

In defining the *state* of a physical system considerable caution must be exercised, for we wish to remain in keeping with the requirements outlined in the introductory paragraphs. First it is well to notice that by state the scientist never means anything not subject to arbitrary fixation; indeed the definition of state is made to conform to the needs of each particular subject. It is quite different, for instance, in classical mechanics from what it is in thermodynamics or in electrodynamics. Hence we need not feel ill at ease when in quantum mechanics a new choice is made. Leaving elucidation until later: *a state is*[1] *a function of certain variables, a function*

---

[1] The reader who dislikes this phrase may substitute " is represented by " for the simple " is." We wish to warn, however, that the spirit of quantum mechanics permits no distinction in meaning between these two expressions.

from which by the rules of quantum theory significant information can be obtained. The variables may be chosen in several ways, each giving rise to a consistent description equivalent to all others; here they will be taken to be space coordinates, for this gives rise to the form of quantum mechanics most commonly used, namely Schrödinger's. By state, or state function, we thus mean a mathematical construct, $\phi(x_1,y_1,z_1; x_2,y_2,z_2; \cdots x_n,y_n,z_n)$. It is possible, as we shall later see, to associate the variables $x_1 \cdots z_n$ with the dimensions of configuration space of the classical analogue of the system in question. In particular, the number of variables needed in ϕ for a complete description of its behavior (at a given instant of time) has always been found to be equal to the number of its classical degrees of freedom. This must indeed be the case in order that large scale bodies be consistently described both by quantum mechanics and by classical mechanics. States may change with time; hence a state in its widest meaning may be written

$$\phi(x_1 y_1 z_1 \cdots z_n, t)$$

Certain restrictions are to be placed upon state functions, restrictions which will take on greater plausibility in view of the postulates of the next section. Most important among them are two: first ϕ, which may be a complex function, must possess an integrable square[2] in the sense that

$$\int \phi^* \phi \, d\tau < \infty \tag{11-1}$$

where $d\tau$ is the "volume of configuration space," i.e., in rectangular coordinates

$$d\tau \equiv dx_1 dy_1 dz_1 \cdots dx_n dy_n dz_n$$

Second,

$$\phi \text{ is single-valued} \tag{11-2}$$

The function ϕ may of course be expressed in any other system of space coordinates by the ordinary geometric transformations of Chapter 5. Condition (2) is particularly important when one of the variables is an angle, say α, for it then requires that

$$\phi(\alpha) = \phi(\alpha + 2n\pi) \tag{11-3}$$

n being an integer.

Finally we must include in our list of definitions another mathematical construct, that of an *operator*. Every specific mathematical operation, like adding 6, or multiplying by c, or extracting the third root, etc., can be

---

[2] This statement requires modification in some cases. See remarks concerning "continuous spectrum," sec. 11.9c. Condition (1) must be rigorously maintained without exception when $\int d\tau$ is finite. It seems best to present the foundations of the theory with this restriction, leaving necessary generalizations for later.

represented by a characteristic symbol which is then called an operator.

Operators are: $6+$, $c\cdot$, $\sqrt[3]{}$, $\dfrac{d}{dx}$, $\displaystyle\int_a^b dt$, $A\dfrac{d^2}{dx^2} + B\dfrac{d}{dx} + C$, and so forth.

In general they act on functions. They can be applied in succession. When they are so applied, the order in which the operators occur is important. For convenience, let us use more general symbols for operators, such as P and Q. If P stands for $a+$ and Q for $c\cdot$, then PQf means $a + cf$ where f is a function; however QPf means $c(a + f)$. Thus

$$QPf = PQf + (c - 1)a \qquad (11\text{-}4)$$

Such an equation is said to be an operator equation. The reader will at once verify that, if P stands for $\partial/\partial x$ and Q for $x\cdot$, the operator equation

$$PQf - QPf = f \qquad (11\text{-}5)$$

holds.

There is an important difference between eqs. (4) and (5); the second is homogeneous in f, the first is not. From the second, f may be canceled symbolically so that it reads

$$PQ - QP = 1 \qquad (11\text{-}5)$$

Only homogeneous operator equations of this kind, usually written in the latter form without explicit insertion of the operand f, are of interest in quantum mechanics.

The formalism of operators is convenient also in other ways. It is possible, for instance, to define a periodic function $\phi(x)$ by writing

$$e^{hD} \phi(x) = \phi(x)$$

D being d/dx; for the left-hand side is, on expansion, simply the Taylor series for $\phi(x + h)$.

Two operators, P and Q, are said to commute when $PQ - QP$ is zero. Thus $c\cdot$ and d/dx commute if c is a constant. Other examples of commuting operators are: $x\cdot$ and $\partial/\partial y$; d/dx and $\displaystyle\int_a^b dx$ if a and b are constants; $a+$ and $(-b)$. Clearly, every operator commutes with itself or any power of itself, provided that by the n-th power we mean the n-fold iteration of the operator.

11.3. Postulates.[3]—a. The fundamental postulates of quantum mechanics are three in number. The first concerns the use of observables.

[3] Henceforth in the present section, and in all subsequent sections up to 11.25, states will be supposed to be independent of the time; i.e., ϕ does not contain t. Such states are known as *stationary* ones, and the part of quantum mechanics dealing with them will be called *quantum statics*. In *quantum dynamics*, introduced in sec. 11.25, a new postulate (Schrödinger's " time " equation) will be needed. This postulate is not included in the present list. Nor do we include the Pauli principle, which is also of axiomatic status, and which will be presented in sec. 11.33. The present limitation is made for pedagogical reasons.

Brief reflection will show that classical physics associates with observables certain definite functions of suitable variables: x, y, z with position, mv with linear momentum, $\frac{1}{2}mv^2$ with kinetic energy, and so forth. These functions are chosen to describe experience most adequately. There is no logical reason which would exclude the use of more abstract mathematical entities in this association. It has indeed been found that, for the description of atomic phenomena, certain *operators* should replace the functions which in classical mechanics represent observables. The first postulate may be stated as follows:

To every observable there corresponds an operator.

The correct operator to be associated with a given observable must be found by trial. In the following table we give a brief summary of the four most important operators of quantum mechanics; the observables in question are understood to refer to systems classically described as groups of mass points having $3n$ degrees of freedom $(j = 1, 2, \cdots, n)$, subject to no *external* forces (total energy constant) and not requiring relativity treatment. The first column gives the name of the observable, the second its classical representation, the third its quantum mechanical representation.

Cartesian coordinate	x_j	$x_{j\cdot}$
Cartesian component of linear momentum of j-th particle	$p_{xj} = m_j \dot{x}_j$	$\dfrac{\hbar}{i}\dfrac{\partial}{\partial x_j}$
X-component of angular momentum of j-th particle	$m_j(y_j \dot{z}_j - z_j \dot{y}_j)$	$\dfrac{\hbar}{i}\left(y_j \dfrac{\partial}{\partial z_j} - z_j \dfrac{\partial}{\partial y_j} \right)$
Total energy	$\frac{1}{2}\sum_j \dfrac{1}{m_j}(p_{xj}^2 + p_{yj}^2 + p_{zj}^2)$ $+ V(x_1 \cdots z_n)$	$-\dfrac{\hbar^2}{2}\sum_{j=1}^{n}\dfrac{1}{m_j}\left(\dfrac{\partial^2}{\partial x_j^2} + \dfrac{\partial^2}{\partial y_j^2} + \dfrac{\partial^2}{\partial z_j^2} \right)$ $+ V(x_1 \cdots z_n)$

m_j is the mass of the j-th particle; $\hbar$ is an abbreviation for Planck's constant, h, divided by 2π.

The operator form of the Cartesian coordinate $x_{\cdot}$, is identical with its classical representation and has been included only for formal reasons. Linear momentum, a differential operator, is basic in the construction of the last two entries in the table.

When the operator corresponding to the linear momentum $\mathbf{p}$ of a single particle is written in the vector form $-i\hbar\nabla$, those corresponding to angular momentum and energy of this particle may be constructed according to classical formulas: Angular momentum $= \mathbf{r} \times \mathbf{p} = -i\hbar\mathbf{r} \times \nabla$, and energy $= (1/2m)p^2 + V = -(\hbar^2/2m)\nabla^2 + V$. These vector forms are valid in all other systems of coordinates and should be used as the basis for transformations.

In view of the table, the reader will easily verify the following operator equations:

Let Q_k stand for the operator " k-th Cartesian coordinate," P_k for the k-th component of linear momentum. Then

$$P_k Q_l - Q_l P_k = -i\hbar\delta_{kl} \tag{11-6}$$

Also, if L_x, L_y and L_z denote the components of the angular momentum operator for a single particle,[4]

$$\begin{aligned}
L_x L_y - L_y L_x &= i\hbar L_z \\
L_y L_z - L_z L_y &= i\hbar L_x \\
L_z L_x - L_x L_z &= i\hbar L_y
\end{aligned} \tag{11-7}$$

Commutation rules, like (6) and (7), are often sufficient to define the operators involved without recourse to their explicit form, but the latter is usually helpful.

b. The second postulate states:

The only possible values which a measurement of the observable whose operator is P can yield are the eigenvalues p_λ of the equation

$$P\psi_\lambda = p_\lambda\psi_\lambda \tag{11-8}$$

provided ψ_λ obeys conditions (1) and (2), namely: $\int \psi_\lambda^*\psi_\lambda d\tau < \infty$ and ψ_λ is single-valued.

The range of integration depends on the particular problem under consideration, as will be seen later.

We illustrate the meaning of this postulate by a few examples. Let us find the measurable values of the linear momentum of a particle, known to be somewhere on the X-axis between the finite points $x = a$ and $x = b$. The operator P is $-i\hbar(\partial/\partial x)$. Eq. (8) therefore becomes a first-order differential equation which can obviously be satisfied if ψ_λ is assumed to be a function of x only. It reads

$$-i\hbar\frac{d\psi_\lambda}{dx} = p_\lambda\psi_\lambda \tag{11-9}$$

and has the solution

$$\psi_\lambda = ce^{(i/\hbar)p_\lambda x}$$

---

[4] L_y and L_z may be obtained from L_x in the table by cyclical permutation of coordinates.

Is this solution satisfactory from the point of view of eqs. (1) and (2)? It is certainly single-valued; moreover, $\int \psi_\lambda^* \psi_\lambda dx = (b - a)c^*c$ is finite for every finite c. Hence no restriction upon p_λ results; *all* values of the linear momentum may be found upon measurement. The eigenvalues of the linear momentum form a continuous spectrum (λ is not a discrete index) and every function of the form $ce^{(i/\hbar)px}$ with constant p is an eigenfunction. As far as measurable values of linear momentum are concerned, quantum mechanics leads to the same result as classical physics.

This is not true for the *angular* momentum of a single particle. Here eq. (8) reads

$$-i\hbar \left(x \frac{\partial}{\partial y} - y \frac{\partial}{\partial x} \right) \psi_\lambda = m_\lambda \psi_\lambda \tag{11-10}$$

provided we consider the z-component and write m_λ for the eigenvalues. Obviously, ψ_λ must be a function of both x and y. But a simple transformation of coordinates reduces the equation to a simpler form. On putting $x = r \cos \theta$ and $y = r \sin \theta$, we have

$$\frac{d}{d\theta} = -r \sin \theta \frac{\partial}{\partial x} + r \cos \theta \frac{\partial}{\partial y} = x \frac{\partial}{\partial y} - y \frac{\partial}{\partial x}$$

Therefore eq. (10) becomes

$$-i\hbar \frac{d\psi_\lambda}{d\theta} = m_\lambda \psi_\lambda$$

and ψ_λ is seen to be a function of θ alone. The solution is

$$\psi_\lambda = ce^{(i/\hbar)m_\lambda \theta}$$

It certainly has an integrable square, because the range of θ extends from 0 to 2π, or more exactly, from $2\pi n$ to $2\pi(n + 1)$, where n is an integer. But ψ_λ violates the condition of single-valuedness which must be imposed in the form (3). To satisfy it we must require that

$$\psi_\lambda(\theta) = \psi_\lambda(\theta + 2\pi)$$

and this implies $e^{(2\pi i/\hbar)m_\lambda} = 1$. This is true only if

$$m_\lambda = \lambda \hbar, \quad \lambda \text{ an integer} \tag{11-11}$$

Hence the only observable values of the angular momentum are given by (11), and the eigenfunctions are $ce^{i\lambda\theta}$. This result is identical with the postulate of the older Bohr theory concerning angular momentum.

Next we consider the possible values of the total energy of a single mass point. The energy operator appearing in the table is often referred to as

the *Hamiltonian* operator and is denoted by the symbol H.   Let us use E_λ
for the eigenvalues.   The operator equation then becomes

$$H\psi_\lambda \equiv -\frac{\hbar^2}{2m}\nabla^2\psi_\lambda + V(x,y,z)\psi_\lambda = E_\lambda\psi_\lambda \tag{11–12}$$

This equation, written perhaps more frequently in the form

$$\nabla^2\psi_\lambda + \frac{2m}{\hbar^2}(E_\lambda - V)\psi_\lambda = 0 \tag{11–12}$$

was found by *Schrödinger* and bears his name.   Its solutions and eigenvalues clearly depend on the functional nature of $V(x,y,z)$; they will be
reserved for detailed consideration in secs. 9 et seq.

A rather peculiar result is obtained when (8) is applied to the coordinate
" operator."   The eigenvalues of " x " are the values ξ_λ for which the
equation

$$x \cdot \psi_\lambda = \xi_\lambda\psi_\lambda$$

an ordinary algebraic one, possesses solutions.   On writing it in the form

$$(x - \xi_\lambda)\psi_\lambda = 0$$

it is evident that either $x = \xi_\lambda$ or $\psi_\lambda = 0$.   In plainer language, ψ_λ as a
function of x vanishes everywhere except at $x = \xi_\lambda$, a constant.   From a
rigorous mathematical point of view such a function is a monstrosity, but
it is useful for certain purposes to introduce it, as Dirac[5] has done.   It is
called $\delta(x - \xi_\lambda)$, the symbol being fashioned after the Kronecker δ, and is
best visualized as something like $\lim\limits_{a\to0} ce^{-(x-\xi_\lambda)^2/a}$.   For later use the constant $c(a)$ will be so chosen that $\int_{-\infty}^{\infty} \delta(x - \xi)dx = 1$, so that

$$\int_{-\infty}^{\infty} f(x)\delta(x - \xi)dx = f(\xi) \tag{11–13}$$

Now it is clear that such a " function " can be formed for every value ξ_λ,
hence every point of the X-axis is an eigenvalue of the x-coordinate.[6]

The significance of the second postulate is best grasped when it is
regarded as furnishing a catalogue of the measurable values of all observables for which operators are known.   It implies no information concerning
the meaning of the eigenfunctions ψ_λ.   These are, of course, states of the

---

[5] Dirac, P. A. M., "Principles of Quantum Mechanics," Third Edition; Clarendon
Press, Oxford, 1947.

[6] The operator $x\cdot$ has a continuous spectrum.   Correspondingly, the integral
$\int \delta^2(x - \xi)dx$ does not exist!   See sec. 11.9c.

system in the sense explained. Their nature will unfold itself when the third postulate has been set forth. For the present we only note that every ψ_λ is indeterminate with respect to a constant multiplier; eq. (8) will also be satisfied by $constant \cdot \psi_\lambda$. On the other hand, $\int \psi_\lambda^* \psi_\lambda d\tau$ exists. We may require, therefore, that ψ_λ is normalized after the manner of sec. 8.2. Henceforth this will be assumed unless a statement to the contrary is made. In this connection it may be recalled, however, that normalization may fail intrinsically when the eigenvalues p_λ form a continuous spectrum. In Chapter 8 this was shown to be the case in instances where the range of the fundamental variable became infinite. These require special treatment.

The ψ_λ will be orthogonal if operator and boundary conditions conform to the circumstances of the Sturm-Liouville theory (sec. 8.5). This theory, as will later be seen, covers most of the cases occurring in quantum mechanics, but must be generalized somewhat to be applicable to complex operators.

c. We turn to the third postulate which states:

When a given system is in a state ϕ, the expected mean of a sequence of measurements on the observable whose operator is P is given by

$$\bar{p} = \int \phi^* P \phi d\tau \tag{11-14}$$

The expected mean is defined as in statistics: If a large number of measurements is made on the system, and the measured values are $p_1, p_2, \cdots, p_N$, then $\bar{p} \equiv 1/N \sum_{i=1}^{N} p_i$. Note that eq. (14) does not predict the outcome of a single measurement.

In writing (14) we are again supposing that ϕ is normalized. This can be brought about in all physical problems by " confining " the system in configuration space, that is, by taking the volume in which it moves to be finite, so that $\int d\tau$ exists. Even if the volume is infinite, $\int \phi^* \phi d\tau$ may still exist, but in general the situation then calls for special treatment involving the use of *eigendifferentials* instead of eigenfunctions.* A more general form of eq. (14), which often works when the volume of configuration space is infinite, is the following

$$\bar{p} = \lim_{\tau \to \infty} \frac{\displaystyle\int_\tau \phi^* P \phi d\tau}{\displaystyle\int_\tau \phi^* \phi d\tau} \tag{11-14'}$$

* See Morse, P. M., and Feshbach, H., "Methods of Theoretical Physics," McGraw-Hill Book Co., Inc., 1953.

We illustrate the meaning of (14) by a few examples. Let a system having one degree of freedom be in a state described by $\phi = (b/\pi)^{1/4} e^{-(b/2)(x-\xi)^2}$. Then the mean value of its position will be:

$$\bar{x} = \int_{-\infty}^{\infty} \phi^2 x\, dx = \xi$$

its mean momentum:

$$\bar{p}_x = -i\hbar \int \phi \phi'\, dx = 0$$

its mean kinetic energy:

$$\bar{E}_{kin} = -\frac{\hbar^2}{2m} \int \phi \phi''\, dx = \frac{\hbar^2}{2m} \int (\phi')^2\, dx = \frac{b}{2} \cdot \frac{\hbar^2}{2m}$$

It is interesting to note that, the more concentrated the function ϕ (the greater b) the larger will be the mean kinetic energy. To calculate the mean total energy we should have to know the form of $V(x)$.

Let us take $\phi = e^{ikx}/(b-a)^{1/2}$. We then find

$$\bar{x} = \int_a^b \phi^* x\phi\, dx = \frac{b+a}{2}$$

$$\bar{p}_x = -i\hbar \int_a^b \phi^* \phi'\, dx = k\hbar$$

$$\bar{E}_{kin} = -\frac{\hbar^2}{2m} \int_a^b \phi^* \phi''\, dx = \frac{k^2\hbar^2}{2m}$$

If in this example the range is extended to infinity, let us say in such a way that $-a = b \to \infty$, the function e^{ikx} can clearly not be normalized. One must then use eq. (14') in the form

$$\bar{p} = \lim_{a \to \infty} \frac{\displaystyle\int_{-a}^{a} \phi^* P \phi\, dx}{\displaystyle\int_{-a}^{a} \phi^* \phi\, dx}$$

which gives the same results as those obtained above.

The three postulates here stated and exemplified do not reveal an *intuitive* meaning of the state function ϕ. It is therefore not unusual in textbooks on quantum mechanics to add another postulate stating that $\phi^*(x)\phi(x)$ signifies the probability that the " particle " whose state is ϕ be found at the point x of configuration space (with suitable generalization for more than one degree of freedom). This is indeed true, and it may be well for the reader to form this basic conception; but this statement is not a further postulate since it may be deduced from those already given. (Cf. sec. 6.)

DEDUCTIONS FROM THE POSTULATES

11.4. Orthogonality and Completeness of Eigenfunctions.—In Chapter 8, orthogonality and completeness of the eigenfunctions belonging to the Sturm-Liouville operator L have been discussed. The proofs there given need to be generalized if they are to be applied to quantum mechanics, for the operators occurring there are not all of the same structure as L. (One of the most important equations encountered, the one-dimensional Schrödinger equation (12), is of the Sturm-Liouville type.) They often involve many variables, they may be differential operators of the first order, they may be complex; in fact they may not be differential operators at all. To simplify the theory we shall assume that the eigenvalues p_λ of eq. (8) are discrete, and that the boundary conditions on acceptable state functions are of the form 1 and 2. Whenever convenient we shall even assume that ϕ vanishes at the boundary of configuration space, over which integrations are to be carried out, in a manner suitable to our needs. Unless these restrictions are made the arguments become involved and in some respects problematic. It would then be necessary to conduct a separate proof for every problem of interest; thus elegance would fall prey to rigor.

We first define what is meant by an *Hermitian operator*. Let u and v be two " acceptable " functions, defined over a certain range of configuration space τ. We then say that the operator P is Hermitian if

$$\int_\tau u^* \cdot Pv d\tau = \int_\tau v \cdot P^*u^* d\tau \qquad (11\text{--}15)$$

All operators of interest in quantum mechanics have this property. As a sample proof we show this for the linear momentum $P_j = -i\hbar(\partial/\partial q_j)$, associated with the j-th Cartesian coordinate:

$$\int u^* Pv d\tau = -i\hbar \int_\tau u^* \frac{\partial v}{\partial q_j} d\tau = -i\hbar \int_\tau u^* \frac{\partial v}{\partial q_j} dq_1 \cdots dq_n$$

First perform the integration over q_j, which yields

$$-i\hbar \int_\sigma u^* v dq_1 \cdots dq_{j-1} dq_{j+1} \cdots dq_n + i\hbar \int_\tau v \frac{\partial}{\partial q_j} u^* d\tau$$

The first integral, a " surface " integral taken only over $n - 1$ coordinates but with u and v evaluated at the end points of the range for q_j, will vanish provided u and v vanish sufficiently strongly for these extreme values of q_j, which is what we are supposing. The remaining integral is indeed identical with $\int vP^*u^* d\tau$.

The Hermitian property of $x\cdot$ is obvious.   To prove it for the Hamiltonian H, two partial integrations are necessary;   the details may be left as an exercise for the reader. .

Hermitian operators have real eigenvalues.   This fact follows at once from eq. (15).   The eigenvalues of P are defined by the equation

$$P\psi_\lambda = p_\lambda\psi_\lambda \tag{11-16}$$

This also implies the validity of the equation

$$P^*\psi_\lambda^* = p_\lambda^*\psi_\lambda^* \tag{11-17}$$

Now multiply (16) by ψ_λ^* and (17) by ψ_λ, and integrate over $d\tau$ obtaining

$$\int \psi_\lambda^* P\psi_\lambda d\tau = p_\lambda \int \psi_\lambda^*\psi_\lambda d\tau$$

$$\int \psi_\lambda P^*\psi_\lambda^* d\tau = p_\lambda^* \int \psi_\lambda^*\psi_\lambda d\tau$$

By (15) the left-hand sides of these two equations are equal, for ψ_λ is certainly an acceptable function in the sense outlined before.   Hence $p_\lambda^* = p_\lambda$;   i.e., p_λ is real.   Since the eigenvalues of operators are measurable values of observables, which must of necessity be real, the physical significance of an operator is assured when it has the Hermitian property.

Let us again consider eq. (16).   If ψ_μ is some other eigenfunction, it is evident that

$$\int \psi_\mu^* P\psi_\lambda d\tau = p_\lambda \int \psi_\mu^*\psi_\lambda d\tau \tag{11-18}$$

But if we start with the equation

$$P^*\psi_\mu^* = p_\mu\psi_\mu^*$$

which is true because p_μ is real, we also conclude that

$$\int \psi_\lambda P^*\psi_\mu^* d\tau = p_\mu \int \psi_\mu^*\psi_\lambda d\tau \tag{12-19}$$

Combining (18) and (19) we find

$$\int \psi_\mu^* P\psi_\lambda d\tau - \int \psi_\lambda P^*\psi_\mu^* d\tau = (p_\lambda - p_\mu) \int \psi_\mu^*\psi_\lambda d\tau$$

If P is Hermitian the left-hand side vanishes.   Hence either $p_\lambda = p_\mu$ or $\int \psi_\mu^*\psi_\lambda d\tau = 0$.   We see that *eigenfunctions of Hermitian operators, belonging to different eigenvalues, are orthogonal.*

The completeness of the eigenfunctions of all operators employed in quantum mechanics is usually assumed.   To the authors' knowledge, a rigorous proof has not been given. Since, however, our main interest will be in the Schrödinger equation which is of the Sturm-Liouville type, this point need not detain us further. In the following we shall assume completeness of all ψ_λ whenever this property is needed.

Problem. Show that the angular momentum operator $L_z = -i\hbar(\partial/\partial\theta)$ is Hermitian.

11.5. Relative Frequencies of Measured Values.—Important consequences can now be deduced from the third postulate, eq. (14). We first note that, if P is Hermitian, every power of P is Hermitian. Moreover, if (14) is true for every operator P, it must certainly hold for the operator P^r. It implies, therefore,

$$\overline{p^r} = \int \phi^* P^r \phi \, d\tau, \quad r = 1, 2, \cdots \tag{11-20}$$

The left-hand side stands, of course, for the r-th moment of the statistical aggregate of the measured values, i.e.,

$$\overline{p^r} = \sum_i \rho_i p_i^r \tag{11-21}$$

provided ρ_i is the relative frequency of the occurrence of the i-th eigenvalue p_i in the set of measurements. In accordance with eq. (20), the state function ϕ predicts not only the *mean*, but all *moments* of the aggregate of measurements.[7] Now eq. (20) may be transformed as follows. Let the eigenfunctions of P be denoted by ψ_λ, so that $P\psi_\lambda = p_\lambda\psi_\lambda$. On allowing P to operate on both sides of this equation, there results $P^2\psi_\lambda = p_\lambda P\psi_\lambda = p_\lambda^2\psi_\lambda$. By continuing this process, the relation

$$P^r\psi_\lambda = p_\lambda^r\psi_\lambda \tag{11-22}$$

is established. If the function ϕ appearing in (20) is expanded in terms of the ψ_λ,

$$\phi = \sum_i a_i\psi_i$$

and this series is substituted, we find

$$\overline{p^r} = \int \sum_{ij} a_i^* a_j \psi_i^* P^r \psi_j \, d\tau = \sum_{ij} a_i^* a_j p_j^r \int \psi_i^* \psi_j \, d\tau$$

$$= \sum_i a_i^* a_i p_i^r$$

by virtue of (22) and the orthogonality of the ψ_i. Comparing this with (21) it is clear that

$$\sum_i \rho_i p_i^r = \sum_i |a_i|^2 p_i^r$$

---

[7] For terminology, see sec. 12.3.

for every integer r.   But this can be true only if

$$\rho_i = | a_i |^2 \tag{11-23}$$

In words: when the system is in the state ϕ, a measurement of the observable corresponding to P will yield the value p_i with a probability (relative frequency) $| a_i |^2$, a_i being the coefficient of ψ_i in the expansion $\phi = \sum_\lambda a_\lambda \psi_\lambda$, and ψ_λ is one of the eigenfunctions of P.   The coefficients a_i are called *probability amplitudes*.

They may be expressed in terms of ϕ and ψ_i by the relation

$$\int \psi_i^* \phi d\tau = \sum_\lambda \int \psi_i^* a_\lambda \psi_\lambda d\tau = a_i \tag{11-24}$$

Consequently, eq. (23) may also be written

$$\rho_i = \left| \int \psi_i^* \phi d\tau \right|^2 \tag{11-25}$$

An interesting result is obtained when, in this equation, we let ϕ be one of the eigenfunctions belonging to the operator P itself, e.g., ψ_j.   It then reads

$$\rho_i = \left| \int \psi_i^* \psi_j d\tau \right|^2 = \delta_{ij}$$

All relative frequencies are zero except the one measuring the occurrence of the eigenvalue p_j, which is unity.   Thus we conclude that an eigenstate ψ_j of an operator P is a state in which the system yields with *certainty* the value p_j when the observable corresponding to P is measured.   Eigenfunctions are simply state functions of this determinate character.

11.6. Intuitive Meaning of a State Function.—Consider now a system, like a simple mass point with one degree of freedom, whose state function is $\phi(x)$.   We wish to know the probability that a measurement of its position will give the value $x = \xi$.   The eigenfunction corresponding to the operator $x \cdot$ for the value ξ has been shown to be

$$\psi_\xi = \delta(x - \xi)$$

Eq. (25) now reads

$$\rho_\xi = \left| \int \delta(x - \xi) \phi(x) dx \right|^2 = | \phi(\xi) |^2 \tag{11-26}$$

by virtue of (13).   The probability (density) of finding the system at ξ is given by the square of its state function.   This fact provides a simple intuitive meaning for the state function.   It can be generalized to several dimensions.

Let $q_1, q_2, \cdots, q_n$ be the coordinates on which ϕ depends. Using the former arguments, the eigenfunction corresponding to the composite coordinate operator $q_1 \cdot q_2 \cdots q_n$ may be shown to be

$$\psi_{\xi_1 \xi_2 \cdots \xi_n} = \delta(q_1 - \xi_1)\delta(q_2 - \xi_2) \cdots \delta(q_n - \xi_n) \qquad (11\text{-}27)$$

If, therefore, we wish to find the probability $\rho_{\xi_1 \xi_2 \cdots \xi_n}$ of finding the system at the point $(\xi_1 \xi_2 \cdots \xi_n)$ of configuration space, we must use eq. (25) with ψ_i replaced by (27). Hence

$$\rho_{\xi_1 \cdots \xi_n} =$$
$$\left| \int\int \cdots \int \delta(q_1 - \xi_1) \cdots \delta(q_n - \xi_n)\phi(q_1 q_2 \cdots q_n)dq_1 dq_2 \cdots dq_n \right|^2$$
$$= \left| \phi(\xi_1 \xi_2 \cdots \xi_n) \right|^2$$

11.7. Commuting Operators.—Let P and R be two operators satisfying the relation $PR - RP = 0$, and let their eigenfunctions be ψ_λ and χ_μ, that is

$$P\psi_\lambda = p_\lambda \psi_\lambda, \quad R\chi_\mu = r_\mu \chi_\mu \qquad (11\text{-}28)$$

We assume the state function to be ψ_i so that, when P is measured, there results with certainty the value p_i. But

$$RP\psi_i = PR\psi_i = p_i R\psi_i$$

Considering only the last two members of this equation, we may say that $(R\psi_i)$ is an eigenfunction of P, namely that belonging to the eigenvalue p_i. But this is possible only if $R\psi_i = \text{const.} \ \psi_i$. Comparison with the second equation (28) shows the constant to be one of the r_μ, and ψ_i to be one of the eigenfunctions χ_μ. We conclude that commuting operators have *simultaneous* eigenstates; i.e., measurements on their observables yield definite values for both; they do not " spread."

The fact that, when P and Q are non-commuting operators and the state of the system is an eigenstate of P, measurements on Q will give a statistical aggregate of values and not a single one with certainty, is usually attributed to the interference of measuring devices. For instance, the measurement of a particle's position disturbs its momentum, and vice versa, so that when one is ascertained with precision, the other quantity loses it. From this point of view, measurements on the observables associated with commuting operators are said to be *compatible*, the procedures of measurement do not conflict with each other.

11.8. Uncertainty Relation.—The proof of the famous Heisenberg uncertainty principle which will now be given requires the use of an inequality, similar to a well known relation due to Schwarz, though not identical with it. (Cf. eq. 3–112.) It states: if u and v are any two " acceptable " functions in the sense specified in connection with the definition of Hermitian operators (sec. 11.4), then

$$\int u^*u d\tau \cdot \int v^*v d\tau \geqq \frac{1}{4}\left[\int (u^*v + v^*u)d\tau\right]^2 \qquad (11\text{-}29)$$

We assume a system to be in a state ϕ, which need not be an eigenstate of any particular operator, and we are interested in the results of measurements on the observables belonging to two operators, P and Q, at present unspecified. Introduce into eq. (29) the following functions

$$u = (P - \bar{p})\phi \quad \text{and} \quad v = i(Q - \bar{q})\phi$$

where $\bar{p}$ and $\bar{q}$ are mean values associated with P and Q through the relation (14). Eq. (29) then reads

$$\int (P - \bar{p})^*\phi^*(P - \bar{p})\phi d\tau \cdot \int (Q - \bar{q})^*\phi^*(Q - \bar{q})\phi d\tau \geqq$$

$$\frac{1}{4}\left[i\int (P - \bar{p})^*\phi^*(Q - \bar{q})\phi d\tau - i\int (Q - \bar{q})^*\phi^*(P - \bar{p})\phi d\tau\right]^2$$

Now P and Q are Hermitian and satisfy eq. (15); $\bar{p}$ and $\bar{q}$ are constants. Therefore the inequality reduces to

$$\int \phi^*(P - \bar{p})^2\phi d\tau \cdot \int \phi^*(Q - \bar{q})^2\phi d\tau \geqq -\frac{1}{4}\left[\int \phi^*(PQ - QP)\phi d\tau\right]^2$$

$$(11\text{-}30)$$

Let us consider the meaning of the quantity $\int \phi^*(P - \bar{p})^2\phi d\tau$. When ϕ is expanded in eigenfunctions ψ_λ of P, $\phi = \sum_\lambda a_\lambda\psi_\lambda$, and the expansion is introduced in the integral, the result is $\sum_\lambda |a_\lambda|^2(p_\lambda - \bar{p})^2$, and this, in view of eq. (23), is nothing other than the *dispersion*[8] of the statistical aggregate of p-measurements about their mean. For this quantity we may introduce the more familiar symbol $\overline{\Delta p^2}$. A similar identification is to be made for $\int \phi^*(Q - \bar{q})^2\phi d\tau$. Inequality (30) then takes the more interesting form

$$\overline{\Delta p^2} \cdot \overline{\Delta q^2} \geqq -\frac{1}{4}\left[\int \phi^*(PQ - QP)\phi d\tau\right]^2 \qquad (11\text{-}31)$$

---

[8] The "dispersion" is the square of the so-called "standard deviation." It is an index of the "spread" of the measurements. See Chapter 12.

Now if P and Q commute, the right-hand side is zero, and it is possible for $\overline{\Delta p^2}$ or $\overline{\Delta q^2}$ to be zero, or even for both to vanish. This state of affairs recalls the result of sec. 7, which was that both p- and q-measurements could yield single values without spread.

When P and Q do not commute, relation (31) sets a lower limit for the product of the dispersions, often called uncertainties. Suppose, for instance, that P is the operator $-i\hbar(\partial/\partial q)$, the linear momentum associated with q, and Q stands for the coordinate q. We then have

$$PQ - QP = i\hbar \qquad (11\text{-}32)$$

When this is put into (31) the result is $\overline{\Delta p^2} \cdot \overline{\Delta q^2} \geqq \hbar^2/4$, or, written in terms of standard deviations, δp and δq,

$$\delta p \cdot \delta q \geqq \frac{\hbar}{2} \qquad (11\text{-}33)$$

This is Heisenberg's uncertainty relation.

Our result need not be cast in the form of an inequality. It is indeed quite possible to calculate both δp and δq separately and exactly when the state function ϕ is given, as the postulates show.

A slight generalization of the present conclusions is also possible. There are other operators, such as L_z and θ (cf. eq. 10 et seq.) which also obey eq. (32). In fact all quantities which are called canonically conjugate in classical physics[9] have operators which satisfy it. (Later we shall see that energy and time belong to this class.) For all these, the uncertainty relation in the form (33) is valid.

Problem. Show that, if the state function ϕ is an eigenfunction of the angular momentum operator L_z corresponding to the eigenvalue l_z, the product of δl_x and δl_y is at least as great as $(\hbar/2)l_z$.

SCHRÖDINGER EQUATIONS

Attention will now be given to the eigenvalues and eigenfunctions of the energy operator, that is, to the solutions of the various forms of the Schrödinger equations, eq. (12).

11.9. Free Mass Point.—The simplest example of a physical system is the free mass point for which the potential energy V may be taken to be zero. In that case eq. (12) reads

$$\nabla^2 \psi + k^2 \psi = 0 \qquad (11\text{-}34)$$

provided we omit the subscript λ and write $k^2 \equiv 2mE/\hbar^2$. This quantity k^2 has a rather simple classical significance which it is well to recognize at

[9] Cf. sec. 9.2.

once. For if E is the total energy of the particle, which is in this case purely kinetic, then $E = \frac{1}{2}mv^2 = p^2/2m$. Hence $k \equiv p/\hbar$, p being the classical momentum of the particle. Note also that k has the dimension of a reciprocal length.

Eq. (34) has already been solved in Chapter 7 (cf. eq. 7–33), where it appeared as the space form of the wave equation. To select the proper solution, we must consider the fundamental domain, τ, of our problem. Here, a great number of possibilities present themselves.

a. *Enclosure is a Parallelepiped.* If the particle is known to be within a parallelepiped of side lengths l_1, l_2, and l_3, then τ is this volume of space. Moreover, since $|\psi(xyz)|^2$ has already been identified as the probability of finding the particle at the point x, y, z, this quantity must certainly be zero everywhere outside τ. For reasons of continuity (which can, by more expanded arguments, be shown to result from our axioms) we require that $|\psi|^2$, and hence ψ itself, shall vanish on the boundaries of τ also. In view of this boundary condition, the solution of (34) in rectangular coordinates, namely eq. 7–36, must be chosen. In more explicit form it reads

$$\psi = (A_1 e^{ik_1 x} + B_1 e^{-ik_1 x})(A_2 e^{ik_2 y} + B_2 e^{-ik_2 y})(A_3 e^{ik_3 z} + B_3 e^{-ik_3 z}),$$
$$k^2 = k_1^2 + k_2^2 + k_3^2$$

The origin of the parallelepiped may be taken in one corner. Vanishing of ψ at the boundary then requires:

$$A_s + B_s = 0, \quad A_s e^{ik_s l_s} + B_s e^{-ik_s l_s} = 0, \quad s = 1, 2, 3$$

The first condition makes each parenthesis of ψ a sine-function; the second implies

$$k_s = \frac{n_s \pi}{l_s}$$

where n_s is an integer. Hence

$$\psi = c \sin\left(\frac{n_1 \pi}{l_1} x\right) \sin\left(\frac{n_2 \pi}{l_2} y\right) \sin\left(\frac{n_3 \pi}{l_3} z\right) \tag{11-35}$$

and

$$k^2 = \left(\frac{n_1^2}{l_1^2} + \frac{n_2^2}{l_2^2} + \frac{n_3^2}{l_3^2}\right)\pi^2$$

so that

$$E = \frac{\pi^2 \hbar^2}{2m}\left[\left(\frac{n_1}{l_1}\right)^2 + \left(\frac{n_2}{l_2}\right)^2 + \left(\frac{n_3}{l_3}\right)^2\right] \tag{11-36}$$

If ψ is to be normalized, $\int \psi^*\psi dx dy dz = 1$, and the constant c has the value

$$c = \left(\frac{8}{l_1 l_2 l_3}\right)^{1/2} = \left(\frac{8}{\tau}\right)^{1/2}$$

The permitted energy values form a denumerably infinite set. Their arrangement is best represented by constructing a lattice of points filling all space, with the " reciprocal " parallelepiped of sides $1/l_1$, $1/l_2$, $1/l_3$ as crystallographic unit. If from a given point lines are drawn to all other points, the squares of the lengths of these lines (multiplied by $\pi^2\hbar^2/2m$) are the energies of our problem. However, not all these lines represent different states. The function ψ changes only its sign when one of the integers n_1, n_2 or n_3 changes sign; it is not thereby converted into a new, linearly independent function. Hence only the lines lying in one octant of the lattice, with the origin of the lines at one corner, will represent different states. If some of the l's are equal there will be degeneracy (cf. sec. 8.6), for then an interchange of the corresponding n's will not produce a different E, while ψ will be changed into a function which is linearly independent from the original one.

b. *Enclosure is a Sphere.* Eq. (34) must now be solved in spherical coordinates. But this has already been done in sec. 8.4 (cf. eq. 8–25), for an acoustical problem. The eigenfunctions are, aside from a normalizing factor, $\psi = Y_l(\theta,\varphi) r^{-1/2} J_{l+1/2}(kr)$. The permitted energies are determined by the condition $J_{l+1/2}(ka) = 0$ where a is the radius of the enclosure. For any integer l, there will be an infinite set of roots of $J_{l+1/2}$ which we shall label r_{ln}, $n = 1, 2, \cdots, \infty$. The permitted k's are therefore

$$k_{ln} = \frac{r_{ln}}{a}$$

and hence E, which will also depend on two indices (quantum numbers) is given by

$$E_{ln} = \frac{\hbar^2}{2ma^2} (r_{ln})^2$$

The simple model treated here is called the " infinite potential hole." It forms the basis for many nuclear quantum mechanical calculations and is one of the favored starting points for considerations leading to nuclear shell structure.[*] A solution of the potential-hole problem with finite walls[†] requires the use of Bessel functions inside, Hankel functions outside the hole. The sequence of the energy values is unaltered, but all levels are depressed.

---

[*] Mayer, M. G. and Jensen, J.H.D., "Elementary Theory of Nuclear Shell Structure," John Wiley and Sons, Inc., New York, 1955.

[†] Margenau, H., *Phys. Rev.* **46**, 613 (1934).

c. *No Enclosure.* When the particle is allowed to exist anywhere in space, the former boundary conditions need not be applied. The simplest way to treat this case is to return to case (a) and permit l_1, l_2, and l_3 to become infinite. Let us first consider the eigenvalues. The lattice of points will condense as the l's increase, until finally it forms a continuum; the energy states (lengths of connecting lines squared) will also move closer and closer together until finally all (positive) energies are permitted. A similar effect may be brought about by increasing the *mass* of the particle, as a glance at eq. (36) will show. Quantum mechanics indicates no quantization of the energy for particles which are not restricted in their motion, or which have an infinite mass.

What happens to the ψ-function, (35), as the l's increase? Clearly, the normalizing constant c tends to zero, causing ψ also to vanish. The meaning of this is quite simple: As the space in which the mass point moves increases indefinitely, the chance of finding it at a given point, $\mid \psi(x,y,z) \mid^2$, approaches zero. The failure of the normalization rule is therefore not merely a mathematical phenomenon, but physically reasonable. To circumvent it, several procedures may be employed. One is to suppose that there is an infinite number of particles in all space, N per unit volume, and accordingly to put $\int \mid \psi \mid^2 d\tau$, taken over a unit of volume, equal to N. This leaves c finite.[10]

When there are no boundary conditions the ψ-function need not be written as a product of sines. In fact in the absence of an enclosure sine, cosine and exponential functions are equally acceptable. Hence we may, if we desire, write

$$\psi_E = c(\mathbf{k})e^{i\mathbf{k}\cdot\mathbf{r}}, \quad E = \frac{\hbar^2}{2m}k^2$$

using the notation explained in connection with eq. (38) of Chapter 7.

Problem. Calculate eigenfunctions and eigenvalues of a free particle enclosed in a cylinder of radius a and length d, obtaining

$$\psi = ce^{i[(n\pi/d)z+m\varphi]}J_m(\alpha\rho)$$

where αa is a root of J_m,

$$E_n = \frac{\hbar^2}{2m}\left(\frac{n^2\pi^2}{d^2} + \alpha^2 a^2\right)$$

11.10. One-Dimensional Barrier Problems.—For a one-dimensional problem the Schrödinger equation is

$$\frac{d^2\psi}{dx^2} + \frac{2m}{\hbar^2}[E - V(x)]\psi = 0$$

---

[10] Another procedure is discussed for instance in Sommerfeld. A., " Atombau und Spektrallinien," Vol. II.

Let us take V to be the step function given by the solid line in Fig. 1, that is: $V = 0$ if $x < 0$, $V = V =$ constant if $x > 0$. The solutions for the two regions are easily written down:

$$\psi_l = A_l e^{ik_l x} + B_l e^{-ik_l x}, \quad x < 0 \quad \text{(left of 0)}$$

$$\psi_r = A_r e^{ik_r x} + B_r e^{-ik_r x}, \quad x > 0 \quad \text{(right of 0)}$$

with

$$k_l = \frac{\sqrt{2mE}}{\hbar} \quad \text{and} \quad k_r = \frac{\sqrt{2m(E - V)}}{\hbar}$$

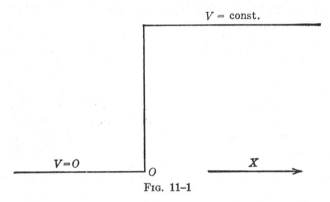

$V = \text{const.}$

$V = 0$

O

X

FIG. 11-1

But how are they to be joined? The differential equation tells us that ψ'' suffers a finite discontinuity as we pass across the discontinuity in V. The increase in ψ' in crossing the origin will be

$$\lim_{\xi \to 0} \int_{-\xi}^{\xi} \psi'' dx = \lim_{\xi \to 0} \xi \, (\psi_l'' + \psi_r'') = 0$$

Hence ψ' (and a fortiori ψ) remains continuous at the origin. The constants A and B must therefore be fixed by requiring

$$\psi_l(0) = \psi_r(0); \quad \psi_l'(0) = \psi_r'(0)$$

In addition to these two we have an equation expressing normalization, *three* relations in all. However, there are four constants (A_l, A_r, B_l, B_r) to be determined. The mathematical situation is therefore such that one of them may be chosen at will. Let us then put B_r equal to zero. The physical meaning of this will at once be clear.

On applying the continuity conditions we have

$$A_l + B_l = A_r; \quad k_l(A_l - B_l) = k_r A_r$$

whence

$$B_l = \frac{k_l - k_r}{k_l + k_r} A_l \tag{11-37}$$

The coefficients A and B have a simple significance. Let us analyze from our fundamental point of view a state function of the form $\psi = Ae^{ikx} + Be^{-ikx}$. In view of the third postulate (eq. 14') it represents a *mean* momentum

$$\bar{p} = -i\hbar \frac{\int \psi^*\psi' dx}{\int \psi^*\psi dx}$$

and a *mean square* momentum

$$\overline{p^2} = -\hbar^2 \frac{\int \psi^*\psi'' dx}{\int \psi^*\psi dx}$$

We have intentionally left the limits of integration indefinite. In evaluating the integrals occurring here we assume that the range of integration is very much larger than the wave length of the particles, $2\pi/k$. The integral over the last two terms of $\psi^*\psi = A^*A + B^*B + AB^*e^{2ikx} + A^*Be^{-2ikx}$ will then vanish, and

$$\int \psi^*\psi dx = (|A|^2 + |B|^2)l$$

l being the range of integration. By a similar procedure,

$$\int \psi^*\psi' dx = ik(|A|^2 - |B|^2)l \text{ and } \int \psi^*\psi'' dx = -k^2(|A|^2 + |B|^2)l$$

Hence

$$\bar{p} = k\hbar \frac{|A|^2 - |B|^2}{|A|^2 + |B|^2}, \text{ while } \overline{p^2} = k^2\hbar^2$$

It will also be observed that ψ is an *eigenstate* of the operator $\left(-i\hbar \frac{\partial}{\partial x}\right)^2$, but not of $-i\hbar \frac{\partial}{\partial x}$.

Translated into particle language, this state of affairs must be expressed as follows. Since *all* particles have a root mean square momentum of magnitude $k\hbar$, and yet the mean momentum along x is smaller than $k\hbar$, some of them must be traveling to the right, others to the left, with momentum $k\hbar$. If a fraction α travels to the right and β to the left,

$$(\alpha - \beta)k\hbar = \bar{p}, \quad (\alpha + \beta)k\hbar = \sqrt{\overline{p^2}}$$

whence

$$\frac{\beta}{\alpha} = \left(1 - \frac{\overline{p}}{k\hbar}\right)\Big/\left(1 + \frac{\overline{p}}{k\hbar}\right) = \frac{|B|^2}{|A|^2}$$

In our problem, β/α is the *reflection coefficient* of the barrier of potential energy V. In view of eq. (37) it is given by

$$R = \frac{|k_l - k_r|^2}{|k_l + k_r|^2}$$

Two cases of interest may be distinguished, (a) $E < V$, (b) $E > V$. In classical mechanics, a particle would certainly be reflected in case a, ($R = 1$), certainly transmitted in case b, ($R = 0$). The matter is not quite so simple in quantum mechanics. In case a, k_l is real but k_r is imaginary. R is thus always 1 in agreement with the classical prediction. But in case b both k_l and k_r are real, and $R < 1$ but not zero. Hence *every* potential barrier reflects particles, even though classically one would expect them to be only retarded.

Before leaving this matter, we must justify the procedure of setting B_r equal to zero. This is now seen to mean omission of a beam of particles travelling to the left in the region to the right of the origin. Had such a beam been included, the physical condition corresponding to ψ would have implied the incidence of *two* beams of particles upon the origin, one from the left and one from the right. In that case, β/α is not the reflection coefficient of the barrier. The ψ-function we have chosen permits that interpretation, for it corresponds to one beam incident from the left, one reflected and one transmitted beam.

Problem. Prove that $\overline{p}$ is the same whether it is computed to the left or to the right of the origin [use conditions (37)].

A study of more complicated barriers, such as that depicted in Fig. 2, reveals a new and striking feature: the " tunnel effect." The energy E of the incident particles is assumed to be greater than V_1 and V_3, but smaller than V_2, so that from the classical point of view every particle would certainly be reflected. If we define

$$k_1^2 = \frac{2m}{\hbar^2}(E - V_1); \quad \kappa^2 = -k_2^2 = \frac{2m}{\hbar^2}(V_2 - E); \quad k_3^2 = \frac{2m}{\hbar^2}(E - V_3)$$

the ψ-functions for the three regions are

$$\psi_1 = A_1 e^{ik_1 x} + B_1 e^{-ik_1 x}, \quad x < 0$$

$$\psi_2 = A_2 e^{\kappa x} + B_2 e^{-\kappa x}, \quad 0 \leq x \leq a$$

$$\psi_3 = A_3 e^{ik_3 x}, \quad x > a$$

The continuity conditions for ψ and ψ' at both $x = 0$ and $x = a$ are seen to be:

$$A_1 + B_1 = A_2 + B_2$$

$$ik_1(A_1 - B_1) = \kappa(A_2 - B_2)$$

$$A_2 e^{\kappa a} + B_2 e^{-\kappa a} = A_3 e^{ik_3 a}$$

$$\kappa(A_2 e^{\kappa a} - B_2 e^{-\kappa a}) = ik_3 A_3 e^{ik_3 a}$$

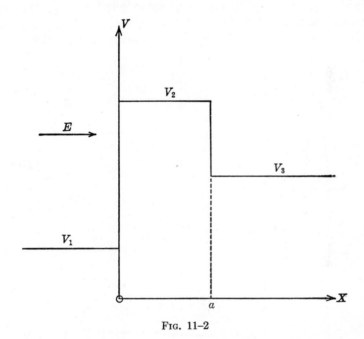

FIG. 11–2

From these, B_1, A_2, and B_2 may be eliminated.   When this is done we obtain the relation

$$A_1 = \tfrac{1}{2} A_3 e^{ik_3 a} \left\{ \left(1 + \frac{k_3}{k_1}\right) \cosh \kappa a + i \left(\frac{\kappa}{k_1} - \frac{k_3}{\kappa}\right) \sinh \kappa a \right\} \quad (11\text{–}38)$$

An argument similar to that which led us to identify the reflection coefficient R with $|B|^2/|A|^2$, shows the *transmission coefficient* of the present barrier to be

$$T = \frac{|A_3|^2 k_3}{|A_1|^2 k_1}$$

This may be computed from (38). In doing so we assume that $\kappa a \gg 1$ so that both cosh κa and sinh κa become $\frac{1}{2}e^{\kappa a}$. Then

$$T = \frac{16k_1k_3}{(k_1 + k_3)^2 + \left(\kappa - \dfrac{k_1k_3}{\kappa}\right)^2} \cdot e^{-2\kappa a}$$

As the width of the barrier increases, the factor $e^{-2\kappa a}$ (sometimes called the " transparency factor ") rapidly diminishes.

The surprising fact is that particles are able to " tunnel " through the barrier although their kinetic energy is not great enough to allow them to pass it. Classically speaking, the kinetic energy of a particle would be

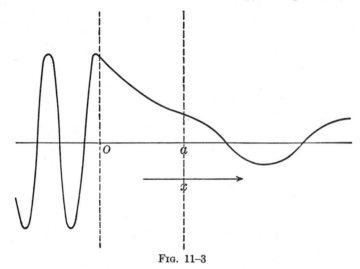

FIG. 11–3

negative while it is in region 2. Quantum mechanically, this statement is devoid of meaning, since it is improper to compute $\overline{E - V}$ for this region alone.[11]

Fig. 3 gives a qualitative plot of the (real part of the) ψ-function in the three regions here considered. It is seen that the barrier attenuates the wave coming from the left, permitting a fraction of its amplitude to pass out at a. The situation is quite analogous to the passage of a wave through an absorbing layer.

11.11. Simple Harmonic Oscillator.—The potential energy, usually expressed in the form $\frac{1}{2}kx^2$, is $\frac{1}{2}m\omega^2x^2$ when written in terms of the mass m and the classical frequency $\omega = 2\pi\nu$ of the oscillator. The meaning of ω is

[11] More complicated barriers are discussed by Condon, E. U., *Rev. Mod. Phys.* **3**, 43 (1931); Eckart, C., *Phys. Rev.* **35**, 1303 (1930).

simply that of a parameter appearing in V; we must no longer expect the oscillator to go back and forth $\omega/2\pi$ times per second. The Schrödinger equation is

$$\frac{d^2\psi}{dx^2} + (\epsilon - \beta^2 x^2)\psi = 0 \qquad (11\text{-}39)$$

if we use the abbreviations

$$\epsilon = \frac{2mE}{\hbar^2}, \quad \beta = \frac{m\omega}{\hbar}$$

The substitution $\xi = \sqrt{\beta}x$ reduces (39) to the form of the differential equation for " Hermite's orthogonal functions,"

$$\frac{d^2\psi}{d\xi^2} + \left[1 - \xi^2 + \left(\frac{\epsilon}{\beta} - 1\right)\right]\psi = 0$$

which was studied in Chapter 2 (cf. eq. 2-66). It was there found that its solution is of the form $e^{-\xi^2/2}H(\xi)$, $H(\xi)$ being a solution of Hermite's equation (2-62). Now $H(\xi)$ is a polynomial if the quantity α, which corresponds to the present $\frac{1}{2}(\epsilon/\beta - 1)$, is an integer. Unless this is true, H is a superposition of the infinite sequences (2-63) and (2-64). But both of these approach infinity like e^{ξ^2}, as closer inspection will show. If they are multiplied by $e^{-\xi^2/2}$, they will not yield a ψ-function which has an integrable square between the limits $-\infty$ and $+\infty$, which we are here assuming to exist. Hence $H(\xi)$ must be chosen in its polynomial form, $H_n(\xi)$. Also, $\frac{1}{2}(\epsilon/\beta - 1) = n$, and this leads to

$$E_n = (n + \tfrac{1}{2})\hbar\omega = (n + \tfrac{1}{2})h\nu \qquad (11\text{-}40)$$

$$\psi_n = ce^{-(\beta/2)x^2}H_n(\sqrt{\beta}x) \qquad (11\text{-}41)$$

If the oscillator has three degrees of freedom, the Schrödinger equation is

$$\nabla^2\psi + (\epsilon - \beta^2 r^2)\psi = 0$$

when the same abbreviations as above are used. The method of separation of variables (Chapter 7) which involves the substitution of $X(x) \cdot Y(y) \cdot Z(z)$ for ψ at once reduces this partial differential equation to three ordinary ones

$$X'' + (\epsilon_1 - \beta^2 x^2)X = 0, \quad Y'' + (\epsilon_2 - \beta^2 y^2)Y = 0$$

$$Z'' + (\epsilon_3 - \beta^2 z^2)Z = 0$$

provided that $\epsilon_1 + \epsilon_2 + \epsilon_3 = \epsilon$. Each of these has a solution of the form (41), so that

$$\psi_{n_1 n_2 n_3} = ce^{-(\beta/2)r^2}H_{n_1}(\sqrt{\beta}x) \cdot H_{n_2}(\sqrt{\beta}y) \cdot H_{n_3}(\sqrt{\beta}z) \qquad (11\text{-}42)$$

and

$$E_{n_1 n_2 n_3} = (n_1 + n_2 + n_3 + \tfrac{3}{2})\hbar\omega \tag{11-43}$$

The orthogonality of the functions (41) has been proved in eq. 3–92. From this formula, the normalizing constant c may also be computed. For if

$$\int_{-\infty}^{\infty} c^2 e^{-\beta x^2} H_n^2(\sqrt{\beta}x)\,dx = \beta^{-1/2}\int_{-\infty}^{\infty} c^2 e^{-\xi^2} H_n^2(\xi)\,d\xi$$

$$= c^2 \cdot 2^n n! \sqrt{\frac{\pi}{\beta}} = 1$$

then

$$c = \left(\frac{\beta}{\pi}\right)^{1/4} (n!\,2^n)^{-1/2}$$

A similar computation, which involves three integrations, yields for the constant c of eq. (42) the value

$$\left(\frac{\beta}{\pi}\right)^{3/4} (n_1!\,n_2!\,n_3!\,2^{n_1+n_2+n_3})^{-1/2}$$

Further mathematical details concerning the functions here encountered, as well as a table of the H_n-polynomials, are given in sec. 3.10.

Problem. The treatment above implied that the 3-dimensional oscillator was isotropic; i.e., bound with equal forces in all directions. Calculate eigenvalues and eigenfunctions for an anisotropic oscillator with potential energy

$$V = \tfrac{1}{2}m(\omega_1^2 x^2 + \omega_2^2 y^2 + \omega_3^2 z^2)$$

11.12. Rigid Rotator, Eigenvalues and Eigenfunctions of L^2.—A rigid rotator is a pair of point masses held together by a rigid, inflexible and inextensible (massless) bond. A diatomic molecule is a fair approximation to a rigid rotator. Before attempting to solve the Schrödinger equation for such a system it is well to digress briefly and consider the eigenvalue equation for an operator which so far we have not introduced, but which is easily constructed. We have seen that the operators corresponding to the components of angular momentum of a particle are

$$L_x = -i\hbar\left(y\frac{\partial}{\partial z} - z\frac{\partial}{\partial y}\right)$$

$$L_y = -i\hbar\left(z\frac{\partial}{\partial x} - x\frac{\partial}{\partial z}\right) \tag{11-44}$$

$$L_z = -i\hbar\left(x\frac{\partial}{\partial y} - y\frac{\partial}{\partial x}\right)$$

From these, we wish to construct the operator

$$L^2 = L_x^2 + L_y^2 + L_z^2 \tag{11-45}$$

It is advantageous to do this in polar (spherical) coordinates.* Putting $x = r \sin \theta \cos \varphi$, $y = r \sin \theta \sin \varphi$, $z = r \cos \theta$, we have

$$\frac{\partial}{\partial x} = \sin \theta \cos \varphi \frac{\partial}{\partial r} + \frac{1}{r} \cos \theta \cos \varphi \frac{\partial}{\partial \theta} - \frac{1}{r} \frac{\sin \varphi}{\sin \theta} \frac{\partial}{\partial \varphi}$$

$$\frac{\partial}{\partial y} = \sin \theta \sin \varphi \frac{\partial}{\partial r} + \frac{1}{r} \cos \theta \sin \varphi \frac{\partial}{\partial \theta} + \frac{1}{r} \frac{\cos \varphi}{\sin \theta} \frac{\partial}{\partial \varphi}$$

$$\frac{\partial}{\partial z} = \cos \theta \frac{\partial}{\partial r} - \frac{1}{r} \sin \theta \frac{\partial}{\partial \theta}$$

When these results are introduced in (44) and (45) is formed, there results

$$L^2 = -\hbar^2 \left\{ \frac{1}{\sin \theta} \frac{\partial}{\partial \theta} \left(\sin \theta \frac{\partial}{\partial \theta} \right) + \frac{1}{\sin^2 \theta} \frac{\partial^2}{\partial \varphi^2} \right\} \tag{11-46}$$

The observable values which the *square* of the angular momentum may assume are the eigenvalues p of the equation

$$L^2 \psi = p \psi \tag{10-47}$$

This equation is easily solved by the method of separation of variables (cf. Chapter 7). Clearly, ψ is a function of θ and φ. Put $\psi = \Theta(\theta) \cdot \Phi(\varphi)$ into (47). This equation will then break up into two ordinary equations (the process is analogous to the construction of eqs. 7–42a and 7–42b):

$$\hbar^2 \left\{ \frac{1}{\sin \theta} \frac{\partial}{\partial \theta} (\sin \theta \Theta') - \frac{m^2}{\sin^2 \theta} \Theta + \frac{p}{\hbar^2} \Theta \right\} = 0$$

$$\Phi'' = -m^2 \Phi$$

The quantity m must be an integer to insure single-valuedness of Φ. The second equation therefore has the solution $\Phi = \text{const.} \; e^{im\varphi}$, m an integer. The first is the equation for associated Legendre functions, (eq. 7.42b), except that the constant $l(l + 1)$ appearing there is here replaced by $p/\hbar^2$. The solution previously obtained is

$$\Theta = \sin^m \theta \frac{d^m}{d(\cos \theta)^m} P_l (\cos \theta)$$

Now the Legendre function P_l was shown to behave singularly at $\cos \theta = \pm 1$ unless l is an integer, in fact it would contain unlimited powers of $x(= \cos \theta)$. The same would be true for Θ if l were arbitrary. But in

* See also the problem at the end of this section.

that case $\int \psi^*\psi d\tau$, which contains the factor

$$\int_0^\pi \Theta^2 \sin \theta d\theta = \int_{-1}^1 \Theta^2 dx$$

would certainly not exist. We conclude, therefore, that l must be an integer, and that the eigenvalues of L^2 are

$$p = l(l+1)\hbar^2$$

On the other hand, the eigenfunctions of L^2 are of the form

$$\sin^m \theta \frac{d^m}{d\theta^m} P_l \,(\cos \theta)e^{im\varphi} = P_l^m \,(\cos \theta)e^{im\varphi} \tag{11-48}$$

in the notation adopted in Chapter 3 (cf. eq. 3–43). Since the eigenvalue p does not depend on m but only on l, functions like (48) with different m will satisfy eq. (47). The most general solution of that equation is therefore,[12]

$$\psi = \sum_{m=-l}^l c_m P_l^m \,(\cos \theta)e^{im\varphi} \tag{11-49}$$

In Chapter 7 this function has already been encountered; it is called a spherical harmonic and denoted by $Y_l(\theta,\varphi)$ (cf. eq. 7–43 et seq.). Hence

$$\psi = Y_l(\theta,\varphi) \tag{11-50}$$

Since $d\tau = \sin \theta d\theta d\varphi$, normalization requires that

$$\int_0^\pi \sin \theta d\theta \int_0^{2\pi} d\varphi \psi^*\psi = 1$$

When (49) is inserted the integral becomes

$$2\pi \sum_{-l}^l c_m^* c_m \int_{-1}^1 [P_l^m(x)]^2 dx = \frac{4\pi}{2l+1} \sum_{-l}^l |c_m|^2 \frac{(l+m)!}{(l-m)!}$$

(cf. eq. 3–62). Hence, for normalization, the constants c_m appearing in (49) must satisfy the relation

$$\sum_{m=-l}^l |c_m|^2 \frac{(l+m)!}{(l-m)!} = \frac{2l+1}{4\pi}$$

and are otherwise arbitrary.

We are now ready to return to the problem of the rigid rotator. In the first place, we shall assume it proper to replace it by a *single* mass, rigidly tied to a center of rotation, and having the same moment of inertia

[12] We define here and elsewhere: $P_l^{-m} = P_l^m$, as in (3–62).

as the original system. The condition upon the state function in accord with this assumption—aside from single-valuedness—is simply $r = a$, a constant. The best procedure is therefore to write down the Schrödinger equation for a particle moving in three dimensions, and then to put $r = a$, $d\psi/dr = 0$. This requires the use of polar (spherical) coordinates. The potential energy V, in this case, is clearly constant and may be taken to be zero.

Schrödinger's equation reads* (cf. Chapter 5 for transformation of ∇^2)

$$\frac{1}{r^2}\frac{\partial}{\partial r}\left(r^2\frac{\partial\psi}{\partial r}\right) + \frac{1}{r^2\sin\theta}\frac{\partial}{\partial\theta}\left(\sin\theta\frac{\partial\psi}{\partial\theta}\right) + \frac{1}{r^2\sin^2\theta}\frac{\partial^2\psi}{\partial\varphi^2} + \frac{2M}{\hbar^2}E\psi = 0 \quad (11\text{-}51)$$

When r is put equal to a the first term on the left vanishes, and the remainder becomes very similar to $L^2\psi$. Indeed if we introduce, a new operator Λ^2 defined as $(1/\hbar^2)L^2$, eq. (51) may be written

$$\Lambda^2\psi = \frac{2Ma^2}{\hbar^2}E\psi \quad (11\text{-}52)$$

But the eigenvalues of Λ^2 are obviously $l(l+1)$, and its eigenfunctions are the same as those of L^2. The constant $(2Ma^2/\hbar^2)E$, must be identified with $l(l+1)$. Hence the eigenvalues and eigenfunctions are

$$E = \frac{\hbar^2}{2Ma^2}l(l+1); \quad \psi_{l,m} = Y_l(\theta,\varphi) \quad (11\text{-}53)$$

Problem. Show by vector algebra that

$$-\Lambda^2 \equiv (\mathbf{r}\times\nabla)^2 = -r^2\nabla^2 + 2r\frac{\partial}{\partial r} + r^2\frac{\partial^2}{\partial r^2}$$

Hint: Note that $(\mathbf{r}\times\nabla)^2 = \mathbf{r}\cdot[\nabla\times(\mathbf{r}\times\nabla)]$. Then use (4-26) for $\nabla\times\mathbf{U}\times\mathbf{V}$.

11.13. Motion in a Central Field.—By central field is meant a field of force in which the potential energy is a function of r only; V is independent of θ and φ. The isotropic three-dimensional oscillator treated in sec. 11 is an example of motion in a central field. Another is the motion of a particle in a Coulomb field. It is to this last example, an electron attracted by a positive point charge (hydrogen atom), that we shall chiefly direct our attention. But before considering this specific case a few general features of the central field problem will be exposed.

It is now clear that the Laplacian, ∇^2, in spherical polar coordinates has the form

$$\nabla^2 = \frac{1}{r^2}\left\{\frac{\partial}{\partial r}\left(r^2\frac{\partial}{\partial r}\right) + \Lambda^2\right\} \quad (11\text{-}54)$$

---

* To avoid confusion, we write M for the electron mass in this section, returning to the symbol m in the next.

where Λ^2 is given by (46) divided by $-\hbar^2$. The eigenvalues of Λ^2 are $l(l+1)$. The Schrödinger equation therefore reads

$$\frac{1}{r^2}\left\{\frac{\partial}{\partial r}\left(r^2\frac{\partial\psi}{\partial r}\right) - \Lambda^2\psi\right\} + \frac{2m}{\hbar^2}[E - V(r)]\psi = 0 \qquad (11\text{-}55)$$

We write ψ as a product of a function $R(r)$ and another, $A(\theta,\varphi)$, which depends only on the angles. The operator Λ^2 acts only on A. Eq. (55), after multiplication by r^2 and subsequent division by $R \cdot A$, has the form

$$\frac{\frac{d}{dr}\left(r^2\frac{dR}{dr}\right)}{R} + \frac{2mr^2}{\hbar^2}[E - V(r)] = \frac{\Lambda^2 A}{A} \qquad (11\text{-}56)$$

The left-hand side of this equation is a function of r alone, the right a function of θ and φ. By the argument which is familiar from Chapter 7, each side must be a constant, say a. Thus

$$\Lambda^2 A = aA$$

But this is simply the eigenvalue equation for Λ^2. We see, then, that

$$a = l(l+1), \quad \text{and} \quad A = Y_l(\theta,\varphi)$$

The left-hand side of (56) becomes

$$\frac{d}{dr}\left[r^2\frac{dR}{dr}\right] + \frac{2mr^2}{\hbar^2}\left[E - V(r) - \frac{l(l+1)}{2mr^2}\hbar^2\right]R = 0 \qquad (11\text{-}57a)$$

and the substitution $U(r) = rR(r)$ reduces this to

$$U'' + \frac{2m}{\hbar^2}\left[E - V(r) - \frac{l(l+1)\hbar^2}{2mr^2}\right]U = 0 \qquad (11\text{-}57b)$$

The development so far has been totally independent of the form of V, except in assuming it to be a function of r alone. The results obtained are therefore valid for any central field. Summarizing them, we may say:

The energy states of a particle in a central field are always of the form

$$\psi = \frac{1}{r}U_l(r)Y_l(\theta,\varphi)$$

and the function U_l is determined by eq. (57b). It was necessary to add a subscript l to U because the differential equation contains l as a parameter. The energies E are obtained solely from eq. (57b).

That equation looks very much like the one-dimensional Schrödinger equation,

$$\psi'' + \frac{2m}{\hbar^2}[E - V(x)]\psi = 0 \qquad (11\text{-}58)$$

but with the term $l(l + 1)\hbar^2/2mr^2$ added to the normal potential energy. What is the meaning of that term? In classical mechanics, the energy of a particle moving in three dimensions differs from that of a one-dimensional particle by the kinetic energy of rotation, $\frac{1}{2}mr^2\omega^2$. This is precisely the quantity $l(l + 1)\hbar^2/2mr^2$, for we have seen that $l(l + 1)\hbar^2$ is the *certain* value of the square of the angular momentum for the state Y_l, in classical language $(mr^2\omega)^2$, which when divided by $2mr^2$, gives exactly the kinetic energy of rotation.

There is, however, one further difference between (57b) and (58). The fundamental range of r in (57b) starts at $r = 0$ and is limited to positive values, whereas the range of x in (58) may include negative values. This fact often has a more important effect on the eigenvalues than the addition of the terms just mentioned.

Let us now solve eq. (57b), assuming a Coulomb field, e.g., $V(r) = -e^2/r$. The energies E will then be the energy levels of the hydrogen atom.[13] For sufficiently large r the solution is determined by

$$U'' - \left(\frac{\alpha}{2}\right)^2 U = 0 \tag{11-59}$$

provided we define

$$\left(\frac{\alpha}{2}\right)^2 = -\frac{2mE}{\hbar^2} \tag{11-60}$$

The solution of (59) is $U_\infty = c_1 e^{(\alpha/2)r} + c_2 e^{-(\alpha/2)r}$, and this represents the behavior of the correct U at ∞. Let us first suppose that α is real, which means that the energy of the particle is *negative*. U will then certainly not have an integrable square (note that the radial integral has the form $\int_0^\infty R^2 r^2 dr = \int U^2 dr$) if the coefficient c_1 fails to vanish. But we cannot simply put it equal to zero because we have boundary conditions to fulfill! Without going further in our analysis at the moment we expect, therefore, that only special values of α will produce acceptable solutions when α is real. If the total energy of the particle is negative (classically speaking, the particle is bound to the attracting center), the energy is expected to be quantized. The following analysis will bear this out.

If α is imaginary, which means that E is positive, U_∞ shows sinusoidal behavior. It has, in fact, the typical form of the state function for a free particle, and the failure of normalization occurs in the milder manner which we have previously found associated with the presence of a continuous spectrum of eigenvalues. There is indeed no way of choosing c_1 or c_2

[13] If e^2 is replaced by Ze^2, $Z = 2$ represents ionized helium, $Z = 3$ doubly ionized lithium, etc.

or α which would make one U_∞ more acceptable than another. We conclude that, when E is positive, the energy spectrum is continuous.

From the point of view of classical physics this result is welcome, for when E is positive the particle is ionized and moves through space, its energy being unrestricted.

We now discuss the bound states in a more rigorous manner. Put $E = -W$, so that W is positive. Our interest will now return to eq. (57a) which forms a more suitable basis for the present discussion. Let $r = x/\alpha$, where α is defined by (60). Eq. (57a) then reads, after some cancellation,

$$x\frac{d^2R}{dx^2} + 2\frac{dR}{dx} + \left[\frac{2me^2}{\hbar^2\alpha} - \frac{x}{4} - \frac{l(l+1)}{x}\right]R = 0 \qquad (11\text{--}61)$$

But this is precisely the differential equation for associated Laguerre functions, which was studied in Chapter 2 (cf. eq. 71). For our immediate purpose we shall write that equation with n^* in place of n, since otherwise our notation would be in conflict with physical convention. To summarize the results of sec. 2.16:

The equation

$$xy'' + 2y' + \left[n^* - \frac{k-1}{2} - \frac{x}{4} - \frac{k^2-1}{4x}\right]y = 0 \qquad (11\text{--}62)$$

has a solution possessing an integrable square[14] of the form

$$y = e^{-x/2}x^{(k-1)/2}L_{n*}^k(x) \qquad (11\text{--}63)$$

provided n^* and k are positive *integers*. Moreover, $n^* - k \geq 0$ since otherwise L_{n*}^k would vanish.

On comparing (61) and (62) we find, in the first place, that $(k^2 - 1)/4 = i(l + 1)$, hence

$$k = 2l + 1$$

Secondly,

$$n^* - \frac{k-1}{2} = n^* - l = \frac{2me^2}{\hbar^2\alpha}$$

When the value of α is inserted here and the relation is solved for W, we find

$$W = \frac{1}{2}\frac{me^4}{(n^* - l)^2\hbar^2}$$

Because of the conditions on n^* and k, the quantity $n^* - l$ cannot be zero. It is usually denoted by n and called the *total quantum* number (after the rôle it played in the Bohr theory). Our conclusion, then, is this:

---

[14] The reader should convince himself of this fact by going back to sec. 2.16.

The energy states of the hydrogen atom are

$$W_n = -E_n = \frac{1}{2}\frac{me^4}{n^2\hbar^2} \tag{11-64}$$

and the corresponding eigenfunctions are, in accordance with (63),

$$R_{n,l} = c_{n,l}e^{-x/2}x^l L_{n+l}^{2l+1}(x) \tag{11-65}$$

the variable x being defined by

$$x = \alpha r = \frac{\sqrt{8mW}}{\hbar}r = \frac{2me^2}{n\hbar^2}r$$

In the Bohr theory of hydrogen, the first orbit has a radius

$$a_0 = \frac{\hbar^2}{me^2} = 0.53 \times 10^{-8} \text{ cm.}$$

It is sometimes convenient to express x in terms of it. Thus $\alpha = 2/na_0$, and

$$x = \frac{2}{n}\frac{r}{a_0} \tag{11-66}$$

It is to be noticed that x represents a *different* variable for each energy state; the quantum number n determining W appears as a scale factor in the dimensionless variable x.

Some integrals involving $R_{n,l}$, which occur frequently in physical and chemical problems, have been evaluated in sec. 3.11. See also the example at the end of sec. 3.11, which is of interest in this connection.

For later use, we write down in explicit form the state function for the normal hydrogen atom. It is

$$R_{1,0} = c_{1,0}e^{-r/a_0}L_1^1 = 2a_0^{-3/2}e^{-r/a_0}$$

For this state $Y_l = \text{constant} = (4\pi)^{-1/2}$ when the function is normalized. Hence the total ground state function is

$$\psi_0 = (\pi a_0^3)^{-1/2}e^{-r/a_0} \tag{11-67}$$

ψ-functions for the higher states are listed in explicit form in Pauling and Wilson.[15]

When the charge on the nucleus is not e but Ze, a_0 must be replaced by a_0/Z, so that

$$\psi_0 = \left(\frac{Z^3}{\pi a_0^3}\right)^{1/2}e^{-Zr/a_0} \tag{11-67a}$$

---

[15] Pauling, L., and Wilson, E. B., Jr., "Introduction to Quantum Mechanics," McGraw-Hill Book Co., 1935.

Problem a. Using the results of Chapter 3, show that the normalizing factor in (65) is

$$c_{n,l} = \left(\frac{2}{na_0}\right)^{3/2} \left\{\frac{(n-l-1)!}{2n[(n+l)!]^3}\right\}^{1/2}$$

Problem b. Work out the problem of the isotropic oscillator using spherical coordinates, and show that the results agree with those obtained in (42) and (43).

11.14. Symmetrical Top.—In dealing with the problem of a rotating rigid body attention must be given to the kinetic energy operator. To obtain it we first observe that its form in rectangular coordinates, for the n particle problem (cf. sec. 11.31) is

$$T\psi = -\frac{\hbar^2}{2}\sum_{i=1}^{n}\frac{\nabla_i^2}{m_i}\psi$$

The position of a rigid body is best expressed in terms of the Eulerian angles, introduced in sec. 9.5. It was there shown that the classical kinetic energy is given by

$$T_c = \frac{1}{2}\sum_{i=1}^{n} m_i(\dot{x}_i^2 + \dot{y}_i^2 + \dot{z}_i^2)$$

$$= \frac{1}{2}A\dot{\beta}^2 + \frac{1}{2}A\dot{\alpha}^2\sin^2\beta + \frac{1}{2}C(\dot{\gamma} + \dot{\alpha}\cos\beta)^2$$

Let us define a *line element* constructed from the Cartesian coordinates

$$\xi_i = \sqrt{m_i}x_i, \quad \eta_i = \sqrt{m_i}y_i, \quad \zeta_i = \sqrt{m_i}z_i$$

as follows:

$$ds^2 = \sum_{i=1}^{n}(d\xi_i^2 + d\eta_i^2 + d\zeta_i^2) \tag{11–68}$$

This is clearly identical with $2T_c dt^2$. From the form of T_c in Eulerian coordinates it is seen that ds^2 in these coordinates is given by

$$ds^2 = A d\beta^2 + A\sin^2\beta\, d\alpha^2 + C(d\gamma + \cos\beta\, d\alpha)^2 \tag{11–69}$$

Now the quantum mechanical form of T is the Laplacian operator corresponding to the line element ds^2, multiplied by $-\hbar^2/2$. The problem is therefore to transform the Laplacian operator from a set of coordinates in terms of which the line element is given by (68), to a new set in terms of which the line element is (69).

This problem has been discussed in sec. 5.17. If

$$ds^2 = \sum_{\lambda,\mu} g_{\lambda\mu}dq_\lambda dq_\mu$$

then

$$\nabla_q^2\psi = \frac{1}{\sqrt{g}}\sum_{\lambda\mu}\frac{\partial}{\partial q_\lambda}\left[\sqrt{g}\, g^{\lambda\mu}\frac{\partial}{\partial q_\mu}\psi\right]$$

On identifying the $g_{\lambda\mu}$ from (69) we find (putting $q_1 = \beta$, $q_2 = \alpha$, $q_3 = \gamma$)

$$(g_{\lambda\mu}) = \begin{pmatrix} A & 0 & 0 \\ 0 & A\sin^2\beta + C\cos^2\beta & C\cos\beta \\ 0 & C\cos\beta & C \end{pmatrix}$$

and hence

$$(g^{\lambda\mu}) = \begin{pmatrix} \dfrac{1}{A} & 0 & 0 \\ 0 & \dfrac{1}{A\sin^2\beta} & -\dfrac{\cos\beta}{A\sin^2\beta} \\ 0 & -\dfrac{\cos\beta}{A\sin^2\beta} & \dfrac{1}{C}+\dfrac{\cos^2\beta}{A\sin^2\beta} \end{pmatrix}, \; g = A^2 C \sin^2\beta$$

When these results are substituted in the expression for $\nabla_q^2\psi$ we have

$$
\begin{aligned}
T\psi = -\frac{\hbar^2}{2}\nabla_q^2\psi = -\frac{\hbar^2}{2\sin\beta}\Bigg\{ &\frac{\partial\psi}{\partial\beta}\left(\frac{\sin\beta}{A}\frac{\partial\psi}{\partial\beta}\right) \\
&+\frac{\partial}{\partial\alpha}\left[\frac{\sin\beta}{A\sin^2\beta}\frac{\partial\psi}{\partial\alpha} - \frac{\sin\beta\cos\beta}{A\sin^2\beta}\frac{\partial\psi}{\partial\gamma}\right] \\
&+\frac{\partial}{\partial\gamma}\left[-\frac{\sin\beta\cos\beta}{A\sin^2\beta}\frac{\partial\psi}{\partial\alpha} + \left(\frac{\sin\beta}{C}+\frac{\sin\beta\cos^2\beta}{A\sin^2\beta}\right)\frac{\partial\psi}{\partial\gamma}\right]\Bigg\} \\
= -\frac{\hbar^2}{2A}\Bigg\{ &\frac{\partial^2\psi}{\partial\beta^2} + \cot\beta\frac{\partial\psi}{\partial\beta} + \frac{1}{\sin^2\beta}\frac{\partial^2\psi}{\partial\alpha^2} \\
&+\left(\cot^2\beta + \frac{A}{C}\right)\frac{\partial^2\psi}{\partial\gamma^2} - \frac{2\cos\beta}{\sin^2\beta}\frac{\partial^2\psi}{\partial\alpha\partial\gamma}\Bigg\}
\end{aligned}
$$

Since the potential energy in this problem is zero, the Schrödinger equation becomes

$$T\psi = E\psi$$

It is separable; for if we put

$$\psi = u(\alpha)\cdot v(\gamma)\cdot w(\beta)$$

the functions u and v are seen to satisfy equations of the form

$$a_2\frac{d^2u}{d\alpha^2} + a_1\frac{du}{d\alpha} + a_0 u = 0, \quad b_2\frac{d^2v}{d\gamma^2} + b_1\frac{dv}{d\gamma} + b_0 v = 0$$

where the coefficients a_0, a_1, a_2 are not functions of α, and the coefficients b_0, b_1, b_2, are not functions of γ. Such equations have solutions

$$u = e^{im\alpha}, \quad v = e^{ik\gamma}$$

m and k being roots of algebraic quadratic equations involving the coefficients a and b. However, these need not be solved here, since the condition of single-valuedness dictates that m and k be integers. We therefore put

$$u = e^{iM\alpha}, \quad v = e^{iK\gamma}$$

$$M,K = 0, \quad \pm 1, \quad \pm 2, \quad \text{etc.}$$

The Schrödinger equation now reduces to the following ordinary differential equation in the independent variable β:

$$w'' + \cot \beta w'$$
$$- \left[\frac{M^2}{\sin^2 \beta} + \left(\cot^2 \beta + \frac{A}{C} \right) K^2 - 2 \frac{\cos \beta}{\sin^2 \beta} KM - \frac{2A}{\hbar^2} E \right] w = 0$$

The substitutions

$$\tfrac{1}{2}(1 - \cos \beta) = x$$

$$w(\beta) = x^{|K-M|/2}(1 - x)^{|K+M|/2} F(x)$$

which are suggested when this equation is examined for its singularities along the lines of Chapter 2, transform it to

$$(x^2 - x)\frac{d^2 F}{dx^2} + [(1 + p)x - q]\frac{dF}{dx} - n(p + n)F = 0$$

the new parameters being defined as follows:

$$p = 1 + |K - M| + |K + M|$$

$$q = 1 + |K - M|$$

$$n(p + n) = A\left(\frac{2E}{\hbar^2} - \frac{K^2}{C} \right) + K^2 - \tfrac{1}{4}(p - 1)^2 - \tfrac{1}{2}(p - 1)$$

This last relation, when rearranged, may be written

$$E = \frac{\hbar^2}{2A}\left[\left(n + \frac{p + 1}{2} \right)\left(n + \frac{p - 1}{2} \right) + \left(\frac{A}{C} - 1 \right)K^2 \right] \qquad (11\text{--}70)$$

Reference to Chapter 2, eq. 56 will show at once that the differential equation for F is none other than the familiar hypergeometric equation defining the Jacobi polynomials, provided n is an *integer*. Unless this condition is satisfied, F will diverge for $x = 1$, i.e., for $\beta = \pi$.

Eq. (70) takes a simpler form when we introduce the new quantum number

$$J = n + \frac{p - 1}{2} = n + \tfrac{1}{2}|K - M| + \tfrac{1}{2}|K + M|$$

which is evidently a positive integer or zero. We then obtain

$$E = \frac{\hbar^2}{2A}\left[J(J+1) + \left(\frac{A}{C} - 1\right)K^2 \right]$$

an equation which determines the energy levels of the symmetrical top. Note that the quantity $\frac{1}{2}\lvert K - M \rvert + \frac{1}{2}\lvert K + M \rvert$ is equal to the larger of the two integers K and M; in consequence of this neither $\lvert K \rvert$ nor $\lvert M \rvert$ can be greater than J.

The energy levels of the *spherical* top $(A = C)$ are those already obtained in sec. 11.12 (cf. eq. 11–53).

MATRIX MECHANICS

11.15. General Remarks and Procedure.—The formulation of quantum mechanics we have given in the foregoing sections was historically preceded by Heisenberg's matrix theory. The latter, while it appears at first glance to be an altogether different mathematical structure, strikingly produced the same results as the former. But when the initial amazement subsided both formulations were recognized as equivalent. In the present text the Schrödinger-Dirac theory was discussed first because its axioms seem perhaps less strange, and because its point of view has been more widely adopted. The terminology of matrix mechanics, however, enjoys great popularity and is often conducive to clarity of expression.

It is possible, and perhaps pedagogically worth while, to derive Heisenberg's theory from the postulates of part of this chapter. But when this is done, the impressive element of uniqueness which attaches to matrix mechanics is completely lost. To preserve it we proceed to state the basic facts of the theory first, to give an example of its application, and then to exhibit its relation to the preceding developments. We can afford to be brief, for when the equivalence of the two theories is once established, no new insight is likely to be gained by deducing former results over again in a different manner. As before, attention will be limited to what we have called quantum statics. The principal facts of Chapter 10 will be used.

Heisenberg associates with every observable a square Hermitian *matrix*. As in the Schrödinger theory, one of the chief concerns of matrix mechanics is the determination of the measurable values of an observable. Let it be desired to find the observable values of a quantity H, which, classically, is a function of the Cartesian coordinates q_i and momenta p_i, $H = H(q_1 \cdots q_n; p_1 \cdots p_n)$. In our example we shall specify H to be the energy, but this restriction is not necessary. Heisenberg's directions are these:

Find a set of matrices $Q_1, Q_2, \cdots, Q_n; P_1, P_2, \cdots, P_n$ which (a) satisfy the commutation rules

$$Q_m Q_n - Q_n Q_m = O; \quad P_m P_n - P_n P_m = O; \quad P_m Q_n - Q_n P_m = -i\hbar\delta_{nm}E$$
(11-71)

where E is the unit matrix; (b) render the matrix

$$H(Q_1 \cdots Q_n; P_1 \cdots P_n) \text{ diagonal} \qquad (11\text{-}72)$$

By $H(Q_1 \cdots Q_n; P_1 \cdots P_n)$ is meant, of course, the matrix which is the same function of the matrices $Q_1 \cdots P_n$ that the ordinary function H is of $q_1 \cdots p_n$. The existence of the matrix H and its uniqueness will be assumed. When such a set of matrices has been found, *the diagonal elements of H will be the measurable values in question.* (It is also true that the squares of the absolute values of the elements $(Q_i)_{\lambda\mu}$ are simply related to spectroscopic transition probabilities, as will be shown later; but this does not concern us here.) We illustrate the power of the method by an example.

11.16. Simple Harmonic Oscillator.—The Hamiltonian function is (cf. sec. 11)

$$H = \frac{p^2}{2m} + \tfrac{1}{2}m\omega^2 q^2$$

Hence, if P and Q are matrices,

$$H(Q,P) = \frac{1}{2m}(P^2 + m^2\omega^2 Q^2)$$

The straightforward way of working this problem would be to select a set of matrices such as, e.g.,

$$Q_{\lambda\mu} = \delta_{\mu,\lambda-1}, \quad P_{\lambda\mu} = -i\hbar\mu\delta_{\mu,\lambda+1} \qquad (11\text{-}73)$$

which satisfy the commutation rule (71):

$$(PQ)_{\lambda\mu} - (QP)_{\lambda\mu} = -i\hbar\delta_{\lambda\mu} \qquad (11\text{-}71a)$$

as the reader may verify. These must then be subjected to a similarity transformation with some other matrix, say S, until the new matrices

$$Q' = S^{-1}QS, \quad P' = S^{-1}PS$$

when substituted in H, make H a diagonal matrix. (Cf. Chap. 10.) This procedure, however, is usually very cumbersome and is rarely used. The success of the matrix method depends frequently on fortunate guesses or on specific properties of the Hamiltonian. In the present instance the following considerations lead most directly to a solution of the problem.

Suppose that the matrices P and Q, which satisfy (71a) and make H diagonal, have already been found. Then

$$H_{kl} = \frac{1}{2m}(P^2 + m^2\omega^2 Q^2)_{kl} = E_k\delta_{kl} \qquad (11\text{--}74)$$

provided we write E_k for the diagonal elements of H.

Now let
$$A \equiv P - im\omega Q$$
and
$$B \equiv P + im\omega Q.$$
Then, because of eq. (71a),
$$AB = 2mH + m\omega\hbar\mathbf{1} \qquad (11\text{--}75)$$
and
$$BA = 2mH - m\omega\hbar\mathbf{1} \qquad (11\text{--}76)$$

Now form ABA from (75) and (76):

$$A(2mH - m\omega\hbar\mathbf{1}) = (2mH + m\omega\hbar\mathbf{1})A$$

$$\sum_\lambda A_{k\lambda}(E_\lambda\delta_{\lambda j} - \tfrac{1}{2}\omega\hbar\delta_{\lambda j}) = \sum_\lambda (E_k\delta_{k\lambda} + \tfrac{1}{2}\omega\hbar\delta_{k\lambda})A_{\lambda j}$$

$$A_{kj}(E_j - E_k - \omega\hbar) = 0.$$

Hence A_{kj} vanishes unless
$$E_j - E_k = \hbar\omega$$

Next, form BAB from (75) and (76):

$$B(2mH + m\omega\hbar\mathbf{1}) = (2mH - m\omega\hbar\mathbf{1})B$$

$$\sum_\lambda B_{k\lambda}(E_\lambda\delta_{\lambda j} + \tfrac{1}{2}\omega\hbar\delta_{\lambda j}) = \sum_\lambda (E_k\delta_{k\lambda} - \tfrac{1}{2}\omega\hbar\delta_{k\lambda})B_{\lambda j}$$

$$B_{kj}(E_j - E_k + \omega\hbar) = 0 \quad \text{or, by changing the subscripts,}$$

$$B_{jk}(E_k - E_j + \omega\hbar) = 0$$

Hence B_{jk} vanishes unless
$$E_j - E_k = \omega\hbar$$

Now take a diagonal element of eq. (76):

$$(BA)_{jj} = 2m(E_j - \tfrac{1}{2}\hbar\omega) \qquad (11\text{--}77)$$

But
$$(BA)_{jj} = \sum_k B_{jk}A_{kj}.$$

Each term in the summation over k vanishes except the *one* for which $E_k = E_j - \omega\hbar$. Suppose E_j is given. Then *either*

$$E_k = E_j - \omega\hbar$$

is another eigenvalue, in which case the right side of eq. (77) is finite. *Or* there is no eigenvalue which is less than E_j by $\hbar\omega$. Then the right side of (77) is zero and

$$E_j = \tfrac{1}{2}\hbar\omega$$

This must be the lowest eigenvalue. From this analysis we may conclude that the sequence of eigenvalues is

$$\tfrac{1}{2}\hbar\omega, \quad \tfrac{3}{2}\hbar\omega, \quad \tfrac{5}{2}\hbar\omega, \quad \text{etc.}$$

in agreement with the results of sec. 11.

11.17. Equivalence of Operator and Matrix Methods.—We first establish a theorem of great importance in quantum mechanics. Consider a differential, Hermitian operator L of the kind discussed in sec. 4, which generates, through the eigenvalue equation

$$L\phi_i = l_i\phi_i$$

a complete set of orthonormal functions ϕ_i. Whether ϕ_i is a function of one or many coordinates is unimportant in this connection. If we introduce other operators M, N which act on the same variables as L we can clearly form two square arrays of numbers, i.e., matrices, by the rule:

$$M_{ij} \equiv \int \phi_i^* M \phi_j d\tau, \quad N_{ij} \equiv \int \phi_i^* N \phi_j d\tau \qquad (11\text{–}78)$$

$d\tau$ being the element of configuration space of the variables of ϕ. The theorem asserts that *equations which hold between the operators M and N, also hold between the matrices formed by the rule* (78). To prove this it is necessary only to establish this parallelism for the two fundamental operations, addition and multiplication:

$$(M + N)_{ij} = M_{ij} + N_{ij} \qquad (11\text{–}79)$$

$$(MN)_{ij} = \sum_\lambda M_{i\lambda} N_{\lambda j} \qquad (11\text{–}80)$$

The first of these is at once evident from (78). To prove the second, let us expand the function $N\phi_j$ in terms of the ϕ_i themselves:

$$N\phi_j = \sum_\lambda a_{\lambda j}\phi_\lambda \qquad (11\text{–}81)$$

By the general procedure of finding the expansion coefficients,[16]

$$a_{ij} = \int \phi_i^* N \phi_j d\tau = N_{ij} \qquad (11\text{–}82)$$

The left side of (80) is, by definition, $\int \phi_i^* MN \phi_j d\tau$. On using (81)

---

[16] Multiply the equation by ϕ_i^* and integrate, using the orthogonality of the ϕ_i.

and (82), this becomes $(MN)_{ij} = \int \phi_i^* M \sum_\lambda a_{\lambda j} \phi_\lambda d\tau = \sum_\lambda \int \phi_i^* M \phi_\lambda d\tau a_{\lambda j} =$ $\sum_\lambda M_{i\lambda} N_{\lambda j}$, in accord with eq. (80).

If, then, we wish to form *matrices* satisfying relations like (71) or (71a), we need only find operators which conform to them, select an orthonormal set of functions ϕ and construct the matrices by means of the rule (78).

Problem a. The operators $Q = e^{ix}$, and $P = -\hbar e^{-ix}(d/dx)$ satisfy

$$PQ - QP = -i\hbar 1$$

Use the functions $\phi_k = \dfrac{1}{\sqrt{2\pi}} e^{ikx}$, $k = 0, \pm 1, \pm 2, \cdots$ to construct the matrices P_{kl}, Q_{kl}. They will be found to be identical with those given in eq. (73).

Problem b. Construct the matrices X_{nm} and P_{nm}, using $X = x$, $P = -i\hbar(d/dx)$, and taking as the orthonormal set the normalized Hermite orthogonal functions discussed in Chapter 3. Note that n and m can only be 0 or positive.

Ans. $X_{nm} = \sqrt{(n+1)/2}\beta \delta_{m,n+1} + \sqrt{n/2}\beta \delta_{m,n-1}$;

$P_{nm} = i\hbar\beta(n - m)X_{nm}$. [$\beta$ is defined after eq. (39).]

Show that these matrices satisfy (71a).

It is interesting to note here that a Hermitian operator, defined by eq. (15), generates a Hermitian matrix (cf. sec. 11.10). For

$$\int \phi_i^* P \phi_j d\tau = \int \phi_j P^* \phi_i^* d\tau$$

simply means

$$P_{ij} = P_{ji}^*$$

in our present notation.

The success of Heisenberg's directions is now easily understood. The *differential* operators which obey relations analogous to those prescribed for Heisenberg's matrices (71) are

$$Q_m = q_m, \quad P_m = -i\hbar \frac{\partial}{\partial q_m}$$

in other words precisely the former, Schrödinger operators.[17] Suppose we select an orthonormal set of functions, ϕ_i, belonging to the operator L, and construct

$$(Q_m)_{ij} = \int \phi_i^* Q_m \phi_j d\tau, \quad \text{etc.}$$

[17] The fact that there are also others, like the ones considered in problem a, need not disturb us here. The Schrödinger equation which results when they are used appears different, to be sure, but reduces to its familiar form when a change of variable is made.

When these matrices are substituted into the functional form H the result is the same as if we had at once formed

$$H_{ij} = \int \phi_i^* H \phi_j d\tau$$

as follows from the theorem we have proved. But the only condition under which this matrix can be diagonal is

$$H\phi_j = \text{const. } \phi_j \qquad (11\text{–}83)$$

that is to say, the ϕ-functions must be chosen to be eigenfunctions of the Hamiltonian H. The problem of making the matrix H diagonal is equivalent to selecting the proper ϕ_i, i.e., to solving the Schrödinger equation. To see that the diagonal elements of H are the permissible energies E_i of the former theory, we need only substitute $H\phi_j = E_j\phi_j$ into (83), obtaining

$$H_{ij} = E_i \delta_{ij}$$

It is easy to extend the Heisenberg theory beyond the limits of the present development. The second postulate, eq. (8), is valid if P is interpreted as a *matrix* and ψ_λ as a *vector*. In the terminology of Chapter 10, the ψ_λ are then the eigenvectors of the matrix P, and the p_λ are its eigenvalues. The relation of the eigenvectors to the state functions is not difficult to see. Suppose we choose a basic orthonormal set of functions, ϕ_i. Expand the eigenfunction ψ_i appearing in the operator equation

$$P\psi_i = p_i\psi_i \qquad (11\text{–}84)$$

in terms of them, viz., $\psi_i = \sum_\lambda a_{i\lambda}\phi_\lambda$.

Now multiply (84) by ϕ_j^* and integrate. We find immediately

$$\sum_\lambda P_{j\lambda} a_{i\lambda} = p_i a_{ij}$$

and conclude that the eigenvector ψ_i has as components the coefficients which appear in its expansion in terms of the basic ϕ. More explicitly,

$$\psi_i = \begin{bmatrix} a_{i1} \\ a_{i2} \\ a_{i3} \\ \cdot \\ \cdot \\ \cdot \end{bmatrix}$$

The last equation then reads $(P\psi_i)_j = P_i(\psi_i)_j$. If the basic set is identical with the eigenfunctions of the operator P, the eigenvector has only one non-vanishing component.

Finally, even the third postulate, (14), may be retained in the Heisen-

377 VARIATIONAL (RITZ) METHOD **11.18**

berg theory if its form is suitably changed. We interpret ϕ as a vector $\boldsymbol{\phi}$ with components a_i, the a_i being the coefficients in the expansion of the *function* $\phi = \sum_\lambda a_\lambda \phi_\lambda$ in terms of our basic ϕ_i (ϕ without subscript here denotes an *arbitrary* state function, not necessarily one of the set ϕ_i), but ϕ^* not as the complex conjugate, but the associate vector:

$$\boldsymbol{\phi}^\dagger = (a_1^* a_2^* a_3^* \cdots)$$

P represents the matrix $P_{ij} = \int \phi_i^* P \phi_j d\tau$. Eq. (14) must then be modified to

$$\bar{p} = \boldsymbol{\phi}^\dagger P \boldsymbol{\phi}$$

which reads, when written more explicitly,

$$\bar{p} = \sum_{\lambda\mu} a_\lambda^* P_{\lambda\mu} a_\mu$$

When the ϕ_i are taken to be the eigenstates of the operator P, the matrix P becomes diagonal, and $\bar{p} = \sum_\lambda a_\lambda^* a_\lambda P_\lambda$, which is the same relation as was found in the Schrödinger theory under these conditions.

Problem. Calculate the integral

$$\int_{-\infty}^{\infty} x^r e^{-x^2} H_n(x) H_m(x) dx$$

by the methods of matrix mechanics. Let $\phi_n = c_n e^{-x^2/2} H_n(x)$, $\phi_m = c_m e^{-x^2/2} H_m(x)$, where c_n, c_m are normalizing factors, and note that, aside from normalizing factors, the integral is the matrix element $(x^r)_{nm}$. Now $x_{\lambda\mu}$ is given by eq. (3–93); this may be used in calculating

$$(x^r)_{nm} = \sum_{\lambda,\mu,\nu,\cdots\sigma} x_{n\lambda} x_{\lambda\mu} x_{\mu\nu} \cdots x_{\sigma m}$$

APPROXIMATION METHODS FOR SOLVING EIGENVALUE PROBLEMS

11.18. Variational (Ritz) Method.—In Chapter 8 we showed that the differential equation $L(u) + \lambda w u = (pu')' - qu + \lambda w u = 0$ is the necessary (though not sufficient!) condition upon u if it is to minimize the integral $\Lambda(u) = \int (pu'^2 + qu^2)dx$. Furthermore, it was seen that $\Lambda(u)$ could be transformed (cf. eq. 8–37) by simple steps to $-\int uL(u)dx$. The theory in this simple form is applicable to every one-dimensional Schrödinger equation, for in that case the Hamiltonian operator $H = -(\hbar^2/2m)(d^2/dx^2) + V(x)$ is of the form $-L$ if only we identify p with $\hbar^2/2m$ and q with V. Hence we may at once say that the Schrödinger equation is the necessary condition upon ψ so that the integral

$$\int \psi H \psi d\tau$$

shall be a minimum. The one-dimensional variation theory may also be applied, though in a somewhat more cumbersome manner, to every ordinary differential equation to which the multi-dimensional Schrödinger equation gives rise on separation of variables. It is possible, however, to prove a far more general theorem which is of utmost utility in numerous problems of applied mathematics, a theorem of which the former statement is a special case.

Let P be a Hermitian operator. We wish to find the normalized function ψ which will make the integral

$$\int_\tau \psi^* P \psi d\tau$$

a minimum. The integration extends, as usual, over configuration space, and we shall assume for the sake of definiteness that τ is a finite portion of configuration space. Certainly, the necessary condition upon ψ is that

$$\delta \left\{ \int \psi^* P \psi d\tau - \lambda \int \psi^* \psi d\tau \right\}$$

shall vanish; λ is an undetermined (Lagrangian) multiplier (cf. sec. 6.5). Now the variation symbol and the integral sign are commutable in this expression because the limits of the integration are supposed finite and fixed. Hence we have

$$\int \delta\psi^* \cdot P\psi d\tau + \int \psi^* \cdot \delta(P\psi)d\tau - \lambda \int \delta\psi^* \cdot \psi d\tau - \lambda \int \psi^* \delta\psi d\tau = 0 \quad (11\text{--}85)$$

The second integral in this expression may be transformed in two steps. First, $\delta(P\psi)$ may be replaced by $P(\delta\psi)$ since the operator P suffers no variation. Second, because P is Hermitian and both ψ and $\delta\psi$ are acceptable functions, $\int \psi^* P(\delta\psi)d\tau = \int \delta\psi \cdot P^*\psi^*d\tau$. Eq. (85) therefore reads

$$\int \delta\psi^*(P\psi - \lambda\psi)d\tau + \int \delta\psi(P^*\psi^* - \lambda\psi^*)d\tau = 0 \quad (11\text{--}86)$$

Here $\delta\psi$ is an entirely arbitrary function. Let us take it to be real, so that $\delta\psi^* = \delta\psi$. Eq. (86) can then be satisfied only if

$$P\psi - \lambda\psi + P^*\psi^* - \lambda\psi^* = 0$$

On the other hand, if we take $\delta\psi$ to be imaginary, so that $\delta\psi^* = -\delta\psi$, we conclude

$$P\psi - \lambda\psi - P^*\psi^* + \lambda\psi^* = 0$$

Addition of the last two equations yields

$$P\psi = \lambda\psi$$

subtraction gives

$$P^*\psi^* = \lambda\psi^*$$

We have shown that, if

$$\delta \int \psi^* P\psi d\tau = 0 \qquad (11\text{-}87)$$

for normalized ψ, this function must satisfy the eigenvalue equation

$$P\psi = \lambda\psi \qquad (11\text{-}88)$$

which also automatically determines λ. Whether, when (88) is satisfied, the minimum of, or indeed the integral, $\int \psi^* P\psi d\tau$, actually exists, is a point we have not investigated. It is customary in physics not to worry about these eventualities, for they are difficult to discuss. The mathematical equivalence of the minimal property of the integral and eq. (88) is usually taken as a matter of faith.

If ψ satisfies eq. (88), then $\int \psi^* P\psi d\tau = \lambda$. From what has been said it follows, therefore, that the integral $\int \varphi^* P\varphi d\tau$ computed with a function different from the minimizing ψ, cannot be smaller than λ. But here a slight complication arises, for there are many eigenvalues λ. All that we can really say is that for a function φ in the " neighborhood " of ψ_i, the integral will not be greater than λ_i. Certainly, however,

$$\int \varphi^* P\varphi d\tau \geqq \lambda_0 \qquad (11\text{-}89)$$

if φ is any analytic and continuous function[18] and λ_0 the lowest eigenvalue.

The Ritz method,[19] named after its inventor, is a systematic procedure, based upon the foregoing variational considerations, for solving the eigenvalue equation (88) by substituting into the integral in (87) a suitable sequence of functions which causes the integral to converge upon the value λ. Instead of presenting the method in its original form, we shall

---

[18] Restriction to functions with a certain number of derivatives is necessary because P is in general a differential operator, and $P\varphi$ must have meaning.

[19] Ritz, W., *J. f. reine und angew. Math.* **135**, 1 (1909); Courant-Hilbert, p. 150.

here work out some of its features in a manner more directly adapted to the needs of quantum mechanics, and with a slight loss of rigor. We are usually interested in finding the energies, particularly the lowest (normal state) energy of physical or chemical systems, hence we identify at once the operator P in (89) with the Hamiltonian H.

The simplest way of finding an approximation to the lowest energy of a system is to use (89) directly. Sometimes a good guess can be made as to the general form of the true state function ψ, a form which may allow the inclusion of one or more arbitrary parameters. The integral in (89) is then computed with this function, and the result is minimized with respect to the parameters. An example will clarify the method.

11.19. Example: Normal State of the Helium Atom.—The helium atom consists of two electrons moving in the field of a nucleus of charge $2e$ and at the same time repelling each other. We consider the nucleus as stationary and denote the distances of the two electrons from it by r_1 and r_2 respectively; r_{12} is the interelectronic distance. The potential energy is $-2e^2(1/r_1 + 1/r_2) + e^2/r_{12}$, and the Schrödinger equation

$$H\psi = \left\{ -\frac{\hbar^2}{2m}(\nabla_1^2 + \nabla_2^2) - 2e^2\left(\frac{1}{r_1} + \frac{1}{r_2}\right) + \frac{e^2}{r_{12}} \right\}\psi = E\psi \quad (11\text{--}90)$$

A subscript on the symbol ∇ indicates that the Laplacian is to be taken with respect to the coordinates labeled by the subscript. If the term e^2/r_{12} were absent eq. (90) would be separable, for then the operator H would be the sum of two helium-ion Hamiltonians, $H = H_1 + H_2$, the first acting on the coordinates of electron 1, the second on those of electron 2. But the equation

$$(H_1 + H_2)\psi = E\psi$$

may be separated on substitution of $\psi = u(1)v(2)$, where $u(1)$ stands for a function of the space coordinates of electron 1, and $v(2)$ is defined similarly. For it becomes, after division by ψ,

$$\frac{H_1 u(1)}{u(1)} + \frac{H_2 v(2)}{v(2)} = E, \quad \text{a constant}$$

which indicates that $H_1 u(1) = E_1 u(1)$; $H_2 v(2) = E_2 v(2)$; $E_1 + E_2 = E$. But the first two of these are simply Schrödinger equations for the singly charged helium ion, whose solutions we already know. (Cf. eq. 67a.) Since we wish to find the *lowest* energy of our system, we identify the functions as follows:

$$u(1) = \left(\frac{Z^3}{\pi a_0^3}\right)^{1/2} e^{-Zr_1/a_0}, \quad v(2) = \left(\frac{Z^3}{\pi a_0^3}\right)^{1/2} e^{-Zr_2/a_0}$$

and ψ is the product of these.

The correct solution of eq. (90) is certainly not of this exact form because of the " interaction term " e^2/r_{12}, whose effect on ψ one would expect to be very complicated indeed. Aside from other changes, it will cause ψ to depend on r_{12} explicitly. But from a physical point of view, the repulsion between the electrons will cause both of them to be, on the average, farther away from the nucleus than if the repulsion were absent. This would mean that the functions u and v are in error with respect to the scale factor Z/a_0. If this were smaller, a more extended probability distribution would result. (For the helium ion $Z = 2$.) It would seem expedient, therefore, that we take as our " trial " function in the variational procedure the function $\phi = u(1)v(2)$ but with an undetermined Z.

In calculating

$$\int \phi H \phi d\tau \tag{11-91}$$

it is well to have available the differential equation whose solutions are u and v:

$$-\frac{\hbar^2}{2m} \nabla_1^2 u(1) = \frac{Ze^2}{r_1} u(1) + Z^2 E_H u(1), \quad v(2) = u(2) \tag{11-92}$$

Here E_H is the energy of the normal hydrogen atom, $E_H = -e^2/2a_0$ ($= -13.53$ e. volts). The differential $d\tau$ in (91) represents, of course, the product of the volume element for the two electrons. When H is taken from (90), we find, using (92) and the fact that u is normalized,

$$\int \phi H \phi d\tau = 2Z^2 E_H + (Z-2)e^2 \int \left(\frac{1}{r_1} + \frac{1}{r_2}\right) \phi^2 d\tau + e^2 \int \frac{\phi^2}{r_{12}} d\tau \tag{11-93}$$

The integral

$$\int \frac{\phi^2}{r_1} d\tau = \int \frac{u^2(1)}{r_1} d\tau_1 \cdot \int u^2(2) d\tau_2 = \int \frac{u^2(1)}{r_1} \cdot r_1^2 dr_1 \sin\theta d\theta d\varphi$$

is easily computed directly. It has, in fact, already been evaluated (cf. sec. 3.11, example) and found to be Z/a_0. The other integral, $\int \frac{\phi^2}{r_2} d\tau$, has the same value. We leave the evaluation of $e^2 \int \frac{\phi^2}{r_{12}} d\tau$ for later; its value is $-\frac{5}{4}ZE_H$. Hence, eq. (93) becomes

$$\int \phi H \phi d\tau = 2Z^2 E_H + (Z-2) \cdot 2Z \frac{e^2}{a_0} - \frac{5}{4} Z E_H$$

$$= Z \left[2Z - 4(Z-2) - \tfrac{5}{4}\right] E_H$$

This expression is to be made as small as possible by choosing Z properly, i.e., the coefficient must take its maximum value because $E_H < 0$. Putting the derivative with respect to Z equal to zero, we find for the minimizing Z the value $27/16$, which is somewhat less than 2 as we expected. Hence the best energy value attainable by adjusting Z in our function is $Z(27/4 - 2Z)E_H = 5.695E_H$. The energy found experimentally is $5.807E_H$. The difference between these two values is to be ascribed to the defects of the simple trial function here chosen.

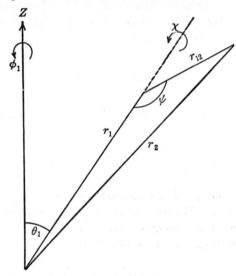

FIG. 11–4

A very interesting summary of the results of the present method as applied to helium is given by Pauling and Wilson.[20] Their table shows how the value of the integral approaches the experimental energy as increasingly refined trial functions are used.

To complete the analysis we indicate how the integral

$$I = e^2 \int \frac{\phi^2}{r_{12}} \, d\tau$$

may be computed. The method is typical of the evaluation of "double volume" integrals involving the variable r_{12}, and hence perhaps of some interest. The volume element

$$d\tau = r_1^2 dr_1 \sin \theta_1 d\theta_1 d\varphi_1 \cdot r_2^2 dr_2 \sin \theta_2 d\theta_2 d\varphi_2$$

may also be expressed as follows: (see Fig. 4)

$$d\tau = r_1^2 dr_1 \sin \theta_1 d\theta_1 d\varphi_1 \cdot r_{12}^2 dr_{12} \sin \psi d\psi d\chi$$

[20] Pauling and Wilson, p. 224.

Now $r_2^2 = r_1^2 + r_{12}^2 - 2r_1r_{12}\cos\psi$, whence $r_2dr_2 = r_1r_{12}\sin\psi\,d\psi$ provided r_1 and r_{12} are held fixed. By means of this relation $\sin\psi\,d\psi$ may be eliminated from the last expression for $d\tau$, and we obtain

$$d\tau = r_1dr_1r_2dr_2r_{12}dr_{12}\sin\theta_1d\theta_1d\varphi_1d\chi$$

Substitute this volume element into I, and integrate at once over the angles, thus introducing the factors $2 \cdot 2\pi \cdot 2\pi$. On using the abbreviation $\alpha = 2Z/a_0$, we obtain

$$I = \frac{\alpha^6 e^2}{8} \iiint e^{-\alpha(r_1+r_2)}r_1dr_1r_2dr_2dr_{12}$$

The ranges of integration are: $0 \leq r_1 \leq \infty, 0 \leq r_2 \leq \infty$; $|r_2 - r_1| \leq r_{12} \leq r_1 + r_2$. The absolute value sign on the limit for r_{12} forces us to split the integration over r_2 into two parts, (a) $r_2 > r_1$, (b) $r_2 < r_1$. In range (a) the lower limit of r_{12} is $r_2 - r_1$, in case (b) it is $r_1 - r_2$. Thus

$$I = \frac{\alpha^6 e^2}{8}\left\{\int_0^\infty e^{-\alpha r_1}r_1dr_1\int_{r_1}^\infty e^{-\alpha r_2}r_2dr_2\int_{r_2-r_1}^{r_2+r_1}dr_{12} + \int_0^\infty e^{-\alpha r_2}r_2dr_2\int_{r_2}^\infty e^{-\alpha r_1}r_1dr_1\int_{r_1-r_2}^{r_1+r_2}dr_{12}\right\}$$

Inspection shows that the two triple integrals are equal. The calculation is now perfectly straightforward; it makes use of the formula

$$\int_0^\infty e^{-px}x^n dx = p^{-(n+1)}n!$$

and leads to the result

$$I = \frac{5}{8}Z\frac{e^2}{a_0} = -\frac{5}{4}ZE_H$$

which was used above.

11.20. The Method of Linear Variation Functions.—It is often convenient to use as the trial function ϕ in $\int \phi^* H\phi d\tau$ a linear combination of definite functions u_i which are judged suitable for the problem at hand. The coefficients appearing in the linear combination may then be treated as variable parameters. Thus, assume

$$\phi = \sum_{\lambda=1}^n a_\lambda u_\lambda \tag{11-94}$$

where the u's need *not* form an orthonormal set. We define

$$\int u_i^* u_j d\tau = \Delta_{ij}, \quad \int u_i^* H u_j d\tau = \mathcal{H}_{ij}, \quad E = \frac{\int \phi^* H\phi d\tau}{\int \phi^* \phi d\tau}$$

The symbol $\mathcal{H}_{ij}$ in place of H_{ij} is to remind the reader of the fact that the matrix $\mathcal{H}$ does not possess the simple properties of H because the former is not constructed with an orthonormal, complete set of functions. The denominator in the expression for E is needed to normalize the function ϕ. According to the variational principle, $E \geq E_0$, the lowest energy state of the system.

We wish to find the condition that E shall be a minimum, and the minimal value of E. Insertion of (94) gives

$$E = \sum_{\lambda,\mu=1}^{n} a_\lambda^* a_\mu \mathcal{H}_{\lambda\mu} \Big/ \sum_{\lambda\mu} a_\lambda^* a_\mu \Delta_{\lambda\mu}$$

This expression will be an extremum, and we hope a minimum, if E is so adjusted that $\partial E/\partial a_k^*$ and $\partial E/\partial a_k$ are zero for every k from 1 to n. Let us take the derivative with respect to a_k^* on both sides of the last equation after it is written in the form

$$E \sum_{\lambda\mu} a_\lambda^* a_\mu \Delta_{\lambda\mu} = \sum_{\lambda\mu} a_\lambda^* a_\mu \mathcal{H}_{\lambda\mu} \qquad (11\text{--}95)$$

The result is

$$\frac{\partial E}{\partial a_k^*} \sum_{\lambda\mu} a_\lambda^* a_\mu \Delta_{\lambda\mu} + E \sum_\mu a_\mu \Delta_{k\mu} = \sum_\mu a_\mu \mathcal{H}_{k\mu}, \quad k = 1, 2, \cdots, n$$

When the first term is omitted $(\partial E/\partial a_k^* = 0)$ the remainder of the equation represents the condition that E shall be a minimum. Differentiation of (95) with respect to a_k leads in a similar way to

$$E \sum_\lambda a_\lambda^* \Delta_{\lambda k} = \sum_\lambda a_\lambda^* \mathcal{H}_{\lambda k}$$

an equation which is simply the conjugate of the former. Both may conveniently be written

$$\sum_\mu a_\mu (\mathcal{H}_{k\mu} - \Delta_{k\mu} E) = 0, \quad k = 1, 2, \cdots, n \qquad (11\text{--}96)$$

If this system of equations is to have a solution different from the trivial one: every $a_\mu = 0$, then the determinant constructed from the coefficients of the a_μ must vanish. Thus

$$\begin{vmatrix} \mathcal{H}_{11} - \Delta_{11}E & \mathcal{H}_{12} - \Delta_{12}E & \cdots & \mathcal{H}_{1n} - \Delta_{1n}E \\ \mathcal{H}_{21} - \Delta_{21}E & \mathcal{H}_{22} - \Delta_{22}E & \cdots & \mathcal{H}_{2n} - \Delta_{2n}E \\ \cdots\cdots\cdots\cdots\cdots\cdots\cdots\cdots\cdots\cdots\cdots\cdots\cdots\cdots\cdots \\ \cdots\cdots\cdots\cdots\cdots\cdots\cdots\cdots\cdots\cdots\cdots\cdots\cdots \\ \mathcal{H}_{n1} - \Delta_{n1}E & \mathcal{H}_{n2} - \Delta_{n2}E & \cdots & \mathcal{H}_{nn} - \Delta_{n\,n}E \end{vmatrix} = 0 \qquad (11\text{--}97)$$

This is an equation of the n-th degree in E and therefore has n roots. The lowest of these will be an approximation to the lowest energy of the system.

The other roots approximate, though in general much more poorly, to the $n - 1$ higher states of the system.

11.21. Example: The Hydrogen Molecular Ion Problem.—The H_2^+-ion consists of two positive charges $+e$, which we shall consider stationary and a distance R apart, and one electron whose distances from the protons will be denoted by r_A and r_B. See Fig. 5.

The Hamiltonian operator is

$$H = -\frac{\hbar^2}{2m}\nabla^2 - \frac{e^2}{r_A} - \frac{e^2}{r_B} + \frac{e^2}{R}$$

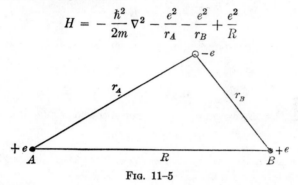

Fɪɢ. 11–5

If the terms $e^2/R - e^2/r_B$ were missing, H would be the Hamiltonian of a hydrogen atom with its proton at A, whose normal state function is (cf. eq. 67)

$$u_A = (\pi a_0^3)^{-1/2} e^{-r_A/a_0}$$

On the other hand, if the terms $e^2/R - e^2/r_A$ were missing, the normal state function would be

$$u_B = (\pi a_0^3)^{-1/2} e^{-r_B/a_0}$$

From a physical point of view one of these solutions is as good as the other: u_A implies that the electron is entirely attached to proton A, u_B that it is attached to proton B. Neither is the case. Let us see what happens if we take for the variation function ϕ a linear combination of u_A and u_B. We put

$$\phi = a_A u_A + a_B u_B$$

using letters as subscripts rather than the number indices which appear in (94).[21] The lowest energy is at once obtained as the lowest root of (97) which takes the simple form

$$\begin{vmatrix} \mathcal{H}_{AA} - \Delta_{AA}E & \mathcal{H}_{AB} - \Delta_{AB}E \\ \mathcal{H}_{BA} - \Delta_{BA}E & \mathcal{H}_{BB} - \Delta_{BB}E \end{vmatrix} = 0 \qquad (11\text{–}98)$$

---

[21] In more complicated molecules it is well to label electrons by numbers, nuclei by letters. We here follow this convention.

Now $\Delta_{AA} = \int u_A^* u_A d\tau = \Delta_{BB} = 1$, because u_A and u_B are normalized.

They are not orthogonal, but $\Delta_{AB} = \Delta_{BA}$. Similarly, $\mathcal{H}_{AB} = \int u_A H u_B d\tau$
$= \mathcal{H}_{BA}$ and $H_{BB} = H_{AA}$ since H is insensitive to an interchange of A and
$'B$. With these simplifications the two roots of (98) are found to be

$$E_1 = \frac{\mathcal{H}_{AA} + \mathcal{H}_{AB}}{1 + \Delta_{AB}} \quad E_2 = \frac{\mathcal{H}_{AA} - \mathcal{H}_{AB}}{1 - \Delta_{AB}}$$

The ϕ-functions corresponding to these energies are obtained from (96):

$$a_A(\mathcal{H}_{AA} - E) + a_B(\mathcal{H}_{AB} - \Delta_{AB}E) = 0$$
$$a_A(\mathcal{H}_{BA} - \Delta_{BA}E) + a_B(\mathcal{H}_{BB} - E) = 0 \tag{11-99}$$

On inserting $E = E_1$ we get $a_A = a_B$; hence the corresponding

$$\phi_1 = c_1(u_A + u_B)$$

If ϕ_1 is to be normalized, $c_1 = [2(1 + \Delta_{AB})]^{-1/2}$. If E_2 is inserted in (99),
we find $a_B = -a_A$, so that

$$\phi_2 = c_2(u_A - u_B)$$

The normalizing factor is in this case $c_2 = [2(1 - \Delta_{AB})]^{-1/2}$.

The remainder of the work is the computation of the three quantities
Δ_{AB}, $\mathcal{H}_{AA}$ and $\mathcal{H}_{AB}$. It involves nothing new and will be left to the
reader. The integrals are most easily evaluated in spheroidal coordinates
(cf. eq. 5-40). $\xi = (r_A + r_B)/R$, $\eta = (r_A - r_B)/R$ and φ, the latter
measured around R. In terms of these

$$d\tau = \frac{R^3}{8}(\xi^2 - \eta^2)d\xi d\eta d\varphi, \quad \text{and} \quad u_A u_B = (\pi a_0^3)^{-1}e^{-(R/a_0)\xi}$$
$$1 \le \xi < \infty; \ -1 \le \eta \le 1$$

The following results will be found:

$$\Delta_{AB} = e^{-\rho}\left(1 + \rho + \frac{\rho^2}{3}\right)$$

$$\mathcal{H}_{AA} = E_H + \frac{e^2}{R} + J, \quad \text{where}$$

$$J = -\int u_A \frac{e^2}{r_B} u_A d\tau = -\frac{e^2}{R}[1 - e^{-2\rho}(1 + \rho)]$$

$$\mathcal{H}_{AB} = \left(E_H + \frac{e^2}{R}\right)\Delta_{AB} + K, \quad \text{where}$$

$$K = -\int u_A \frac{e^2}{r_A} u_B d\tau = -\frac{e^2}{R}e^{-\rho}(\rho + \rho^2)$$

$$\left.\begin{array}{c} \\ \\ \\ \\ \\ \\ \\ \\ \\ \\ \\ \end{array}\right\} \tag{11-100}$$

The parameter $\rho \equiv R/a_0$; E_H is defined as in sec. 19. The quantities J and K are of interest. According to its definition, J represents the Coulomb attraction energy between a negative charge of density u_A^2 and the proton B. The integral K has no such simple interpretation; it is called an *exchange integral*. Its importance is best appreciated if E_1 and E_2 are written more explicitly with the use of (100):

$$E_1 = E_H + \frac{e^2}{R} + \frac{J + K}{1 + \Delta_{AB}}$$

$$E_2 = E_H + \frac{e^2}{R} + \frac{J - K}{1 - \Delta_{AB}}$$

Because K is negative, E_1 is the lower root. Had we omitted the function u_B from our trial function ϕ, the variational result would have been

$$E = E_H + \frac{e^2}{R} + J$$

E_1 is lower than this by virtue of the presence of K (and of course Δ_{AB}). But in classical parlance, a lower energy must be regarded as due to the presence of additional attractive forces between the constituents of the system, i.e., a hydrogen atom and a proton. These forces would be given by $\partial K/\partial R$; they are commonly called *exchange forces*. They possess no classical interpretation; their significance is rooted entirely in the variational method through which they arise.

Of course E_1 is only an approximation to the true energy, which is lower for every R.[3] Its most important feature is that it possesses a *minimum*, which explains the stability of the H_2^+ ion. Classical mechanics would yield no minimum and is therefore incompetent to account for the existence of this ion. A detailed comparison of E_1 with the experimental energy is given in Pauling and Wilson.[22]

Problem. Let u_0, u_1, u_2 be the three lowest energy states of the simple harmonic oscillator, H_0 its Hamiltonian. The Hamiltonian for an oscillator in an electric field is $H = H_0 + kx$, where k is a constant. Calculate by the variational method the lowest energy of this system, using as trial functions (a) u_0, (b) $a_0u_0 + a_1u_1$, (c) $a_0u_0 + a_1u_1 + a_2u_2$.

Ans. (a) $\frac{1}{2}h\nu$, (b) $h\nu - \sqrt{k^2x_{01}^2 + \frac{1}{4}(h\nu)^2} \approx \frac{1}{2}h\nu - k^2x_{01}^2/h\nu$, (c) $\frac{1}{2}h\nu - h\nu k^2x_{01}/[(h\nu)^2 - k^2x_{12}^2]$ (approximately). Here x_{ij} is defined as $\int u_i x u_j d\tau$, as usual.

11.22. Perturbation Theory.—The following problem is frequently met in quantum mechanics. We know the energy states of a given system, say an atom, and also its eigenfunctions. A small perturbation, such as an

[22] Pauling and Wilson, loc. cit.

electric or magnetic field, is now imposed; this changes, presumably by slight amounts, both energies and state functions. Mathematically, the situation is described in this way. We know the solutions and eigenvalues of

$$H^0\psi_i = E_i^0\psi_i \tag{11-101}$$

where H^0 is the " unperturbed " Hamiltonian. We wish to find solutions and eigenvalues of

$$H\phi_i = E_i\phi_i, \quad H = H^0 + H' \tag{11-102}$$

H' being considered as a " small " addition to H^0. (By a small operator we mean one whose matrix elements, formed with the functions ψ_i, are all small compared with the diagonal elements of H^0.)

To solve the problem we use the method of linear variation functions, using as our trial function

$$\phi = \sum_\lambda a_\lambda\psi_\lambda \tag{11-103}$$

If we allow an infinite number of terms in this summation and choose the coefficients properly, we expect ϕ to be the *correct* solution of (102), for the ψ_λ of (101) form a complete set. But since the ψ_i are orthonormal, the energies are given as the roots of (97) with every Δ_{ij} replaced by a Kronecker δ_{ij}, so that E appears only in the principal diagonal. Moreover,

$$\mathcal{H}_{ij} = H_{ij} = (H^0)_{ij} + H'_{ij} = E_i^0\delta_{ij} + H'_{ij}$$

$$H'_{ij} = \int \psi_i^* H'\psi_j d\tau$$

Hence the determinant reads

$$\begin{vmatrix} H'_{11}-(E-E_1^0) & H'_{12} & H'_{13} & H'_{14} & \cdots \\ H'_{21} & H'_{22}-(E-E_2^0) & H'_{23} & H'_{24} & \cdots \\ H'_{31} & H'_{32} & H'_{33}-(E-E_3^0) & H'_{34} & \cdots \\ H'_{41} & H'_{42} & H'_{43} & H'_{44}-(E-E_4^0)\cdots \\ \cdots & \cdots & \cdots & \cdots \\ \cdots & \cdots & \cdots & \cdots \end{vmatrix} = 0$$

$$\tag{11-104}$$

If all its roots could actually be found they would indeed be the exact energies of our problem. But in the case we are visualizing certain simplifying approximations are in order. Suppose we are interested in the energy E_1, that is, the energy to which E_1^0 is changed by the perturbation. (E_1 need not be the lowest energy of our system, for the states may be labeled in an arbitrary order.) *If E_1^0 is a non-degenerate level, then E_1*

will lie much closer to E_1^0 than to any other unperturbed E_i^0. This suggests the following approximations:

a. Put $E = E_1^0$ in all diagonal elements except the first.

b. Since every difference $E_1^0 - E_i^0$ for $i \neq 1$ is large compared to H'_{ii}, the latter may be omitted in all diagonal elements except the first.

c. Neglect all non-diagonal elements except those in the first row and the first column, since they affect E_1 only in a secondary way.

When this is done, the determinant reads (we now write ΔE_1 for the perturbation $E - E_1^0$ we are seeking)

$$\begin{vmatrix} H'_{11} - \Delta E_1 & H'_{12} & H'_{13} & H'_{14} & \cdots \\ H'_{21} & E_2^0 - E_1^0 & 0 & 0 & \cdots \\ H'_{31} & 0 & E_3^0 - E_1^0 & 0 & \cdots \\ H'_{41} & 0 & 0 & E_4^0 - E_1^0 & \cdots \\ \cdots\cdots\cdots\cdots\cdots\cdots\cdots\cdots\cdots\cdots \end{vmatrix} = 0 \qquad (11\text{-}105)$$

It may be evaluated by the usual process of adding multiples of rows or columns. In this instance, multiply the second row by $H'_{12}/(E_2^0 - E_1^0)$, and then subtract it from the first. The element H'_{12} will then disappear from the first row, but the first element is converted into

$$H'_{11} - \Delta E_1 - \frac{H'_{12}H'_{21}}{E_2^0 - E_1^0}$$

Next, multiply the third row by $H'_{13}/(E_3^0 - E_1^0)$ and subtract it from the first. The result will be disappearance of H'_{13} and addition of $-H'_{13}H'_{31}/(E_3^0 - E_1^0)$ to the first element. This process is continued until all non-diagonal elements of the first row have disappeared. We now have

$$\left(H'_{11} - \Delta E_1 - \sum_{\lambda=2}^{\infty} \frac{H'_{1\lambda}H'_{\lambda 1}}{E_\lambda^0 - E_1^0} \right) \ (E_2^0 - E_1^0)(E_3^0 - E_1^0) \cdots = 0$$

If E_1^0 is non-degenerate, as we are supposing, none of the parentheses except the first can be zero. We therefore conclude

$$\Delta E_1 = H'_{11} - \sum_{\lambda=2}^{\infty} \frac{H'_{1\lambda}H'_{\lambda 1}}{E_\lambda^0 - E_1^0} \qquad (11\text{-}106)$$

and this is the Rayleigh-Schrödinger perturbation formula. The quantity H'_{11} is often called the *first-order* perturbation, the sum on the right is called the *second-order* perturbation. By retaining more elements in (104) third and higher orders may be computed, but these are rarely used. When the approximation (106) is not sufficient it is generally preferable to return to the variation scheme, or to find a more successful way of evaluating the determinant (104).

Formula (106) may, of course, be used to calculate the perturbation in any energy level which is non-degenerate; to show this fact it may be written in the form

$$\Delta E_k = H'_{kk} - \sum_\lambda{}' \frac{|H'_{k\lambda}|^2}{E^0_\lambda - E^0_k} \qquad (11\text{-}107)$$

where we have also used the Hermitian property of $H'_{k\lambda}$. The prime on the summation symbol indicates that the term in which $\lambda = k$ should be omitted.

Next, let us find the coefficients a_λ in (103). They are obtained from (96) which now reads

$$\sum_\mu a_\mu (E^0_k \delta_{k\mu} + H'_{k\mu} - E\delta_{k\mu}) = 0, \quad k = 1, 2, \cdots$$

In accordance with the approximations which led to eq. (106) we put $E = E^0_1$ and neglect every $H'_{k\mu}$ unless one of the subscripts is 1. We then find

$$a_1 H'_{21} + a_2 (E^0_2 - E^0_1) = 0 \quad \text{if} \quad k = 2$$

$$a_1 H'_{31} + a_3 (E^0_3 - E^0_1) = 0 \quad \text{if} \quad k = 3, \text{ etc.}$$

Hence

$$a_\lambda = \frac{H'_{\lambda 1}}{E^0_1 - E^0_\lambda} a_1, \quad \lambda \neq 1$$

or in general, if we are interested not in E_1 but in E_k,

$$a_\lambda = \frac{H'_{\lambda k}}{E^0_k - E^0_\lambda} a_k, \quad \lambda \neq k \qquad (11\text{-}108)$$

The coefficient a_k must be chosen so that ϕ is normalized. Since all other a_λ are small, its value is very nearly unity and may be taken as such.

Formulas (107) and (108) have been derived by assuming that the level, k, whose perturbation is being calculated, was non-degenerate. For degenerate levels both formulas obviously fail, for they contain terms with vanishing denominators (several E^0_λ being equal to E^0_k). To deal with the case of degeneracy we have to return to the fundamental determinant (104). If the functions $u_1, u_2, \cdots, u_n$ all belong to the same energy E^0_1 (we then say that the level E^0_1 has an n-fold degeneracy), these functions are equally concerned in the perturbation, and if we formerly retained all matrix elements of the form $H'_{1\lambda}$, we must now retain $H'_{2\lambda}, H'_{3\lambda}, \cdots, H'_{n\lambda}$ also. But for most purposes sufficient accuracy results if we neglect all elements connecting a state of the degenerate group with all states not belonging to that group. Eq. (104) reduces in this case to

$$\begin{vmatrix} H'_{11} - \Delta E & H'_{12} & H'_{13} & \cdots & H'_{1n} \\ H'_{21} & H'_{22} - \Delta E & H'_{23} & \cdots & H'_{2n} \\ H'_{31} & H'_{32} & H'_{33} - \Delta E & \cdots & H'_{3n} \\ \cdots\cdots\cdots\cdots\cdots\cdots\cdots\cdots\cdots\cdots\cdots\cdots\cdots\cdots \\ H'_{n1} & H'_{n2} & H'_{n3} & \cdots & H'_{nn} - \Delta E \end{vmatrix} = 0 \quad (11\text{–}109)$$

the n roots of this equation (of which some may coincide) are the energies into which E_1^0 will " split " as the result of the perturbation. They cannot, of course, be represented by a general formula.

These energies are said to represent the first-order perturbation. If greater accuracy is desired the work may be continued in this way. By substituting the first-order energies into eqs. (96) and neglecting all states not belonging to the degenerate group, n sets of coefficients $a_1, a_2, \cdots, a_n$ are found, each set belonging to a single first-order energy. This yields n functions

$$v_i = \sum_{\lambda=1}^{n} a_{i\lambda} u_\lambda$$

If now we construct matrix elements with the v-functions, $H'_{ij} = \int v_i^* H v_j d\tau$, these will be *diagonal;* for solving (109) is the well-known procedure for diagonalizing the matrix $\boldsymbol{H'}$. (See Chapter 10.) Hence, when the v-functions are chosen to represent the n degenerate states, the second order perturbation can be computed by formula (107), from which the terms with vanishing denominator are now absent because every $H'_{k\lambda}$ corresponding to them is zero.

11.23. Example: Non-Degenerate Case. The Stark Effect.—Let H^0 represent the Hamiltonian operator for any one-electron system, and let $\psi_1, \psi_2,$ be its eigenfunctions. When a uniform electric field along X is applied, the term $H' = -eFx$ is added to H^0, e being the electronic charge and F the field strength. The normal state of the system is non-degenerate, hence formula (107) may be used. Denoting the normal state by the subscript zero, we find

$$\Delta E_0 = -eF x_{00} - e^2 F^2 \sum_\lambda{}' \frac{|x_{0\lambda}|^2}{E_\lambda^0 - E_0^0} \qquad (11\text{–}110)$$

Here $x_{0\lambda} = \int \psi_0^* x \psi_\lambda d\tau$. The first term on the right is usually zero because $|\psi_0|^2$ is an even function of x; thus the " first-order Stark effect " is absent.

In classical physics, the increment in energy of an atom due to a static electric field is expressed in terms of the polarizability α in the form

$$\Delta E = -\tfrac{1}{2}\alpha F^2$$

On comparing this with (110) we find for the polarizability of the normal state of our system

$$\alpha = 2e^2 \sum_\lambda{}' \frac{|x_{0\lambda}|^2}{E_\lambda^0 - E_0^0}$$

For an oscillator, this takes a particularly simple form, since all $x_{0\lambda}$ vanish with the exception of $x_{01} = \sqrt{1/2\beta}$ (cf. Chapter 3, eqs. 92 and 93). Also, $E_\lambda^0 = (\lambda + \frac{1}{2})h\nu$. Thus

$$\alpha = 2e^2 \frac{x_{01}^2}{h\nu}$$

Comparison with the problem of sec. 21 shows that second-order perturbation theory gives in this instance the same result as the variational method with the trial function $a_0\psi_0 + a_1\psi_1$. In general, however, the use of a simple variation function yields a much poorer result for the polarizability than the method of sec. 22.

11.24. Example: Degenerate Case. The Normal Zeeman Effect.— The energy states of the hydrogen atom were found to be

$$R_{n,l}(r)Y_l(\theta,\varphi)$$

To a given l, there belong $2l + 1$ spherical harmonics of the form

$$Y_l = \sum_{m=-l}^{l} c_m P_l^m (\cos\theta)e^{im\varphi}, \quad (P_l^{-m} = P_l^m)$$

and each such combination with its own set of coefficients c_m, forms a proper eigenfunction when multiplied by $R_{n,l}$. The energy does not depend on m; the state under consideration has therefore a $(2l + 1)$-fold degeneracy.

Let us choose the $2l + 1$ functions in the simplest possible way, namely by letting each Y_l contain only one term, as follows:

$$R_{n,l} \cdot c_{-l}P_l^l e^{-il\varphi}, \quad R_{n,l} \cdot c_{-l+1}P_l^{l-1}e^{-i(l-1)\varphi}, \quad \cdots, \quad R_{nl} \cdot c_l P_l^l e^{il\varphi}$$

and label them $\phi_1, \phi_2, \cdots, \phi_{2l+1}$, in that order.

The Zeeman effect is the splitting of the energy levels of an atom in a magnetic field. When a uniform magnetic field along the Z-axis and of strength F is applied to the hydrogen atom, its unperturbed Hamiltonian takes on the extra term[23]

$$H' = -\frac{i\hbar e}{2Mc} F \frac{\partial}{\partial \varphi} \equiv -iA \frac{\partial}{\partial \varphi}$$

Each matrix element $H'_{ij} = \int \phi_i^* H' \phi_j d\tau$ contains the factor $\int R_{n,l}^2 r^2 dr$

[23] See Van Vleck, J. H., " The Theory of Electric and Magnetic Susceptibilities," Oxford, 1932. We write here M for the electron mass to avoid conflict with the summation index m (magnetic quantum number). Note that H' is the quantum representation of $(e/2M_c)\mathbf{F} \cdot \mathbf{L}$, where $\mathbf{L}$ is the angular momentum vector of sec. 11.3.

which, by virtue of the normalization of the radial functions, is unity.   If
we form, e.g., H'_{12} we obtain an integral over θ times $\displaystyle\int_0^{2\pi} e^{il\varphi}\frac{\partial}{\partial\varphi}e^{-i(l-1)\varphi}d\varphi$,
and this vanishes   In the same way all other non-diagonal matrix elements are seen to be zero.   The diagonal element H'_{11} is

$$-iA\cdot(-il).\int \phi_1^*\phi_1 d\tau = -Al$$

and the others are similarly constructed.

When these elements are substituted into (109) we have

$$\begin{vmatrix} -lA-\Delta E & 0 & 0 & 0 \\ 0 & -(l-1)A-\Delta E & 0 & 0 \\ 0 & 0 & -(l-2)A-\Delta E & 0 \\ \multicolumn{4}{c}{\dotfill 0} \\ 0 & 0 & 0\cdots\cdots lA-\Delta E \end{vmatrix} = 0$$

The determinant is already diagonal, our choice of functions was a fortunate one.   The perturbed energies are clearly

$$\Delta E = mA = m\frac{\hbar eF}{2Mc}, \quad m = -l,\ -l+1,\ \cdots 0,\ 1\cdots l$$

Classically, an electron in a magnetic field F performs a uniform precession of angular frequency $\omega_L = eF/2Mc$, known as the *Larmor frequency*. Thus we see that $\Delta E = m\hbar\omega_L$.

Problem. Calculate the Stark effect of the rigid rotator (cf. sec. 11.12), for the state $l = 3$, adopting the same choice for the spherical harmonics as above.   Here $H' = -eaF\cos\theta$, provided the electric field F is along Z.   The determinant will not be diagonal.   To calculate the matrix elements, use formulas (3–48 and 53).   Include in your calculation successively more states: $l = 2,3,4$; $l = 1,2,3,4,5$.

TIME-DEPENDENT STATES. SCHRÖDINGER'S TIME EQUATION

11.25. General Considerations.—In all preceding considerations we have assumed that the states of the systems in question were stationary ones, that the time coordinate could be disregarded in describing them.   In generalizing the theory so as to make it applicable to states which change in time it is well to look back and see why a time-free description was possible thus far.

It is important to note that the time, t, in classical mechanics is canonically conjugate to the energy, E, in the same sense that x is conjugate to p_x.   Let us then for the moment consider the operator $P_x = -i\hbar(\partial/\partial x)$. Its eigenstates were seen to be (cf. eq. 9) $\psi_p = ce^{(i/\hbar)px}$.   What do they tell us about the distribution of the system in x?   The answer is, it is uniform.

Whatever is true at the point x_1, is also true at the point x_2. This is the meaning of the uncertainty principle applied to the case at hand: if the momentum is known with certainty, the state function is entirely non-committal with regard to x. If in the calculation of the mean value of an operator Q,

$$\bar{q} = \int \psi_p^* Q \psi_p d\tau$$

Q did not depend on x, we could have afforded to neglect the factor $e^{(i/\hbar)px}$ of ψ_p altogether. It had to be included, however, because most operators of interest do depend on x.

But this trivial situation existed with regard to the time coordinate in all the Schrödinger problems considered heretofore. The states were those in which the energy was known with certainty, and for this reason the state functions were completely indiscriminate in respect to t. What was true at t_1 was also true at t_2. Moreover, the other operators used were independent of t. This condition will always be present as long as we are dealing with *closed* systems, for the energy will then be constant in time.

When the system is an open one, the present method must clearly fail. But the last remarks contain the hint that we should, perhaps, associate with E the operator $-i\hbar(\partial/\partial t)$. This would lead to the eigenvalue equation

$$-i\hbar \frac{\partial}{\partial t} \psi = E\psi$$

which is certainly too simple because the energy depends on other things beside the time. The example above gives us no definite lead at this point because p_x does possess the single dependence on x. There is, however, only one reasonable way to include these other variables, namely, to put them into E, which thereby ceases to be an eigenvalue: E must be replaced by the Hamiltonian operator H. We then arrive at *Schrödinger's time equation*

$$-i\hbar \frac{\partial}{\partial t} \psi = H\psi \tag{11-111}$$

H is to be constructed as before by replacement of every Cartesian coordinate p_i by $-i\hbar(\partial/\partial q_i)$ and the dependence on t is to be introduced explicitly.

It is immaterial, of course, whether we choose eq. (111) or its complex conjugate equation. The latter choice has certain advantages and will here be made. Furthermore, we shall use the symbol u (more or less generally) for time-dependent state functions and thus record Schrödinger's time equation in the form

$$i\hbar \frac{\partial u(q_1 \cdots q_n, t)}{\partial t} = H\left(-i\hbar \frac{\partial}{\partial q_1}, \cdots - i\hbar \frac{\partial}{\partial q_n} \; ; \; q_1 \cdots q_n; \; t\right) u(q_1 \cdots q_n; \; t)$$

$$(11\text{-}112)$$

This equation, being of the first order in t, permits prediction of the state u at any future (or past) time when u is known as a function of the coordinates at present. Although it is closely related to the preceding developments, eq. (112) is a new postulate not derivable from those already given.

The present theory must be valid also in the special case when H does not contain t. When that is true eq. (112) is separable. On writing $u = \psi(q_1 \cdots q_n) \cdot f(t)$ it becomes equivalent to the equation

$$\frac{H\psi}{\psi} = i\hbar \frac{\frac{\partial f}{\partial t}}{f}$$

each side of which must represent a constant. But in view of the form of the left-hand side, that constant must be one of the eigenvalues of the operator H, say E_λ, so that

$$\frac{\partial f}{\partial t} = \frac{-iE_\lambda}{\hbar} f$$

Hence

$$f_\lambda = ce^{-(iE_\lambda/\hbar)t}$$

The general solution of eq. (112) for the special case in which H is independent of the time is

$$u = \sum_\lambda c_\lambda \psi_\lambda e^{-(iE_\lambda/\hbar)t} \qquad (11\text{-}113)$$

We have formerly said that any state function, such as u, could be expanded in the orthonormal system of functions ψ_λ. This expansion was written as

$$u = \sum_\lambda a_\lambda \psi_\lambda$$

We now see that this is indeed true even when the analysis is made on the basis of eq. (112), but the coefficients a_λ are always functions of the time: $a_\lambda = c_\lambda e^{-(iE_\lambda/\hbar)t}$. The mean value of E, computed for the state (113), is

$$\bar{E} = \sum_\lambda |c_\lambda|^2 E_\lambda = \sum_\lambda |a_\lambda|^2 E_\lambda$$

It is independent of t. But the probability of finding the system at the point $q_1 \cdots q_n$ of configuration space, $u^*u = \sum_{\lambda\mu} c_\lambda^* c_\mu \psi_\lambda^* \psi_\mu e^{(i/\hbar)(E_\lambda - E_\mu)t}$, is a superposition of oscillating functions of the time. The only way for this time dependence to be obliterated would be to have $c_\lambda = \delta_{\lambda j}$ in (113), in which case

$$u^*u = \psi_j^*\psi_j$$

Thus, whenever a state is formed by superposition of energy eigenstates, the mean energy of the system remains constant, but the configuration of the system changes in time. The reader should note, of course, that the solution of the Schrödinger equation (12) when multiplied by $e^{-(iE_\lambda/\hbar)t}$ is also a solution of (112), but that the solution of (112) does not in general satisfy (12).

Problem. Let the time-dependent Hamiltonian be $H = H_0 + V(t)$, where H_0 acts only on space coordinates and has eigenfunctions ψ_λ, eigenvalues E_λ. Show that

$$u = \sum_\lambda c_\lambda \psi_\lambda e^{-(i/\hbar)(E_\lambda t + \int V dt)}$$

11.26. The Free Particle; Wave Packets.—The eigenfunctions of the energy of a free mass point (cf. sec. 11.9) moving in one dimension without restriction are $\psi_k = e^{ikx}$, its energies $E_k = \dfrac{\hbar^2}{2m} k^2$, and there is no quantization. The general solution of eq. (112) for the free particle is therefore,

$$u = \int_{-\infty}^{\infty} c(k)e^{i[kx - (\hbar/2m)k^2 t]}\, dk \qquad (11\text{--}114)$$

a function constructed after the manner of (113) but with an integral instead of a sum. An integral very similar to this has been already encountered in the mathematical formulation of waves (cf. eq. 7–38) and of diffusion phenomena (eq. 7–53). It is interesting to inquire what form u will have at some time t if at $t = 0$ it is given by $u = u_0(x)$. The coefficient $c(k)$ may be determined by Fourier analysis. We have

$$u_0 = \int_{-\infty}^{\infty} c(k)e^{ikx}dk$$

whence by eq. 8–13

$$c(k) = \frac{1}{2\pi}\int_{-\infty}^{\infty} u_0(\xi)e^{-ik\xi}d\xi$$

Eq. (114) therefore reads

$$u(x,t) = \frac{1}{2\pi}\int\int_{-\infty}^{\infty} u_0(\xi)e^{i[k(x-\xi)-(\hbar/2m)k^2 t]}d\xi dk$$

In this instance, the integration over k cannot be performed (as it could in the diffusion problem, sec. 7.14). To proceed further it is necessary to introduce the function u_0 explicitly.

Assume that $u_0 = e^{-x^2/2a^2}$. Then, with the use of the formula

$$\int_{-\infty}^{\infty} e^{-\lambda x^2 + i\mu x}\,dx = \sqrt{\frac{\pi}{\lambda}}\, e^{-\mu^2/4\lambda} \tag{11-115}$$

we find

$$c(k) = \frac{a}{\sqrt{2\pi}}\, e^{-a^2 k^2/2}$$

Hence

$$u = \frac{a}{\sqrt{2\pi}} \int_{-\infty}^{\infty} e^{-k^2[a^2/2 + i(\hbar/2m)t] + ikx}\,dk$$

$$= \left(1 + i\frac{\hbar}{ma^2}t\right)^{-1/2} \exp - \left[\frac{x^2}{2\left(a^2 + \dfrac{i\hbar}{m}t\right)}\right] \tag{11-116}$$

again with the aid of (115).

Eq. (114) represents a superposition of *waves* of wave length $2\pi/k$ and frequency $\nu = (\hbar/4\pi m)k^2$. The form of u_0 here chosen describes a concentration of waves about the origin, a phenomenon called a " wave packet." Such a wave packet does not retain its spatial distribution; eq. (116) is characteristic of the manner in which it diffuses.

From the point of view of quantum mechanics, u_0^2 is the probability density of the particle at $t = 0$. It represents a Gauss error function of " width " a. At time t,

$$u^* u = \left[1 + \left(\frac{\hbar}{ma^2}t\right)^2\right]^{-1/2} \exp - \left[\frac{x^2}{a^2 + \dfrac{\hbar^2}{m^2 a^2}t^2}\right]$$

The probability density is still a Gauss function, but of smaller maximum and of width $[a^2 + (\hbar^2/m^2 a^2)t^2]^{1/2}$.

Problem a. Compute how long it would take an electron, localized within $a = 10^{-10}$ cm., to diffuse through twice that distance.

b. How long would it take an object weighing one gram, localized within 1 cm., to diffuse through twice that distance?

c. Show that if $u_0 = ce^{iKx}$, where K is a constant, the wave will be of the form $u = e^{iKx - (\hbar/2m)K^2 t}$.

If our particle is free to move in three dimensions, then as shown in sec. 11.9,

$$\psi_{\mathbf{k}} = e^{i\mathbf{k}\cdot\mathbf{r}}, \quad \text{and} \quad E_{\mathbf{k}} = \frac{\hbar^2}{2m}k^2$$

Hence (114) has the form

$$u = \int c(\mathbf{k})e^{i[\mathbf{k}\cdot\mathbf{r} - (\hbar/2m)k^2 t]}\,d\mathbf{k} \tag{11-117}$$

Again, if $u = u_0(x,y,z)$ at $t = 0$

$$u_0 = \int c(\mathbf{k})e^{i\mathbf{k}\cdot\mathbf{r}}d\mathbf{k}$$

whence by 3-dimensional Fourier analysis

$$c(\mathbf{k}) = \frac{1}{8\pi^3}\int u_0(\xi,\eta,\zeta)e^{-i\mathbf{k}\cdot\rho}d\rho$$

the vector ρ having components ξ, η, ζ.

Assume now, in analogy with the one-dimensional case, that

$$u_0 = e^{-r^2/2a^2}$$

At $t = 0$ the wave packet is a spherical concentration of waves centered about the origin; the probability packet has a similar shape and a width a. On inserting u_0 into the relation for c we have

$$c(\mathbf{k}) = \frac{1}{8\pi^3}\int_{-\infty}^{\infty} e^{-(\xi^2/2a^2)-ik_1\xi}d\xi \int e^{-(\eta^2/2a^2)-ik_2\eta}d\eta \int e^{-(\zeta^2/2a^2)-ik_3\zeta}d\zeta$$

$$= \left(\frac{a}{\sqrt{2\pi}}\right)^3 e^{-(a^2/2)k^2}$$

This gives

$$u = (2\pi)^{-3/2}a^3 \int e^{-[(a^2/2)+i(\hbar^2/2m)t]k_1^2+ik_1x}dk_1$$

times two similar integrals with k_1 replaced by k_2 and k_3, x by y and z. Hence

$$u = \left(1 + \frac{i\hbar}{ma^2}t\right)^{-3/2} \exp -\left[\frac{r^2}{2\left(a^2 + i\frac{\hbar}{m}t\right)}\right]$$

The interpretation of this result is not different from that of (116).

Before leaving the subject of " particle waves," we should remark that every component wave of the packet (117), being of the form $e^{i(\mathbf{k}\cdot\mathbf{r}-2\pi\nu t)}$, travels in a positive direction along $\mathbf{k}$. Had we chosen the sign as in eq. (111) and not as in (112), the waves would have been of the form $e^{i(\mathbf{k}\cdot\mathbf{r}+2\pi\nu t)}$, which implies that they travel along $-\mathbf{k}$. Since $k\hbar$ represents the momentum of the particle, the latter choice is an unsuitable one. We also note that the wave length $\lambda = 2\pi/k = 2\pi\hbar/mv = h/mv$ conforms to the *De Broglie* formula. The *phase* velocity of the waves is $\nu\lambda = \hbar k/2m = mv/2m = v/2$, but their group velocity,[24] defined as $2\pi(d\nu/dk) = v$, is equal to the classical speed of the particle.

[24] For a discussion of group velocity, see Sommerfeld, A., " Wellenmechanischer Ergänzungsband," Friedr. Vieweg & Sohn, Braunschweig, 1929, p. 46.

11.27. Equation of Continuity, Current.—If the state function changes in time in accordance with the Schrödinger equation

$$Hu = i\hbar \dot{u} \tag{11-118}$$

will it remain normalized? If it does not, there occurs a destruction or creation of probability; while initially there was certainty of finding the particle somewhere in space, there might later be uncertainty, a situation which would clearly be physically untenable. Permanence of normalization, however, follows immediately from (118). For

$$\frac{\partial}{\partial t} \int u^* u d\tau = \int [\dot{u}^* u + u^* \dot{u}] d\tau = \frac{i}{\hbar} \int [uH^* u^* - u^* Hu] d\tau$$

because of (118), and the last expression is zero on account of the Hermitian character of H.

Having shown that $u^* u$ is conserved we can define a probability *current* by subjecting $u^* u$, which we will call ρ for the moment, to the equation of continuity

$$\frac{\partial \rho}{\partial t} + \nabla \cdot \mathbf{I} = 0 \tag{11-119}$$

Whatever $\mathbf{I}$ turns out to be must be regarded as the current corresponding to the " flow " of the quantity $u^* u$. We shall limit our consideration to the case of a *single* particle so that

$$H = -\frac{\hbar^2}{2m} \nabla^2 + V(x,y,z)$$

although generalization to many-dimensioned configuration space is easy. Again because of (118)

$$\frac{\partial \rho}{\partial t} = \dot{u}^* u + u^* \dot{u} = \frac{i}{\hbar} (uH^* u^* - u^* Hu)$$

$$= \frac{i\hbar}{2m} (u^* \nabla^2 u - u \nabla^2 u^*) = \nabla \cdot \left[\frac{i\hbar}{2m} (u^* \nabla u - u \nabla u^*) \right]$$

To satisfy (119) we must put[25]

$$\mathbf{I} = -\frac{i\hbar}{2m} (u^* \nabla u - u \nabla u^*) \tag{11-120}$$

It is interesting to observe that a state u which has no *complex dependence* on a space variable has no current associated with it. Thus, in the

[25] This form of I is correct so long as the potential energy V is of the scalar form here used. When H contains a vector potential, $\mathbf{A}$, the term $(e/c)\mathbf{A}$ must be added to the expression for the current here given.

free particle problem, $\cos kx$ and $\sin kx$ represent stationary states, but e^{ikx} and e^{-ikx} have currents.

Problem. Compute I for the various regions of the barrier problems considered in sec. 11.10.

11.28. Application of Schrödinger's Time Equation. Simple Radiation Theory.—The cases in which eq. (118) can be solved exactly are not numerous and not very interesting. When the time equation (118) is not separable, resort must be taken to approximation methods, the most useful of which will now be illustrated.

Let an atom, whose normal Hamiltonian function, free from all perturbations, is H_0, be suddenly subjected to a light wave which adds a perturbing energy

$$V(x,t) = -eF_0 x \sin \omega t \qquad (11\text{-}121)$$

to H. Physically, this means the light wave is monochromatic and has frequency $\nu = \omega/2\pi$; its electric vector is along X and of amplitude F_0.[26] If V did not contain x and $\sin \omega t$ in product form, eq. (118) with $H = H_0 + V$ would be separable; the fusion of x and t into V spoils separability.

In solving (118) we use the following initial condition: At $t = 0$, when the atom was exposed to the perturbation V, the atom was certainly in an eigenstate of the operator H_0, say in the state ψ_1 corresponding to the energy E_1 which we shall take to be the lowest energy of the system. Or, if we wish to include the trivial time dependence of the state, we take

$$u = \psi_1 e^{-(iE_1/\hbar)t} \qquad (11\text{-}122)$$

The solution of

$$(H_0 + V)v = i\hbar\dot{v} \qquad (11\text{-}123)$$

which we desire, is certainly available in the form

$$v = \sum_{\lambda} c_\lambda \psi_\lambda e^{-(iE_\lambda/\hbar)t} \qquad (11\text{-}124)$$

---

[26] Eq. (121) is a valid approximation for the purpose at hand. It neglects the energy due to the magnetic vector of the light wave whose contribution is small compared to (121) in the ratio v/c, where v is the velocity of the charge composing the atom and c the velocity of light. For hydrogen, v/c is $1/137$. Furthermore, eq. (121) implies that the wave length of the light is large compared with the size of the atom. Correctly, $V = -eF_0 x \sin\left(\omega t - \dfrac{2\pi z}{\lambda}\right)$, and we are omitting the term z/λ. The legitimacy of this will be clear from the following analysis.

provided we let the coefficients c be functions of the time.   This follows immediately from the completeness of the ψ_λ with respect to functions of the space coordinates.   When (124) is substituted into (123), there results

$$\sum_\lambda c_\lambda (H_0\psi_\lambda + V\psi_\lambda)e^{-(iE_\lambda/\hbar)t} = \sum_\lambda (c_\lambda E_\lambda\psi_\lambda + i\hbar\dot{c}_\lambda\psi_\lambda)e^{-iE_\lambda t/\hbar}$$

wherein each term $H_0\psi_\lambda$ on the left cancels $E_\lambda\psi_\lambda$ on the right.   Let us now multiply the remaining terms of the equation by ψ_k^* and integrate over configuration space, remembering the orthogonality of the ψ_λ.   Then, after simple rearrangement,

$$\dot{c}_k = -\frac{i}{\hbar}\sum_\lambda c_\lambda V_{k\lambda}e^{i[(E_k-E_\lambda)/\hbar]t}, \quad k = 1, 2, 3, \cdots \quad (11\text{-}125)$$

where, as usual,

$$V_{k\lambda} = \int \psi_k^* V \psi_\lambda d\tau$$

If the unperturbed atom has an infinite number of states, (125) represents an infinite set of linear differential equations, which in general can not be solved.   But we now recall that at $t = 0$, $v = u$; which means that all c_k except c_1 were zero at that time.   Thereafter c_1 decayed from 1 to some smaller value, while all other c's grew from 0 to various finite values. We now limit our inquiry to times so small that c_1 is still sensibly unity, and the other c's are small compared with it, although $\dot{c}_1$ may be quite comparable with the time derivatives of other c's.   This permits the approximation of replacing every c_λ on the right-hand side of (125) by its value at $t = 0$, while retaining every $\dot{c}_k$.   The equation then beomes

$$\dot{c}_k = -\frac{i}{\hbar} V_{k1}e^{(i/\hbar)(E_k-E_1)t}$$

To simplify writing we introduce the abbreviation

$$\frac{E_k - E_1}{\hbar} \equiv \omega_k$$

and observe that every $\omega_k > 0$, since, as we are assuming, E_1 is the lowest energy state.   In view of (121),

$$V_{k1} = -eF_0 x_{k1} \sin \omega t = \tfrac{1}{2}ieF_0 x_{k1}(e^{i\omega t} - e^{-i\omega t})$$

so that

$$\dot{c}_k = \frac{eF_0}{2\hbar} x_{k1}\left[e^{i(\omega_k+\omega)t} - e^{i(\omega_k-\omega)t} \right]$$

On integration,

$$c_k = \frac{ieF_0}{2\hbar} x_{k1} \left[\frac{e^{i(\omega_k - \omega)t} - 1}{\omega_k - \omega} - \frac{e^{i(\omega_k + \omega)t} - 1}{\omega_k + \omega} \right], \quad k \neq 1$$

where we have at once adjusted the constant of integration so that $c_k = 0$ when $t = 0$. For physical reasons, only the first term in the square parenthesis need be retained because it alone can attain appreciable magnitude. (Both ω and $\omega_k > 0$.) In fact c_k is large only when $\omega \approx \omega_k$, and this fact is accentuated when c_k is squared:

$$|c_k|^2 = \frac{e^2 F_0^2}{2\hbar^2} |x_{k1}|^2 \frac{1 - \cos(\omega_k - \omega)t}{(\omega_k - \omega)^2} = \frac{e^2 F_0^2}{4\hbar^2} |x_{k1}|^2 \frac{\sin^2\left[\left(\frac{\omega_k - \omega}{2}\right)t\right]}{\left(\frac{\omega_k - \omega}{2}\right)^2}$$

(11-126)

We now interpret this result. The coefficient c_k is, in view of (124), the k-th probability amplitude in the expansion of the state function v at time t in terms of energy eigenstates of the normal atom. Hence because of sec. 5, $|c_k|^2$ is the probability that at time t the k-th energy level of the atom be excited; it is the "transition probability" from state 1 to state k when the atom has been exposed to monochromatic light of frequency $\omega/2\pi$ for t seconds.

Many interesting conclusions of a physical nature can be drawn from eq. (126), of which only two will here be mentioned. First, the transition probability is proportional to the square of the matrix element connecting the states in question. Whenever x_{1k} vanishes, $|c_k|^2 = 0$. Hence the vanishing of x_{1k} is the criterion of a "forbidden" transition. In the second place, the transition probability is small unless $\omega \approx \omega_k$, which is the Bohr frequency condition.

Problem. The reader may be surprised to find that $|c_k|^2$ is not a linear function of t, as might be expected on physical grounds. Show that, when the incident light forms a *continuous* spectrum of uniform intensity, $|c_k|^2$ is proportional to t. (For this purpose, (126) must be integrated over ω from 0 to ∞; but the integration may without appreciable error be taken from $-\infty$ to $+\infty$.)

ELECTRON SPIN. PAULI THEORY

11.29. Fundamentals of the Theory.—The theory so far developed describes the general behavior of atomic and molecular systems surprisingly well, but it makes some false predictions, particularly with regard to the finer details of the energy states of atoms, the Zeeman effect, and the magnetic properties of electrons. It was soon apparent that the state of a single electron could not be represented as a function of three space coordinates alone, but that another parameter was required whose interpretation was for some time in doubt.   Most decisive in clarifying the situation was the spectroscopic observation of the doubling of the energy levels of a single electron: In all alkali atoms, for instance, two levels are found where the Schrödinger equation permits only one.   The energy difference between these levels was such as would be produced by a small magnet of magnetic moment $\hbar e/2mc$ setting itself once parallel and then opposite to the magnetic field present in the atom on account of the electron's revolution.   Also, the angular momentum corresponding to these two energy states was known to be different; it was equal to that caused by the electron's orbital motion, plus $\hbar/2$ in one, minus $\hbar/2$ in the other state.

Uhlenbeck and Goudsmit suggested that the electron behaves like a spinning top having a "spin" angular momentum of magnitude $\hbar/2$ which, however, can only add or subtract its whole amount, in quantum fashion, to any angular momentum the electron already possesses as a result of its orbital motion.   Correspondingly, the electron generates by its spin a magnetic moment of magnitude $\hbar e/2mc$ (m is the electron mass, c the velocity of light), and this also communicates itself in *toto*, either parallel or in opposition, to any magnetic moment already present.

To describe the electron spin as an angular momentum of the usual kind and to associate with it an operator like L (eq. 44) proved a fruitless undertaking, chiefly because L would have more than two eigenstates.   The most successful procedure of including the spin in the quantum mechanical formalism, aside from Dirac's relativistic treatment of the electron, is that of Pauli which will now be described.   What follows will refer only to the spin states of a *single* electron; some applications to several electrons may be found in secs. 34 and 35.

Since the three space coordinates are insufficient to specify the complete state of an electron, we introduce a fourth, the "spin coordinate," and denote it by s_z.   It corresponds, in classical language, to the cosine of the angle between the axis of the spin angular momentum and the Z-axis of coordinates.   This visual interpretation, while in no way dictated by the mathematical formalism, will be found a useful mental aid.   Thus the state function of an electron has the form

$$\phi(x,y,z,s_z)$$

Since in all that follows, the hypothetical spin coordinates s_x and s_y are never needed, we shall henceforth delete the subscript z on s, but retain the above interpretation.   Hence $\phi = \phi(x,y,z,s)$.   Finally, it is well for the moment to abstract attention entirely from the space dependent part of the wave function, i.e., to consider x, y, z as fixed, concentrating our inquiry solely upon the electron spin.   Then $\phi = \phi(s)$.

If s, like x, y and z, were permitted to assume a continuous range of values, difficulties would result. Pauli therefore postulates—in a manner admittedly *ad hoc* and designed to force success of the theory—that the range of s consists of only two points: $s = \pm 1$ (classical meaning: spin vector is parallel or in opposition to Z). A function of s is therefore defined only at these two points. The most general spin function is, accordingly,

$$\phi(s) = a\delta_{s,+1} + b\delta_{s,-1} \tag{11-127}$$

where the δ's are Kronecker symbols.

Our postulates involved certain integrals over configuration space. But an integral over configuration space consisting of two points vanishes. It becomes necessary to redefine the integral as a summation over the two points:

$$\int F(s)ds \equiv F(-1) + F(1)$$

If $\phi(s)$ is to be normalized,

$$\int (|a|^2\delta_{s,+1}^2 + |b|^2\delta_{s,-1}^2 + (a^*b + b^*a)\delta_{s,+1}\delta_{s,-1})ds$$

$$= |a|^2 + |b|^2 = 1 \tag{11-128}$$

In a very trivial sense, eq. (127) represents an expansion of a function $\phi(s)$ in a complete orthonormal set of functions, $\delta_{s,+1}$ and $\delta_{s,-1}$. To what operator do these two functions belong as eigenstates? The answer is suggested by intuition and will be justified by its complete success; it is the operator S_z which is associated with the observable: spin angular momentum along Z. We must now give thought to the mathematical structure of this operator.

Empirical evidence cited in the introductory paragraphs demands that its two eigenvalues be $\pm\hbar/2$. Hence it must satisfy the two equations

$$S_z\delta_{s,+1} = \frac{\hbar}{2}\delta_{s,+1}$$

$$S_z\delta_{s,-1} = -\frac{\hbar}{2}\delta_{s,-1} \tag{11-129}$$

It is possible to show that no differential operator of the type encountered previously can satisfy these equations without giving rise to an infinite number of other eigenstates. But why search for the operator? The simplest point of view, and that here taken, is to regard eqs. (129) as a *definition* of the operator S_z.[27]   The result of applying S_z to the most general function of s (eq. 126), can be constructed on the basis of (129), hence (129) exhausts the meaning of S_z and is its definition.

To simplify the notation, and to be in accord with custom, we now introduce the symbol $\alpha(s)$ for $\delta_{s,+1}$, and $\beta(s)$ for $\delta_{s,-1}$. Furthermore, we define a new operator

$$\sigma_z = \frac{2}{\hbar} S_z$$

which has eigenvalues ± 1, for the simple expedient to save writing.   Then, in view of (129),

$$\sigma_z \alpha(s) = \alpha(s), \quad \sigma_z \beta(s) = -\beta(s) \tag{11-130}$$

It is indeed possible and often useful to find an explicit operator in form of a *matrix* which will satisfy these equations.   This matrix is easily formed by means of the principles outlined in sec. 17.   Our eigenstates are $\psi_1 = \alpha$, $\psi_2 = \beta$, and we construct $(\sigma_z)_{ij} = \int \psi_i^* \sigma_z \psi_j d\tau$ with the integral replaced by a summation.   We thus obtain the two-square matrix

$$\sigma_z = \begin{pmatrix} 1 & 0 \\ 0 & -1 \end{pmatrix} \tag{11-131}$$

To let it operate on what was formerly the function $\phi(s)$ the latter has to be regarded as a vector whose components are its expansion coefficients: If the *function* ϕ is given by

$$\phi(s) = a\alpha + b\beta$$

a and b being numbers, then the *vector* $\phi(s)$ is

$$\phi = \begin{pmatrix} a \\ b \end{pmatrix}$$

Thus, in the matrix representation,

$$\sigma_z \phi \equiv \begin{pmatrix} 1 & 0 \\ 0 & -1 \end{pmatrix} \begin{pmatrix} a \\ b \end{pmatrix} \tag{11-132}$$

and the reader will easily verify by the rules of Chapter 10 that the two *eigenvectors* of σ_z are $\phi = \begin{pmatrix} a \\ 0 \end{pmatrix}$ and $\phi = \begin{pmatrix} 0 \\ b \end{pmatrix}$, where the values of both a

---

[27] An operator P is in general uniquely determined when the result of its action upon each member of an orthonormal set of functions is known.   This method of defining an operator is ordinarily not useful because an infinite number of relations like (129) would be required.

and b must be unity because of (128). The *eigenvalues* are, respectively, $+1$ and -1. But the *functions* ϕ corresponding to the vectors $\begin{pmatrix} 1 \\ 0 \end{pmatrix}$ and $\begin{pmatrix} 0 \\ 1 \end{pmatrix}$ are clearly α and β, which takes us back to the scheme (130).

It is seen that there is a complete isomorphism between the two descriptions of the operator S_z and its eigenstates ϕ: One in terms of matrices and eigenvectors, where the rule of operations is (132); the other in terms of *linear substitution operators* and eigenfunctions, where the rule of operations is (130).

The question now arises as to the structure of the operators S_x and S_y, associated with the other two components of the spin.[28]   In endeavoring to construct them it is important to recall one significant fact concerning the ordinary angular momentum **L**: its components do not commute with one another.   In fact (see eq. 7)

$$L_x L_y - L_y L_x = i\hbar L_z, \quad L_y L_z - L_z L_y = i\hbar L_x$$

$$L_z L_x - L_x L_z = i\hbar L_y$$

Let us assume that the components of the spin **S**, this being an angular momentum operator, must be subject to the same commutation rules.   In terms of σ rather than **S**, we postulate

$$\sigma_x \sigma_y - \sigma_y \sigma_x = 2i\sigma_z; \quad \sigma_y \sigma_z - \sigma_z \sigma_y = 2i\sigma_x; \quad \sigma_z \sigma_x - \sigma_x \sigma_z = 2i\sigma_y \quad (11\text{--}133)$$

These relations imply that an eigenstate of S_z, e.g., $\alpha(s)$ or $\beta(s)$, cannot be a simultaneous eigenstate of S_x or S_y (sec. 7).

The construction of σ_x and σ_y, σ_z being given, is more easily performed in the matrix scheme.   If we set ourselves the problem of determining two matrices σ_x and σ_y, which, when combined with σ_z of eq. (131), obey (133), we easily find that the answer is not unique.   But certainly the solution

$$\sigma_x = \begin{pmatrix} 0 & 1 \\ 1 & 0 \end{pmatrix} \quad \sigma_y = \begin{pmatrix} 0 & -i \\ i & 0 \end{pmatrix} \quad (11\text{--}134)$$

is a possible one.   The ambiguity here encountered permits just enough freedom to make possible a rotation of coordinate axes (see Chap. 15).

Let us, then, accept (134) as our solution in matrix form.   Clearly, σ_x has eigenvalues ± 1, eigenvectors $\sqrt{\frac{1}{2}} \begin{pmatrix} 1 \\ 1 \end{pmatrix}$ and $\sqrt{\frac{1}{2}} \begin{pmatrix} 1 \\ -1 \end{pmatrix}$; σ_y has eigenvalues ± 1, eigenvectors $\sqrt{\frac{1}{2}} \begin{pmatrix} 1 \\ i \end{pmatrix}$ and $\sqrt{\frac{1}{2}} \begin{pmatrix} 1 \\ -i \end{pmatrix}$.   The observable values

---

[28] While we need only one spin *coordinate*, s_z, all three components of the *operator* must be introduced because they appear in the Hamiltonian and other operators.

of *all three* components S_x, S_y and S_z are therefore $\pm\hbar/2$. When these results are translated into the function language they read as follows.

The equation $\sigma_x\phi(s) = \lambda\phi(s)$ has two possible (normalized) solutions:

$$\left.\begin{array}{ll} \lambda = 1, & \phi(s) = \sqrt{\tfrac{1}{2}}[\alpha(s) + \beta(s)] \\ \lambda = -1, & \phi(s) = \sqrt{\tfrac{1}{2}}[\alpha(s) - \beta(s)] \end{array}\right\} \tag{a}$$

The equation $\sigma_y\phi(s) = \lambda\phi(s)$ has two possible solutions:

$$\left.\begin{array}{ll} \lambda = 1, & \phi(s) = \sqrt{\tfrac{1}{2}}[\alpha(s) + i\beta(s)] \\ \lambda = -1, & \phi(s) = \sqrt{\tfrac{1}{2}}[\alpha(s) - i\beta(s)] \end{array}\right\} \text{(b)} \quad (11\text{–}135)$$

The equation $\sigma_z\phi(s) = \lambda\phi(s)$ has two possible solutions:

$$\left.\begin{array}{ll} \lambda = 1, & \phi(s) = \alpha(s) \\ \lambda = -1, & \phi(s) = \beta(s) \end{array}\right\} \tag{c}$$

If now we write the eqs. (135a) in the simpler form

$$\sigma_x\alpha + \sigma_x\beta = \alpha + \beta, \quad \sigma_x\alpha - \sigma_x\beta = -(\alpha - \beta)$$

and solve these by adding and subtracting, we find

$$\sigma_x\alpha = \beta, \quad \sigma_x\beta = \alpha$$

The same procedure applied to (135b) and (135c) yields similar relations. Summarizing these results: The operators σ_x, σ_y, σ_z may be represented either by the set of linear substitutions

$$\sigma_x\alpha = \beta, \quad \sigma_y\alpha = i\beta, \quad \sigma_z\alpha = \alpha, \tag{11–136}$$
$$\sigma_x\beta = \alpha; \quad \sigma_y\beta = -i\alpha; \quad \sigma_z\beta = -\beta$$

or by the matrices

$$\sigma_x = \begin{pmatrix} 0 & 1 \\ 1 & 0 \end{pmatrix}, \quad \sigma_y = \begin{pmatrix} 0 & -i \\ i & 0 \end{pmatrix}, \quad \sigma_z = \begin{pmatrix} 1 & 0 \\ 0 & -1 \end{pmatrix} \tag{11–137}$$

For practical use, the set of substitutions is to be preferred.

Note that the operators

$$\sigma^+ \equiv \tfrac{1}{2}(\sigma_x + i\sigma_y)$$

and
$$\sigma^- \equiv \tfrac{1}{2}(\sigma_x - i\sigma_y)$$

satisfy the convenient relations

$$\sigma^+\alpha = 0 \qquad \sigma^-\alpha = \beta$$
$$\sigma^+\beta = \alpha \qquad \sigma^-\beta = 0$$

They are sometimes called " displacement operators."

We return to the consideration of the general state function of an electron, which includes x, y, z and s as arguments. Such a function may certainly be expanded in eigenfunctions of σ_z, i.e.,

$$\phi(x,y,z,s) = \phi_+(x,y,z)\alpha(s) + \phi_-(x,y,z)\beta(s)$$

Normalization now requires

$$\int \phi^*\phi d\tau \equiv \sum_s \int \phi^*\phi dx dy dz = \int (\phi_+^*\phi_+ + \phi_-^*\phi_-)dx dy dz = 1$$

The operators σ_x, σ_y, σ_z do not act on ϕ_+ and ϕ_- which are only functions of x, y, z; in other words, they commute with space coordinates. Thus, for instance,

$$\sigma_y\phi(x,y,z,s) = \sigma_y\phi_+\alpha + \sigma_y\phi_-\beta = \phi_+\sigma_y\alpha + \phi_-\sigma_y\beta = i\phi_+\beta - i\phi_-\alpha$$

In the matrix scheme, $\phi(x,y,z,s)$ is represented by the vector

$$\phi = \begin{pmatrix} \phi_+(x,y,z) \\ \phi_-(x,y,z) \end{pmatrix}$$

In the sense of this analysis it may be said that the introduction of the spin in the Pauli manner causes all Schrödinger functions to become two-component functions.

Problem. Carry out the algebra involved in finding the two Hermitian matrices (134).

11.30. Applications.—a. *Atom in a Magnetic Field.* Our interest here is not in a complete solution of this problem, which may be found worked out in most books on quantum mechanics, but in its salient mathematic features. We wish to find the energies of a one-electron atom (e.g., hydrogen or, with good approximation, the alkalis) when it is placed in a uniform magnetic field. The Hamiltonian consists of two parts, one acting on the electron's space coordinates and one acting on the spin coordinate. The former will be called H_0; the latter is the " spin energy." If the magnetic field $\mathcal{H}$ is taken along the Z-axis, the classical energy of a particle of magnetic moment μ would be $\mu \cdot \mathcal{H} = \mu_z\mathcal{H}_z$. But empirically, the magnetic moment associated with the spin is $(\hbar e/2mc)\sigma$. We shall here write μ for the constant $\hbar e/2mc$. In quantum mechanical transcription, then, the " spin energy " is $\mu\mathcal{H}_z\sigma_z$ where σ_z is interpreted as the operator (130) or (131). The Schrödinger equation becomes

$$(H_0 + \mu\mathcal{H}_z\sigma_z)\Psi = E\Psi \tag{11-138}$$

Let

$$\Psi(x,y,z,s) = \psi_+(x,y,z)\alpha(s) + \psi_-(x,y,z)\beta(s)$$

and substitute, obtaining

$$\alpha(s)[H_0 + \mu\mathcal{H}_z - E]\psi_+ + \beta(s)[H_0 - \mu\mathcal{H}_z - E]\psi_- = 0$$

provided relations (136) are used. Since α and β are linearly independent, orthogonal functions of s, their coefficients in the last equation must separately vanish.[29] Hence we have

$$H_0\psi_+ = (E - \mu\mathcal{H}_z)\psi_+, \quad H_0\psi_- = (E + \mu\mathcal{H}_z)\psi_- \quad (11\text{-}139)$$

Now let E_0 be an eigenvalue of H_0, ψ_0 the corresponding eigenfunction. The first of eqs. (139) (which is nothing more than an eigenvalue equation for the operator H_0) then says $E - \mu\mathcal{H}_z = E_0$, or $E = E_0 + \mu\mathcal{H}_z$, $\psi_+ = \psi_0$. On substituting this value of E into the second equation it reads $H_0\psi_- = (E_0 + 2\mu\mathcal{H}_z)\psi_-$, and this can only be satisfied by putting $\psi_- = 0$ because $E_0 + 2\mu\mathcal{H}_z$ is not an eigenvalue of H_0. Thus we obtain as one solution of (138)

$$E = E_0 + \mu\mathcal{H}_z, \quad \Psi = \psi_0(x,y,z)\alpha(s) \quad (11\text{-}140a)$$

But we can also start with the second of eqs. (139) and assume ψ_- to be ψ_0, $E + \mu\mathcal{H}_z$ to be E_0. Then $\psi_+ = 0$ and we have

$$E = E_0 - \mu\mathcal{H}_z, \quad \Psi = \psi_0(x,y,z)\beta(s) \quad (11\text{-}140b)$$

How does the inclusion of the spin modify the eigenvalues and eigenfunctions of the Schrödinger equation when there is no magnetic field? The answer is obtained by letting $\mathcal{H}_z$ vanish in (140a, b). Both values of E coalesce to E_0 which now represents the ordinary Schrödinger energy in the absence of a field, but the functions Ψ remain distinct. The spin thus introduces a *degeneracy* into the Schrödinger representation of states. Formulas (140) account—in a primitive way—for the doubling of the alkali energy levels, the field $\mathcal{H}_z$ being caused in that case by the electron's orbital motion, and not by external agencies.

Problem. Solve eq. (138) by the method of separation of variables, i.e., by putting $\Psi = \psi(x,y,z)\phi(s)$, and show that (140) is the solution obtained by that method also.

b. *A Spin Problem.* Having shown how spin and coordinate functions cooperate in the description of the state of an electron, let us omit further reference to space coordinates and inquire what are the energies which an electron, placed in a uniform magnetic field of arbitrary direction, may assume regardless of its translational motion. The only energy of interest is that due to the spin. Let $\mathcal{H}$ be the magnetic field strength. The Schrödinger equation reads

$$\mu\mathcal{H}\cdot\sigma\psi = \mu(\mathcal{H}_x\sigma_x + \mathcal{H}_y\sigma_y + \mathcal{H}_z\sigma_z)\psi(s) = E\psi(s) \quad (11\text{-}141)$$

If $\mathcal{H}$ is taken along Z, the equation reduces to

$$\mu\mathcal{H}\sigma_z\psi(s) = E\psi(s) \quad (11\text{-}142)$$

---

[29] This can be seen explicitly if the equation is multiplied by either $\alpha(s)$ or $\beta(s)$ and then " integrated " over s.

The operator on the left is but a constant multiple of σ_z and must therefore have the same eigenfunctions as σ_z, i.e., α and β. The corresponding eigenvalues are at once seen to be $E = \pm\mu\mathcal{H}$. We shall show that eq. (141) has the same eigenvalues, but different eigenfunctions.

Make the substitution $\psi = a\alpha(s) + b\beta(s)$ in eq. (141). On using, subsequently, relations (136) the result will be

$$\mu\{\mathcal{H}_x(a\beta + b\alpha) - i\mathcal{H}_y(a\beta - b\alpha) + \mathcal{H}_z(a\alpha - b\beta)\} - E(a\alpha + b\beta) = 0$$

As before, the coefficients of α and β may be put equal to zero separately, so that

$$\left.\begin{aligned}
\mu(\mathcal{H}_x a - i\mathcal{H}_y a - \mathcal{H}_z b) &= Eb \\
\mu(\mathcal{H}_x b + i\mathcal{H}_y b + \mathcal{H}_z a) &= Ea
\end{aligned}\right\} \tag{11-143}$$

If the equations are to have solutions a, b, which are different from zero, the determinant of the coefficients of a, b must vanish, whence $E = \pm\mu\mathcal{H}$. On substituting $E = +\mu\mathcal{H}$ into the first of eqs. (143) and then taking the square of its absolute value, we have

$$(\mathcal{H}_x^2 + \mathcal{H}_y^2)|\,a\,|^2 = (\mathcal{H} + \mathcal{H}_z)^2|\,b\,|^2$$

Let us call the angle between $\mathcal{H}$ and the Z-axis, θ, so that $\mathcal{H}_x^2 + \mathcal{H}_y^2 = \mathcal{H}^2 \sin^2 \theta$, and $\mathcal{H}_z = \mathcal{H} \cos \theta$. Furthermore, in view of (128), $|\,b\,|^2 = 1 - |\,a\,|^2$. When these substitutions are made and the last equation is solved, the squares of the absolute values of a, b are found to be $\cos^2 \theta/2$ and $\sin^2 \theta/2$, respectively. Let us then put $a = \cos \theta/2$, $b = e^{i\delta} \sin \theta/2$, treating δ as a phase constant. With the further substitutions $\mathcal{H}_x = H \sin \theta \cos \phi$, $\mathcal{H}y = H \sin \theta \sin \phi$, where ϕ is the azimuth of the field, we find from (143) that $\delta = -\phi$.

In a similar way, when $E = -\mu H$, $\delta = \pi - \phi$, $a = \sin \theta/2$, $b = -e^{-i\phi} \cos \theta/2$.

We conclude that eq. (141) has the eigenvalues $E_1 = \mu\mathcal{H}$, $E_2 = -\mu\mathcal{H}$, and the corresponding eigenfunctions

$$\left.\begin{aligned}
\psi_1 &= \cos \frac{\theta}{2} \cdot \alpha(s) + \sin \frac{\theta}{2} e^{-i\phi}\beta(s) \\
\psi_2 &= \sin \frac{\theta}{2} \cdot \alpha(s) - \cos \frac{\theta}{2} e^{-i\phi}\beta(s)
\end{aligned}\right\} \tag{11-144}$$

Notice that, when the field $\mathcal{H}$ is reversed in direction (i.e., $\theta \to \pi - \theta$, $\phi \to \phi + \pi$), ψ_1 and ψ_2 exchange their roles.

Problem. Solve eq. (141) by diagonalizing the matrix

$$\mathcal{H}_x\sigma_x + \mathcal{H}_y\sigma_y + \mathcal{H}_z\sigma_z = \begin{pmatrix} \mathcal{H}_z & \mathcal{H}_x - i\mathcal{H}_y \\ \mathcal{H}_x + i\mathcal{H}_y & -\mathcal{H}_z \end{pmatrix}$$

and show that it leads to the same results.

THE MANY–BODY PROBLEM AND THE EXCLUSION PRINCIPLE

11.31. Separation of the Coordinates of the Center of Mass.—In classical mechanics, a system containing many particles and subject only to internal forces behaves in such a way that its center of mass moves uniformly on a straight line. As a corollary of this theorem every classical two-body problem may be reduced to a one-body problem.[30] A similar fact may be proved in quantum theory.

The Schrödinger equation for a system of n particles of masses $m_1, \cdots, m_n$ reads:

$$\left(-\sum_1^n \frac{\hbar^2}{2m_i} \nabla_i^2 + V \right)\psi = E\psi \tag{11-145}$$

where $\nabla_i^2 = \partial^2/\partial x_i^2 + \partial^2/\partial y_i^2 + \partial^2/\partial z_i^2$. The potential energy, V, is to be regarded as a function of the *relative* coordinates $x_j - x_i$, $y_j - y_i$, $z_j - z_i$. We first transform to a new set of coordinates, defined as follows:

$$X = \frac{1}{M}\sum_1^n m_i x_i, \quad M = \sum_1^n m_i$$
$$x_2' = x_2 - X, \quad x_3' = x_3 - X, \cdots \quad x_n' = x_n - X \tag{11-146}$$

with similar relations for the y and z components. Note that x_1' is missing; the coordinates of one particle have been eliminated by the introduction of the center of mass coordinates X, Y, Z. In computing the sum of the Laplacian operators occurring in (145) in terms of the new coordinates we observe:

$$\frac{\partial X}{\partial x_i} = \frac{\partial Y}{\partial y_i} = \frac{\partial Z}{\partial z_i} = \frac{m_i}{M} ; \quad \frac{\partial x_j'}{\partial x_i} = \frac{\partial y_j'}{\partial y_i} = \frac{\partial z_j'}{\partial z_i} = \delta_{ij} - \frac{m_i}{M}$$

Using these relations, simple differentiation yields

$$\frac{\partial^2\psi}{\partial x_1^2} = \frac{m_1^2}{M^2}\left(\frac{\partial^2\psi}{\partial X^2} - 2\sum_{i=2}^n \frac{\partial^2\psi}{\partial X \partial x_i'} + \sum_{i,j=2}^n \frac{\partial^2\psi}{\partial x_i'\partial x_j'} \right)$$

$$\frac{\partial^2\psi}{\partial x_i^2} = \frac{m_i^2}{M^2}\left(\frac{\partial^2\psi}{\partial X^2} - 2\sum_2^n \frac{\partial^2\psi}{\partial X \partial x_j'} + \sum_{j,k=2}^n \frac{\partial^2\psi}{\partial x_j'\partial x_k'} \right)$$

$$+ 2\frac{m_i}{M}\left(\frac{\partial^2\psi}{\partial X \partial x_i'} - \sum_{j=2}^n \frac{\partial^2\psi}{\partial x_i'\partial x_j'} \right) + \frac{\partial^2\psi}{\partial x_i'^2}$$

[30] So long as relativity effects are neglected.

and similar expressions for the derivatives with respect to y and z. When these are combined we obtain, in place of (145), the equation

$$\left\{ -\frac{\hbar^2}{2M}\nabla^2 - \sum_2^n \frac{\hbar^2}{2m_i}\nabla'^2_i + \frac{\hbar^2}{2M}\sum_{i,j=2}^n \left(\frac{\partial^2}{\partial x'_i \partial x'_j} + \frac{\partial^2}{\partial y'_i \partial y'_j} + \frac{\partial^2}{\partial z'_i \partial z'_j} \right) + V \right\}\psi$$
$$= E\psi \qquad (11\text{--}147)$$

Here ∇^2 is the Laplacian with respect to the center of mass coordinates, ∇'^2_i with respect to the primed coordinates. While V is not directly a function of the primed coordinates, it may be expressed in terms of them because $x_j - x_i = x'_j - x'_i$. A difficulty might seem to appear in connection with $x_i - x_1$ because x'_1 is absent from the primed set. But it is easily seen that $m_1 x'_1 = -\sum_2^n m_i x'_i$, whence $x_i - x_1 = x'_i + \frac{1}{m_i}\sum_j^n m_j x'_j$. Therefore V, when expressed in terms of the new coordinates, will not contain X, Y, or Z.

As a result, eq. (147) is separable; therefore ψ may be written as $\Psi(X,Y,Z) \cdot \phi(x'_2 \cdots z'_n)$.

Correspondingly, $E = E_c + E'$, where E_c is the energy associated with $\Psi(X,Y,Z)$, determined by

$$-\frac{\hbar^2}{2M}\nabla^2\Psi = E_c\Psi$$

This is the Schrödinger equation of a free particle of mass M, it produces, as we know, no quantization. The remainder of (147) describes the internal motion of the particles:

$$\left(-\sum_2^n \frac{\hbar^2}{2m_i}\nabla'^2_i + \frac{\hbar^2}{2M}\sum_{i,j=2}^n \nabla'_i \cdot \nabla'_j + V \right)\phi = E'\phi \qquad (11\text{--}148)$$

It differs from the normal form of Schrödinger's equation by the presence of the terms in $\nabla'_i \cdot \nabla'_j$ and by the fact that V has a different functional form in the primed coordinates than in the unprimed ones.

The coordinates (146) measure the position of the i-th particle relative to the center of mass. It is also possible to use a less symmetrical but physically more useful set of coordinates, which is closely related to (146). If we put

$$X = \frac{1}{M}\sum_1^n m_i x_i, \quad M = \sum_1^n m_i$$
$$x'_2 = x_2 - x_1, \quad x'_3 = x_3 - x_1, \cdots, \quad x'_n = x_n - x_1 \qquad (11\text{--}149)$$

thus measuring all coordinates relative to that one which has been eliminated (x_1), we obtain in the same manner the equation

$$\left\{ -\frac{\hbar^2}{2M}\nabla^2 - \sum_2^n \frac{\hbar^2}{2m_i}\nabla'^2_i - \frac{\hbar^2}{2m_1}\sum_{i,j=2}^n \nabla'_i \cdot \nabla'_j + V \right\}\psi = E\psi \qquad (11\text{--}150)$$

This form is particularly useful when it is desired to calculate the energy of a many-electron atom, for particle 1 may then be taken to be the nucleus and the summations in (150) are extended only over the electrons. The equation remaining after separation of the motion of the center of mass is now

$$\left\{ -\frac{\hbar^2}{2}\left(\sum_i \frac{1}{m}\nabla_i'^2 + \frac{1}{m_1}\sum_{i,j}\nabla_i' \cdot \nabla_j' \right) + V \right\} \phi = E'\phi$$

where m is the mass of an electron, m_1, that of the nucleus. It may be written in terms of the *reduced mass*

$$\mu = \frac{mm_1}{m + m_1}$$

as follows:

$$\left\{ -\frac{\hbar^2}{2\mu}\sum_i \nabla_i'^2 - \frac{\hbar^2}{2m_1}\sum_{i\neq j}\nabla_i' \cdot \nabla_j' + V \right\} \phi = E'\phi \qquad (11\text{-}151)$$

The terms in the double summation play an important role in the isotope effect of heavy atoms.[31] They are present whenever the number of electrons is greater than one. For the case of hydrogen, eq. (151) has the same form as Schrödinger's equation for a *stationary* nucleus, except for the replacement of the electron mass by μ. Hence the true energies of the hydrogen atom are not exactly given by eq. (64), but by that equation with μ written for m.

Note that the function V is different in (148) and (151), and that the terms of the double summation have opposite signs. Nevertheless the equivalence of these two equations for the two-body problem may be seen as follows. Write for the potential energy in (151)

$$V = V(x',y',z'), \quad \text{where } x' = x_2 - x_1, \quad \text{etc.}$$

The V-function of (148) must then be expressed in terms $x_2 - X, y_2 - Y, z_2 - Z$. Now $x_2 - x_1 = \dfrac{(m_1 + m_2)}{m_1}(x_2 - X)$. Therefore we must use in (148)

$$V = V\left(\frac{m_1 + m_2}{m_1}x', \ \frac{m_1 + m_2}{m_1}y', \ \frac{m_1 + m_2}{m_1}z' \right)$$

and the equation reads

$$\left[-\frac{\hbar^2}{2}\left(\frac{1}{m_2} - \frac{1}{m_1 + m_2} \right)\nabla'^2 + V(\alpha x', \alpha y', \alpha z') \right]\psi(x',y',z') = E'\psi(x',y',z')$$

[31] See Hughes, A. L., and Eckart, C., *Phys. Rev.* **36**, 694 (1930).

where $\alpha = (m_1 + m_2)/m_1$. If here we put $\alpha x' = x''$, $\alpha y' = y''$, $\alpha z' = z''$, it becomes

$$\left[-\frac{\hbar^2}{2} \alpha^2 \left(\frac{1}{m_2} - \frac{1}{m_1 + m_2} \right) \nabla''^2 + V(x'', y'', z'') \right] \psi = E' \psi$$

which is identical with eq. (151).

11.32. Independent Systems.—Physical systems are independent, or isolated from one another, if the Hamiltonian operator of one contains no terms referring to another system. There is then no interaction between them. Consider n independent systems, and let the coordinates of the r-th system (including the spin coordinate) be symbolized by the single letter q_r. If its Hamiltonian operator is H_r, its Schrödinger equation will be

$$H_r \psi_i^{(r)}(q_r) = E_i^{(r)} \psi_i^{(r)}(q_r) \tag{11-152}$$

$E_i^{(r)}$ being the i-th eigenvalue of the r-th system.

The state function describing the entire assemblage of n systems will satisfy the equation

$$(H_1 + H_2 + \cdots H_n)\Psi(q_1, q_2, \cdots q_n) = E\Psi(q_1, q_2, \cdots q_n) \tag{11-153}$$

To find its solutions we put $\Psi(q_1, q_2, \cdots q_n) = \psi^{(1)}(q_1)\psi^{(2)}(q_2) \cdots \psi^{(n)}(q_n)$ tentatively. Substitution in (153) and use of the fact that H_1 acts only on q_1, etc., leads at once to the equation

$$\frac{H_1 \psi^{(1)}}{\psi^{(1)}} + \frac{H_2 \psi^{(2)}}{\psi^{(2)}} + \cdots \frac{H_n \psi^{(n)}}{\psi^{(n)}} = E$$

which shows that each term $H_r \psi^{(r)}/\psi^{(r)}$ is separately a constant, say $E^{(r)}$, and that the sum of all these constants is E. But if $H_r \psi^{(r)}/\psi^{(r)} = E^{(r)}$, then $\psi^{(r)}$ must be one of the set of functions defined by (152), and $E^{(r)}$ one of the energies $E_i^{(r)}$. Therefore

$$\Psi(q_1, q_2, \cdots q_n) = \psi_i^{(1)}(q_1) \cdot \psi_j^{(2)}(q_2) \cdots \psi_s^{(n)}(q_n)$$
$$E = E_i^{(1)} + E_j^{(2)} + \cdots E_s^{(n)} \tag{11-154}$$

This result is indeed what intuition would lead us to expect. For clearly the total energy of a number of isolated systems is the sum of the individual energies. Furthermore, if w_1 is the probability that system 1 be found at q_1, w_2 that system 2 be found at q_2, then the probability that both of these statements be true simultaneously is the product $w_1 w_2$. Hence the individual ψ-functions, whose squares are these probabilities, must likewise combine as factors.

This latter circumstance is dictated also by the time dependence of the Schrödinger states eq. (113). For only the product of the individual functions $\psi^{(1)}e^{-(i/\hbar)E^{(1)}t}$, $\psi^{(2)}e^{-(i/\hbar)E^{(2)}t}$, etc., will have the factor $e^{-(i/\hbar)\sum_r E^{(r)}\cdot t}$ required in $\Psi(q_1, q_2, \cdots q_n)e^{-(i/\hbar)Et}$.

11.33. The Exclusion Principle.—When two independent systems occupy the energy states $E_i^{(1)}$ and $E_j^{(2)}$ respectively, the combined system has an energy

$$E = E_i^{(1)} + E_j^{(2)}$$

and a state function

$$\Psi = \psi_i^{(1)}(q_1) \cdot \psi_j^{(2)}(q_2) \tag{11-155}$$

We shall suppose for the moment that the individual states $\psi_i^{(1)}$ and $\psi_j^{(2)}$ are non-degenerate. Then, unless there happen to be two energies $E_l^{(1)}$ and $E_k^{(2)}$ whose sum is precisely the same as $E_i^{(1)} + E_j^{(2)}$, the combined state (155) will also be non-degenerate. This will generally be the case when the two systems are *different* in a physical sense.

But if they are similar, e.g., both electrons, or both hydrogen atoms, another situation arises. We may then drop all superscripts in the description of the states, and write (155)

$$E = E_i + E_j, \quad \Psi = \psi_i(q_1) \cdot \psi_j(q_2) \tag{11-156}$$

This state is degenerate, although ψ_i and ψ_j are not; for if we interchange the indices i and j, or what is the same, interchange the coordinates q_1 and q_2 in Ψ, there results a different Ψ-function but not a different energy. This degeneracy, which is peculiar to the description of any aggregate of similar systems, is known as *exchange degeneracy*. Classically it implies that the energy of the total system is unaltered when two individual constituents *exchange places and spins*.

In the more general case where E_i has g_i and E_j has g_j linearly independent functions associated with it, the number of Ψ's corresponding to E will be, not $g_i g_j$, but $2g_i g_j$.

Returning to the case of non-degeneracy of ψ_i and ψ_j we note that the two functions

$$\Psi_I = \psi_i(q_1)\psi_j(q_2), \quad \Psi_{II} = \psi_j(q_1)\psi_i(q_2)$$

which are linearly independent, are equally good representatives of the state in which $E = E_i + E_j$. Moreover, any linear combination of the two satisfies the Schrödinger equation for this value of E, and has just claim to be considered. Of course, only two such combinations can be linearly independent. Let us then consider the function

$$a\Psi_I + b\Psi_{II}$$

where we shall assume $|a|^2 + |b|^2 = 1$ to assure normalization. On " exchanging " the two systems, $\Psi_I \to \Psi_{II}$ and $\Psi_{II} \to \Psi_I$, hence the function above transforms itself into

$$b\Psi_I + a\Psi_{II}$$

the numerical value of which for any given configuration (q_1, q_2) will in general be different from $a\Psi_I + b\Psi_{II}$. Physically, this implies that the configuration which results when the two systems exchange places has an altogether different probability than the original, a consequence that is clearly objectionable.

However, among all linear combinations there are two which avoid this dilemma. They are the symmetric[32] combination

$$\Psi_S(q_1, q_2) = \sqrt{\tfrac{1}{2}}(\Psi_I + \Psi_{II})$$

and the " antisymmetric " one

$$\Psi_A(q_1, q_2) = \sqrt{\tfrac{1}{2}}(\Psi_I - \Psi_{II})$$

They are independent and indeed orthogonal; the first remains unaltered on exchange of systems, the second changes its sign. Both, therefore, yield probabilities $\left|\psi\right|^2$ which are insensitive to exchange.

Consider now, not two, but n independent similar systems, in states $\psi_i, \psi_j, \cdots \psi_s$. The assemblage has the energy $E = E_i + E_j + \cdots E_s$, and is described by the state function

$$\Psi(q_1, q_2, \cdots q_n) = \psi_i(q_1)\psi_j(q_2) \cdots \psi_s(q_n) \tag{11-157}$$

But every permutation of the q's among the ψ's on the right will produce a new function belonging to the same E, provided the subscripts, $i, j, \cdots s$ are all different (which we shall assume for the moment). Hence, if P represents any one of the $n!$ possible permutations of the q's and $\Psi_P(q_1, q_2, \cdots q_n)$ the function which results from (157) when this permutation is made, then

$$\Psi(q_1, q_2, \cdots q_n) = \sum_P a_P \Psi_P \tag{11-158}$$

where the a_P are arbitrary constants, one for each permutation (arbitrary except for the normalization condition), represents an acceptable state function for the energy E. Since there were originally $n!$ linearly independent functions, there will also be $n!$ linearly independent combinations of the type (158).

Fortunately, most of these are uninteresting, for they cause

$$\left|\Psi(q_1, q_2, \cdots q_n)\right|^2$$

---

[32] A function is said to be *symmetric* with respect to a given operation if the operation leaves it unchanged; it is said (in quantum mechanics) to be *antisymmetric* if the operation changes its sign without altering it in any other way.

to change when an exchange is made among any of the q's. There are certainly two combinations, however, which preserve probabilities on exchange. One is the symmetrical, the other the antisymmetrical combination. The symmetrical one is formed by making all the coefficients a_P in (158) equal:

$$\Psi_S (q_1,q_2,\cdots q_n) = (n!)^{-1/2}\sum_P \Psi_P \qquad (11\text{-}159)$$

the antisymmetric one by giving opposite signs to *even* and *odd* permutations (cf. Chapter 15):

$$\psi_A (q_1,q_2,\cdots q_n) = (n!)^{-1/2}\sum_P (-1)^P \psi_P \qquad (11\text{-}160)$$

A practical way of constructing (160) is to write the determinant

$$\Psi_A = (n!)^{-1/2} \begin{vmatrix} \psi_i(q_1)\psi_i(q_2)\psi_i(q_3)\cdots\psi_i(q_n) \\ \psi_j(q_1)\psi_j(q_2)\psi_j(q_3)\cdots\psi_j(q_n) \\ \cdots\cdots\cdots\cdots\cdots\cdots \\ \cdots\cdots\cdots\cdots\cdots\cdots \\ \psi_s(q_1)\psi_s(q_2)\psi_s(q_3)\cdots\psi_s(q_n) \end{vmatrix} \qquad (11\text{-}160')$$

which the reader will easily recognize as equivalent to the expansion (160).

It is to these two functions, Ψ_S and Ψ_A, that we must confine our attention. Lest the simplicity of our formalism obscure significant details, we recall that q_r stands for *all* coordinates of the r-th system. Thus, if the systems were electrons, $\psi_j(q_r)$ would be an abbreviation for a combination of space and spin functions:

$$\varphi_{j+}(x_r,y_r,z_r)\alpha(s_r) + \varphi_{j-}(x_r,y_r,z_r)\beta(s_r)$$

in the notation of sec. 29, and an interchange of q_r and q_p means that x_r is to be exchanged against x_p, y_r against y_p, z_r against z_p and s_r against s_p.

There is no *a priori* way of deciding which of the two functions, (159) or (160), is preferable. But here the exclusion principle, early recognized by Pauli, creates simplicity in a most effective way. It states that *if the individual systems belong to a certain class* (see below), *only antisymmetric functions may be used in describing the assemblage.* This principle is of the nature of a postulate; it has not yet been deduced from more fundamental axioms, although one might hope, from a mathematical point of view, that this will prove possible.[33] Why nature insists upon antisymmetric states for some and symmetric states for others among its creatures is at present a puzzle.

The elementary systems to which Pauli's principle is known to apply

---

[33] A very searching and interesting examination of the principle in the light of other fundamental issues has been given by Pauli, *Phys. Rev.* **58**, 716 (1940).

are: electrons, positrons, protons, neutrons, neutrinos and mu-mesons; photons, on the other hand, and several kinds of meson, are described by symmetrical state functions.

Perhaps the most important consequence of the exclusion principle is this. Suppose our assemblage consists of electrons, two of which are described by the same function ψ_i (i.e., the functions are identical with respect to positional and spin factors). The determinant $(160')$ will then have two equal rows, and hence will vanish. We may therefore say: *two systems obeying the Pauli principle cannot be in the same state.* This fact governs the structure of atoms and molecules; each electron added to the shell of an atom must have its own set of quantum numbers.

The exclusion principle makes it impossible to distinguish two states which differ only by an interchange of two constituent systems, a fact which has already been noted.

Photons, which are described by the symmetrical function (159), may exist in identical states, because that function does not vanish when two sets of indices like i and j, contained in Ψ_P become equal.

11.34. Excited States of the Helium Atom.—To show how the Pauli principle is applied we treat some of the excited states of the helium atom. The latter is to be regarded as a simple assemblage of 2 electrons moving in the Coulomb field of the nucleus (and under their mutual repulsion), hence the considerations of the foregoing section apply. However, in the first part of our treatment we shall ignore both the electron spin and the exclusion principle.

The Schrödinger equation has already been given (eq. 90); it is

$$\left(H_1 + H_2 + \frac{e^2}{r_{12}} \right) \Psi = E\Psi \qquad (11\text{--}161)$$

where

$$H_i = -\frac{\hbar^2}{2m} \nabla_i^2 - \frac{2e^2}{r_i}$$

If the term e^2/r_{12} were absent the two electrons would be independent, and Ψ would be a product of the form $\psi_i(q_1) \cdot \psi_j(q_2)$, E being $E_i + E_j$. Moreover ψ_i and ψ_j would be hydrogen eigenfunctions with atomic number $Z = 2$, for H_1 and H_2 are Hamiltonian operators for a single electron in a Coulomb field. To retain the notation of sec. 19 we shall now write u for the individual electron functions, so that, in the absence of the interaction term,

$$\Psi = u_i(x_1 y_1 z_1) u_j(x_2 y_2 z_2) \qquad (11\text{--}162)$$

Functions of this type will be used as *variation* functions with the complete Hamiltonian (161). Let us first give thought to the proper choice of the individual functions u.   The state corresponding to the lowest energy of a single electron is (cf. eq. 67a)

$$u_{10} = \left(\frac{2^3}{\pi a_0^3}\right)^{1/2} e^{-2r/a_0} \tag{11-163}$$

We are writing here, in place of the single subscript i, the values of the two quantum numbers $n = 1$ and $l = 0$.   The first excited state is either

$$u_{20} = R_{20} Y_0(\theta,\varphi)$$

or

$$u_{21} = R_{21} Y_1(\theta,\varphi)$$

The spherical harmonic Y_0 is a constant, but Y_1 is any linear combination of the three functions $P_1^1(\cos\theta)e^{i\varphi}$, $P_1^0(\cos\theta)$ and $P_1^1(\cos\theta)e^{-i\varphi}$.   It will be convenient to choose the following normalized combinations

$$Y_x = \sqrt{\frac{3}{16\pi}}[P_1^1(\cos\theta)e^{i\varphi} + P_1^1(\cos\theta)e^{-i\varphi}] = \sqrt{\frac{3}{4\pi}}\sin\theta\cos\varphi = \sqrt{\frac{3}{4\pi}}\frac{x}{r}$$

$$Y_y = -i\sqrt{\frac{3}{16\pi}}[P_1^1(\cos\theta)e^{i\varphi} - P_1^1(\cos\theta)e^{-i\varphi}] = \sqrt{\frac{3}{4\pi}}\sin\theta\sin\varphi$$

$$= \sqrt{\frac{3}{4\pi}}\frac{y}{r}$$

$$Y_z = \sqrt{\frac{3}{4\pi}}P_1^0(\cos\theta) = \sqrt{\frac{3}{4\pi}}\cos\theta = \sqrt{\frac{3}{4\pi}}\frac{z}{r}$$

and to define[34]

$$\left.\begin{aligned} u_{20} &= R_{20}Y_0 \\ u_{2x} &= R_{21}Y_x \\ u_{2y} &= R_{21}Y_y \\ u_{2z} &= R_{21}Y_z \end{aligned}\right\} \tag{11-164}$$

as the four independent, orthonormal functions describing the first excited state of the one-electron system.   The product (162) can be formed by combining u_{10} with any one of the four functions (164); furthermore, the arguments can be interchanged in each of the functions thus constructed. We are therefore concerned with the following eight functions, each of which is a solution of eq. (161) with the term e^2/r_{12} deleted, and belongs to the energy

$$E_0 = -\frac{2e^2}{a_0}(1 + \tfrac{1}{4}) = 5E_H \tag{11-165}$$

[34] R_{21} is given in eq. (65); its explicit form will not be needed here.

$$\psi_1 = u_{10}(1)u_{20}(2) \qquad\qquad \psi_2 = u_{20}(1)u_{10}(2)$$
$$\psi_3 = u_{10}(1)u_{2x}(2) \qquad\qquad \psi_4 = u_{2x}(1)u_{10}(2)$$
$$\psi_5 = u_{10}(1)u_{2y}(2) \qquad\qquad \psi_6 = u_{2y}(1)u_{10}(2) \qquad (11\text{-}166)$$
$$\psi_7 = u_{10}(1)u_{2z}(2) \qquad\qquad \psi_8 = u_{2z}(1)u_{10}(2)$$

In writing them we have indicated the arguments $(x_1y_1z_1)$ and $(x_2y_2z_2)$ simply by (1) and (2). A combination of these functions

$$\Phi = \sum_{\lambda=1}^{8} a_\lambda \psi_\lambda$$

will be used as a variation function in the sense of sec. 20. The best energies of the system are given by (97), and this reduces at once to the form (104) because the ψ_λ are orthonormal and belong to the operator $H^0 = H_1 + H_2$. The perturbing term is

$$H' = \frac{e^2}{r_{12}}$$

The next step in the solution of our problem is the calculation of the matrix elements $\int \psi_i^* H' \psi_j dx_1 dy_1 dz_1 dx_2 dy_2 dz_2$ using the functions (166), the details of which may be left for the reader.[35] Symmetry arguments may be used to show that

$$H_{11}' = H_{22}', \quad H_{33}' = H_{44}', \quad H_{55}' = H_{66}', \quad H_{77}' = H_{88}'$$

and that only functions in the same line of (166) give non-vanishing elements. Furthermore the volume element adopted in the evaluation of I (sec. 19) is convenient in proving:

$$H_{33}' = H_{55}' = H_{77}'; \quad H_{34}' = H_{56}' = H_{78}'$$

Since the ψ_λ are real, $H_{ij}' = H_{ji}'$. We are left, therefore, only with the following matrix elements:

$$H_{11}' = \int u_{10}^2(1)u_{20}^2(2)\, \frac{e^2}{r_{12}}\, d\tau \equiv J$$

$$H_{12}' = \int u_{10}(1)u_{20}(2)\, \frac{e^2}{r_{12}}\, u_{20}(1)u_{10}(2)d\tau \equiv K$$

$$H_{33}' = \int u_{10}^2(1)u_{2x}^2(2)\, \frac{e^2}{r_{12}}\, d\tau \equiv J'$$

$$H_{34}' = \int u_{10}(1)u_{2x}(2)\, \frac{e^2}{r_{12}}\, u_{2x}(1)u_{10}(2) \equiv K'$$

---

[35] See Heisenberg, W., *Z. Phys.* **39**, 499 (1926).

In a sense previously defined, (see sec. 11.21) J and J' are Coulomb integrals, K and K' exchange integrals.

The determinantal eq. (97) becomes

$$\begin{vmatrix} J - \epsilon & K & 0 & 0 \\ K & J - \epsilon & 0 & 0 \\ 0 & 0 & J' - \epsilon & K' \\ 0 & 0 & K' & J' - \epsilon \\ & & & & J' - \epsilon & K' & 0 & 0 \\ & & & & K' & J' - \epsilon & 0 & 0 \\ & & & & 0 & 0 & J' - \epsilon & K' \\ & & & & 0 & 0 & K' & J' - \epsilon \end{vmatrix} = 0 \quad (11\text{-}167)$$

provided we write ϵ for $E - E_0$. All elements not written are zeros. The determinant has two *single* roots: $\epsilon_1 = J - K$, $\epsilon_2 = J + K'$ and two *triple* roots: $\epsilon_3 = J' - K'$, $\epsilon_4 = J' + K'$. The perturbation e^2/r_{12} may

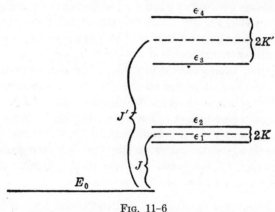

Fig. 11-6

therefore be said to change the one unperturbed level E_0 into four perturbed levels: $E_0 + \epsilon_1$, $E_0 + \epsilon_2$, $E_0 + \epsilon_3$, $E_0 + \epsilon_4$, as indicated qualitatively in the diagram (Fig. 6).

To find the functions corresponding to the eight roots ϵ we must return to equations (96):

$$a_1(J - \epsilon) + a_2 K = 0$$

$$a_1 K + a_2(J - \epsilon) = 0$$

$$a_3(J' - \epsilon) + a_4 K' = 0$$

$$a_3 K' + a_4(J' - \epsilon) = 0 \quad \text{etc.}$$

On substituting ϵ_1 for ϵ we find $a_2 = -a_1$, $a_3 = a_4 = \cdots = a_8 = 0$. On substituting $\epsilon = \epsilon_2$, we find $a_2 = a_1$, $a_3 = a_4 = \cdots = a_8 = 0$, and so forth.

We thus obtain the set of energies and normalized variation functions given in the first two columns of Table 1.

TABLE 1

E	ψ		
$E_0 + J - K$	$\sqrt{\tfrac{1}{2}}(\psi_1 - \psi_2)$	a	Triplet
$E_0 + J + K$	$\sqrt{\tfrac{1}{2}}(\psi_1 + \psi_2)$	s	Singlet
$E_0 + J' - K'$	$\sqrt{\tfrac{1}{2}}(\psi_3 - \psi_4)$	a	Triplet
	$\sqrt{\tfrac{1}{2}}(\psi_5 - \psi_6)$	a	Triplet
	$\sqrt{\tfrac{1}{2}}(\psi_7 - \psi_8)$	a	Triplet
$E_0 + J' + K'$	$\sqrt{\tfrac{1}{2}}(\psi_3 + \psi_4)$	s	Singlet
	$\sqrt{\tfrac{1}{2}}(\psi_5 + \psi_6)$	s	Singlet
	$\sqrt{\tfrac{1}{2}}(\psi_7 + \psi_8)$	s	Singlet

It now becomes necessary to include the spin into our analysis. To do this accurately would require a modification of the Hamiltonian operator (161), for the magnetic moments of the spinning electrons produce an interaction with the magnetic field due to their orbital motions and this interaction has not been included in (161). We shall omit this spin-orbit interaction and refer the reader to the literature for the more accurate treatment.[36] In other words, we shall suppose that the Hamiltonian does not act on the spin coordinates. The state function is then separable and appears as the product of an *orbital* (any of the functions in the table) and a spin function, and the latter may be taken as an eigenfunction of σ_z for each electron. Let us consider these spin functions more closely. For the two electrons, we have four functions:

$$\alpha(s_1)\alpha(s_2), \quad \alpha(s_1)\beta(s_2), \quad \beta(s_1)\alpha(s_2), \quad \text{and} \quad \beta(s_1)\beta(s_2)$$

These, however, do not have convenient exchange properties, for when s_1 and s_2 are interchanged, the first and last remain unaltered, the second transforms into the third and the third into the second. But it is possible to construct from the second and third two other, equivalent functions, which are symmetrical and antisymmetrical with respect to an exchange of spin coordinates. They are, when normalized, $\sqrt{\tfrac{1}{2}}[\alpha(s_1)\beta(s_2) + \beta(s_1)\alpha(s_2)]$ and $\sqrt{\tfrac{1}{2}}[\alpha(s_1)\beta(s_2) - \beta(s_1)\alpha(s_2)]$. We have in this way obtained four spin functions

$$\Sigma_1 = \alpha(s_1)\alpha(s_2), \quad \Sigma_2 = \sqrt{\tfrac{1}{2}}[\alpha(s_1)\beta(s_2) + \beta(s_1)\alpha(s_2)], \quad \Sigma_3 = \beta(s_1)\beta(s_2);$$
$$A = \sqrt{\tfrac{1}{2}}[\alpha(s_1)\beta(s_2) - \beta(s_1)\alpha(s_2)] \tag{11-168}$$

[36] Condon, E. U., and Shortley, G. H., " The Theory of Atomic Spectra," Macmillan Co., New York, 1935.

the first three of which are symmetrical, only the last being antisymmetrical. Furthermore, this set of functions is orthogonal (and complete).

To include the spin we need only multiply each one of the functions in Table 1 by one of the spin functions Σ_1 to A, a procedure which yields 32 different functions of position and spin coordinates. But here the exclusion principle effects a great simplification. It says that only functions which are antisymmetrical when *all* coordinates, i.e., position and spin coordinates, of the two electrons are interchanged, are to be permitted. Hence a function of Table 1 which is symmetrical can only be combined with A, and a function which is antisymmetrical only with Σ_1, Σ_2 and Σ_3.

Now the functions marked a in the table are antisymmetric; they can be multiplied by any one of the three Σ-functions. Each of them corresponds, therefore, to *three* states. For this reason the energy states $E_0 + J' - K'$ and $E_0 + J - K$ are said to be *triplet* states. If spin-orbit interaction had been included in our calculation each of these levels would have appeared as three closely adjacent levels, while the other energies, marked singlets, would have remained single.

It is true that the functions in Table 1 are only approximate solutions of eq. (161). Nevertheless what we have said about their symmetry with respect to exchange of electrons may be shown to hold rigorously. The structure of the helium energy spectrum, and in particular the singlet-triplet character of the states, are therefore correctly given by the simple theory of this section; the numerical values of the energy levels will be in error.

The normal state of the helium atom, whose energy was computed approximately in sec. 19 of this chapter, is given in the present notation by $u_{10}(1)u_{10}(2)$, if we neglect the spin. It is clearly symmetrical and can only be multiplied by A when the spins are introduced. Hence it is a singlet state. When the helium atom is in a singlet state, its probability of passing into a triplet state under emission or absorption of radiation is very small, as may be shown by an extension of the methods used in sec. 11.28. Hence triplet and singlet levels do not " combine," and helium may be said to have two distinct spectra, the triplet spectrum to which spectroscopists apply the term " orthohelium " spectrum, and the singlet spectrum called " parhelium " spectrum.

Problem a. Instead of using the 8 functions (166) as linear variation functions, start with the 32 functions obtained from (166) by multiplying each of them by Σ_1, Σ_2, Σ_3, A. Show that, if these 32 functions are suitably arranged, the determinantal equation is a four-fold repetition of the one obtained above, and that it yields the same results in regard to both energies and functions.

Problem b. The following spin operators for two electrons may be defined:

$$\sigma_z = \sigma_{z1} + \sigma_{z2}$$
$$\sigma^2 = (\sigma_1 + \sigma_2)^2 = \sigma_{x1}^2 + \sigma_{y1}^2 + \sigma_{z1}^2 + \sigma_{x2}^2 + \sigma_{y2}^2 + \sigma_{z2}^2 + 2(\sigma_{x1}\sigma_{x2} + \sigma_{y1}\sigma_{y2} + \sigma_{z1}\sigma_{z2})$$

where σ_{x1} is the operator σ_x acting on spin coordinate s_1, etc. Show that Σ_1, Σ_2, Σ_3 and A are all eigenstates with respect to both of these operators, in particular that

$$\sigma_z\Sigma_1 = 2\Sigma_1, \quad \sigma_z\Sigma_2 = 0, \quad \sigma_z\Sigma_3 = -2\Sigma_3, \quad \sigma_z A = 0$$
$$\sigma^2\Sigma_1 = 8\Sigma_1, \quad \sigma^2\Sigma_2 = 8\Sigma_2, \quad \sigma^2\Sigma_3 = 8\Sigma_3, \quad \sigma^2 A = 0$$

Are these results consistent with the classical interpretation according to which Σ_1 is the state in which both spins are parallel and along Z,

Σ_2 is the state in which both spins are parallel and perpendicular to Z,

Σ_3 is the state in which both spins are parallel and along $-Z$,

A is the state in which both spins are opposed and yield no resultant angular momentum?

11.35. The Hydrogen Molecule.—One of the stumbling blocks of pre-quantum chemistry was the phenomenon of homo-polar binding; it is impossible to explain on the basis of classical dynamics the union of two hydrogen atoms to form a molecule. The only attraction which two neutral structures like H-atoms could possibly exhibit was due to quadrupole forces, and these were known to be too weak to account for molecular binding. It was shown by Heitler and London that the homo polar bond is caused by a typical quantum-mechanical effect: the "exchange" of the two electrons. Its meaning will be clear from the following discussion.

The method of calculation[37] to be employed is a simple one which lays little claim to quantitative accuracy[38] but exposes the significant facts in a beautiful way. It is similar to the treatment of the H_2^+-ion, from which it differs by the presence of two electrons instead of one. The coordinate system to be used will be clear from Fig. 7; particles 1 and 2 are electrons,

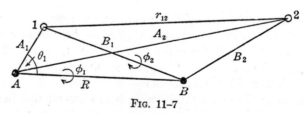

Fig. 11–7

A and B are the protons whose positions are regarded as fixed. In connection with Fig. 7, we also wish to outline the use of a coordinate system and a volume element which are very convenient in the numerical work involved in this problem.

The coordinate system for the *two* electrons will contain the six variables A_1, B_1, B_2, r_{12}, φ_1, φ_2,

$$| B_2 - B_1 | \le r_{12} \le B_1 + B_2, \qquad 0 < B_2 < \infty$$
$$| B_1 - R | \le A_1 \le B_1 + R, \qquad 0 < B_1 < \infty$$

[37] Heitler, W., and London, F., *Z. Phys.* **44**, 455 (1927).

[38] The most elaborate and accurate calculation, also employing the variational method was made by James, H. M., and Coolidge, A. S., *J. Chem. Phys.* **1**, 825 (1933).

The volume element $d\tau = d\tau_1 d\tau_2$, where

$$d\tau_1 = A_1^2 dA_1 \sin\theta_1 d\theta_1 d\varphi_1$$

Now

$$B_1^2 = A_1^2 + R^2 - 2A_1 R \cos\theta_1$$

whence

$$2B_1 dB_1 = 2A_1 R \sin\theta_1 d\theta_1$$

On eliminating $\sin\theta_1 d\theta_1$ from $d\tau_1$ by means of this last relation, we find

$$d\tau_1 = \frac{1}{R} A_1 dA_1 B_1 dB_1 d\varphi_1$$

The element $d\tau_2$ is obtained by writing down an expression similar to $d\tau_1$, but using B_1 as base line:

$$d\tau_2 = \frac{1}{B_1} r_{12} dr_{12} B_2 dB_2 d\varphi_2$$

Hence the product $d\tau_1 d\tau_2$ is

$$d\tau = \frac{1}{R} A_1 dA_1 B_2 dB_2 r_{12} dr_{12} dB_1 d\varphi_1 d\varphi_2 \tag{11-169}$$

Several similar volume elements can be constructed by the same method.

After this excursion, let us consider the Schrödinger equation of the H_2-problem. It is

$$H\psi = \left\{ -\frac{\hbar^2}{2m}(\nabla_1^2 + \nabla_2^2) - e^2\left(\frac{1}{A_1} + \frac{1}{B_2} + \frac{1}{A_2} + \frac{1}{B_1} - \frac{1}{r_{12}} - \frac{1}{R}\right) \right\}\psi$$
$$= E\psi \tag{11-170}$$

We endeavor to solve it by the method of linear variation functions, choosing as constituents of the trial function simple but reasonable approximations to the correct ψ. If H did not contain the last four items in the parenthesis multiplying e^2 it would simply be the sum of two hydrogen-atom Hamiltonians, and

$$\psi = u_A(1)u_B(2)$$

where

$$u_A(1) = (\pi a_0^3)^{-1/2} e^{-A_1/a_0}, \quad u_B(2) = (\pi a_0^3)^{-1/2} e^{-B_2/a_0}$$

are hydrogen functions centered about A and B respectively. On the other hand, if the terms $1/A_1 + 1/B_2 - 1/r_{12} - 1/R$ were missing from the parenthesis, H would also be the sum of two hydrogen-atom Hamiltonians, but $\psi = u_B(1)u_A(2)$. Both of these ψ's are equally good approximations, and both must be included in the trial function. Note that they differ with respect to an exchange of the electrons (or, what amounts in this problem to the same thing, the protons). Hence we adopt

$$\Phi = c_1 u_A(1)u_B(2) + c_2 u_B(1)u_A(2) \tag{11-170a}$$

as variation function in minimizing $\int \Phi H \Phi d\tau$. As explained in sec. 20, the process leads to the secular equations

$$\left. \begin{array}{l} c_1(\mathcal{K}_{11} - \Delta_{11}E) + c_2(\mathcal{K}_{12} - \Delta_{12}E) = 0 \\ c_1(\mathcal{K}_{21} - \Delta_{21}E) + c_2(\mathcal{K}_{22} - \Delta_{22}E) = 0 \end{array} \right\} \tag{11-171}$$

and E is given by

$$\begin{vmatrix} \mathcal{K}_{11} - \Delta_{11}E & \mathcal{K}_{12} - \Delta_{12}E \\ \mathcal{K}_{21} - \Delta_{21}E & \mathcal{K}_{22} - \Delta_{22}E \end{vmatrix} = 0 \tag{11-172}$$

Here

$$\Delta_{11} = \int u_A^2(1)u_B^2(2)d\tau_1 d\tau_2 = \Delta_{22} = 1$$

$$\Delta_{12} = \int u_A(1)u_B(2)u_B(1)u_A(2)d\tau_1 d\tau_2 = \left(\int u_A(1)u_B(1)d\tau_1\right)^2 = \Delta_{21}$$

The latter integral is familiar from sec. 21, it is the quantity there called Δ_{AB}. Hence

$$\Delta_{12} = \Delta_{21} = e^{-2\rho}\left(1 + \rho + \frac{\rho^2}{3}\right)^2, \quad \rho = \frac{R}{a_0}$$

Next, we turn to

$$\mathcal{K}_{11} = \int u_A(1)u_B(2)Hu_A(1)u_B(2)d\tau_1 d\tau_2$$

The ∇^2-terms in H need not be calculated; their effect upon $u_A(1)$ and $u_B(2)$ is at once obtainable from the differential equations which these functions satisfy:

$$-\frac{\hbar^2}{2m}\nabla_1^2 u_A(1) = \left(E_H + \frac{e^2}{A_1}\right)u_A(1),$$

$$-\frac{\hbar^2}{2m}\nabla_2^2 u_B(2) = \left(E_H + \frac{e^2}{B_2}\right)u_B(2)$$

In this way we find

$$\mathcal{K}_{11} = 2E_H + 2J + J' + \frac{e^2}{R}$$

where

$$J = -e^2 \int u_A^2(1)u_B^2(2)B_1^{-1}d\tau_1 d\tau_2 = -e^2 \int \frac{u_A^2(1)}{B_1} d\tau_1 \quad (11\text{–}173)$$

and

$$J' = e^2 \int \frac{u_A^2(1)u_B^2(2)}{r_{12}} d\tau_1 d\tau_2 \quad (11\text{–}174)$$

J is given in sec. 21, eq. (100), and J' has the value

$$J' = \frac{e^2}{R}\left[1 - e^{-2\rho}\left(1 + \frac{11}{8}\rho + \frac{3}{4}\rho^2 + \frac{\rho^3}{6}\right)\right]$$

Problem. Prove this result, using the system of coordinates and the volume element (169).

Furthermore,

$$\mathcal{H}_{22} = \mathcal{H}_{11}$$

as the reader will easily verify. In a similar way,

$$\mathcal{H}_{12} = \mathcal{H}_{21} = 2E_H\Delta_{12} + 2K\Delta_{12}^{1/2} + K' + \frac{e^2}{R}\Delta_{12}$$

where

$$K = -e^2 \int u_A(1)u_B(1)B_1^{-1}d\tau_1 \quad (11\text{–}175)$$

and

$$K' = e^2 \int \frac{u_A(1)u_B(1)u_A(2)u_B(2)}{r_{12}} d\tau_1 d\tau_2 \quad (11\text{–}176)$$

The value of K is given in eq. (100), and

$$K' = \frac{e^2}{5a_0}\left\{e^{-2\rho}\left(\frac{25}{8} - \frac{23}{4}\rho - 3\rho^2 - \frac{1}{3}\rho^3\right)\right.$$
$$\left. + \frac{6}{\rho}\left[\Delta(\gamma + \ln \rho) - 2\sqrt{\Delta\Delta'}Ei(-2\rho) + \Delta'Ei(-4\rho)\right]\right\}$$

where $\gamma = 0.5772$ (Euler-Mascheroni constant),

$$\Delta = \Delta_{12}, \quad \Delta' = e^{2\rho}\left(1 - \rho + \frac{\rho^2}{3}\right)^2$$

and $Ei(x)$ is an abbreviation for the exponential integral

$$Ei(x) = \int_{-\infty}^{x} \frac{e^u}{u} du,$$

which is tabulated and discussed, for instance, in "Tables of Sine, Cosine and Exponential Integrals," Federal Works Agency, New York, 1940.

Problem. Evaluate K'. See in this connection, Sugiura, Y., *Z. f. Phys.* **45**, 484 (1927).

The two roots of (172) are

$$E_1 = \frac{\mathcal{K}_{11} + \mathcal{K}_{12}}{1 + \Delta} = 2E_H + \frac{e^2}{R} + \frac{2J + J' + 2K\Delta^{1/2} + K'}{1 + \Delta}$$

$$E_2 = \frac{\mathcal{K}_{11} - \mathcal{K}_{12}}{1 - \Delta} = 2E_H + \frac{e^2}{R} + \frac{2J + J' - 2K\Delta^{1/2} - K'}{1 - \Delta} \qquad (11\text{-}177)$$

Substitution into (171) shows that to E_1 there corresponds the function

$$\Phi_1 = [2(1 + \Delta)]^{-1/2}[u_A(1)u_B(2) + u_B(1)u_A(2)] \qquad (11\text{-}178)$$

and to E_2 the function

$$\Phi_2 = [2(1 - \Delta)]^{-1/2}[u_A(1)u_B(2) - u_B(1)u_A(2)] \qquad (11\text{-}179)$$

The energies E_1 and E_2 are plotted against R, the internuclear distance, in Pauling and Wilson.[39] It will be seen that E_1 has a minimum in the neighborhood of the experimental internuclear distance of the H_2-molecule; at this minimum E_1 is negative and equal in order of magnitude to the experimentally known minimum which causes the stability of the molecule. On the other hand, E_2 is positive for all R, decreasing in monotone fashion with increasing R. It, therefore, corresponds to repulsion between the atoms. Comparison of E_1 and E_2 shows the difference in their behavior as functions of R to be predominantly due to the presence of the K and K' integrals. These would have been missing if electron exchange had not been taken account of by introducing the *two* functions constituting the Φ of eq. (170). In that case also, there would have been only one energy and not two. Now while (170) may be a crude approximation, the fact that two equivalent functions, differing only with respect to electron exchange, will compose the correct solution of (170) is beyond doubt, hence the qualitative aspects here obtained cannot be questioned. The integrals K and K' are called *exchange integrals*.

Let us now include the spin and apply the Pauli principle. The spin functions are those already encountered in the helium problem, eq. (168). If the resultant function is to be antisymmetrical, Φ_1, which is symmetrical in the position coordinates of electrons 1 and 2 must be multiplied by an antisymmetrical function of the spins, of which there is only one, namely A. However, Φ_2 may be multiplied by one of the three functions Σ_1, Σ_2 or Σ_3. It represents a triplet state while Φ_1 is a singlet.

To the energy E_2, therefore, there correspond three times as many quantum mechanical states as to E_1. From this fact may be drawn the

[39] Loc. cit., p. 344.

conclusion that when two H-atoms approach they will, ceteris paribus, be three times as likely to repel as to attract each other.

REFERENCES

To begin with source material, there are: Schrödinger's charming volume " Wave Mechanics " (Blackie and Son, London, 1928) which is a collection of his epoch-making papers of 1926 and 1927; Heisenberg's more popular " The Physical Principles of the Quantum Theory " (Chicago University Press, Chicago, 1930); De Broglie and Brillouin's " Selected Papers on Wave Mechanics " (Blackie and Sons, London, 1928); and Born and Jordan's " Elementare Quantenmechanik " (J. Springer, Berlin, 1930). The foundations of the subject, both mathematical and philosophical, are treated most thoroughly but also most abstractly by Dirac in his " Principles of Quantum Mechanics" (Clarendon Press, Oxford, Third Edition, 1947) and by J. v. Neumann in " Mathematische Grundlagen der Quantenmechanik " (J. Springer, Berlin, 1932).

General treatises are:

Condon, E. U., and Morse, P. M., " Quantum Mechanics," McGraw-Hill Book Co., Inc., New York, 1929.

Ruark, A. E., and Urey, H. C., " Atoms, Molecules, and Quanta," McGraw-Hill Book Co., Inc., New York, 1930.

De Broglie, L., " Théorie de la Quantification," Hermann et Cie, Paris, 1932.

Frenkel, J., " Wave Mechanics," Vols. I and II, Clarendon Press, Oxford, 1932, 1934.

" Handbuch der Physik," Vol. XXIV, Parts I and II (numerous authors), Julius Springer, Berlin, 1933.

Pauling, L., and Wilson, E. B., " Introduction to Quantum Mechanics," McGraw-Hill Book Co., Inc., New York, 1935.

Jordan, P., " Anschauliche Quantenmechanik," J. Springer, Berlin, 1936.

Kemble, E. C., " The Fundamental Principles of Quantum Mechanics," McGraw-Hill Book Co., Inc., New York, 1937.

Dushman, S., " Elements of Quantum Mechanics," John Wiley and Sons, Inc., New York, 1938.

Sommerfeld, A., " Atombau und Spektrallinien," Vol. II, Vieweg und Sohn, Braunschweig, 1939.

Rojanski, V., " Introductory Quantum Mechanics," Prentice-Hall, Inc., New York, 1939.

Mott, N. F., and Sneddon, I. N., " Wave Mechanics and its Applications " Oxford Press, 1948.

Schiff, L. I., " Quantum Mechanics," McGraw-Hill Book Co., Inc., New York, 1949.

Bohm, D., " Quantum Theory," Prentice-Hall, Inc., New York, 1951.

Slater, J. C., " Quantum Theory of Matter," McGraw-Hill Book Co., Inc., New York, 1951.

Houston, W. V., " Principles of Quantum Mechanics," McGraw-Hill Book Co., Inc., New York, 1951.

Landé, A., " Quantum Mechanics," Pitman Publishing Corp., New York, 1951.

A list of books in which quantum mechanics is applied to special problems follows.

Van Vleck, J. H., " The Theory of Electric and Magnetic Susceptibilities," Clarendon Press, Oxford, 1932.

Condon, E. U., and Shortley, G. H., " The Theory of Atomic Spectra," The Macmillan Co., New York, 1935.

Seitz, F., " Modern Theory of Solids," McGraw-Hill Book Co., Inc., New York, 1940.

Pauling, L., " The Nature of the Chemical Bond," Cornell University Press, Ithaca, N. Y., 1940.

Eyring, H., Walter, J. and Kimball, G. E., " Quantum Chemistry," John Wiley and Sons, Inc., New York, 1944.

Glasstone, S., " Theoretical Chemistry," D. Van Nostrand Co., Inc., New York, 1944.

Flügge, S., and Marschall, H., " Rechenmethoden der Quantentheorie," Part 1, Springer, Berlin, 1947. This book contains many problems on elementary quantum mechanics and the details of the solution for each.

Mott, N. F., and Massey, H. S. W., " The Theory of Atomic Collisions," Second Edition, Clarendon Press, Oxford, 1948.

Corson, E. M., " Perturbation Methods in the Quantum Mechanics of n-Electron Systems," Hafner Publishing Co., New York, 1950.

Coulson, C. A., " Valence," Clarendon Press, Oxford, 1952.

Pitzer, K. S., " Quantum Chemistry," Prentice-Hall, Inc., New York, 1953.

Corson, E. M., " Introduction to Tensors, Spinors, and Relativistic Wave-Equations," Blackie and Son Ltd., London, 1953.

Finkelnburg, W., " Einführung in die Atomphysik," Third Edition, Springer, Berlin, 1954.

CHAPTER 12

STATISTICAL MECHANICS

12.1. Permutations and Combinations.—The purpose of the present chapter is not primarily an exposition of the ideas of statistical mechanics, which is available in several modern texts,[1] but a brief and summary review of the chief analytical techniques used in the treatment of this subject. We begin by discussing the principal formulas of the theory of combinations.

a. The number of possible permutations of n different (distinguishable) objects is $n!$

The proof is simple: the first object can be put in n different positions. When its place is fixed, $n - 1$ different positions are left open for the second. Hence these two objects can be arranged in $n(n - 1)$ different ways without disturbing the relative order of the remaining $(n - 2)$ objects. But the third can occupy $n - 2$ different places, and so on. The total number of possible arrangements is therefore $n(n - 1)(n - 2) \cdots 2 = n!$

b. Suppose we wish to arrange the n objects in r piles, the number in each pile being prescribed. Let the number of objects in the first pile be n_1, that in the second n_2, etc., so that $\sum_{i=1}^{r} n_i = n$. It is desired to find the number, M, of possible arrangements of this kind. If M is multiplied by the number of possible permutations of all objects in the first pile, then by the number of possible permutations of the objects in the second pile and so on for all the piles, we must obtain the total number of permutations of n objects. Thus

$$Mn_1!n_2! \cdots n_r! = n!$$

whence

$$M = \frac{n!}{n_1!n_2! \cdots n_r!} \tag{12-1}$$

There is another combinatorial problem which leads to the same result. Suppose the n objects fall into r classes, the objects in each class being alike

[1] Tolman, R. C., " The Principles of Statistical Mechanics," Clarendon Press, Oxford, 1938. Chapman, S. and Cowling, T. G., " The Mathematical Theory of Non-Uniform Gases," University Press, Cambridge, 1939. Mayer, J. E. and Mayer, M. G., " Statistical Mechanics," John Wiley and Sons, 1940. Lindsay, R. B., " Physical Statistics," John Wiley and Sons, 1941.

(indistinguishable). Let the first class contain n_1 objects, the second n_2, etc. The number of possible *distinguishable* arrangements of the n objects will then be seen to be obtainable by the reasoning employed above. Hence M represents also the number of arrangements of n things groupable into r *classes*, the members of each class being alike.

c. The number of ways in which m objects can be selected from a set of n objects is $n!/[m!(n - m)!]$. This follows at once from (1), for a withdrawal of m individuals is equivalent to an arrangement of the n objects into two piles, one containing m, the other $(n - m)$ objects. We note that this number

$$\frac{n!}{m!(n - m)!} \equiv \binom{n}{m} \tag{12-2}$$

It is often referred to as the number of *combinations* of n things taken m at a time. We observe that, since $\binom{n}{m} = \binom{n}{n - m}$, it is equal to the number of combinations of n things taken $n - m$ at a time.

Eq. (2) also provides the answer to another, apparently different question. Assume that we have n boxes, and a smaller number, m, of *indistinguishable* objects to be placed in them in such a way that no box contains more than one object. The number of ways in which this can be done is given by (2), for the assignment of m objects to n boxes is entirely equivalent to the selection of m objects from a set of n objects.

d. When in accordance with theorem (c), a certain selection of m objects has been made, a permutation among these m objects does not produce a new *combination*. It does, however, produce a new *arrangement*. Thus, to every combination given by eq. (2), there correspond $m!$ arrangements of the m objects. The total number of arrangements of n things taken m at a time is therefore

$$\binom{n}{m} m! = \frac{n!}{(n - m)!} \tag{12-3}$$

If, in the problem of placing m objects into n boxes $(n \geq m)$ discussed in (c), the objects are assumed to be *distinguishable*, so that our interest is no longer merely in the individual boxes each of which contains an object, but also in the arrangement of the individual objects placed in them, eq. (3) is applicable. It expresses the number of ways in which m distinguishable objects can be placed in n boxes, zero or one object per box.

e. Let us now determine the number of ways in which m *in*distinguishable particles can be put into n boxes. Suppose that the m particles were placed along a line in any manner whatever and that $(n - 1)$ partitions were used to separate the particles. If one more partition were then placed at the end of the line, the particles could be regarded as having been placed into n boxes. If, therefore, we consider the m particles and the $(n - 1)$ partitions, visualized as walls, as a set of $(m + n - 1)$ objects, our problem becomes tantamount to finding the number of ways in which $(n - 1)$ walls can be arranged among the totality of $(m + n - 1)$ objects. This number, from sec. c and eq. (2) is

$$\binom{n + m - 1}{n - 1} = \binom{n + m - 1}{m} \tag{12-4}$$

The preceding result is obtainable in several other ways, among which the following is sometimes given. Suppose that there are n boxes and m objects, as before. The first box can be selected in n ways, leaving $(n + m - 1)$ boxes and objects which can be arranged in $(n + m - 1)!$ ways or a total number of $n(n + m - 1)!$ arrangements. However, permutations of boxes or particles among themselves do not correspond to recognizably different arrangements. Since this last number is $n!m!$, the desired number is again given by eq. (4).

In the mathematical literature, the result of eq. (4) is sometimes known as the number of "combinations with repetitions." We note that it equals the number of combinations of $(n + m - 1)$ things taken m at a time, where repetitions are *not* allowed.

A recursion formula for the case with repetition is sometimes useful. If there are three objects, taken two at a time, it is found that there are six possibilities: (aa, ab, ac, bb, bc, cc); if taken three at a time, there are ten cases: $(aaa, aab, aac, abb, acc, bbb, bbc, bcc, ccc)$. By mathematical induction, it is easy to show that for n objects taken m at a time

$$C_m(n) = \frac{n + m - 1}{m} C_{m-1}(n) \tag{12-5}$$

If m is given the successive values $1, 2, 3, \cdots, k$ and the equations multiplied together, the result is the now familiar one of eq. (4).

f. The number of ways in which m *distinguishable* objects may be placed in n boxes is clearly n^m for the first object can be put into n places; with each of these dispositions of the first object can be combined n dispositions of the second object, and so on.

12.2. Binomial Coefficients.—The coefficients $\binom{n}{t}$ appear in Newton's famous binomial expansion

$$(a + b)^n = \sum_{t=0}^{n} \binom{n}{t} a^t b^{n-t} \tag{12-6}$$

where t is an integer. Its proof is fairly obvious, since the number of ways in which t factors a and $(n - t)$ factors b can be selected from n factors $(a + b)$ is $\binom{n}{t}$ ways, by virtue of (2). We note that

$$\binom{n}{0} = 1, \quad \binom{n}{1} = n, \quad \binom{n}{n} = 1$$

also

$$\binom{n}{t} = 0 \quad \text{if } t > n, \quad \text{this because } (n - t)! = \infty$$

Most of the relations to be studied here are valid for non-integral values of n provided we define

$$\binom{x}{t} = \frac{x(x - 1) \cdots (x - t + 1)}{1 \cdot 2 \cdots t}$$

An important series in binomial coefficients may be obtained as follows. In view of (6), $\binom{n + k}{r}$ is the coefficient of $a^r b^{n+k-r}$ in the expansion of $(a + b)^{n+k}$. But

$$(a + b)^n (a + b)^k = \left[\sum_{t=0}^{n} \binom{n}{t} a^t b^{n-t} \right] \left[\sum_{s=0}^{k} \binom{k}{s} a^s b^{k-s} \right]$$

$$= \sum_{s,t} \binom{k}{s} \binom{n}{t} a^{t+s} b^{n+k-t-s}$$

The coefficient of $a^r b^{n+k-r}$ in this double sum is obtained by putting $t + s = r$ and summing over t. Hence

$$\binom{n + k}{r} = \sum_{t=0}^{r} \binom{n}{t} \binom{k}{r - t} \tag{12-7}$$

This is known as the *addition theorem* of the binomial coefficients. From it, numerous other relations can be derived.

On putting $k = 1$, we have

$$\binom{n + 1}{r} = \binom{n}{r} + \binom{n}{r - 1}$$

If $k = r = n$,

$$\binom{2n}{n} = \sum_{t=0}^{r} \binom{n}{t} \binom{n}{n - t} = \sum_{t=0}^{n} \binom{n}{t}^2$$

If we observe that

$$\binom{-n}{r} = (-1)^r \binom{n + r - 1}{r}$$

we may also put $k = -1$ in (7), obtaining

$$\binom{n - 1}{r} = \sum_{t=0}^{r} (-1)^{r-t} \binom{n}{t}$$

If, in Newton's formula, we let $a = b = 1$, we find

$$\sum_{t=0}^{n} \binom{n}{t} = 2^n$$

but if $a = -b$,

$$\sum_{t=0}^{n} (-1)^t \binom{n}{t} = 0$$

12.3. Elements of Probability Theory.—An aggregate of *elements*, such as a set of observations, a sequence of results of some operation (e.g., throwing a die), is called a *probability aggregate* if it is permissible to apply the rules of the probability calculus to the aggregate. Whether or not this application is proper is usually decided on the basis of intuition: it seems clear that the decimal expansion of the fraction $\frac{1}{7}$ does not form an aggregate of digits to which probability considerations may validly be applied; on the other hand, no hesitation is felt in subjecting the outcome of a series of throws of a die to probability reasoning. In the former sequence (.142857142857, etc.) the digits occur with too much regularity to be regarded as " distributed at random." The criteria for randomness, which decide whether an aggregate is a probability aggregate, may be stated with considerable precision[2] but will be omitted here.

Every element is regarded as having one of a number, s, of distinguishable *properties*. (Each throw of a die is an element, the number appearing uppermost is a property; $s = 6$. In measuring a physical quantity, each measurement is an element, each measured value a property; s may be infinite in this example.) If n_i is the number of times the i-th property occurs and n the total number of elements,

$$\frac{n_i}{n}$$

is defined as the *relative frequency* of the i-th property. By the *probability* of the i-th property is meant the limit

$$\lim_{n \to \infty} \frac{n_i}{n} = w_i \tag{12-8}$$

[2] See, Lindsay, R. B., and Margenau, H., " Foundations of Physics," John Wiley and Sons, 1936.

The existence of this limit is a matter which has given rise to considerable discussion; it will here be assumed.[3] The totality of the w_i is called the *distribution* of the probability aggregate. Obviously, $\sum_i w_i = 1$.

The properties may be *discrete* (throwing dice) or *continuous* (value of a physical quantity, such as position of a particle). In the former case the distribution is sometimes said to be *arithmetical*, in the latter case, *geometrical*. In the continuous case a different formulation of probability is more convenient. Let x denote the continuous property. The probability w_x, defined by (8) is clearly zero, but the probability that x shall lie between x and $x + \Delta x$ is finite and is, moreover, usually proportional to the range Δx provided this range is sufficiently small. Hence we may write for this probability

$$w(x)\Delta x$$

and the function $w(x)$, which does not have the physical dimension of a probability (a pure number) is called the *probability density*. Clearly,

$$\int w(x)dx = 1$$

if the integral is taken over the entire range of properties.

When a distribution w_i or $w(x)$ is given, certain expressions frequently occurring in statistical theories can be calculated. We present the most important of these, using parallel formulations for the arithmetical and geometrical cases. To make this possible, we write $w(x_i)$ for the former w_i, thus letting x_i represent the i-th property.

If $f(x)$ is a function defined for every x_i (or x) which has a non-vanishing probability, the *mean* of $f(x)$ with respect to the distribution $w(x)$ is given by

$$\bar{f} = \begin{cases} \sum_i f(x_i)w(x_i) \\ \int f(x)w(x)dx \end{cases} \qquad (12\text{-}9)$$

The *dispersion* of $f(x)$ with respect to $w(x)$ is defined by

$$D(f) = \begin{cases} \sum_i [f(x_i) - \bar{f}]^2 w(x_i) \\ \int [f(x) - \bar{f}]^2 w(x)dx \end{cases} \qquad (12\text{-}10)$$

On taking for the function $f(x)$ the variable x itself there results

$$\bar{x} = \begin{cases} \sum_i x_i w(x_i) \\ \int x w(x)dx \end{cases} \qquad (12\text{-}11)$$

[3] For further remarks see Lindsay and Margenau, loc. cit., Chapter 4.

and

$$D(x) \equiv \sigma^2 = \begin{cases} \sum_i (x_i - \bar{x})^2 w(x_i) \\ \int (x - \bar{x})^2 w(x) dx \end{cases} \tag{12-12}$$

The quantity σ^2 is called the *dispersion* of the distribution $w(x)$, σ is known as the *standard deviation*. As is clear from its definition, σ is a measure of the spread of $w(x)$ about its mean. If $w(x)$ were regarded as a distribution of mass, σ would represent its radius of gyration. By the *r*-th *moment* of the distribution is meant the quantity

$$\overline{x^r} = \begin{cases} \sum x_i^r w(x_i) \\ \int x^r w(x) dx \end{cases}$$

For distributions with an infinite range of properties, higher moments do not always exist. The dispersion of $w(x)$ may be expressed in terms of its first and second moments. In view of (12),

$$\sigma^2 = \overline{x^2} - 2\bar{x}^2 + \bar{x}^2 = \overline{x^2} - \bar{x}^2$$

Under certain conditions it is possible to expand a geometrical distribution in terms of its moments, provided these exist. For simplicity we shall take these moments about $\bar{x}$ as origin, so that $\bar{x} = 0$, $\overline{x^2} = \sigma^2$, etc. One can then prove[4] that

$$w(x) = \frac{1}{\sigma\sqrt{2\pi}} e^{-x^2/2\sigma^2} \left\{ 1 + \sum_{i=3}^{\infty} \frac{c_i}{i!} H_i\left(\frac{x}{\sigma}\right) \right\}$$

where H_i is the i-th Hermite polynomial, and

$$c_3 = \frac{\overline{x^3}}{\sigma^3}, \quad c_4 = \frac{\overline{x^4}}{\sigma^4} - 3, \quad c_5 = \frac{\overline{x^5}}{\sigma^5} - 10\frac{\overline{x^3}}{\sigma^3}, \quad c_6 = \frac{\overline{x^6}}{\sigma^6} - 15\frac{\overline{x^4}}{\sigma^4} + 30$$

This expansion is particularly useful when $w(x)$ does not depart too greatly from a normal " Gauss " distribution: $w(x) = e^{-x^2/2\sigma^2} \big/ \sigma\sqrt{2\pi}$.

Problems. Two geometrical distributions of considerable interest in physics and chemistry are

$$w_1(x) = \frac{h}{\sqrt{\pi}} e^{-h^2(x-a)^2}$$

$$w_2(x) = \frac{a}{\pi} \frac{1}{a^2 + x^2}$$

[4] See Zernike, F., " Handbuch der Physik," Vol. III, J. Springer, 1928, p. 448.

a. Show that, for w_1, $\bar{x} = a$ and $\sigma^2 = 1/2h^2$.

b. Show that the r-th moment of w_1 is $1 \cdot 3 \cdot 5 \cdots (r-1)/2^{r/2}h^r$ if r is even; for odd r it is zero. (Take $a = 0$). All moments of w_1 are finite.

c. Show that, for w_2, all odd moments are zero and no even moment (except the zero-th) exists.

12.4. Special Distributions.—A problem which is basic in statistical mechanics and in the theory of errors will here be discussed in some detail. It is of considerable historical interest, its solution being connected with the names of Newton, Bernoulli, Laplace, Poisson and Gauss. Consider n boxes, each containing P black balls and Q white balls. We wish to find the probability $w_n(m)$, that in drawing one ball from each of the n boxes, m of them will be white.

The probability of drawing a black ball from a given box is clearly $P/(P+Q) \equiv p$, that of drawing a white ball is $Q/(P+Q) \equiv q$. Thus $w_1(0) = p$, $w_1(1) = q$. If $n = 2$, the probability aggregate has the following properties: bb, bw, wb, ww (b = black, w = white), and these occur with the probabilities p^2, pq, pq, q^2; hence $w_2(0) = p^2$, $w_2(1) = 2pq$, $w_2(2) = q^2$. In general, the probability that m white balls will be drawn from n *specified* boxes and $n - m$ black ones from the remaining boxes will be

$$q^m p^{n-m}$$

But in view of eq. (2) there are $\binom{n}{m}$ ways of selecting m boxes from a total number of n boxes. Hence the answer to the problem, first found by Newton, is

$$w_n(m) = \binom{n}{m} p^{n-m} q^m \tag{12-13}$$

It is clear from (6) that

$$\sum_{m=0}^{n} w_n(m) = 1$$

since $q + p = 1$. Eq. (13) has of course a more general significance than the one here particularized: it represents the probability of m successes in n independent trials if the probability of success in a single trial is q.

To calculate the mean of m and the dispersion of the arithmetical distribution $w_n(m)$ we consider the identity

$$(p + qy)^n = \sum_{m=0}^{n} w_n(m)y^m$$

where y is a variable. On differentiation with respect to y this reads

$$n(p + qy)^{n-1}q = \sum_{m=0}^{n} mw_n(m)y^{m-1} \tag{12-14}$$

When we let $y = 1$ in this equation, the right hand side becomes $\overline{m}$, so that

$$\overline{m} = nq \qquad (12\text{--}15)$$

The mean number of successes is equal to the probability of success in a single trial, multiplied by the number of trials.

To find the dispersion, we differentiate (14) once more, and then set $y = 1$. The result is:

$$n(n-1)q^2 = \sum_{m=0}^{n} m(m-1)w_n(m) = \overline{m^2} - \overline{m}$$

To obtain the dispersion we must add to the right hand side the quantity $\overline{m} - \overline{m}^2$ which, according to (15), equals $nq - n^2q^2$. Hence

$$\sigma^2 = \overline{m^2} - \overline{m}^2 = nq(1 - q) = nqp \qquad (12\text{--}16)$$

Especially interesting is the case where $q \ll p$, so that $p \sim 1$. For then the dispersion is numerically equal to the mean number of successes, a criterion which can sometimes be used to determine whether the successes are due entirely to chance. For applications of the formulas here developed, particularly to the case of radioactive emission, the reader is referred to Lindsay's Physical Statistics. (See also the problem of the random walk at the end of this section.)

For large values of n and m expression (13) is difficult to use because of the inconvenience in dealing with factorials of large numbers. We shall now prove that in this case $w_n(m)$ can be approximated by the Gauss error law. Let us first see what happens to $w_n(m)$ as $n \to \infty$. It is clear from (15) and (16) that both $\overline{m}$ and σ^2 tend to infinity, that is to say, if we were to plot $w_n(m)$ against m, the mean (which for sufficiently large m is also the maximum of $w_n(m)$) would move outward from the origin and the distribution would broaden out indefinitely. However, the quantity x, defined as the deviation from the mean and measured on a proper scale which contracts as n increases, namely

$$x = \frac{m - \overline{m}}{\sqrt{n}} \qquad (12\text{--}17)$$

will remain finite. We shall try to convert $w_n(m)$ into $w(x)$, assuming that $n \to \infty$.

First compute

$$\ln w_n(m) = \ln n! - \ln m! - \ln (n-m)! + (n-m)\ln p + m\ln q$$

Now by Stirling's formula, which is valid for large numbers,

$$\ln n! = (n + \tfrac{1}{2})\ln n - n + \tfrac{1}{2}\ln 2\pi + \frac{1}{12n} - \left(\text{terms of order } n^{-3}\right)$$

Hence[5]

$$-\lim_{n \to \infty} \ln w_n(m) = \tfrac{1}{2} \ln \frac{2\pi m(n-m)}{n} + m \ln \frac{m}{nq} + (n-m) \ln \frac{n-m}{np}$$
(12–18)

In view of (17) and (15)

$$m = nq\left(1 + \frac{x}{n^{1/2}q}\right), \quad n - m = np\left(1 - \frac{x}{n^{1/2}p}\right)$$

When these expressions are introduced in (18) there results

$$-\ln w(x) = \tfrac{1}{2} \ln 2\pi npq + \frac{1}{2}\frac{x^2}{q} + \frac{1}{2}\frac{x^2}{p}$$

provided we use the expansion of the logarithm

$$\ln (1 + x) = x - \frac{x^2}{2} + \cdots$$

and retain no terms in negative powers of n. Thus we have, since $p + q = 1$,

$$-\ln w(x) = \tfrac{1}{2} \ln 2\pi npq + \frac{x^2}{2pq}$$

whence

$$w(x) = \frac{1}{\sqrt{2\pi npq}} e^{-x^2/2pq}$$
(12–19)

When written again in terms of m it is

$$\lim_{n \to \infty} w_n(m) = \frac{1}{\sqrt{2\pi p\overline{m}}} e^{-(m-\overline{m})^2/2p\overline{m}} = \frac{1}{\sqrt{2\pi}\sigma} e^{-(m-\overline{m})^2/2\sigma^2}$$
(12–20)

These results have a special significance with respect to errors of measurement, as can be seen from the following (oversimplified) argument. Suppose that the true value of a measured quantity is A, but that there are n causes of error, each of which will add to A the amount ΔA or $-\Delta A$ with equal probability. If m of these n causes contribute ΔA then the resulting error is $r\Delta A = [m - (n - m)]\Delta A$, and therefore the probability of this error is $w_n(m)$ with $m = (n + r)/2$. For large n the distribution of errors is then given by (20):[6]

$$w(r) = \frac{1}{\sqrt{2\pi}\sigma} e^{-(r-\overline{r})^2/8\sigma^2}$$

---

[5] In arriving at this result, it is convenient to add to the literal expansion of the logarithm by Stirling's formula the quantity $(m \ln n - m \ln n)$.

[6] Note however, that σ^2 is no longer the dispersion with respect to the r-distribution. Furthermore, $\displaystyle\int_{-\infty}^{\infty} w(r)dr \neq 1$.

If $p = q = \frac{1}{2}$, $\overline{m} = n/2$ and $\overline{r} = 0$. If we denote $r\Delta A$ by ϵ_r, Gauss' error law

$$w(\epsilon_r) = \text{const.}\ e^{-h^2\epsilon_r^2}$$

immediately results; the constant h, which depends on ΔA and is always determined empirically, is often called the "measure or index of precision."

In the analysis leading to (19) quantities of the order $1/nq$ and $1/npq$ were neglected, the assumption being that p and q are numbers not greatly different from unity. Under these conditions the mean of m, nq, is a large number. This, then, is a criterion for the applicability of eq. (19).

It may happen, however, that q is small, so small indeed that nq is of order unity in a given application. In this case the distribution (13) has, to be sure, spread out indefinitely ($n \to \infty$) but the mean has remained small; the resulting distribution is quite asymmetrical. To deal with this situation we put

$$q = \frac{a}{n}, \quad p = 1 - \frac{a}{n}$$

and treat $a = \overline{m}$ as a number of order unity while n tends to infinity. Thus

$$w_n(m) = \frac{n(n-1)\cdots(n-m+1)\,a^m}{m!}\ \frac{\left(1 - \dfrac{a}{n}\right)^n}{n^m\left(1 - \dfrac{a}{n}\right)^m}$$

$$= \left(1 - \frac{a}{n}\right)^n \frac{a^m}{m!} \cdot \frac{1\left(1 - \dfrac{1}{n}\right)\cdots\left(1 - \dfrac{m-1}{n}\right)}{\left(1 - \dfrac{a}{n}\right)^m}$$

As $n \to \infty$, the last fraction takes on the value 1. Hence, under these conditions,

$$\lim_{n\to\infty} w_n(m) = \frac{a^m e^{-a}}{m!} \tag{12-21}$$

since

$$\lim_{n\to\infty}\left(1 + \frac{a}{n}\right)^n = e^a$$

Formula (21) was first derived by Poisson and bears his name. It is used in the theory of radioactivity.

Problem a. Plot $w_n(m)$ for $q = \frac{1}{3}$; $n = 5, 10, 50$. Observe the change from an asymmetrical to a symmetrical distribution. Compare Poisson's formula with the plot for $n = 5$.

Problem b. " *Random walk.*" A person, making steps of length l, is just as likely to step forwards as backwards ($p = q = \frac{1}{2}$). Prove that, after taking n steps, he will have gone forward a distance rl with a probability

$$(\tfrac{1}{2})^n \binom{n}{\dfrac{n+r}{2}}$$

Show also that $\bar{r} = 0, \overline{r^2} = n$.

12.5. Gibbsian Ensembles.

—It is the main purpose of statistical mechanics to provide a formalism by means of which the facts of thermodynamics (cf. Chapter 1) can be deduced. This may be done in several different ways, that is, on the basis of several distinct sets of fundamental axioms. Two of these stand out for their success and clarity. One, the system of Gibbs, is particularly suited to a development of the classical laws of thermodynamics, i.e., those relations whose understanding is possible without the use of quantum mechanics. Gibbs' statistical mechanics will be summarized in this section and the next. The remainder of the present chapter will be devoted to the method of Darwin and Fowler with the aid of which the subject of quantum statistics is most satisfactorily discussed.

The central concept of Gibbs'[7] theory is the ensemble, the meaning of which will now be discussed. Statistical mechanics deals with certain properties of physical objects, as for instance a given body of gas, or liquid, or a solid. Such an object will be called a *system*, or, more specifically, a thermodynamic system. If it has n degrees of freedom, then its complete mechanical state can be specified in terms of n generalized coordinates and n generalized momenta, a total of $2n$ numbers. Mathematically, these $2n$ numbers may be said to define a point in a space of $2n$ dimensions, and this space is called the *phase space* of the system. At any instant of time, the system is represented by one point in its phase space, and in the course of time, this point will move, describing a certain trajectory in phase space. When the position of the representative point at any instant is known, it is theoretically possible by the laws of mechanics to calculate its position at any other time, but such a prediction is practically not feasible. Other, less detailed methods of description must be chosen.

In the simplest instance of a thermodynamic system, the ideal gas consisting of ν molecules, $n = 3\nu$, and the phase space has 6ν dimensions. A representative point would correspond to an exact assignment of 3 com-

---

[7] There is no more lucid and careful exposition of J. W. Gibbs' ideas than his own, "Elementary Principles of Statistical Mechanics"; C. Scribner's Sons, 1902; Collected Works, vol. II, Longmans, Green & Co., 1928, New York. See also "Commentary on the Scientific Writings of J. Willard Gibbs," Yale University Press, 1936.

ponents of momentum and position to each of the ν molecules, and the path of the point would portray the changes which the values of all these quantities undergo in time. In this case, another picture is often useful. One may regard the phase space of the system as being composed of ν subspaces, one for each molecule. Such a subspace is called a μ-space (" molecule space ") in order to distinguish it from the entire phase space which is often designated as γ-space (" gas space "). In the case of molecules regarded as mass points, μ-space has 6 dimensions, although in general for molecules having internal degrees of freedom the number of dimensions is greater. Use of the μ-space is often very convenient, but it loses its significance except as an approximate description when strong interaction exists between the molecules.

We shall denote the n generalized coordinates of our system by $q_1 q_2 \cdots q_n$, the generalized momenta by $p_1 \cdots p_n$. Out of these we construct an element $d\phi$ of phase space in which the p's and q's are taken as Cartesian coordinates:

$$d\phi = dq_1 dq_2 \cdots dq_n dp_1 dp_2 \cdots dp_n$$

It is possible to show[8] that any point transformation

$$q_i' = q_i'(q_1 \cdots q_n)$$

$$p_i' = p_i'(q_1 \cdots q_n, p_1 \cdots p_n)$$

leaves $d\phi$ invariant; thus

$$d\phi = dq_1' \cdots dq_n' dp_1' \cdots dp_n'$$

Since the system is assumed to obey mechanical laws, Hamilton's canonical equations must be valid (cf. 9–13):

$$\dot{p}_i = -\frac{\partial H}{\partial q_i}, \quad \dot{q}_i = \frac{\partial H}{\partial p_i}, \quad i = 1, 2, \cdots n \qquad (12\text{--}22)$$

From these equations it follows at once that through every point in phase space there passes but one trajectory; for when every p and q is given, equations (22) determine uniquely the rate of change of every coordinate in phase space. Hence the representative point can never cross its previous path. Whether the motion of the point will ultimately carry it through all regions of phase space has not been proved completely; such behavior is tentatively asserted by the so-called ergodic hypothesis[9] which, however, is not needed in Gibbs' formulation of statistical theory.

It would seem that the values of thermodynamic quantities such as temperature, pressure, etc., could be regarded as time averages over the

[8] See Gibbs, loc. cit. or Lindsay, loc. cit.
[9] See Tolman, loc. cit.

motion of the representative point of the system in its own phase space. The development of this conjecture, however, is fraught with rather formidable difficulties and is not usually attempted. Instead, Gibbs introduces what he calls an *ensemble* of systems, by which is meant a very large set of imagined replicas of the one real system under consideration. These systems are not in identical states, but the state of each is represented by a phase point in its own phase space. Since all imaginary systems of the ensemble are similar as to number of molecular constituents and Hamiltonian function, all points can be plotted in the *same* phase space, in which they will be distributed with a certain density, D.

This density will in general be different in different parts of phase space, and it will change in time. Hence

$$D = D(p_1 \cdots p_n; q_1 \cdots q_n; t)$$

Nothing has as yet been said about the initial distribution of points in phase space which, in view of the meaning of the ensemble, is quite arbitrary. Whatever the functional form of D, we must require

$$\int D d\phi = N \quad \text{for every } t$$

if N is the (very large) number of systems in the ensemble. It is convenient also to introduce a " probability of phase "

$$P = \frac{D}{N} \tag{12-23}$$

such that

$$\int P d\phi = 1$$

12.6. Ensembles and Thermodynamics.—By virtue of Liouville's theorem, proved in almost all books on statistical mechanics (also known as the principle of conservation of density-in-phase, a name due to Gibbs), the representative points move in phase space as though they constituted an incompressible fluid of varying density. A group of points filling a certain region of phase space at a time t_0 can neither contract nor expand during its motion; it will continue to occupy the same volume but with altered shape. Mathematically these statements are expressed as follows:

$$\frac{\partial D}{\partial t} + \sum_{i=1}^{n} \frac{\partial D}{\partial p_i} \dot{p}_i + \sum_{i=1}^{n} \frac{\partial D}{\partial q_i} \dot{q}_i = 0 \tag{12-24}$$

Thus it is seen that phase space possesses no intrinsic property of accumulating phase points in some regions or not admitting them to others; Liouville's theorem shows phase space to be *indifferent* to the motion of the points. This fact suggests the following fundamental postulate, by means of which contact is established between an ensemble and thermodynamic experience:

The probability that, at any instant t, a given real system be found in the state characterized by $q_1 \cdots q_n$, $p_1 \cdots p_n$ is the same as the probability $P(p_1 \cdots p_n; q_1 \cdots q_n; t)$ that a system selected at random from the corresponding ensemble shall have the phase $q_1 \cdots q_n$, $p_1 \cdots p_n$ at the instant t. The probability that the values of the p's and q's shall lie within a small extension of phase $\Delta\phi$ is proportional to $\Delta\phi$. We are thus attributing equal intrinsic probabilities to equal volumes of phase space, a procedure suggested, though not made necessary, by Liouville's principle.

In accordance with this postulate we may calculate mean values of dynamical quantities of the real system by computing mean values over the individuals composing the ensemble. If R is such a quantity, expressible as a function of momenta and coordinates, then

$$\overline{R}(t) = \int R(p_1 \cdots p_n; q_1 \cdots q_n) P(p_1 \cdots p_n; q_1 \cdots q_n, t) d\phi \quad (12\text{--}25)$$

And by $\overline{R}(t)$ is meant in general the *expected mean value* of the quantity R which would be obtained when R is actually measured at the time t. It can be shown that deviations from this expected mean are extremely small when the system in question has many degrees of freedom, so that the expected mean may be identified for practical purposes with the value of R actually measured in a single observation. Moreover, we shall see at once that under equilibrium conditions P is not a function of t, so that $\overline{R}$, also, will not be a function of t. One may then think of $\overline{R}$ as the mean value of the quantity R in a *temporal* sense, i.e., $\overline{R} = \int_0^T \dfrac{R dt}{T}$ for sufficiently large T, without violating the spirit of the postulate.

If the thermodynamic system is in equilibrium, the number of representative points in any given extension in phase, $\Delta\phi$, must remain constant in time. The condition of equilibrium may therefore be stated in the form

$$\frac{\partial D}{\partial t} = 0$$

For the equilibrium case, in which we are chiefly interested (a reversible thermodynamic change consists of a sequence of equilibrium states), Liouville's theorem states

$$\sum_{i=1}^n \left(\frac{\partial D}{\partial p_i} \dot{p}_i + \frac{\partial D}{\partial q_i} \dot{q}_i \right) = 0 \quad (12\text{--}26)$$

Let us now give thought to the initial form of the function $D(p_1 \cdots p_n; q_1 \cdots q_n)$. We know that if it satisfies eq. (26), then it implies $\dot{D}(t) = 0$ and corresponds to an equilibrium condition of the thermodynamic system. Hence D will forever be independent of t. But we note that if we put $D = D(H)$,.where H is the Hamiltonian function of the system, D will certainly satisfy eq. (26); for the left-hand side of that equation will read

$$\frac{\partial D}{\partial H}\left[\sum_{i=1}^{n}\left(\frac{\partial H}{\partial p_i}\dot{p}_i + \frac{\partial H}{\partial q_i}\dot{q}_i\right)\right]$$

and this vanishes because of (22). Hence we take $D = D(H)$.

Further restrictions on D cannot be imposed on the basis of mechanical or statistical reasoning, except for the obvious facts that D must be everywhere positive and must satisfy $\int D d\phi = N$. However, the choice of the function must be such as to lead to the thermodynamic formulas when thermodynamic quantities are computed by eq. (25). The important choices by which this success can be achieved, as Gibbs has shown, are these:

$$\left.\begin{array}{l}D(H) = \text{const. when } E_0 \leq H \leq E_0 + \Delta E \\ D(H) = 0 \text{ for all other values of } H\end{array}\right\} \tag{12-27}$$

$$D(H) = Ce^{-H/\theta} \tag{12-28}$$

where C and θ are positive constants. The first is called the *microcanonical* or *energy shell* ensemble, the second the *canonical* ensemble. The energy shell ensemble seems most reasonable from the physical point of view, for a system in equilibrium is one of fixed total energy, i.e., fixed within an interval of error ΔE, and systems not having an energy within this range are excluded from consideration. However, the canonical ensemble, although it assigns a finite density to points corresponding to those members of the ensemble which do not satisfy the requirement of constant energy, also leads to the correct thermodynamic relations. Since it is mathematically easier to handle, it enjoys greater popularity than the former, and was indeed preferred by Gibbs.

The connection between the two types of ensemble may be exhibited in the following way. Consider a gas whose phase density in γ-space is represented by a microcanonical ensemble. Let it consist of molecules with μ-spaces, μ_1, μ_2, etc., with probability distribution P_i in space μ_i. Denote the element of extension in μ_i by $d\phi_i$. Since energy exchanges may take place between the molecules, P_i cannot be represented by a microcanonical distribution; it must indeed be finite for all energies, H_i, of the i-th molecule. Nevertheless, the probability that molecule 1 be within the element $d\phi_1$ of its μ-space, molecule 2 within $d\phi_2$ of its μ-space, etc., simultaneously, equals the probability that the whole gas be in the element $d\phi = d\phi_1 d\phi_2 \cdots d\phi_\nu$ of γ-space. Hence

$$P_1(H_1)d\phi_1 P_2(H_2)d\phi_2 \cdots P_\nu(H_\nu)d\phi_\nu = P(H)d\phi \qquad (12\text{-}29)$$

so that

$$P_1(H_1) \cdot P_2(H_2) \cdots P_\nu(H_\nu) = P(H_1 + H_2 + \cdots H_\nu)$$

We wish this functional equation to be satisfied for every value of the total energy $H = \sum H_i$ although, of course, for any given H the constant $P(H)$ may be described by the microcanonical distribution. The solution of eq. (29) therefore leads to a very natural extension of this distribution.

Eq. (29) holds for every ν. If the gas consisted of only 2 molecules,

$$P_1(H_1) \cdot P_2(H_2) = P(H_1 + H_2)$$

Hence it follows that $P_i(0) = 1$ for every i. If we denote $\log P_i$ by f_i, we have

$$f_1(H_1) + f_2(H_2) = f(H_1 + H_2) \qquad (12\text{-}30)$$

On putting $H_2 = 0$, this reads

$$f_1(H_1) + f_2(0) = f(H_1)$$

and since $f_2(0) = 0, f_1 = f$. Thus all f_i are seen to be the same function, f. We are thus led to consider the equation

$$f(x) + f(y) = f(x + y)$$

When y is taken equal to x, we have $2f(x) = f(2x)$, and so by induction,

$$f(nx) = nf(x)$$

for every integer n. From this relation,

$$f\left[n\left(\frac{x}{n}\right)\right] = nf\left(\frac{x}{n}\right), \quad \text{whence} \quad f\left(\frac{x}{n}\right) = \frac{1}{n}f(x)$$

and

$$f\left(\frac{m}{n}x\right) = \frac{m}{n}f(x)$$

where m is another integer. Finally,

$$f(x) = f(x \cdot 1) = xf(1) = \text{const. } x$$

We have shown that the only function which satisfies eq. (30) is $f_i = cH_i$, whence

$$P_i(H_i) = e^{cH_i}$$

But P_i, being a probability, must remain finite for every H_i, a quantity which may tend to $+\infty$, though not to $-\infty$. Hence c is a negative constant. Following Gibbs, we write for it $-1/\theta$, so that finally

$$P_i(H_i) = e^{-H_i/\theta} \tag{12-31}$$

This defines the canonical ensemble, in accordance with eq. (28). The constant C in (28) has been introduced to insure normalization in γ-space: $\int P d\phi = 1$; the functions P_i in (29) are not properly normalized in each μ-space, as is evident on closer inspection.

Problem. Consider as system a single particle of mass m in a constant gravitational field. Note that the microcanonical ensemble is given by Fig. 1, where all points not lying between the two parabolas A and B, corresponding to $H = E_0$ and $H = E_0 + \Delta E$, have zero density. Show that the group of points lying between p_1 and p_2 at $t = 0$, will lie between p_1' and p_2' at time t, such that $p_1' = p_1 + mgt$, $p_2' = p_2 + mgt$. Prove also the invariance of the element of phase volume, i.e., area ϕ_1 = area ϕ_2. (Liouville's theorem.)

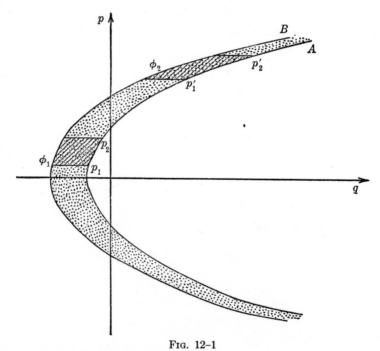

Fig. 12-1

12.7. Further Considerations Regarding the Canonical Ensemble.—
As an illustration regarding the use of the canonical ensemble we derive the Maxwell law for the distribution of velocities in an ideal gas. In

accordance with eq. (28) we put

$$Pd\phi = Ce^{-H/\theta}dp_1dp_2\cdots dp_ndq_1dq_2\cdots dq_n$$

$$= Ce^{-H/\theta}\prod_{i=1}^{\nu} dp_{ix}dp_{iy}dp_{iz}dx_idy_idz_i$$

The constant C must be so chosen as to make $\int Pd\phi = 1$. For an ideal gas,

$$H = \sum_{i=1}^{\nu}\left[\frac{p_{ix}^2 + p_{iy}^2 + p_{iz}^2}{2m} + V(x_iy_iz_i)\right] = \sum_{i=1}^{\nu}H_i$$

where $V(x,y,z)$ is the potential energy of a particle in an external field if such a field is present. The probability that particle i have an energy H_i corresponding to p_{ix}, p_{iy}, p_{iz}; x_i, y_i, z_i, regardless of the states of all other particles is clearly given by the integral $\int Pd\phi$ extended over the momenta and positions of all particles except i:

$$P_idp_{ix}dp_{iy}dp_{iz}dx_idy_idz_i = c'e^{-H_i/\theta}d\phi_i \qquad (12\text{--}32)$$

c' being some other constant. This relation, often called the Maxwell-Boltzmann law, is really nothing more than eq. (31). When no external field V is present, it may be written in more explicit form, for the constant c' can then be determined. Since

$$\int c'e^{-H_i/\theta}d\phi_i = 1$$

we have

$$\frac{1}{c'} = \int\int\int\int\int\int e^{-(p_x^2+p_y^2+p_z^2)/2m\theta}dp_xdp_ydp_zdxdydz$$

$$= \tau\int\int\int e^{-(p_x^2+p_y^2+p_z^2)/2m\theta}dp_xdp_ydp_z$$

where τ is the volume of the gas. Thus

$$\frac{1}{c'} = \tau\left[\int_{-\infty}^{\infty}e^{-u^2/2m\theta}du\right]^3 = \tau\cdot(2m\theta)^{3/2}\cdot\pi^{3/2}$$

When (32) is now expressed in terms of velocities instead of momenta and an integration over the volume is carried out on both sides, the result is

$$P(v_x,v_y,v_z)dv_xdv_ydv_z = (2\pi m\theta)^{-3/2}e^{-(v_x^2+v_y^2+v_z^2)m/2\theta}d(mv_x)d(mv_y)\cdot d(mv_z)$$

$$= \left(\frac{m}{2\pi\theta}\right)^{3/2}e^{-(v_x^2+v_y^2+v_z^2)m/2\theta}dv_xdv_ydv_z \qquad (12\text{--}33)$$

To find the meaning of the parameter θ we compute the mean energy of the i-th particle, $\overline{H_i}$, which, as we know from simple statistical theory, must be equal to $\frac{3}{2}kT$. We have

$$\overline{H_i} = \frac{\tau}{2m} \int \int \int_{-\infty}^{\infty} (p_x^2 + p_y^2 + p_z^2)c'\, e^{-(p_x^2+p_y^2+p_z^2)/2m\theta} dp_x dp_y dp_z$$

$$= \tau(2m)^{3/2}\theta^{5/2}c' \int \int \int_{-\infty}^{\infty} (u^2 + v^2 + w^2)e^{-(u^2+v^2+w^2)} du\,dv\,dw$$

$$= \tau(2m)^{3/2}\theta^{5/2}c' \cdot \frac{3}{2}\pi^{3/2} = \frac{3}{2}\theta$$

If this is to be equal to $\frac{3}{2}kT$, we must put

$$\theta = kT \tag{12-34}$$

Making this substitution in (33) we obtain Maxwell's law for the distribution of velocities

$$P(v_x,v_y,v_z)dv_x dv_y dv_z = \left(\frac{m}{2\pi kT}\right)^{3/2} e^{-(v_x^2+v_y^2+v_z^2)m/2kT} dv_x dv_y dv_z \tag{12-35}$$

The probability that the absolute value of v shall lie between v and $v + dv$ is derived from this expression by transforming the " volume " element $dv_x dv_y dv_z$ to spherical coordinates, where it takes the form $v^2 dv \sin\theta d\theta d\varphi$, and then integrating over θ and φ. Thus

$$P(v)dv = 4\pi v^2 \left(\frac{m}{2\pi kT}\right)^{3/2} e^{-(m/2kT)v^2} dv \tag{12-36}$$

According to its derivation $P(v)$ denotes the probability that *one* molecule shall have a speed about v. It is then clear that $\nu P(v)$ represents the number of molecules having this speed. It is this last interpretation which is usually given to Maxwell's law.

For most purposes it is convenient to write the canonical distribution law, eq. (28), in a slightly different form. When we put C, the positive constant occurring in that equation, equal to $Ne^{\psi/\theta}$, where ψ is a new parameter depending on θ, we have

$$P(H) = e^{(\psi-H)/\theta} \tag{12-37}$$

the standard form used by Gibbs.

In conclusion, let us attempt to correlate the quantities H, ψ, and θ with thermodynamic quantities. This can be done through the thermodynamic relations, the most important of which are:

$$dU = TdS - \sum_i f_i d\xi_i \tag{12-38}$$

and

$$-dA = SdT + \sum_i f_i d\xi_i \tag{12-39}$$

Here U stands for the internal energy of the system, S for its entropy, A for the Helmholtz free energy. The force f_i is defined by $f_i = -\partial H/\partial \xi_i$, it is called into play when the i-th *external* coordinate is changed. The summation $\sum f_i d\xi_i$ represents, therefore, the total work done by the thermodynamic system when it undergoes a (reversible) change involving variation of the ξ_i.

We now return to the ensemble whose distribution is given by (37). This distribution will change in detail as the condition of the system changes. But it will change in such a way that

$$\int e^{(\psi-H)/\theta}d\phi = 1$$

On differentiating this relation (permitting the external parameters ξ_i as well as θ and hence ψ to be altered) we have

$$d\int e^{(\psi-H)/\theta}d\phi = \int e^{(\psi-H)/\theta}\left[\frac{d\psi}{\theta} - \frac{\psi-H}{\theta^2}d\theta - \frac{1}{\theta}\sum_i \frac{\partial H}{\partial \xi_i}d\xi_i\right]d\phi$$

$$= \frac{d\psi}{\theta} - \frac{\overline{\ln P}}{\theta}d\theta + \frac{1}{\theta}\sum_i \bar{f}_i d\xi_i = 0 \qquad (12\text{--}40)$$

provided we use eq. (37) and indicate averages over the ensemble by horizontal bars, i.e., $\bar{Q} \equiv \int PQ d\phi$. The last equation may be written

$$-d\psi = -\overline{\ln P}\,d\theta + \sum \bar{f}_i d\xi_i \qquad (12\text{--}41)$$

But since $\overline{\ln P} = \dfrac{\psi - \bar{H}}{\theta}$, we also have

$$\psi = \theta \cdot \overline{\ln P} + \bar{H} \qquad (12\text{--}42)$$
$$d\psi = d\bar{H} + \theta\,d(\overline{\ln P}) + \overline{\ln P}\,d\theta$$

When this is substituted in (41), the result is

$$d\bar{H} = -\theta\,d(\overline{\ln P}) - \sum_i \bar{f}_i d\xi_i \qquad (12\text{--}43)$$

Now it is clear that $d\bar{H}$ must be identified with the increase in total energy of the thermodynamic system, dU. We have already established the relation $\theta = kT$. Furthermore, the $\bar{f}_i$ can hardly be anything other than the actual forces acting on the real system. We then see that (43) is the *exact* analogue of the thermodynamic relation (38), provided we interpret $-\overline{\ln P}$ as entropy divided by Boltzmann's constant, k.

When eq. (41) is now compared with (39), ψ is at once seen to be the Helmholtz free energy, A. With this additional interpretation, eq. (42) becomes the familiar

$$A = U - TS$$

Further pursuit of this matter shows that *all* thermodynamic relations are satisfied if we correlate thermodynamic with statistical quantities as follows:

$$
\left.\begin{aligned}
&\text{Total energy } (U) \text{ corresponds to } \overline{H}\\
&\text{Absolute temp. } (T) \text{ corresponds to } \theta/k\\
&\text{Entropy } (S) \text{ corresponds to } -k\ \overline{\ln P}\\
&\text{Helmholtz free energy } (A) \text{ corresponds to } \psi
\end{aligned}\right\}
\qquad (12\text{–}44)
$$

When A is given, many thermodynamic properties of the system at hand are known (cf. Chapter 1). It is therefore important to know how to compute the free energy, i.e., ψ, statistically. Since $\int e^{(\psi - H)/\theta} d\phi = 1$, we have $e^{-\psi/\theta} = \int e^{-H/\theta} d\phi$. The integral $\int e^{-H/\theta} d\phi \equiv I$, which is thus seen to be basic in the evaluation of ψ, is often called the *phase integral* (also " sum of state " and " partition function "). In terms of it

$$\psi = -kT \ln I$$

Problem. Using (36), verify the following relations for the first and second moments of the velocity distribution of an ideal gas:

$$\overline{|v|} = \frac{2}{\sqrt{\pi}} \left(\frac{2kT}{m}\right)^{1/2}$$

$$\overline{v^2} = \frac{3kT}{m}$$

Show also that the most probable velocity is $(2kT/m)^{1/2}$.

12.8. The Method of Darwin and Fowler.—A statistical method different from that of Gibbs but also leading to the correct thermodynamic laws, and more adaptable to the needs of quantum mechanics, has been introduced by Darwin and Fowler.[10] We shall first describe its fundamental features and then use it to derive quantum mechanical distribution laws. Consider a system made up of ν similar particles. No reference to an ensemble will here be made; all arguments concern this single, real system. If the particles are *independent*, as will now be assumed, each individual particle may be said to be in a definite energy state ϵ_i, this ϵ_i being an eigenvalue of the Schrödinger equation (see Chapter 11) for the single particle, with boundary conditions corresponding to the volume of the total system if the latter is a fluid, or other suitable conditions if it is a

[10] *Phil. Mag.* **44**, 450, 823 (1922); **45**, 1, 497 (1923). For a general and more recent treatment see Fowler, R. H., " Statistical Mechanics," Second Edition, Cambridge University Press, 1936.

crystal. By the *state of the total system* we mean the aggregate of single particle states, that is the assignment of individual particles to the various energies ϵ_i. But a *microscopic* assignment of particles to energies, as in the statement: particle 1 has energy ϵ_i, particle 2 has energy ϵ_j, particle 3 has energy ϵ_k, etc., has no meaning in quantum mechanics in view of the exclusion principle. The best that can be done in specifying a state is therefore to say that a_1 particles have energy ϵ_1, a_2 particles have energy ϵ_2, $\cdots a_i$ particles have energy ϵ_i, etc. Thus a state is defined when a system of " occupation numbers," $a_1, a_2, \cdots a_s$ is given. These must obviously satisfy the relation

$$\sum_i a_i = \nu \qquad (12\text{–}45)$$

We shall also prescribe that the system shall have a fixed total energy E, so that

$$\sum_i a_i \epsilon_i = E \qquad (12\text{–}46)$$

Now it is possible, as will be shown in the next section, to assign a *statistical weight*, $w(a_1 \cdots a_s)$ to each state $a_1 \cdots a_s$. The average of a quantity $Q(a_1 \cdots a_s)$ which takes on different values for different states is then defined by

$$\overline{Q} = \frac{\sum w(a_1 \cdots a_s) Q(a_1 \cdots a_s)}{\sum w(a_1 \cdots a_s)} \qquad (12\text{–}47)$$

The summations in this expression are understood to be taken over all values of a_1, a_2, a_3, etc., which satisfy conditions (45) and (46); the index s is in general very large; it is given by $\epsilon_s \leqq E$, $\epsilon_{s+1} > E$. Contact with experience is made in the Darwin-Fowler theory by assuming that $\overline{Q}$ is the observed value of the quantity Q when a measurement is made on the system. The Gibbsian ensemble average is here replaced by an average over the states of a single thermodynamic system. The fact that they agree is rather noteworthy from a logical point of view. To carry through the calculation of an average like (47) it is necessary (a) to construct a suitable weighting function, w; (b) to devise means for evaluating the restricted sums appearing in that equation.

12.9. Quantum Mechanical Distribution Laws.—In quantum mechanics, the weight of an energy state is defined as its degree of degeneracy: it is equal to the number of linearly independent state functions belonging to the eigenstate in question. This postulate will here be invoked. Our system, however, is one containing ν similar particles; hence it is necessary to apply all the considerations of secs. 11.32 and 11.33, in particular the exacting demands of the exclusion principle. But for the moment it seems well to consider the number of eigenstates of E belonging to the statistical state $(a_1 \cdots a_s)$ when the exclusion principle is left out of account.

As an example, consider the simple case of 5 particles, with energy partition $a_1 = 2$ and $a_2 = 3$. Let the single-particle function belonging to the single-particle energy ϵ_1 be ψ_1, that belonging to ϵ_2, ψ_2. The Schrödinger equation for the 5 particles corresponding to $E = 2\epsilon_1 + 3\epsilon_2$ will then be satisfied by the simple product

$$\psi_1(1)\psi_1(2)\psi_2(3)\psi_2(4)\psi_2(5) \qquad (12\text{–}48)$$

as well as by any function obtained from this product through permutation of the arguments 1 to 5 (each numeral designates all coordinates, including the spin, of the corresponding particle). But not all the 5! products thus obtained are independent. For instance a transposition of 1 and 2, or a permutation among particles 3, 4 and 5 causes no change in the function. The number of *different* combinations is obviously equal to the number of ways in which 5 objects can be arranged in 2 piles, one containing 2 the other 3 objects. This, according to eq. (1), is

$$\frac{5!}{2!3!}$$

The generalization of this result is immediate; the number of different energy eigenfunctions belonging to $(a_1 a_2 \cdots a_s)$ is given by

$$w(a_1 \cdots a_s) = \frac{\nu!}{a_1! a_2! \cdots a_s!} \qquad (12\text{–}49)$$

This is true so long as each individual particle function, ψ_i, is non-degenerate. Suppose now that the energy ϵ_i itself can be realized by g_i different functions. Each ψ_i then has a weight g_i, and the product of a_i such functions has a weight $g_i^{a_i}$. We then obtain, in place of (49), the more general result

$$w(a_1 \cdots a_s) = \nu! \frac{g_1^{a_1} g_2^{a_2} \cdots g_s^{a_s}}{a_1! a_2! \cdots a_s!} \qquad (12\text{–}50)$$

To see this in detail, let us return to the example (48) and assume $g_1 = 3$, $g_2 = 2$. It is then necessary to introduce new functions, e.g., b, c, d in place of ψ_1; e and f in place of ψ_2. Instead of $\psi_1(1)\psi_1(2)$ we can now have

$$b(1)b(2), \quad c(1)c(2), \quad d(1)d(2), \quad b(1)c(2), \quad b(2)c(1), \quad b(1)d(2),$$
$$b(2)d(1), \, c(1)d(2), \, c(2)d(1)$$

and in place of $\psi_2(3)\psi_2(4)\psi_2(5)$

$$e(3)e(4)e(5), \, f(3)f(4)f(5), \, e(3)f(4)f(5), \, e(4)f(3)f(5), \, e(5)f(3)f(4),$$
$$e(3)e(4)f(5), \quad e(3)e(5)f(4), \quad e(4)e(5)f(3)$$

Eq. (50) would be the statistical weight of the state $a_1 \cdots a_s$ if no symmetry requirements, no Pauli principle, had to be respected.

How many different functions can be constructed out of the individual ψ_i when overall antisymmetry is demanded? We have seen that the only antisymmetric function available has the determinantal form eq. (11-160'). This, however, vanishes when any two particles are described by the same ψ_i, for then two rows of the determinant are equal. Hence, if there is no degeneracy in the individual particle functions (all g_i are 1), only one function is constructible; in other words,

$$w(a_1 \cdots a_s) = \begin{cases} 1 \text{ if every } a_i \leqq 1 \\ 0 \text{ if any } a_i > 1 \end{cases}$$

On the other hand, if the i-th state has a degeneracy $g_i > 1$, the number of non-vanishing determinants which can be constructed is equal to the number of ways in which a_i arguments can be distributed among g_i different functions, and this, by virtue of eq. (2) is $\binom{g_i}{a_i}$. Thus we obtain in general, when the exclusion principle is applied,

$$w(a_1 \cdots a_s) = \binom{g_1}{a_1}\binom{g_2}{a_2} \cdots \binom{g_s}{a_s} \tag{12-51}$$

Note that this vanishes when any a_i is greater than its corresponding g_i, so that the preceding equation is a special case of this.

Finally, we consider the case in which the total function is symmetrical. As was shown in sec. 11.33, this is of the form $\sum_P \Psi_P$, where Ψ_P is a function constructed like (48), with a particular permutation of the arguments 1 to ν. But if any permutation of arguments is made in $\sum_P \Psi_P$, this function is transformed into itself. Hence $w(a_1 \cdots a_s)$ is always 1 provided all g_i are 1. *But if this is not true, then the degeneracy of ϵ_i gives rise to as many different combinations of functions $\psi_i^{(1)}\psi_i^{(2)} \cdots \psi_i^{(gi)}$ as there are ways of distributing the a_i arguments amongst them, without regard to the number of arguments associated with the same function.* This number, by eq. (4), is $\binom{g_i + a_i - 1}{a_i}$. We have thus determined w for the symmetrical case to be

$$w(a_1 \cdots a_s) = \binom{g_1 + a_1 - 1}{a_1} \cdots \binom{g_s + a_s - 1}{a_s} \tag{12-52}$$

Assemblies of particles whose motion is governed by the Pauli principle must be described by antisymmetric functions. Their statistical weights are given by (51). The formulas which ensue from its use are characteristic of *Fermi-Dirac* statistics, the type of statistics to which electrons, neutrons, protons are subject. Henceforth we refer to (51) as $w_{F.D.}$. On the other hand, photons, nuclei and atoms containing an even number of elementary particles (e.g., He^4) are known to require symmetrical state functions for their collective description. Their statistical states have weights given by (52); they are said to obey *Einstein-Bose* statistics. Henceforth we write $w_{E.B.}$ in place of (52). No known constituents of material bodies are described by (50), although it is precisely that assignment of weights which leads to the Maxwell-Boltzmann law. It will be shown, however, that both quantum formulations, (51) and (52), give rise to distribution laws which under many thermodynamic circumstances are practically identical with the Maxwell-Boltzmann law. For this reason, and for the sake of generality, we continue to include eq. (50) in our consideration, and refer to it as $w_c{}^{11}$ (classical assignment of statistical weights).

It is to be noted that all three statistical weights may be written in the form

$$w = \prod_j \gamma(a_j)$$

if we put

$$\left. \begin{aligned} \gamma_c(a_j) &= \frac{(g_j)^{a_j}}{a_j!} \\ \gamma_{F.D.}(a_j) &= \binom{g_j}{a_j} \\ \gamma_{E.B.}(a_j) &= \binom{g_j + a_j - 1}{a_j} \end{aligned} \right\} \qquad (12\text{-}53)$$

In proceeding thus the factor $\nu!$ in eq. (50) is being omitted. However, since this factor is independent of the a's and hence constant for all statistical states, it will cancel when averages are computed after the manner of eq. (47). This is the principal use which will here be made of the weight function. In many other problems the omission of $\nu!$ is not permissible. (See the remarks after eq. 72.)

The quantity Q whose average we wish to calculate is a_r, the number of particles having a given energy ϵ_r. It is necessary, therefore, to evaluate

$$\bar{a}_r = \frac{\sum\limits_{(a)} a_r \prod\limits_j \gamma(a_j)}{\sum\limits_{(a)} \prod\limits_j \gamma(a_j)} = \frac{A}{W} \qquad (12\text{-}54)$$

W is here written for the sum of all statistical weights compatible with the fixed energy E, A for $\bar{a}_r W$. The summations appearing in (54) are

[11] The formula for w_c can also be derived as follows, without reference to state functions. Divide phase space into cells according to the energies of the individual particles: in the i-th cell a particle has energy ϵ_i. If the i-th cell has a fundamental weight g_i, then the number of ways in which the state $a_1, a_2 \cdots a_s$ can be realized by assignments of specific particles to cells is given by eq. (50).

taken over all values of a_1, a_2, $a_3 \cdots a_s$ which satisfy both (45) and (46). We first calculate W.

For this purpose consider the expansion

$$M = \sum_{a_1 a_2 \cdots a_s} \left[\prod_j \{\gamma(a_j)\} x^{\sum_j a_j} z^{\sum_j a_j \epsilon_j} \right] \tag{12-55}$$

in which the summation over the a's is entirely unrestricted, each one of the many a's taking all values from 0 to ∞. M may be regarded as a function of x and z, depending parametrically upon the eigenvalues ϵ_j characteristic of the particles in question. A moment's reflection will show that W is the coefficient of $x^\nu z^E$ in the expansion M; in other words, because of the theorem of residues eq. (3-3),

$$W = \left(\frac{1}{2\pi i}\right)^2 \oint \oint \frac{M\, dx\, dz}{x^{\nu+1} z^{E+1}} \tag{12-56}$$

the integrals being taken counter-clockwise about the poles of the integrand, i.e., about $x = 0$ and $z = 0$, x and z being considered as complex variables.

Now M may be evaluated rather simply. First note that it can be written

$$M = \sum_{a_1 \cdots a_s} \prod_j \{\gamma(a_j) x^{a_j} z^{a_j \epsilon_j}\}$$

$$= \prod_j \left\{ \sum_{n=0}^{\infty} \gamma(n) x^n z^{n\epsilon_j} \right\}$$

The summation in $\{ \quad \}$ can be performed for all three of the functions γ listed in (53). Let us put $xz^{\epsilon_j} \equiv r$, $\sum_{n=0}^{\infty} \gamma(n) r^n \equiv f(r)$. We then obtain

$$\left.\begin{aligned} f_c &= \sum_{n=0}^{\infty} \frac{(g_j r)^n}{n!} = \exp\,(g_j r) = \exp\,\{g_j x z^{\epsilon_j}\} \\[2mm] f_{F.D.} &= \sum_{n=0}^{\infty} \binom{g_j}{n} r^n = (1 + r)^{g_j} = (1 + xz^{\epsilon_j})^{g_j} \\[2mm] f_{E.B.} &= \sum_{n=0}^{\infty} \binom{g_j + n - 1}{n} r^n = (1 - xz^{\epsilon_j})^{-g_j} \end{aligned}\right\} \tag{12-57}$$

The last result, which is perhaps not so obvious, is easily verified by writing down the MacLaurin expansion of

$$(1 - r)^{-g} = 1 + \frac{g}{1!} r + \frac{g(g + 1)}{2!} r^2 + \frac{g(g + 1)(g + 2)}{3!} r^3 + \cdots$$

which is identical with the summation in $f_{E.B.}$.

Thus we see that

$$M = \prod_j f(xz^{\epsilon_j}) \qquad (12\text{-}58)$$

and W is related to M by eq. (56). The effect of the degeneracy factors g_j in (57) is rather interesting: they merely appear as exponents in the f-functions in all three cases.

Let us now consider the numerator of eq. (54). If we calculate the quantity $(1/\ln z)(\partial M/\partial \epsilon_r)$, we find, using eq. (55),

$$\frac{1}{\ln z}\frac{\partial}{\partial \epsilon_r} \sum_{a_1 a_2 \cdots a_s} \left\{ \prod_j \gamma(a_j)x^{a_j}(e^{\ln z})^{a_j \epsilon_j} \right\} = \sum_{a_1 a_2 \cdots a_s} a_r \left\{ \prod_j \gamma(a_j)x^{a_j}z^{a_j \epsilon_j} \right\}$$

On comparing this with (54) we see that A is the coefficient of $x^\nu z^E$ in the expansion of $(1/\ln z)(\partial M/\partial \epsilon_r)$. But this last quantity can be put in a form more suitable for our purposes. In view of (58),

$$\frac{1}{\ln z}\frac{\partial M}{\partial \epsilon_r} = \frac{1}{\ln z}\left\{ \frac{M}{f(xz^{\epsilon_r})}\frac{df(xz^{\epsilon_r})}{d(xz^{\epsilon_r})} \cdot xz^{\epsilon_r} \cdot \ln z \right\}$$

$$= Mx\frac{\partial}{\partial x}\ln f(xz^{\epsilon_r})$$

Summarizing these steps, we note:

$$A = \left(\frac{1}{2\pi i}\right)^2 \oint \oint x\frac{\partial}{\partial x}\ln f(xz^{\epsilon_r}) \frac{M\,dx\,dz}{x^{\nu+1}z^{E+1}} \qquad (12\text{-}59)$$

If now it were possible to find a path of integration around the origin with respect to both x and z, such that the function $Mx^{-\nu-1}z^{-E-1}$ were practically zero everywhere along that path except in the immediate neighborhood of two definite points, say $x = \xi$ and $z = \vartheta$, the evaluation of $\bar{a}_r = A/W$ would be very simple. For it would then be permissible to take the factor $x(\partial/\partial x)\ln f$, multiplying the integrand in (59), in front of the integral sign and give to x and z the values ξ and ϑ, and the integrals themselves would cancel. We should then have

$$\bar{a}_r = \xi\frac{\partial}{\partial \xi}\ln f(\xi\vartheta^{\epsilon_r}) \qquad (12\text{-}60')$$

This procedure is indeed justified, as the following section will show. If the f-functions are identified in accordance with (57), the result is seen to be

$$(\bar{a}_r)_c = \xi g_r \vartheta^{\epsilon_r}$$

$$(\bar{a}_r)_{F.D.} = \frac{g_r \vartheta^{\epsilon_r}}{\xi^{-1} + \vartheta^{\epsilon_r}}$$

$$(\bar{a}_r)_{E.B.} = \frac{g_r \vartheta^{\epsilon_r}}{\xi^{-1} - \vartheta^{\epsilon_r}}$$

The physical significance of the parameters ξ and ϑ can be fixed most simply by the subsequent plausibility argument which we offer in lieu of more detailed considerations[12] adducible to establish this meaning more completely.   The first of the expressions above must be the Maxwell-Boltzmann law which reads, for the situation here considered, $\bar{a}_r = A_0 g_r e^{-\epsilon_r/kT}$, the factor A_0 being so determined that $\sum_r a_r = \nu$. Hence ξ must be identified with A_0, and ϑ with $e^{-1/kT}$.   In the other two relations ξ must also act as a normalizing factor, while ϑ has the same meaning as in $(\bar{a}_r)_c$.   Hence we conclude

$$
\left.
\begin{aligned}
(\bar{a}_r)_c &= \xi g_r e^{-\epsilon_r/kT} & \text{(a)} \\[2mm]
(\bar{a}_r)_{F.D.} &= \frac{g_r e^{-\epsilon_r/kT}}{\xi^{-1} + e^{-\epsilon_r/kT}} & \text{(b)} \\[2mm]
(\bar{a}_r)_{E.B.} &= \frac{g_r e^{-\epsilon_r/kT}}{\xi^{-1} - e^{-\epsilon_r/kT}} & \text{(c)}
\end{aligned}
\right\} \quad (12\text{--}60)
$$

It is easily seen in a qualitative way that ξ must increase when ν increases (the volume of the system being fixed) if $\sum a_r$ is to remain equal to ν.   In (a), ξ is in fact proportional to ν, but this simple dependence fails in (b) and (c).   Nevertheless, if ν is very small, $\xi^{-1} \gg 1 > e^{-\epsilon_r/kT}$. In this case both (b) and (c) reduce to the classical form (a).   Hence for sufficiently small densities all assemblies show an essentially classical behavior.   Closer investigation (see any of the references at the beginning of this chapter) indicates that this is true for all ordinary molecules at ordinary temperatures, thus justifying the use of classical statistics. The main instances in which quantum distribution laws are needed are the motion of electrons in metals (b), the photon gas, and helium at very low temperatures (c).

All thermodynamic relations can be deduced by the method here described provided the following associations between thermodynamic quantities and elements appearing in the Darwin-Fowler scheme are made; the first is obvious, the second has already been obtained; the others will be derived later on (cf. eqs. 71 and 72):

$$
\left.
\begin{aligned}
&U \text{ corresponds to } E \\
&T \text{ corresponds to } -(k \ln \vartheta)^{-1} \\
&S \text{ corresponds to } k \ln W
\end{aligned}
\right\} \quad (12\text{--}61)
$$

$$
\nu A \text{ corresponds to } -kT\left(\sum_j \ln f(\xi \vartheta^{\epsilon_j})\right) - \nu \ln \xi
$$

12.10. The Method of Steepest Descents.—The evaluation of $\bar{a}_r$ in the last section depended for its validity on our ability to find a point $x = \xi$ in

[12] Cf. Fowler, loc. cit.

the integration with respect to x, and another point $z = \vartheta$ in the integration with respect to z, at which the integrand of (56) would be large, and where its value would descend very steeply on both sides along the path of integration. Such a point has interesting properties which we shall first investigate in connection with a simpler but more general example.

Let $\varphi(z)$ be a function in the complex plane, z being $x + iy$. We wish to find a point in the X, Y-plane such that, as we cross that point in the direction of steepest descent of φ, φ will be a maximum at the point. To be more specific we will refer in this inquiry not to φ, but to the *real part of φ*. Suppose the point thus defined is $z = \vartheta$.

On writing

$$\varphi(z) = g(x,y) + ih(x,y) \tag{12-62}$$

where g and h are real functions, it is clear that these must satisfy the Riemann relations:

$$g_x = h_y, \quad g_y = -h_x{}^{13} \tag{12-63}$$

Our specification amounts to this: $dg = g_x dx + g_y dy$ shall be zero along the path on which g decreases most rapidly. The direction of this path is the direction of the negative gradient of g, namely $-\nabla g$. Since this is $-(ig_x + jg_y)$, this direction is defined by $dy/dx = g_y/g_x$. But by reason of eq. (63), this is $-h_x/h_y$. On the other hand, if $dy/dx = -h_x/h_y$, then

$$h_x dx + h_y dy = dh = 0$$

in the same direction. Now the vanishing of both dg and dh at ϑ is possible only if $\varphi'(\vartheta) = 0$. We may conclude, therefore, that the point in question, if it exists at all, satisfies the condition

$$\varphi' = 0$$

If φ' has a real root, the point ϑ will obviously lie on the real axis.

Next, it will be shown that ϑ is a "*saddle point*," i.e., that the curvature on the path of steepest descent is *opposite* to that along a path at right angles to this direction. For any direction dy/dx

$$d^2g = g_{xx}dx^2 + 2g_{xy}dxdy + g_{yy}dy^2$$

---

[13] By g_x is meant $\partial g/\partial x$, etc. To prove these well known relations we observe

$$\varphi_x = \varphi' = g_x + ih_x$$

$$\varphi_y = i\varphi' = g_y + ih_y$$

with $\varphi' = d\varphi/dz$. Hence $\varphi' = g_x + ih_x = -ig_y + h_y$, which is equivalent to eqs. (63) when real and imaginary parts are equated separately. Note also that (63) implies: $g_{xx} = -h_{yy}$.

The direction at right angles to dy/dx is fixed by the substitution of $-dy$ for dx, and of dx for dy. Hence, on writing $d^2g_\perp$ for the curvature at right angles to the direction dy/dx,

$$d^2g_\perp = g_{xx}dy^2 - 2g_{xy}dxdy + g_{yy}dx^2$$

so that

$$d^2g + d^2g_\perp = (g_{xx} + g_{yy})(dx^2 + dy^2) = 0$$

in view of eqs. (63).

With this general knowledge, let us return to the calculation of (56):

$$W = (2\pi i)^{-2} \oint \oint \frac{M(x,z)}{x^\nu z^E} \frac{dx}{x} \frac{dz}{z} \tag{12-64}$$

$$M = \prod_j f(r_j), \quad r_j = xz^{\epsilon_j}$$

Let us further put

$$Mx^{-\nu}z^{-E} = e^{Y(x,z)}$$

so that

$$Y = \sum_j \ln f(r_j) - \nu \ln x - E \ln z \left.\right\} \tag{12-65,}$$

A saddle point of the integrand of (64) is then determined by

$$\frac{\partial Y}{\partial x} = 0$$

in the integration with respect to x, and by

$$\frac{\partial Y}{\partial z} = 0$$

in the integration with respect to z. The first of these leads to

$$\sum_j \frac{f'(\rho_j)}{f(\rho_j)} \vartheta^{\epsilon_j} - \frac{\nu}{\xi} = 0 \tag{12-66}$$

the second to

$$\frac{1}{\vartheta} \sum_j \frac{f'(\rho_j)}{f(\rho_j)} \rho_j \epsilon_j - \frac{E}{\vartheta} = 0 \tag{12-67}$$

where ρ_j has been written for $\xi\vartheta^{\epsilon_j}$. Eqs. (66) and (67) define the saddle point (ξ,ϑ) in $X - Z$ space:

$$\xi\sum_j \frac{f'}{f} \vartheta^{\epsilon_j} = \nu \tag{12-66'}$$

$$\xi\sum_j \frac{f'}{f} \vartheta^{\epsilon_j} = E \tag{12-67'}$$

These results at once take on a more interesting form when we insert eq. (60′):

$$\bar{a}_r = \xi \frac{f'}{f} \vartheta^{\epsilon_r}$$

The saddle point conditions are then seen to be nothing more than the conservation conditions of our problem:

$$\sum_j \bar{a}_j = \nu$$

$$\sum_j \bar{a}_j \epsilon_j = E$$

In classical statistics we may also write

$$\xi \sum_j g_j \vartheta^{\epsilon_i} = \nu$$

$$\xi \sum_j g_j \vartheta^{\epsilon_i} \epsilon_j = E$$

In quantum statistics, the equations become

$$\xi \sum_j g_j \frac{\vartheta^{\epsilon_i}}{1 \pm \xi \vartheta^{\epsilon_i}} = \nu$$

$$\xi \sum_j g_j \frac{\vartheta^{\epsilon_i} \epsilon_j}{1 \pm \xi \vartheta^{\epsilon_i}} = E$$

where the positive sign is to be taken in the Fermi-Dirac case and the negative sign in the Bose-Einstein case.

In the following we shall also need the values of $\partial^2 Y / \partial x^2$, $\partial^2 Y / \partial z^2$, and $\partial^2 Y / \partial x \partial z$ at the saddle point. To save writing the discussion will be limited for the present to F.D. statistics. Here one finds [14]

$$\frac{\partial^2 Y}{\partial x^2}\bigg|_{\xi,\vartheta} = \frac{1}{\xi^2} \sum_j g_j \frac{\xi \vartheta^{\epsilon_i}}{(1 + \xi \vartheta^{\epsilon_i})^2} \equiv \frac{A}{\xi^2}$$

$$\frac{\partial^2 Y}{\partial z^2}\bigg|_{\xi,\vartheta} = \frac{1}{\vartheta^2} \sum_j g_j \xi \vartheta^{\epsilon_i} \left(\frac{\epsilon_j}{1 + \xi \vartheta^{\epsilon_i}} \right)^2 \equiv \frac{B}{\vartheta^2}$$

$$\frac{\partial^2 Y}{\partial x \partial z}\bigg|_{\xi,\vartheta} = \frac{1}{\xi \vartheta} \sum_j g_j \xi \vartheta^{\epsilon_i} \frac{\epsilon_j}{(1 + \xi \vartheta^{\epsilon_i})^2} \equiv \frac{C}{\xi \vartheta}$$

The quantities A, B, and C are to be defined by these relations; it is easily seen that, for a gas consisting of many particles, they are very large

[14] Note that the symbol A has a different meaning than in the last section! Neither is it to be confused with the Helmholtz free energy.

numbers.[15]   Moreover, the reader will be able to show that

$$AB > C^2 \tag{12-68}$$

always.

Having located the saddle point, we expand $Y(x,z)$ in its neighborhood.

$$Y(x,z) = Y(\xi,\vartheta) + \frac{\partial Y}{\partial \xi}(x - \xi) + \frac{\partial Y}{\partial \vartheta}(z - \vartheta) \tag{12-69}$$

$$+ \frac{1}{2}\left[\frac{\partial^2 Y}{\partial \xi^2}(x - \xi)^2 + 2\frac{\partial^2 Y}{\partial \xi \partial \vartheta}(x - \xi)(z - \vartheta) + \frac{\partial^2 Y}{\partial \vartheta^2}(z - \vartheta)^2\right] + \cdots$$

The first derivatives vanish at the saddle point, which, as may be shown from eqs. (66) and (67), lies on the real axis of both x and z. In order to carry out the integrations in (64), it is suggested that the paths be taken across ξ and ϑ, and this is done with greatest convenience by choosing the circuits

$$x = \xi e^{i\alpha}, \quad -\pi < \alpha \leq \pi; \quad z = \vartheta e^{i\beta}, \quad -\pi < \beta \leq \pi$$

When these substitutions are made in (69), this expression becomes

$$Y(x,z) = Y(\xi,\vartheta) - \tfrac{1}{2}(A\alpha^2 + B\beta^2 + 2C\alpha\beta)$$

in the neighborhood of ξ, ϑ, since for small α and β

$$x - \xi = i\xi\alpha \quad \text{and} \quad z - \vartheta = i\vartheta\beta$$

Therefore, in view of (65),

$$Mx^{-\nu}z^{-E} = e^{Y(\xi,\vartheta)}\, e^{-(1/2)(A\alpha^2 + B\beta^2 + 2C\alpha\beta)}$$

and

$$W = (2\pi i)^{-2} \int \int_{-\pi}^{\pi} e^{Y(\xi,\vartheta)\,-\,(1/2)(A\alpha^2+B\beta^2+2C\alpha\beta)}\,(id\alpha)\cdot(id\beta) \tag{12-70}$$

This result shows with impressive clarity how rapidly the integrand " descends " from its saddle point: its " half width " with respect to α for example, is approximately given by $A^{-1/2}$. But A is of the same order of magnitude as ν, the total number of particles. The procedure of the foregoing section was therefore proper.

The question arises as to the behavior of the integrand at points on the contour *not* in the neighborhood of the saddle point, for we hardly have reason thus far to expect that it is small everywhere else. This, however, is not difficult to prove. When written in terms of α and β, the function f of (57) takes the form

$$f_{F.D.} = (1 + \xi\vartheta^{\epsilon_j}\, e^{iu_j})^{\theta_j}, \quad u_j = \alpha + \beta\epsilon_j$$

---

[15] The ϵ_j must here be regarded as dimensionless and of order of magnitude unity or greater. This can be achieved by measuring kT and ϵ_j in the same conveniently chosen unit, in which case kT and hence ϑ, also, become pure numbers.

and M becomes

$$M = \prod_j (1 + 2\xi\vartheta^{\epsilon_j} \cos u_j + \xi^2\vartheta^{2\epsilon_j})^{g_j/2} \exp\left\{ ig_j \tan^{-1} \frac{\xi\vartheta^{\epsilon_j} \sin u_j}{1 + \xi\vartheta^{\epsilon_j} \cos u_j} \right\}$$

The product of all exponential terms, which are purely imaginary, can never exceed unity, the value which it has at the saddle point. Each term in the first parenthesis attains its maximum when $u_j = 0$, or in general $2n\pi$. If $u_j = 2n\pi$ for a given choice of α and β there will be very many u_k, $k \neq j$, for which the parenthesis will not assume its maximum value, so that the product, having a great number of factors, will be much smaller than its value at ξ, ϑ. The only way to insure that M will be a maximum is to make $u_j = 0$ for *all* j, and this requires that both α and β be zero. This maximum will be very strong provided the number of energy states, ϵ_j, is large.

It is of some interest, finally, to conclude the explicit calculation of W. The integral in (70) is easily performed, for the limits may clearly be replaced by $+\infty$ and $-\infty$. Remembering the formula

$$\int_{-\infty}^{\infty} e^{ax^2 \pm bx} \, dx = \sqrt{\frac{\pi}{a}}\, e^{b^2/4a}$$

we find

$$W = \frac{e^{Y(\xi,\vartheta)}}{2\pi\sqrt{AB - C^2}}$$

which, in view of (68), is real and positive. The entropy, defined in (61), becomes therefore

$$S = kY(\xi,\vartheta) - k \ln\left[2\pi\sqrt{AB - C^2}\right] \tag{12-71}$$

In chemistry, it is customary to neglect the second term of S because it is much smaller than the first when the number of particles is large.[16] Now

$$Y(\xi,\vartheta) = \ln M(\xi,\vartheta) - \nu \ln \xi - E \ln \vartheta$$

$$= \sum_j \ln f(\xi\vartheta^{\epsilon_j}) - \nu \ln \xi + \frac{E}{kT}$$

In view of (71), then,

$$E - ST = -kT(\sum_j \ln f - \nu \ln \xi) \tag{12-72}$$

This justifies the identification of the free energy made in (61).

Finally let us endeavor to make contact with classical statistics again. Here it must be remembered that a factor $\nu!$ was omitted in the evaluation of W. We must therefore add to (71) the quantity $k \ln \nu!$, so that we have

---

[16] The full expression must be used when attention is given to the entropy of a nucleus, which contains relatively few neutrons and protons.

in place of (72) for the classical free energy A_{class}.

$$\nu \cdot A_{\text{class}} = -kT(\sum_j \ln f - \nu \ln \xi + \ln \nu!)$$

But

$$\sum_j \ln f = \nu$$

by eq. (66') and

$$\ln \nu! \approx \nu \ln \nu - \nu$$

by Stirling's theorem. Hence

$$\nu \cdot A_{\text{class}} = -kT\left(\nu \ln \frac{\nu}{\xi}\right)$$

Again by (66'), $\nu/\xi = \sum_j g_j \vartheta^{\epsilon_j}$, a quantity to be denoted by Z and often called the *partition function*. In terms of Z,

$$A_{\text{class}} = -kT \ln Z$$

Comparison with the last equation of sec. 12.7 shows that Z is the quantum mechanical analogue of Gibbs' phase integral.[17]

The computational aspects of the statistical method can be simply summarized as follows. Given a system of ν particles whose total energy is E. Each particle has energies ϵ_j obtainable by solving the Schrödinger equation with boundary conditions corresponding to the volume in which the particles are enclosed. For instance, if the volume is a parallelepiped, the ϵ_j are given by eq. (11–36). Thus they depend on the volume of the container.

The thermodynamic properties of the system then depend on two parameters, ξ and ϑ, defined by eqs. (66') and (67'). When these are solved simultaneously and ξ and ϑ are known for the given ν and E, the quantities T, S and A can be calculated from (61). In F.D. and E.B. statistics, eqs. (66') and (67') are such that there is no general method for obtaining explicit solutions for ξ and ϑ. Recourse must then be had to approximations, valid in different ranges of the parameter ξ.[18]

Problem. Find the values of A, B, C in classical and in E.B. statistics. Note that the classical values are obtainable from those derived in the preceding section by letting $\xi \to 0$.

[17] Partition functions for specific substances can often be computed from spectroscopic data. Such calculations are becoming increasingly important in applied thermodynamics. See Taylor, H. S., and Glasstone, S., "A Treatise on Physical Chemistry," Third Edition, Vol. 1, D. Van Nostrand Co., Inc., New York, 1942

[18] See for instance Fowler, loc. cit.

REFERENCES

Tolman, R. C., "The Principles of Statistical Mechanics," Clarendon Press, Oxford, 1938.

Chapman, S., and Cowling, T. G., "The Mathematical Theory of NonUniform Gases," University Press, Cambridge, 1939.

Mayer, J. E. and Mayer, M. G., "Statistical Mechanics," John Wiley and Sons, Inc., 1940.

Lindsay, R. B., "Physical Statistics," John Wiley and Sons, Inc., 1941.

Schrödinger, E., "Statistical Thermodynamics," Cambridge Press, 1948.

Rushbrooke, G. S., "Introduction to Statistical Mechanics," Oxford Press, 1949.

Ter Haar, D., "Elements of Statistical Mechanics," Rinehart and Company, NewYork, 1954.

CHAPTER 13

NUMERICAL CALCULATIONS

13.1. Introduction.—We describe here certain types of numerical calculations which are often required. No theory[1] is presented, but the methods are explained and illustrated by means of worked examples. The reader will find that such computations are usually tedious and time-consuming, hence he should exercise his ingenuity in devising means of reducing the labor involved. Before starting a calculation, he should always consider the possibility of using graphical methods, for these are often simpler than the numerical ones. He should also remember that there is some advantage in representing numerical data by equations of empirical or theoretical form. Such equations, obtained by the method of least squares or otherwise (see sec. 13.37) are generally easier to use for interpolation, differentiation or integration than the methods of this chapter. Finally, he should note that when alternative procedures are given for a particular operation, the special problem at hand may often suggest which of these is the most suitable.

It is assumed that the reader is familiar with the elementary facts concerning significant figures, rounding off and number of significant figures to be retained in addition, multiplication, etc.[2]

For convenience, we divide this chapter into three separate parts. The first deals with methods primarily based on interpolation formulas; the second, with miscellaneous algebraic calculations and the third with a discussion of errors and related problems.

PART 1. NUMERICAL METHODS BASED ON INTERPOLATION FORMULAS

INTERPOLATION

13.2. Interpolation for Equal Values of the Argument.—It often happens that data are given in tabular form with values of x and $y = f(x)$ at certain intervals of x. Suppose a value of y is needed for an x, which is not

---

[1] See references at end of chapter.

[2] Retention of an unnecessary number of significant figures should be carefully avoided, especially in physical and chemical calculations. If $(n + 1)$-digits are carried along in the intermediate stages of the calculations, the final result, obtained by rounding-off and thus containing n significant figures, will be uncertain by one or two units in its last digit. This practice is customary in treating scientific data.

listed in the table. Usually, the simplest procedure is to plot y against x, draw a smooth curve through the points and read from the graph the required value of y. The same result may be obtained by the use of *interpolation formulas.* Provided the given values of x are equidistant, we first form a *difference table*[3] as shown in Table 1 where the first, second, third and r-th *differences* are given by

$$\Delta y_0 = y_1 - y_0, \quad \Delta y_1 = y_2 - y_1, \cdots, \quad \Delta y_{n-1} = y_n - y_{n-1},$$

$$\Delta y_n = y_{n+1} - y_n$$

$$\Delta^2 y_0 = \Delta y_1 - \Delta y_0 = y_2 - 2y_1 + y_0, \cdots$$

$$\Delta^2 y_n = \Delta y_{n+1} - \Delta y_n = y_{n+2} - 2y_{n+1} + y_n$$

$$\Delta^3 y_n = \Delta^2 y_{n+1} - \Delta^2 y_n = y_{n+3} - 3y_{n+2} + 3y_{n+1} - y_n$$

$$\cdots \quad \cdots \quad \cdots \quad \cdots \quad \cdots \quad \cdots \quad \cdots \quad \cdots$$

$$\Delta^r y_n = \Delta^{r-1} y_{n+1} - \Delta^{r-1} y_n = y_{n+r} - r y_{n+r-1} + \frac{r(r-1)}{2!} y_{n+r-2} + \cdots$$

$$+ (-1)^r y_n$$

$$= \sum_{m=0}^{r} (-1)^m \binom{r}{m} y_{n+r-m} \tag{13-1}$$

TABLE 1

x	y	Δ	Δ^2	Δ^3	Δ^4	Δ^5	Δ^6
x_0	y_0						
x_1	y_1	Δy_0					
x_2	y_2	Δy_1	$\Delta^2 y_0$				
x_3	y_3	Δy_2	$\Delta^2 y_1$	$\Delta^3 y_0$			
x_4	y_4	Δy_3	$\Delta^2 y_2$	$\Delta^3 y_1$	$\Delta^4 y_0$		
x_5	y_5	Δy_4	$\Delta^2 y_3$	$\Delta^3 y_2$	$\Delta^4 y_1$	$\Delta^5 y_0$	
x_6	y_6	Δy_5	$\Delta^2 y_4$	$\Delta^3 y_3$	$\Delta^4 y_2$	$\Delta^5 y_1$	$\Delta^6 y_0$

In forming such a table of differences, care must be taken to maintain the correct signs; the subtractions must all be performed in the order given in (1). A convenient check may be obtained by noting that the sum of the entries in any column equals the difference between the first and last entries in the preceding column. It also happens in most cases that the differences of some order will be zero or will vary (perhaps with alternating signs) only in the last few figures of the numbers retained. This is the basis for all of the methods described in the first part of this chapter, for if the unknown $f(x)$ were a polynomial of the n-th degree, the n-th differences would be constant and the $(n + 1)$-th differences zero.

[3] Many different notations and forms of the difference table will be found in books on numerical methods but it will usually be simple to find the relations between the various symbols used.

Now if x_k and y_k are values given in such a table, h is the common interval of x, $h = x_1 - x_0 = x_2 - x_1 = \cdots = x_n - x_{n-1}$, and

$$x = x_k + hu; \quad u = \frac{(x - x_k)}{h} \tag{13-2}$$

then a value of y for an x not contained in the table is given by *Newton's interpolation formula*,

$$y = y_k + u\Delta y_k + \frac{u(u-1)}{2!}\Delta^2 y_k + \frac{u(u-1)(u-2)}{3!}\Delta^3 y_k + \cdots$$

$$+ \frac{u(u-1)(u-2)\cdots(u-r+1)}{r!}\Delta^r y_k \tag{13-3}$$

A second useful form of this equation may also be obtained:

$$y = y_k + u\Delta y_{k-1} + \frac{u(u+1)}{2!}\Delta^2 y_{k-2} + \frac{u(u+1)(u+2)}{3!}\Delta^3 y_{k-3} + \cdots$$

$$+ \frac{u(u+1)(u+2)\cdots(u+r-1)}{r!}\Delta^r y_{k-r} \tag{13-4}$$

It will be noticed that (3) involves differences lying on a *diagonal* line in the table, starting from y_k, while (4) uses differences on a *horizontal* line from y_k. Thus (3) should be used for interpolation near the beginning of a difference table and (4) for interpolation near the end. Summation should be continued until the desired number of significant figures is obtained. These two formulas may also be used to extrapolate at both ends of the difference table but due caution should be used in such cases unless it is known that the function is continuous beyond the tabulated values.

Example 1. Interpolate in Table 2, to find $y = e^{-x^2}$ for $x = 0.0477$. We take $x_k = 0.05$, thus $h = 0.05$, $u = -0.046$. Using (3),

$$y = 0.99750 + 4.6 \times 7.45 \times 10^{-5} - \frac{4.6 \times 1.046 \times 4.85}{2} \times 10^{-5}$$

$$- \frac{4.6 \times 1.046 \times 2.046 \times 1.9}{6} \times 10^{-6}$$

$$= 0.99750 + 0.00034 - 0.00012 = \underline{0.99772}$$

It will be noticed that the third and fourth differences are too small for consideration. The result is correct to the last figure given as may be found by expanding e^{-x^2} in a power series. In this case, the calculations may easily be performed with a slide-rule.

TABLE 2[4]

x	$y = e^{-x^2}$	Δ	Δ^2	Δ^3	Δ^4
0	1.00000				
0.05	0.99750	$-$ 250			
0.10	0.99005	$-$ 745	-495		
0.15	0.97775	-1230	-485	$+10$	
0.20	0.96079	-1696	-466	$+19$	$+9$
0.25	0.93941	-2138	-442	$+24$	$+5$
0.30	0.91393	-2548	-410	$+32$	$+8$

Example 2.   Calculate $y = e^{-x^2}$ for $x = 0.2862$. Since this value is near the end of the table, it is better to use (4) with $x_k = 0.30$, $u = -0.276$. Then,

$$y = 0.91393 + 2.548 \times 2.76 \times 10^{-3} + \frac{4.1 \times 2.76 \times 7.24}{2} \times 10^{-5}$$

$$- \frac{3.2 \times 2.76 \times 7.24 \times 1.724}{6} \times 10^{-6}$$

$$= 0.91393 + 0.00703 + 0.00041 - 0.00002 = \underline{0.92135}$$

This result is also correct to the last significant figure.

An arrangement of tabulated data, somewhat different from that of Table 1, leads to central difference formulas, notably those of Stirling and Bessel.   While these converge faster than Newton's formula, this advantage in most cases is of no practical importance.[5]

Problem.   Interpolate or extrapolate from the data of Table 2 to find $y = e^{-x^2}$ for $x = 0.045$; 0.2775; 0.3018.

13.3. Interpolation for Unequal Values of the Argument.—When the values of x are given for *unequal* intervals, (3) and (4) do not apply, but it is possible to use *divided differences* or the *interpolation formula of Lagrange*. Both methods are tedious to apply and not very precise, hence it is usually better to interpolate from a suitable graph.   We give Lagrange's formula only;   for the method of divided differences, Whittaker and Robinson (loc. cit.) may be consulted.   Suppose $x_0, x_1, \cdots, x_n$ and $y_0, y_1, \cdots, y_n$

---

[4] Following the usual custom, we omit zeros after the decimal point in the various differences.

[5] For details concerning central differences, see references cited at end of chapter.

are known, then for some other value of x,

$$y = f(x) = \frac{(x - x_1)(x - x_2) \cdots (x - x_n)}{(x_0 - x_1)(x_0 - x_2) \cdots (x_0 - x_n)} y_0$$

$$+ \frac{(x - x_0)(x - x_2) \cdots (x - x_n)}{(x_1 - x_0)(x_1 - x_2) \cdots (x_1 - x_n)} y_1 + \cdots$$

$$+ \frac{(x - x_0)(x - x_1) \cdots (x - x_{n-1})}{(x_n - x_0)(x_n - x_1) \cdots (x_n - x_{n-1})} y_n \qquad (13\text{-}5)$$

Example 3. The following data were obtained in the calibration of a platinum-rhodium thermocouple. Find the temperature corresponding to a reading of 9.000 millivolts.

t, °C.	630.5	960.5	1063.0
e, millivolts	5.535	9.117	10.301

With $x = 9.000$, $x_0 = 5.535$, $x_1 = 9.117$, $x_2 = 10.301$, $y_0 = 630.5$, $y_1 = 960.5$, $y_2 = 1063.0$,

$$y = \frac{(-0.117)(-1.301)(630.5)}{(-3.582)(-4.766)} + \frac{(3.465)(-1.301)(960.5)}{(3.582)(-1.184)}$$

$$+ \frac{(3.465)(-0.117)(1063.0)}{(4.766)(1.184)} = \underline{950.4°C.}$$

The value obtained from a carefully constructed curve is 950.2°C.

13.4. Inverse Interpolation.—The problem of inverse interpolation, as the name implies, is that of finding a value of x corresponding to a given value of $y = f(x)$. From Lagrange's formula, it is seen that the roles of x and y may be interchanged so that (5) may be used for inverse interpolation by rewriting it to give $x = \phi(y)$. An illustration of this application of (5) is shown in the following problem. Inverse interpolation may also be effected by *reversion of the series* (3) or (4) to find u as a function of y and Δy. The unknown x is then obtained from (2) or by a method of successive approximations. Full details of both procedures are given by Scarborough (loc. cit.).

Problem. From the data of Example 3, sec. 13.3, find the electromotive force of the thermocouple when the temperature is 750°C.

13.5. Two-way Interpolation.—Suppose the tabulated quantity is given as a function of two independent variables, for example, the index of refraction of water as it varies with both temperature and wavelength. Interpolation to give a value of y for two variables not contained in such tables is best performed by using Newton's formula to interpolate for each

variable separately.   Series, similar to Newton's for direct two-way interpolation, are given by Scarborough (loc. cit.).

NUMERICAL DIFFERENTIATION

13.6. Differentiation Using Interpolation Formula.—In order to determine the numerical derivative of a function of x at a given point, the slope
of the curve of the function may be obtained by graphical means or the
data may be fitted to an empirical equation which is then differentiated.
We may also write

$$\frac{dy}{dx} = \left(\frac{\partial y}{\partial u}\right)\left(\frac{\partial u}{\partial x}\right) \tag{13-6}$$

and if we use (2) and (3) we get

$$\frac{dy}{dx} = \frac{1}{h}\frac{dy}{du} = \frac{1}{h}\left[\Delta y_k + \frac{(2u-1)}{2!}\Delta^2 y_k + \frac{(3u^2-6u+2)}{3!}\Delta^3 y_k + \cdots\right]$$

(13-7)

At the point $x = x_k$, $u = 0$, so we have

$$\left(\frac{dy}{dx}\right)_{x=x_k} = \frac{1}{h}[\Delta y_k - \tfrac{1}{2}\Delta^2 y_k + \tfrac{1}{3}\Delta^3 y_k - \tfrac{1}{4}\Delta^4 y_k + \cdots]$$

$$\left(\frac{d^2 y}{dx^2}\right)_{x=x_k} = \frac{1}{h^2}[\Delta^2 y_k - \Delta^3 y_k + \tfrac{11}{12}\Delta^4 y_k - \cdots] \tag{13-8}$$

More terms and higher order derivatives may be readily found.   Since the
lower order differences disappear upon differentiation, the convergence of
(8) is slower than that of (3) or (4), therefore derivatives obtained in this
way are not very precise.

Maxima or minima in a tabulated function may be found by substituting the differences in (7), equating the derivative to zero and solving for u
and then for x from the relation $x = x_k + hu$.

Example 4.   Find dy/dx and $d^2 y/dx^2$ for $y = e^{-x^2}$ at the point $x = 0.05$
from the data of Table 2.

$$\left(\frac{dy}{dx}\right)_{x=0.05} = \frac{1}{0.05}\left[-0.00745 + \frac{0.00485}{2} + \frac{0.00019}{3} - \frac{0.00005}{4}\right]$$

$$= \underline{-0.09980}$$

$$\left(\frac{d^2 y}{dx^2}\right)_{x=0.05} = \frac{1}{(0.05)^2}\left[-0.00485 - 0.00019 + \frac{0.00055}{12}\right]$$

$$= \underline{-2.00000}$$

The values found by differentiation are

$$dy/dx = -2xy = -0.099750$$
$$d^2 y/dx^2 = 2y(2x^2 - 1) = -1.985025$$

13.7. Differentiation Using a Polynomial.—Another method of finding the derivative has been described by Rutledge.[6] It does not depend on differences but assumes that the given data can be fitted to a polynomial of the fourth or lower degree. Five points must be known, that is, five values of x and y. If h is the equal interval between successive values of x, the derivative of $y = f(x)$ at the point $x = x_k$ is given by the three following *approximately equivalent* expressions.

$$\left(\frac{dy}{dx}\right)_{x=x_k} = \frac{1}{12h}\left[3y_{k+1} + 10y_k - 18y_{k-1} + 6y_{k-2} - y_{k-3}\right]$$

$$= \frac{1}{12h}\left[(y_{k-2} - y_{k+2}) - 8(y_{k-1} - y_{k+1})\right]$$

$$= \frac{1}{12h}\left[y_{k+3} - 6y_{k+2} + 18y_{k+1} - 10y_k - 3y_{k-1}\right] \quad (13\text{-}9)$$

These equations are particularly suitable for solution by one continuous operation with a calculating machine. The method may be extended to apply to polynomials of degree higher than four or to derivatives of higher order.

Example 5. Find dy/dx at $x = 0.15$ for $y = e^{-x^2}$ using the data of Table 2 and the method of this section.

$$\left(\frac{dy}{dx}\right)_{x=0.15} = \frac{1}{12 \times 0.05}[3 \times 0.96079 + 9.7775 - 18 \times 0.99005 +$$
$$6 \times 0.99750 - 1] = -\underline{0.02934}$$

$$= \frac{1}{0.6}[(0.99750 - 0.93941) - 8(0.99005 - 0.96079)]$$

$$= -\underline{0.02933}$$

$$= \frac{1}{0.6}[0.91393 - 6 \times 0.93941 + 18 \times 0.96079 - 9.7775 -$$
$$3 \times 0.99005] = -\underline{0.02933}$$

By direct differentiation, $dy/dx = -0.0293325$.

Problem. Use the data of Table 5 to find dy/dx and d^2y/dx^2 at $x = 0.75$ by the methods of secs. 13.6 and 13.7.

NUMERICAL INTEGRATION

13.8. Introduction.—Suppose $f(x)$ is known to be continuous over an interval of x from a to b but that either the explicit form of $f(x)$ is unknown or it is such a function that its definite integral cannot be determined

[6] Rutledge, G., *Phys. Rev.* **40**, 262 (1932).

conveniently in terms of other known functions.   Numerical evaluation of such integrals, a process called *approximate quadratures*, depends on replacing the integral $\int_a^b f(x)dx$ by another integral $\int_a^b \phi(x)dx$ where $\phi(x)$ can be determined in a simple way.   If $f(x)$ is known to have the $(n + 1)$ values $y_0, y_1, \cdots, y_n$ at $(n + 1)$ points within the interval (a,b), the latter integral may be expressed as

$$\int_a^b \phi(x)dx = A_0y_0 + A_1y_1 + \cdots + A_ny_n \qquad (13\text{-}10)$$

where the $(n + 1)$ quantities A_m are independent of the $(n + 1)$ values of the y_m.   It follows that if $f(x)$ is a polynomial of degree $\leq n$, the error made in replacing $\int_a^b f(x)dx$ by $\sum A_my_m$ may be made to vanish by the proper choice of the A_m.   If $f(x)$ is a polynomial of degree $> n$, the difference between the true value of the integral and (10) may still be small enough to make this procedure useful.   We first consider the methods where the y_m are known at equal intervals.

13.9. The Euler-Maclaurin Formula.—If the explicit form of $f(x)$ is known and it has finite derivatives at the upper and lower limits of the integral or if these derivatives may be determined by numerical methods, the *Euler-Maclaurin formula* may be used to evaluate the integral.   Indicating the values of $f(x)$ at $x = a$ and at $x = b$ by y_0 and y_n and the intermediate values by $y_1, y_2, y_3, \cdots$, this formula is written

$$\int_a^b f(x)dx = h\left[\frac{y_0}{2} + y_1 + y_2 + \cdots + \frac{y_n}{2}\right] - \sum_{\text{odd } r} \frac{h^{r+1}}{(r + 1)!} B_{r+1}[y_n^{(r)} - y_0^{(r)}]$$

$$(13\text{-}11)$$

where $y_n^{(r)}$ and $y_0^{(r)}$ are the r-th derivatives of $f(x)$ at the points b and a. The numerical coefficients B_r are the *Bernoulli numbers*, defined by the relation

$$\frac{x}{e^x - 1} = \sum_{n=0}^{\infty} \frac{B_nx^n}{n!} \qquad (13\text{-}12)$$

which may be rearranged to give the identity

$$\sum_{n=0}^{\infty} \frac{x^n}{(n + 1)!} \cdot \sum_{n=0}^{\infty} \frac{B_nx^n}{n!} \equiv 1$$

Successive values of B_r are obtained from this equation by equating the coefficients of equal powers of x to zero or more simply as follows.   Expand the equation

$$(B + 1)^n = B^n \qquad (13\text{-}13)$$

for a given value of n. Now set $B^r = B_r$ and if $B_0 = 1$, B_{n-2}, B_{n-3}, $\cdots$ are known B_{n-1} may be found. A few of the Bernoulli numbers[7] are $B_1 = -\frac{1}{2}$; $B_3 = B_5 = B_7 = \cdots = 0$; $B_2 = \frac{1}{6}$, $B_4 = -\frac{1}{30}$, $B_6 = \frac{1}{42}$, $B_8 = -\frac{1}{30}$, $B_{10} = \frac{5}{66}$. Putting these numbers in (11) and using the notation (later to be clarified)

$$I_T = h\left[\frac{y_0}{2} + y_1 + y_2 + \cdots + y_{n-1} + \frac{y_n}{2}\right]$$

we obtain the Euler-Maclaurin formula in expanded form

$$\int_a^b f(x)dx = I_{EM} = I_T - \frac{h^2}{12}\Delta + \frac{h^4}{720}\Delta^3 - \frac{h^6}{30240}\Delta^5$$

$$+ \frac{h^8}{1209600}\Delta^7 + \cdots \qquad (13\text{-}14)$$

where

$$\Delta^r = [y_n^{(r)} - y_0^{(r)}]$$

TABLE 3

x	$1/x$	Δ	Δ^2	Δ^3	Δ^4
1.0	1.000000				
1.2	0.833333	-166667			
1.4	0.714286	-119047	$+47620$		
1.6	0.625000	$-\ 89286$	$+29761$	-17859	
1.8	0.555556	$-\ 69444$	$+19842$	$-\ 9919$	$+7940$
2.0	0.500000	$-\ 55556$	$+13888$	$-\ 5954$	$+3965$

Example 6. Divide the interval between 1.0 and 2.0 into five equal parts and evaluate the integral $I = \displaystyle\int_{1.0}^{2.0} (dx/x)$ by the Euler-Maclaurin formula. The required values of $f(x) = 1/x$ are given in Table 3. We also need the derivatives of odd order which are

$$f(x) = 1/x; \quad f^{(n)}(x) = \frac{(-1)^n n!}{x^{n+1}}$$

Then $f'(1) = -1$; $f'''(1) = -6$; $f^V(1) = -120$; $f'(2) = -0.25$; $f'''(2) = -0.375$; $f^V(2) = -1.875$. Since $h = 0.2$, we also find $h^2/12 = 0.003333$; $h^4/720 = 2.222 \times 10^{-6}$; $h^6/30240 = 2.4 \times 10^{-9}$; $\Delta = 0.75$; $\Delta^3 = 5.625$; $\Delta^5 = 118.1$; $I_T = 0.695635$. Hence,

$$I_{EM} = 0.695635 - 0.002500 + 0.000012 - (2.8 \times 10^{-7})$$
$$= 0.693147$$

[7] Those with even subscript only are required in the Euler-Maclaurin formula.

The fifth derivatives contribute to the final result only after six significant figures. A more exact value of the integral is $\ln 2 = 0.69314719$.

13.10. Gregory's Formula.—In case the explicit form of $f(x)$ is unknown, we may rewrite (11), using (8) in place of the derivatives and obtain *Gregory's formula*

$$\int_a^b f(x)dx = I_{Gr} = h\left[\frac{y_0}{2} + y_1 + y_2 + \cdots + \frac{y_n}{2}\right]$$
$$- \frac{h}{12}(\Delta y_{n-1} - \Delta y_0) - \frac{h}{24}(\Delta^2 y_{n-2} + \Delta^2 y_0)$$
$$- \frac{19h}{720}(\Delta^3 y_{n-3} - \Delta^3 y_0) - \frac{3h}{160}(\Delta^4 y_{n-4} + \Delta^4 y_0) - \cdots$$

$$(13\text{-}15)$$

It should be observed that the contents of the parentheses are alternately differences and sums. Additional coefficients of $-h(\Delta^r y_{n-r} \pm \Delta^r y_0)$ may be found by evaluating the definite integral

$$\frac{(-1)^r}{(r+1)!}\int_0^1 z(z-1)(z-2)\cdots(z-r)dz \qquad (13\text{-}16)$$

Example 7. Evaluate the integral of Example 6 by means of Gregory's formula. We find $h/12 = 0.01667$; $h/24 = 0.008333$; $19h/720 = 0.0053$; $3h/160 = 0.0038$. Hence,

$$I_{Gr} = 0.69635 - 0.01667(-0.055556 + 0.166667)$$
$$- 0.008333(0.013888 + 0.047620) - 0.0053(-0.005954 + 0.017859)$$
$$- 0.0038(0.003965 + 0.007940) = \underline{0.693163}$$

The result is not as precise as that obtained in Example 6 because of the small number of available differences.

Problem. Evaluate the integral of Example 8, sec. 13.11, by the Euler-Maclaurin and Gregory formulas. Divide the interval into five equal parts.

13.11. The Newton-Cotes Formula.—Instead of using differences, it is possible to rewrite (11) or (15) in terms of the y_m since the $\Delta^r y_m$ may be reduced to sums of y_m by means of (1). The resulting equation, called the *Newton-Cotes formula* is of the form of (10) where

$$A_m = \frac{(-1)^{n-m}h}{m!(n-m)!}\int_0^n \frac{z(z-1)(z-2)\cdots(z-n)}{(z-m)}dz \qquad (13\text{-}17)$$

Table 4 gives the A_m for several values of n. The values found in this way may be easily checked since it is necessary that

$$A_0 + A_1 + A_2 + \cdots + A_n = nh \qquad (13\text{-}18)$$

When this method is used, it is simpler to divide the interval from a to b into a number of sub-intervals. The number of y_m's in each of these determines the appropriate A_m from Table 4. The value of the integral then equals the sum of the separate terms obtained by applying (10) to each sub-interval. We give a few special cases of the Newton-Cotes formula.

TABLE 4

$n = 1$	$A_0 = A_1 = h/2$
2	$A_0 = A_2 = h/3$; $A_1 = 4h/3$
3	$A_0 = A_3 = 3h/8$; $A_1 = A_2 = 9h/8$
4	$A_0 = A_4 = 14h/45$; $A_1 = A_3 = 64h/45$; $A_2 = 8h/15$
5	$A_0 = A_5 = 95h/288$; $A_1 = A_4 = 125h/96$
	$A_2 = A_3 = 125h/144$
6	$A_0 = A_6 = 41h/140$; $A_1 = A_5 = 54h/35$
	$A_2 = A_4 = 27h/140$; $A_3 = 204h/105$

a. The Trapezoidal Rule. If each sub-interval contains two values of y_m, $n = 1$, $A_0 = A_1 = h/2$ and[8]

$$I_T = \int_a^b f(x)dx = h\left[\frac{y_0}{2} + y_1 + y_2 + \cdots + y_{n-1} + \frac{y_n}{2}\right] \quad (13\text{–}19)$$

This result is exact if the first differences of $f(x)$ are constant. It will be noticed that (19) forms the principle term in both the Euler-Maclaurin and Gregory formulas.

b. Simpson's Rule. If there are an *even* number of y_m and we divide each sub-interval in two parts, we obtain, with $n = 2$ from Table 4, *Simpson's One-Third Rule*:

$$I_S = \int_a^b f(x)dx =$$

$$\frac{h}{3}[y_0 + 4(y_1 + y_3 + \cdots + y_{n-1}) + 2(y_2 + y_4 + \cdots + y_{n-2}) + y_n] \quad (13\text{–}20)$$

This is exact if second differences of $f(x)$ are constant. It is probably the most generally useful of all quadrature formulas.

[8] In order to avoid confusion, it should be noted that n has been taken with two meanings. In Table 4, it refers to the number of intervals between the lower and upper limits of the integral. It now refers to the number of divisions of the sub-intervals. As a subscript in (19), (20) and (21) it indicates the last available value of y_m as in previous equations.

c. *Weddle's Rule.* Taking $n = 6$ from Table 4 and neglecting all differences above the sixth, we obtain *Weddle's Rule*:

$$I_W = \int_a^b f(x)dx = \frac{3h}{10} [y_0 + 5y_1 + y_2 + 6y_3 + y_4 + 5y_5$$

$$+ 2y_6 + 5y_7 + y_8 + 6y_9 + y_{10} + 5y_{11} + 2y_{12} + \cdots$$

$$+ 2y_{n-6} + 5y_{n-5} + y_{n-4} + 6y_{n-3} + y_{n-2} + 5y_{n-1} + y_n] \qquad (13\text{-}21)$$

This is the most accurate of the formulas[9] in this section but it has the disadvantage that the interval must be divided into a number of parts equal to six or some multiple of it.

Various other special cases of the Newton-Cotes equation may be developed. The best known of these, generally called Simpson's Three-Eighth's Rule is obtained from (10) and Table 4 with $n = 3$. As shown by Scarborough (loc. cit.), it is inferior to the One-Third Rule and should never be used.

Example 8. Evaluate $\displaystyle\int_0^{1.5} \frac{x^3}{e^x - 1}\, dx$ by the three preceding methods. This integral is of importance in the Debye theory of the heat capacity of solids;[10] it cannot be evaluated in terms of other known functions. Values of the integral between 0 and n, with n from 0.01 to 24 in steps of 0.01 have been given to six places by Beattie;[11] from his table, $I = 0.615495$. Dividing the interval 0 to 1.50 into six equal parts, we obtain Table 5. Since $h = 0.25$, we find

$$I_T = 0.25 \times (1.991643 + 0.484678)$$

$$= \underline{0.619082}$$

$$I_S = \frac{0.25}{3} \times (4 \times 1.216979 + 2 \times 0.774664 + 0.969357)$$

$$= \underline{0.615550}$$

$$I_W = \frac{3 \times 0.25}{10} \times (5 \times 0.839293 + 1.744021 + 6 \times 0.377686)$$

$$= \underline{0.615495}$$

[9] Note that the last term in (21) has the coefficient unity if $n = 6$ or some multiple of 6. In deriving this formula, the coefficient of the term $\Delta^6 y_0$ is $41/140$, which is taken to be $3/10$ in order to make the final form of the equation as simple as possible. The resulting error is negligible.

[10] See, for example, Taylor, H. S. and Glasstone, S., " A Treatise on Physical Chemistry", Vol. 1, Third Edition, Chapter IV, D. Van Nostrand Co., Inc., New York, 1942.

[11] Beattie, *J. Math. Phys.* **6**, 1 (1926).

It is thus seen that the trapezoidal rule is the least accurate of these three equations while Weddle's rule and Simpson's rule give nearly the same results.

TABLE 5

x	$f(x) = x^3/(e^x - 1)$
0	0
0.25	0.055013
0.50	0.192687
0.75	0.377686
1.00	0.581977
1.25	0.784280
1.50	0.969357

Problem a. Compute some of the coefficients of Table 4.

Problem b. Divide the interval of the integral of Table 5 into twelve equal parts and perform the integrations by the three methods of this section.

13.12. Gauss' Method.—The *method of Gauss* not only determines the $(n + 1)$ values of A_m but also fixes the $(n + 1)y_m$'s of (10) in such a way that the difference between $\int \phi(x)dx$ and $\int f(x)dx$ is a minimum. Since there are now $(2n + 2)$ constants available, it follows that if $f(x)$ is a polynomial of degree $\leq (2n + 1)$, the method will give an exact result for the integral. It will be remembered that the Newton-Cotes method will be exact under similar conditions, if the degree of the polynomial $\leq n$, hence Gauss' method will give a more nearly exact result than the Newton-Cotes method with the same number of values of y_m, or conversely the Newton-Cotes method requires a larger number of known values of the function than Gauss' method for the same allowed error. This is a matter of some importance especially when the given values of y_m are limited in number as they are likely to be when they result from experimental measurements.

In applying the method, it is convenient to change the limits of the integral $\int_a^b f(x)dx$ by making the substitution

$$x = a + (b - a)v \qquad (13\text{-}22)$$

hence in terms of the new variable v, the limits are 0 and 1. Then,

$$f(x) = f[a + (b - a)v] = F(v)$$

$$dx = (b - a)dv \qquad (13\text{-}23)$$

and

$$I_G = \int_a^b f(x)dx = (b - a)\int_0^1 F(v)dv \qquad (13\text{-}24)$$

Developing $F(v)$ in a series similar to (10), we have

$$I_G = R_0 F_0 + R_1 F_1 + R_2 F_2 + \cdots + R_n F_n \qquad (13\text{--}25)$$

where F_m means the numerical value of $F(v_m)$. Now it can be shown[12] that the difference between $\int f(x)dx$ and I_G of (25) is made a minimum provided v_m and R_m are determined by the relations

$$\sum_{m=0}^{n} R_m = 1; \quad \sum_{m=0}^{n} R_m v_m = \tfrac{1}{2}; \quad \sum_{m=0}^{n} R_m v_m^2 = \tfrac{1}{3}; \quad \cdots$$

$$\sum_{m=0}^{n} R_m v_m^r = \frac{1}{(r+1)} \qquad (13\text{--}26)$$

Since there are $(2n+2)$ constants to be evaluated, the most direct procedure would be to solve simultaneously $(2n+2)$ equations like (26). This, however, is very laborious even for small values of n but the v_m alone may be found in the following way. Let $z_0, z_1, z_2, \cdots, z_n$ be the $(n+1)$ real roots of the Legendre polynomial[13] P_{n+1} of degree $(n+1)$ obtained from the equation $P_{n+1}(z) = 0$. Then,

$$v_0 = \tfrac{1}{2}(1 + z_0), \quad v_1 = \tfrac{1}{2}(1 + z_1), \quad \cdots, \quad v_n = \tfrac{1}{2}(1 + z_n) \qquad (13\text{--}27)$$

With the $(n+1)$ values of v_m determined in this way, it is a simple matter to find the remaining constants R_m, $(n+1)$ in number for it is only necessary to solve simultaneously $(n+1)$ relations like (26). Values of both v_m and R_m are given in Table 6.[14]

Some writers make the substitution

$$x = \frac{(a+b)}{2} + \frac{(b-a)}{2} w$$

which changes the limits of the integral to ± 1. In this case,

$$I_G = \frac{(b-a)}{2} \int_{-1}^{+1} g(w)dw = \frac{b-a}{2} \sum_{m=0}^{n} T_m g(w_m)$$

where $g(w)$ corresponds to the former $F(v)$ and

$$T_m = 2R_m; \quad w_m = 2v_m - 1$$

[12] The proof is given by Hobson, E. W., " The Theory of Spherical and Ellipsoidal Harmonics," Cambridge Press, 1931.

[13] See sec. 3.3.

[14] More extensive lists may be found in " Tables of Lagrangian Interpolation Coefficients," Columbia University Press, New York, 1944.

It will be observed that in Gauss' method, the interval is not subdivided equally as in the preceding cases but it is divided symmetrically about the mid-point.

TABLE 6

$n = 2$	$v_0 = 0.11270166$ $v_1 = 0.5$ $v_2 = 0.88729833$	$R_0 = R_2 = \frac{5}{18}$ $R_1 = \frac{4}{9}$
3	$v_0 = 0.06943184$ $v_1 = 0.33000948$ $v_2 = 0.66999052$ $v_3 = 0.93056816$	$R_0 = R_3 = 0.17392742$ $R_1 = R_2 = 0.32607258$
4	$v_0 = 0.04691008$ $v_1 = 0.23076534$ $v_2 = 0.5$ $v_3 = 0.76923466$ $v_4 = 0.95308992$	$R_0 = R_4 = 0.11846344$ $R_1 = R_3 = 0.23931434$ $R_2 = \quad\ 0.28444444$
5	$v_0 = 0.03376524$ $v_1 = 0.16939531$ $v_2 = 0.38069041$ $v_3 = 0.61930959$ $v_4 = 0.83060469$ $v_5 = 0.96623476$	$R_0 = R_5 = 0.08566225$ $R_1 = R_4 = 0.18038079$ $R_2 = R_3 = 0.23395697$

Example 9. Apply Gauss' method to the integral of Examples 6 and 7, subdividing the interval into four parts. From (22) and (44), we find $x = 1 + v$ and $I_G = \int F(v)dv$. From Table 6, with $n = 3$, we obtain

$F_0 = 1/1.069432 = 0.935076$; $F_1 = 1/1.330009 = 0.751875$; $F_2 = 1/1.669990 = 0.598806$; $F_3 = 1/1.930568 = 0.517982$. Then,

$$I_G = 0.173927 \times (0.935076 + 0.517982) +$$
$$0.326072 \times (0.751875 + 0.598806)$$
$$= \underline{0.693145}$$

The result is as precise as that obtained by the Euler-Maclaurin or Gregory formulas but entails much less work.

Problem a. Find values of v_m and R_m for $n = 2$. *Hint:* $P_{n+1}(z) = \dfrac{d^3(z^2 - 1)^3}{dz^3} = 0$.

Problem b. Evaluate the integral of Example 6, sec. 13.9 by Gauss' method. Use the limits 1.0 and 3.0, subdividing this interval into four divisions.

13.13. Remarks Concerning Quadrature Formulas.—The selection of the most suitable quadrature formula to use in a specific case is a matter for which no general rules can be given. When the explicit form of $f(x)$ is known and the differentiations easily made, the Euler-Maclaurin formula has the advantage of giving a result to any required number of figures. When the explicit form of $f(x)$ is not known or if it cannot be differentiated easily, Gregory's formula is useful. As previously stated, the Newton-Cotes formula and its special cases such as the trapezoidal rule, Simpson's and Weddle's rules are approximations to the Euler-Maclaurin and Gregory formulas; they have the advantage of requiring less labor to apply than the two former but result in a loss of accuracy. Gauss' method is apparently not used as often as might be expected in chemical and physical calculations. Since calculating machines are commonly used in such work, the application of it is not laborious and the resulting precision should recommend it.

The reader should remember that in approximate quadratures, the integrand is being replaced by a polynomial, the latter instead of the original function then being integrated. It thus follows that the reliability of the result is determined by the fidelity with which the approximating polynomial matches the given function. Since Gauss' formula fits a polynomial of given degree with fewer known points than any of the other formulas, it should be preferred when the function is of such a form that it can be used. Even if the explicit form of $f(x)$ is unknown, Gauss' formula may still be applied but it requires interpolation between the given y_m to find the proper $F(v)$. When the y_m are the results of experiment and can be arranged at will, Gauss' formula in fact prescribes their optimum positions as those determined by the v_m.

One caution regarding quadrature formulas should be mentioned. If the graph of $f(x)$ is such that the area under one portion of the integral is much larger than that under another portion, the integral should be evaluated separately for each area. The value of h for the sub-interval contributing the least amount to the final result may then be taken as a larger quantity than the h-value for the remaining sub-intervals. If nothing is known of the behavior of $f(x)$, a graph should always be drawn.

NUMERICAL SOLUTION OF DIFFERENTIAL EQUATIONS

13.14. Introduction.—One often encounters differential equations which cannot be solved by any of the methods described in Chapter 2, except that of solution in terms of a series, and this method may be difficult to apply in certain cases. Even when an analytical solution is available, it is sometimes not easy to find numerical values of corresponding pairs of the dependent and independent variables. For example, if the initial conditions $x_0 = 0$, $y_0 = 1$ are given for $dy/dx = (y - x)/(y + x)$, the solutions is $\frac{1}{2} \ln (x^2 + y^2) + \tan^{-1} y/x = \pi/2$ but the labor of finding values of x for given values of y will be very great. In cases of this kind, it is possible to proceed by graphical[15] or numerical methods. The object of the latter is to obtain a table of x and y over the range of x required by the particular problem at hand. When a few such values are known, the table may be extended rather easily, as will be shown. Special methods are required for finding the first few values of x. We present four different ways of starting the solution of a differential equation by numerical methods, and then show how the solution may be continued by extrapolation.

13.15. The Taylor Series Method.—Suppose a differential equation of the first order is given:

$$\frac{dy}{dx} = f(x,y) \qquad (13\text{-}28)$$

with initial values $x = x_0$, $y = y_0$. We may then write the Taylor series

$$y = y_0 + (x - x_0)y_0' + \frac{(x - x_0)^2}{2!}\, y_0'' + \frac{(x - x_0)^3}{3!}\, y_0'''$$

$$+ \cdots + \frac{(x - x_0)^n}{n!}\, y_0^{(n)} \qquad (13\text{-}29)$$

If it is possible to find the various derivatives, the calculations may be extended to as many values of x as desired.

Example 10. Start the solution of the differential equation

$$\frac{dy}{dx} = e^{-x} - y \qquad (13\text{-}30)$$

with initial conditions, $x_0 = 0$, $y_0 = 0$. The exact solution of (30) is found by the methods of Chapter 2 to be $y = xe^{-x}$; the reader will recognize that it is of the form of the differential equation occurring in the study of radioactive disintegration and in the kinetics of chemical reactions involving consecutive first order decompositions. Since $y' = e^{-x} - y$, $y'' = -e^{-x} - y'$, $\cdots$, $y^{(n)} = (-1)^{n-1}e^{-x} - y^{(n-1)}$, it follows that $y_0^{(n)} = (-1)^{n-1}n$ and from (29),

$$y = x - x^2 + \frac{x^3}{2} - \frac{x^4}{6} + \frac{x^5}{24} - \frac{x^6}{120} + \cdots$$

[15] For graphical methods, see Levy, H., and Baggott, E. A., " Numerical Studies in Differential Equations," Vol. 1, Watts and Co., London, 1934, or Sherwood and Reed, " Applied Mathematics in Chemical Engineering," McGraw-Hill Book Co., New York, 1939.

Taking x as 0.1, 0.2 and 0.3, we find the results which appear in Table 7.

TABLE 7

x	y	Exact Values of y
0	0	0
0.1	0.0905	0.09048
0.2	0.1637	0.16375
0.3	0.2222	0.22224

While the method is very simple, it is often tedious to apply as the successive derivatives may become difficult to handle and even at $x = 0.3$ in this case, the fifth derivative is needed. However, it would appear that this procedure is preferable to any other in finding the first few values of y when it is possible to use it.

13.16. The Method of Picard (Successive Approximations or Iteration).—From (28), we see that a solution may be found in the form of an integral equation

$$y = y_0 + \int_{x_0}^{x} f(x,y)dx = y_0 + \int_{x_0}^{x} \left(\frac{dy}{dx}\right)dx \qquad (13\text{--}31)$$

An approximate solution of this equation may be made by assuming that $y = y_0$ under the integral sign. The integral may then be evaluated (by quadratures, if necessary) since it is only a function of x and the constant y_0. Denoting this first approximation to y by ^1y,

$$^1y = y_0 + \int_{x_0}^{x} f(x,y_0)dx \qquad (13\text{--}32)$$

The process may be repeated to give

$$^2y = y_0 + \int_{x_0}^{x} f(x,{}^1y)dx \qquad (13\text{--}33)$$

and so on.

Example 11. Start the solution of Example 10, sec. 13.15 by this method.

$$^1y_1 = y_0 + \int_{0}^{x} (e^{-x} - y_0)dx$$

$$= \int_{0}^{x} e^{-x}dx = 1 - e^{-x}$$

$$^2y_1 = y_0 + \int_0^x (e^{-x} - \,^1y)dx$$

$$= \int_0^x (2e^{-x} - 1)dx = 2(1 - e^{-x}) - x$$

$$^3y_1 = 3(1 - e^{-x}) - \frac{x(4 - x)}{2}$$

With $x = 0.1$, $^3y_1 = \underline{0.0906}$; the next approximation, $^4y_1$ is the same as $^3y_1$, hence we proceed to calculate y_2 at $x = 0.2$ from the relations

$$^1y_2 = \,^3y_1 + \int_{0.1}^x (e^{-x} - \,^3y_1)dx$$

$$= \,^3y_1 + e^{-0.1} + 0.1\,^3y_1 - e^{-x} - 0.0906x$$

$$= 1.0045 - e^{-x} - 0.0906x$$

$$^2y_2 = \,^3y_1 + \int_{0.1}^x (e^{-x} - \,^1y_2)dx$$

$$^3y_2 = \,^3y_1 + \int_{0.1}^x (e^{-x} - \,^2y_2)dx = \underline{0.1639}$$

The next value, $^4y_2$ is the same as $^3y_2$ so we go on in the same way to find $^1y_3$, etc., at $x = 0.3$. The results by the Picard method are seen from Table 7 to be not quite as good as those obtained in Example 10. Moreover, the disadvantages here are similar to those of the method of sec. 13.15, for the successive integrals may become more and more difficult to determine.

13.17. The Modified Euler Method.—If the intervals between successive values of x are small enough we may write $\Delta x = h$ and

$$\Delta y = \left(\frac{dy}{dx}\right) \Delta x \qquad (13\text{--}34)$$

An approximate value of y_1 at $x_1 = x_0 + h$ is then given by

$$^1y_1 = y_0 + \Delta y = y_0 + \left(\frac{dy}{dx}\right)_0 h \qquad (13\text{--}35)$$

An approximation to dy/dx at x_1, may be obtained by the relation

$$^1\left(\frac{dy}{dx}\right)_1 = f(x_1, {}^1y_1) \qquad (13\text{--}36)$$

which leads to an improved value of y_1

$$^2y_1 = y_0 + \frac{h}{2}\left[{}^1\left(\frac{dy}{dx}\right)_1 + \left(\frac{dy}{dx}\right)_0\right] \qquad (13\text{--}37)$$

This in turn will give a better approximation to dy/dx

$$^2\left(\frac{dy}{dx}\right)_1^\cdot = f(x_1, {}^2y_1) \tag{13-38}$$

which may be used to compute the third approximation to y_1. The process is repeated until there is no further change in the results. The values of y and dy/dx at x_2 are found in a similar way.

Example 12. Start the solution of (30) by this method. Since $x_0 = y_0 = 0$, $(dy/dx)_0 = 1$. With $h = 0.1$,

$$^1y_1 = 1 \times 0.1 = 0.1$$

$$^1(dy/dx)_1 = e^{-0.1} - 0.1 = 0.8048$$

$$^2y_1 = 0.1(1.0 + 0.8048)/2 = 0.0902$$

$$^2(dy/dx)_1 = 0.9048 - 0.0902 = 0.8146$$

$$^3y_1 = 0.1(1.0 + 0.8146)/2 = 0.0907$$

$$^3(dy/dx)_1 = 0.9048 - 0.0907 = 0.8141$$

$$^4y_1 = 0.05(1.0 + 0.8141) = \underline{0.0907}$$

No further improvement results by continuing the approximations, so we proceed to $x = 0.2$ with $y_1 = 0.0907$, $(dy/dx)_1 = 0.8141$. Then,

$$^1y_2 = 0.0907 + 0.8141 \times 0.1 = 0.1721$$

$$^1(dy/dx)_2 = e^{-0.2} - 0.1721 = 0.6466$$

and finally,

$$^4y_2 = \underline{0.1641}, \quad ^3(dy/dx)_2 = 0.6546$$

This method is tedious in application but perhaps less complicated than either of the preceding methods since neither differentiation nor integration is required.

13.18. The Runge-Kutta Method.—In this method it is necessary to calculate the four quantities

$$k_1 = f(x_0, y_0)h$$

$$k_2 = f\left(x_0 + \frac{h}{2}, \quad y_0 + \frac{k_1}{2}\right)h$$

$$k_3 = f\left(x_0 + \frac{h}{2}, \quad y_0 + \frac{k_2}{2}\right)h \tag{13-39}$$

$$k_4 = f(x_0 + h, \quad y_0 + k_3)h$$

Then,

$$x_1 = x_0 + h, \quad y_1 = y_0 + \Delta y$$

$$\Delta y = \tfrac{1}{6}(k_1 + 2k_2 + 2k_3 + k_4) \tag{13-40}$$

It will be noted that if $f(x,y)$ is independent of y, (40) reduces to Simpson's rule. The same formulas are used to compute y at x_2, substituting x_1 and y_1 for x_0 and y_0 in (39).

Example 13. With the same differential equation as before (Example 10, sec. 13.15), we find

$$k_1 = (e^{-x_0} - y_0)h = 0.1$$
$$k_2 = (e^{-0.05} - 0.05)0.1 = 0.0901$$
$$k_3 = (e^{-0.05} - 0.0450)0.1 = 0.0906$$
$$k_4 = (e^{-0.1} - 0.0906)0.1 = 0.0814$$

Hence,

$$y_1 = y_0 + \Delta y = \tfrac{1}{6}(0.1 + 0.1802 + 0.1812 + 0.0814) = \underline{0.0905}$$

For the next interval, we find in a similar way, $k_1 = 0.0814$, $k_2 = 0.0730$, $k_3 = 0.0734$, $k_4 = 0.0655$, $\Delta y = 0.0733$, $y_2 = 0.09055 + 0.0733 = \underline{0.1638}$.

The error in the Runge-Kutta method is of the order of h^5. It will be seen that its use is reasonably simple; it is probably the most generally useful of the four methods given here.

13.19. Continuing the Solution.—When the first few values (three or four) of y have been found by one of the preceding methods, the solution may be continued by extrapolation. For this purpose, it is appropriate to use Newton's interpolation formula (4), rewriting it in terms of $y' = dy/dx$, y_k' and the differences $\Delta y_{k-1}'$, $\Delta^2 y_{k-2}'$, $\cdots$, $\Delta^r y_{k-r}'$. Upon substituting this expression in the equation

$$y = \int_{x_1}^{x_2} y' \, dx \tag{13-41}$$

and performing the integration, several useful formulas may be obtained by changing the limits of the definite integral.

$$(\Delta y)_k^{k+1} = h(y_k' + \tfrac{1}{2}\Delta y_{k-1}' + \tfrac{5}{12}\Delta^2 y_{k-2}' + \tfrac{3}{8}\Delta^3 y_{k-3}' + \tfrac{251}{720}\Delta^4 y_{k-4}') \tag{13-42}$$

$$(\Delta y)_{k-1}^{k} = h(y_k' - \tfrac{1}{2}\Delta y_{k-1}' - \tfrac{1}{12}\Delta^2 y_{k-2}' - \tfrac{1}{24}\Delta^3 y_{k-3}' - \tfrac{19}{720}\Delta^4 y_{k-4}') \tag{13-43}$$

$$(\Delta y)_{k-2}^{k-1} = h(y_k' - \tfrac{3}{2}\Delta y_{k-1}' + \tfrac{5}{12}\Delta^2 y_{k-2}' + \tfrac{1}{24}\Delta^3 y_{k-3}' + \tfrac{11}{720}\Delta^4 y_{k-4}') \tag{13-44}$$

$$(\Delta y)_{k-3}^{k-2} = h(y_k' - \tfrac{5}{2}\Delta y_{k-1}' + \tfrac{23}{12}\Delta^2 y_{k-2}' - \tfrac{3}{8}\Delta^3 y_{k-3}' - \tfrac{19}{720}\Delta^4 y_{k-4}') \tag{13-45}$$

$$(\Delta y)_{k-4}^{k-3} = h(y_k' - \tfrac{7}{2}\Delta y_{k-1}' + \tfrac{53}{12}\Delta^2 y_{k-2}' - \tfrac{55}{24}\Delta^3 y_{k-3}' + \tfrac{251}{720}\Delta^4 y_{k-4}') \tag{13-46}$$

The meaning of a symbol such as $(\Delta y)_k^{k+1}$ should be clear. It is the increment to be added to the k-th value of y in the difference table to obtain the next value beyond, that is, the value of y at x_{k+1}. Equation (42) is thus to be used for extending the table to larger values of x while the remaining formulas are useful in checking the values of y already found.

Example 14. Extend the integration of the differential equation of Examples 10, 11, 12 and 13, using the values of y' found in Example 12. We first collect the data as shown in Table 8. To check y at $x = 0.1$, let us use (44). Then since $k = 3$,

$$(\Delta y)_0^1 = 0.1(0.6546 + \tfrac{3}{2}0.1595 + \tfrac{5}{12}0.0264) = \underline{0.0905}$$

Thus, $y_1 = y_0 + (\Delta y)_0^1 = 0.0905$, which shows that the result in Table 8 is in error by 2 units in the last place. Similarly, to check y_2, we use (43) to obtain

$$(\Delta y)_1^2 = 0.1(0.6546 + 0.0798 - 0.0022) = 0.0732$$

and

$$y_2 = y_1 + (\Delta y)_1^2 = 0.0905 + 0.0732 = \underline{0.1637}$$

We now make a new table (Table 9) to include our corrected values of $y, y', \Delta y'$, etc. To find y_3, we use (42) to obtain

$$(\Delta y)_2^3 = 0.1(0.6550 - 0.0796 + 0.0110) = 0.0586$$

$$y_3 = 0.1637 + 0.0586 = \underline{0.2223}$$

A check on y_3 may be found from (43)

$$(\Delta y)_2^3 = 0.1(0.5185 + 0.0682 - 0.0011 + 0.0005) = 0.0586$$

TABLE 8

x	y	y'	$\Delta y'$	$\Delta^2 y'$
0	0	1.0000		
0.1	0.0907	0.8141	−0.1859	
0.2	0.1641	0.6546	−0.1595	+0.0264

TABLE 9

x	y	Δy	y'	$\Delta y'$	$\Delta^2 y'$	$\Delta^3 y'$
0.00	0		1.0000			
0.10	0.0905	0.0905	0.8143	−0.1857		
0.20	0.1637	0.0732	0.6550	−0.1593	+0.0264	
0.30	0.2223	0.0586	0.5185	−0.1365	+0.0128	−0.0136

Since this is the same result as that found previously, we proceed to the next value of x. Moreover, since the preceding y was correct at the first trial, we suspect that the value of h might be increased, say to 0.20. We thus obtain y for $x = 0.40$ in the same manner as before, then rewrite the table for $x = 0, 0.20$ and 0.40. From the new table, we go on to $x = 0.60$, etc.

13.20. Milne's Method.—One further method of continuing the solu-tion of a differential equation is often useful. Supposing the first four values of y and y' have been found by some of the previous methods, we continue as follows.

1. Find a first approximation to the next y by using the formula

$$^1y_k = y_{k-4} + \frac{4h}{3}(2y'_{k-1} - y'_{k-2} + 2y'_{k-3}) \qquad (13\text{-}47)$$

2. Substitute this in the original differential equation (28) to find y'_k.
3. Use the value of y'_k to calculate ^2y_k from the relation

$$^2y_k = y_{k-2} + \frac{h}{2}(y'_k + 4y'_{k-1} + y'_{k-2}) \qquad (13\text{-}48)$$

If ^1y_k and ^2y_k agree to the desired number of figures, we may proceed to the next interval in the same way. If they do not agree, the size of the interval must be decreased. The error due to the use of (48) is

$$E = \tfrac{1}{29}\,|\,{}^2y_k - {}^1y_k\,|$$

Eqs. (47) and (48) are obtained by integrating Newton's interpolation formula (3), after expressing it in terms of y'. Both formulas are exact when fourth differences of y' vanish.

Example 15. Use Milne's method to continue the solution of the differential equation of the previous examples. For $x = 0.4$, we find using Table 9 and (47),

$$^1y_4 = \frac{0.40}{3}(2 \times 0.5185 - 0.6550 + 2 \times 0.8143) = 0.2681$$

From the original differential equation (30)

$$y'_4 = (0.6703 - 0.2681) = 0.4022$$

From (48),

$$^2y_4 = 0.1637 + \frac{0.10}{3}(0.4022 + 4 \times 0.5185 + 0.6550) = \underline{0.2681}$$

Problem. Use the various methods of this chapter to obtain the solutions, cor-rect to four decimal places, of the differential equation $dy/dx = (x - y)$ between 0 and 0.25, with $x_0 = 0$, $y_0 = 1$. The exact solution is $y = (x - 1) + 2e^{-x}$.

13.21. Simultaneous Differential Equations of the First Order.—Sup-pose the given equations are

$$\frac{dy}{dx} = f_1(x,y,z)$$

$$\frac{dz}{dx} = f_2(x,y,z) \qquad (13\text{-}49)$$

where x is the independent variable and y, z are dependent variables. Provided initial values of x, y and z are given, the first increments in y and z due to an increment Δx_1 in x may be found by any of the methods given in the preceding sections. The procedure should be obvious, but it is particularly necessary to check the results carefully at each stage of the solution. If the Runge-Kutta method is used, the following equations replace (39) and (40)

$$k_1 = f_1(x_0, y_0, z_0)h$$

$$k_2 = f_1\left(x_0 + \frac{h}{2}, \quad y_0 + \frac{k_1}{2}, \quad z_0 + \frac{m_1}{2}\right)h$$

$$k_3 = f_1\left(x_0 + \frac{h}{2}, \quad y_0 + \frac{k_2}{2}, \quad z_0 + \frac{m_2}{2}\right)h$$

$$k_4 = f_1(x_0 + h, \, y_0 + k_3, \, z_0 + m_3)h$$

$$m_1 = f_2(x_0, y_0, z_0)h$$

$$m_2 = f_2\left(x_0 + \frac{h}{2}, \quad y_0 + \frac{k_1}{2}, \quad z_0 + \frac{m_1}{2}\right)h \tag{13-50}$$

$$m_3 = f_2\left(x_0 + \frac{h}{2}, \quad y_0 + \frac{k_2}{2}, \quad z_0 + \frac{m_2}{2}\right)h$$

$$m_4 = f_2(x_0 + h, \quad y_0 + k_3, \quad z_0 + m_3)h$$

$$x_1 = x_0 + h; \quad y_1 = y_0 + \Delta y; \quad z_1 = z_0 + \Delta z$$

$$\Delta y = \tfrac{1}{6}(k_1 + 2k_2 + 2k_3 + k_4)$$

$$\Delta z = \tfrac{1}{6}(m_1 + 2m_2 + 2m_3 + m_4) \tag{13-51}$$

13.22. Differential Equations of Second or Higher Order.—Any differential equation of second or higher order is reduced to a system of simultaneous equations by the introduction of new variables. Consider the equation

$$\frac{d^n y}{dx^n} = f(x, y, y', \cdots, y^{(n-1)}) \tag{13-52}$$

where $y' = dy/dx$, $y'' = d^2y/dx^2$, etc. Make the substitutions

$$z_1 = \frac{dy}{dx}; \quad z_2 = \frac{dz_1}{dx}; \quad \cdots, \quad z_{n-1} = \frac{dz_{n-2}}{dx} \tag{13-53}$$

then,

$$\frac{d^n y}{dx^n} = z_n = \frac{dz_{n-1}}{dx} = f(x, y, z_1, z_2, \cdots, z_{n-1}) \tag{13-54}$$

Provided initial values of x, y, z_1, z_2, $\cdots$, z_{n-1} are given, the problem is equivalent to the solution of a system of simultaneous first order differential equations which may be effected as described in sec. 13.21.

In physical problems, differential equations of the type

$$\frac{d^2y}{dx^2} + f(x,y) = 0$$

or

$$\frac{d^2y}{dx^2} + f\left(x,y,\frac{dy}{dx}\right) = 0$$

often arise, with the requirement that the variables satisfy certain boundary conditions, say $x = x_0$, $y = y_0$ and $x = x_1$, $y = y_1$ with the initial value of dy/dx unknown. For example, in the Thomas-Fermi theory[16] of the atom, the equation $d^2y/dx^2 = (y^3/x)^{1/2}$ occurs with the boundary conditions, $x = 0$, $y = 1$, $x = \infty$, $y = 0$. In cases of this kind, a tentative value of dy/dx is assumed and a rough integration is made over the range of x. This first approximation will usually suggest a better guess for the initial value of dy/dx. After several attempts are made, the value of dy/dx may usually be found to the desired accuracy.

Example 16. Find y and dy/dx for the equation

$$\frac{d^2y}{dx^2} + 4x\frac{dy}{dx} - 4y = 0$$

Let $dy/dx = z$, then the second order equation is equivalent to the first order equations

$$\frac{dy}{dx} = z; \quad \frac{dz}{dx} + 4xz - 4y = 0$$

which may be solved by the previous methods. If the Runge-Kutta method is used, $f_1(x,y,z) = z$ and $f_2(x,y,z) = -4xz + 4y$. In this case, f_1 does not depend on x and y, a situation which makes the evaluation of the k's in (50) somewhat simpler than in the general case. The differential equation of this problem may be solved exactly by the substitution $y = ve^{-x^2}$.

PART 2. ALGEBRAIC CALCULATIONS

13.23. Numerical Solution of Transcendental Equations.—No general method exists for finding the roots of transcendental equations such as $xe^x = 1$ or $x^2 = \sin x$. Approximate values may always be found by graphical means; where more precise results are required several analytical procedures are available.

a. *The Method of "Regula Falsi."* Suppose the given equation is $f(x) = 0$, then it is obvious that the plot of $y = f(x)$ will give the required

---

[16] The differential equation and its solution are discussed in more detail by Gombas, P., "Theorie und Lösungsmethoden des Mehrteilchenproblems der Wellenmechanik," Birkhäuser, Basel, 1950.

root when $y = 0$, that is when the graph crosses the x-axis. Two values of x, say x_0 and x_1 with the corresponding values of y are selected from a graph or otherwise. Then if x_0 is near the root desired, a better approximation for the root is given by

$$x = x_0 + \Delta x$$

where

$$\Delta x = \frac{(x_1 - x_0)|\, y_0\,|}{|\, y_0\,| + |\, y_1\,|} \tag{13–55}$$

The process is continued until the required number of figures is obtained.

Example 17. Find the solution[17] of $f(x) = (5 - x)e^x - 5 = 0$ near $x = 5$. One solution is clearly $x = 0$; to find the other let $x_0 = 4.5$, $x_1 = 5.0$; $y_0 = 40.00$, $y_1 = -5.00$, hence

$$^1\Delta x = \frac{0.5 \times 40.00}{45.00} = 0.44; \quad ^1x = 4.50 + 0.44 = 4.94$$

A second approximation with $x_0 = 4.94$; $y_0 = 3.382$ gives

$$^2\Delta x = \frac{0.06 \times 3.382}{8.382} = 0.024; \quad ^2x = 4.94 + 0.024 = 4.964$$

The third approximation with $x_0 = 4.964$; $y_0 = 0.1516$ gives

$$^3\Delta x = \frac{0.036 \times 0.1516}{5.1516} = 0.001; \quad ^3x = 4.964 + 0.001 = \underline{4.965}$$

Further repetition of the calculations show that this result is correct to four significant figures. The value 4.965114 has been obtained by Birge.[18]

b. *The Newton-Raphson Method.* When the derivative of $f(x)$ is easily evaluated numerically, the real roots of $f(x) = 0$ may be determined in the following way. Suppose x_0 is an approximate value of one of the roots, then an improved value of the root is given by

$$x = x_0 + \Delta x; \quad \Delta x = -\frac{f(x_0)}{f'(x_0)} \tag{13–56}$$

The next approximation is found by substituting x in place of x_0 to get a new value of Δx, continuing in this way as long as necessary. In practice, it will be found that after a few approximations, the value of the derivative will change very little with succeeding values of x hence f' need not be recomputed.

[17] This equation occurs in the theory of black-body radiation, see, for example, Taylor, H. S., and Glasstone, S., "A Treatise on Physical Chemistry," Vol. 1, D. Van Nostrand Co., Inc., New York, 1942.

[18] Birge, R. T., *Revs. Mod. Phys.* **13**, 233 (1941).

Example 18. Find x of example 17, starting with $x_0 = 4.9$. Substitution gives $f(x_0) = 8.43$; $f'(x_0) = -120.87$; $\Delta x = 8.43/120.87 = 0.07$; $^1x = 4.97$. The second approximation is obtained from $f(4.97) = -0.677$; $f'(4.97) = -139.78$; $\Delta x = -0.677/139.78 = -0.005$; $^2x = 4.965$.

c. *The Method of Iteration.* If we rewrite our equation $f(x) = 0$ in the form

$$x = \phi(x) \tag{13-57}$$

we may substitute an approximate value of x, say x_0 on the right of (57) to get $^1x = \phi(x_0)$ and repeat to get

$$^2x = \phi(x_1); \quad ^3x = \phi(x_2); \quad \text{etc.} \tag{13-58}$$

It is often possible to write $f(x) = 0$ in the form $x = \phi(x)$ in several different ways, in which case, it is better to start with the simplest such arrangement. A few approximations will indicate whether the chosen form is suitable but if the succeeding values of x do not converge rapidly, one of the alternative functions should be tried. The condition for convergence is found to be that $\phi'(x)$, the derivative of x, be less than unity in the neighborhood of the desired root. As this derivative becomes smaller, the convergence becomes more rapid.

Example 19. Find x of the function in Examples 17 and 18 by the method of iteration. Writing the equation in the form $x = 5e^{-x}(e^x - 1)$ we find with $x_0 = 4.9$; $e^x = 134.3$; $^1x = (5 \times 133.3)/134.3 = 4.963$. The next approximation gives $e^x = 143.1$; $^2x = (5 \times 142.1)/143.1 = 4.965$.

Problem. Solve the equation $x \log x = 1.5334$ by the methods of this section. *Ans.:* $x = \underline{3.1110}$.

13.24. Simultaneous Equations in Several Unknowns.—The real roots of simultaneous algebraic or transcendental equations may be found by the methods of secs. 13.23b or 13.23c. In the Newton-Raphson method, when two equations are given

$$f(x,y) = 0; \quad g(x,y) = 0 \tag{13-59}$$

(56) is replaced by

$$x = x_0 + \Delta x; \quad y = y_0 + \Delta y$$

where

$$\Delta x = \frac{1}{\Delta} \begin{vmatrix} -f(x_0,y_0) & f_y(x_0,y_0) \\ -g(x_0,y_0) & g_y(x_0,y_0) \end{vmatrix}$$

$$\Delta y = \frac{1}{\Delta} \begin{vmatrix} f_x(x_0,y_0) & -f(x_0,y_0) \\ g_x(x_0,y_0) & -g(x_0,y_0) \end{vmatrix}$$

$$\Delta = \begin{vmatrix} f_x(x_0,y_0) & f_y(x_0,y_0) \\ g_x(x_0,y_0) & g_y(x_0,y_0) \end{vmatrix}$$

In the method of iteration, we rewrite (59) as

$$x = \phi(x,y); \quad y = \psi(x,y)$$

then

$$^1x = \phi(x_0,y_0); \quad {}^1y = \psi({}^1x,y_0)$$
$$^2x = \phi({}^1x,{}^1y); \quad {}^2y = \psi({}^2x,{}^1y); \quad \text{etc.} \tag{13-60}$$

Both methods are readily extended to cases of more than two unknowns.

13.25. Numerical Determination of the Roots of Polynomials.—Any of the methods of sec. 13.23 may be applied to determine the real roots of a polynomial. When all of the roots are not required, the Newton-Raphson method is probably more rapid than the others.[19] In order to evaluate $f(x)$ and $f'(x)$ for $x = x_0$, the following procedure will be found useful. Suppose the polynomial is $y(x) = c_0 x^n + c_1 x^{n-1} + \cdots + c_n$. Write the coefficients in a line, supplying zeros if any powers of x are missing. Multiply the number c_0 by x_0 and add the result to c_1; multiply this sum (d_1) by x_0 and add to c_2 continuing until the last sum is obtained; its value equals $y(x)$ for $x = x_0$. The scheme is illustrated in Table 10. In actual computation with a calculating machine nothing need be written down since with proper care to locate the decimal point and due regard to sign, the whole process may be performed as a continuous operation. The use of this method is illustrated further in the last part of Example 20.

TABLE 10

c_0	c_1	c_2	c_3	$\cdots$	c_n
	$c_0 x_0$	$d_1 x_0$	$d_2 x_0$	$\cdots$	$d_{n-1} x_0$
	d_1	d_2	d_3	$\cdots$	d_n

Graeffe's root-squaring method will be found to involve little more labor than the preceding method with the added advantage that it gives all of the roots of the polynomial at once. No initial approximation is required and complex as well as real roots may be found. It is convenient to divide by the coefficient of x^n if necessary so that the polynomial appears in the form $y(x) = x^n + a_1 x^{n-1} + a_2 x^{n-2} + \cdots + a_n = 0$. Using detached coefficients, Table 11 is calculated. Care must be taken with the signs of the doubled cross-products. The new coefficients b_1, b_2, $\cdots$, are then squared and the cross-products of the b's determined in a similar way. As the squaring process is continued, it will be found that the doubled cross-products become progressively smaller, eventually contributing nothing to the next squared terms. When this point is reached, there will be n coeffi-

---

[19] Horner's method does not appear to have any advantages over the Newton-Raphson method. It is described by Mellor, J. W., " Higher Mathematics for Students of Chemistry and Physics," Longmans, Green and Co., New York, 1902, and in most elementary algebra texts.

cients, say $m_1, m_2, \cdots, m_n$. Then if $x_1, x_2, \cdots, x_n$ are the n real roots of the polynomial

$$\left| x_1 \right|^p = m_1; \quad \left| x_2 \right|^p = \frac{m_2}{m_1}; \quad \cdots, \quad \left| x_n \right|^p = \frac{m_n}{m_{n-1}}$$

or,

$$\log \left| x_1 \right| = \frac{1}{p} \log m_1$$

$$\log \left| x_2 \right| = \frac{1}{p} (\log m_2 - \log m_1)$$

$$\log \left| x_3 \right| = \frac{1}{p} (\log m_3 - \log m_2)$$

$$\cdots\cdots\cdots\cdots\cdots\cdots\cdots\cdots\cdots\cdots$$

$$\log \left| x_n \right| = \frac{1}{p} (\log m_n - \log m_{n-1}) \qquad (13\text{--}61)$$

where $p = 2^s$ and s is the number of times the squaring operation has been performed. The signs of the roots must be determined by some rule of signs but this may often be done by inspection.

TABLE 11

1	a_1	a_2	a_3	a_4	$\cdots$
1	a_1^2	a_2^2	a_3^2	a_4^2	$\cdots$
	$-2a_2$	$-2a_1a_3$	$-2a_2a_4$	$-2a_3a_5 + 2a_2a_6$	$\cdots$
		$+2a_4$	$+2a_1a_5$	$-2a_1a_7$	$\cdots$
			$-2a_6$	$+2a_8$	$\cdots$
1	b_1	b_2	b_3	b_4	$\cdots$

In practice, it is best to carry only four or five figures in the calculations, hence tables of squares and four-place logarithms may be used if a calculating machine is unavailable. If more figures are required in the roots, the use of the Newton-Raphson method serves both to give these additional figures and to check the previous calculations.

When two (or more) roots of the polynomial are real and equal, one of the doubled cross-products will not decrease in magnitude as the squaring proceeds; in fact it will always be equal to one-half of the squared term which stands just above it. The squaring in this case is stopped when the other cross-products no longer contribute to the next coefficients.

The presence of complex roots in a polynomial expression is revealed by the fact that the doubled cross-products do not disappear and the signs of some of the sums alternate as the squaring proceeds. The method of finding the complex roots as well as pairs of real roots is described in detail by both Scarborough and by Whittaker and Robinson (loc. cit.).

TABLE 12

1	-5.600×10	4.900×10^2	1.111×10^4	-1.175×10^5
	3.136×10^3	2.401×10^5	1.234×10^8	1.381×10^{10}
	-0.980	12.44	1.152	
		-2.350		
1	2.156×10^3	1.250×10^6	2.386×10^8	1.381×10^{10}
	4.648×10^6	1.562×10^{12}	5.693×10^{16}	
	-2.500	-1.029	-3.452	
		0.028		
1	2.148×10^6	5.610×10^{11}	2.241×10^{16}	1.907×10^{20}
	4.614×10^{12}	3.147×10^{23}	5.022×10^{32}	
	-1.122	-0.963	-2.140	
		0.004		
1	3.492×10^{12}	2.188×10^{23}	2.882×10^{32}	3.637×10^{40}
	1.219×10^{25}	4.787×10^{46}	8.306×10^{64}	
	-0.044	-0.201	-1.591	
1	1.175×10^{25}	4.586×10^{46}	6.715×10^{64}	1.323×10^{81}
1	1.381×10^{50}	2.103×10^{93}	4.388×10^{129}	1.750×10^{162}
1	1.904×10^{100}	4.414×10^{186}	1.925×10^{259}	3.062×10^{324}

Example 20. Find the four real roots of the polynomial[20]

$$y(x) = x^4 - 56x^3 + 490x^2 + 11{,}112x - 117{,}495 = 0$$

The method is apparent from Table 12. It will be seen that the second row of doubled cross-products may be neglected after the eighth power terms and the first row after the thirty-second power terms, hence the squaring is stopped after raising the coefficients to the sixty-fourth power. We then find that

$$\log |x_1| = 100.2797/64 = 1.5669$$

$$\log |x_2| = (186.6448 - 100.2797)/64 = 1.3494$$

$$\log |x_3| = (72.6396)/64 = 1.1350$$

$$\log |x_4| = (65.2016)/64 = 1.0188$$

so that $|x_1| = 36.89$; $|x_2| = 22.36$; $|x_3| = 13.65$; $|x_4| = 10.45$. Inspection shows that all signs are positive except that of x_3. With these

[20] Solution of similar equations is needed to calculate the energy levels of the asymmetric top in quantum mechanics; see, for example, Herzberg, G., "Infrared and Raman Spectra of Polyatomic Molecules," D. Van Nostrand Co., Inc., New York, 1945.

NUMERICAL SOLUTION OF SIMULTANEOUS LINEAR EQUATIONS **13.26**

values, the sum of the roots is 56.05, in approximate agreement with the coefficient of x^3 in the original equation.

In order to improve these values, we make use of the Newton-Raphson method. With $x_1 = 36.89$, we find

$$y(x_1) = [(1 \times 36.89) - 56] + [(-19.11 \times 36.89) + 490]$$
$$+ [(-214.97 \times 36.89) + 11,112] + [(3,181.76 \times 36.89)$$
$$- 117,495] = -120$$

In the same way, from

$$y'(x) = 4x^3 - 168x^2 + 980x + 11,112$$

we find

$$y'(x_1) = [(4 \times 36.89) - 168] + [-(20.44 \times 36.89) + 980]$$
$$+ [(225.97 \times 36.89) + 11,112] = 19,448$$

Then,

$$\Delta x_1 = 120/19,448 = 0.0062$$

and

$$^1x_1 = 36.89 + 0.0062 = 36.8962$$

Repeating the calculations, we obtain $y(^1x_1) = 3.57$; $y'(^1x_1) = 19,478$; $\Delta^1x_1 = -0.0002$; $^2x_1 = \underline{36.8960}$. This value is correct to five significant figures. The same procedure applied to the other roots gives $\underline{22.3410}$; $-\underline{13.6669}$; $\underline{10.4302}$. The sum of these values which is 56.0005 gives a further check on the results.

Problem. Find the roots of $x^3 - 15x^2 + 74x - 120 = 0$, by the Graeffe method. *Ans.*: $x = \underline{4}, \underline{5}, \underline{6}$.

13.26. Numerical Solution of Simultaneous Linear Equations.— Systems of the form

$$\sum_{k=1}^{n} a_{ki}x_k = g_i; \quad (i = 1, 2, \cdots, n) \tag{13-62}$$

where the a_{ki} and g_i are numbers and the x_k are sought, often occur in physical problems, particularly in the solution of the normal equations resulting from a least squares treatment of numerical data (see sec. 13.37b). Several methods of solving such equations are given by Whittaker and Robinson (loc. cit.) but none of these are particularly suitable for machine calculation (see also sec. 10.9). When $a_{ki} = a_{ik}$, which is usually true, the determinantal method described there offers certain advantages but in general when the number of unknowns is greater than four or five the labor of evaluating the determinants becomes prohibitive. The following systematic procedure[21] which is well adapted for machine calculation will be found useful in such cases.

Using detached coefficients, the numbers in (62) are written down as in Table 13. For convenience we assume that there are only four unknowns;

TABLE 13

g_1	g_2	g_3	$-$		
a_{11}	a_{12}	a_{13}	$-a_{22}/a_{21}$	$-a_{23}/a_{21}$	
a_{21}^*	a_{22}	a_{23}	1	0	
a_{31}	a_{32}	a_{33}	0	1	
	g_2'	g_3'	$-$	$-$	
	b_{11}	b_{12}	$-b_{22}/b_{21}$		
	b_{21}^*	b_{22}	1		
	g_3''		$-$		
	c_{11}^*				

(A) on the left, (B) on the right.

extension of the method to a larger number may be made without difficulty. Choose some unknown, say x_2 for elimination. Divide the numbers of the corresponding row of (A) by the first number in that row (we indicate it with a star) and add one's and zero's as shown to form (B). Now consider g_1, g_2, g_3 as a row matrix and multiply the columns of (B) by this row (see sec. 10.6). The results are g_2' and g_3'. For example, $g_2' = g_1 \times (-a_{22}/a_{21}) + g_2 \times 1 + g_3 \times 0$ and $g_3' = g_1 \times (-a_{23}/a_{21}) + g_2 \times 0 + g_3 \times 1$. Multiply rows of ($A$) by columns of ($B$), omitting the starred row of (A). This gives the numbers b_{ij}. Again star an element and repeat the process until the last unknown is eliminated. The values of x are then given by

$$x_1 = g_3''/c_{11}$$
$$x_2 = (g_2' - b_{11}x_1)/b_{21}$$
$$x_3 = (g_1 - a_{11}x_1 - a_{31}x_3)/a_{21} \qquad (13\text{-}63)$$

Some care must be exercised in the order of elimination of the x's, especially if they are of widely different magnitudes. It is always advisable to begin with the smallest one, proceeding with the elimination in order of increasing magnitude. If this is not done, the cumulative errors in the calculations will produce unsatisfactory values of the unknowns.

[21] See Frazer, R. A., Duncan, W. J., and Collar, A. R., "Elementary Matrices," Cambridge University Press, 1938; Jeffreys, H. and Jeffreys, B. S., "Methods of Mathematical Physics", Second Edition, Cambridge University Press; 1950 and Milne, loc. cit.

Example 21. Fit the data of Example 3, sec. 13.3 to an equation of the form $e = x_1 + x_2 t + x_3 t^2$. The three simultaneous equations become

$$x_1 + 630.5 x_2 + 3.975 \times 10^5 x_3 = 5.535$$
$$x_1 + 960.5 x_2 + 9.226 \times 10^5 x_3 = 9.117$$
$$x_1 + 1063.0 x_2 + 11.300 \times 10^5 x_3 = 10.301 \qquad (13\text{-}64)$$

TABLE 14

5.535	9.117	10.301	–	–
1	1	1	−2.32101	−2.84277
630.5	960.5	1063.0	1	0
$3.975 \times 10^{5*}$	9.226×10^5	11.300×10^5	0	1
	−3.72979	−5.43373	–	–
	−1.32101	−1.84277	−1.45033	
	−502.897*	−729.366	1	
		−0.02430	–	
		+0.07313*		

Since the magnitude of the x's is probably $x_3 < x_2 < x_1$, we choose the starred numbers in that order. If we desire four significant figures in the final results, we note that we must carry six figures in the calculations, since two figures disappear in one of the steps. The scheme is shown in Table 14. Then,

$$x_1 = -0.02430/0.07313 = \underline{-0.3323}$$
$$x_2 = -(-3.730 - 1.321 \times 0.3323)/502.9 = \underline{0.00829}$$
$$x_3 = (5.535 + 0.3323 - 630.5 \times 0.00829)/3.975 \times 10^5$$
$$= \underline{1.611 \times 10^{-6}}$$

Substitution of these results in the original equations gives as a check, 5.535, 9.116, 10.300.

13.27. Evaluation of Determinants.—The procedure just outlined is also applicable to the evaluation of determinants, the scheme being similar to that shown in Table 14 except for the fact that the g's are omitted. If the starred elements are taken in the first column and row, that is, in the order a_{11}, b_{11}, c_{11}, etc., the value of the determinant equals the product of all of these starred elements. If some other order is chosen as in Example 21, the determinant still equals this product but it must be multiplied by $(-1)^n$ where n is the number of interchanges required to bring the starred elements into the position of the element which stands first in the corresponding array. If it is convenient to choose starred elements that are not in the first column the necessary modification of the procedure will be found described by Frazer, Duncan and Collar (loc.cit.). A method for determinants, with suitable checking procedures, is also given by Milne (loc. cit.).

Example 22. Evaluate the determinant of the coefficients of the x's in Example 21. Since two interchanges are required to bring the first starred element to the position a_{11} and one interchange to bring b_{21} to b_{11}, the value of the determinant Δ is given by

$$\Delta = (-1)^3 \times (3.975 \times 10^5) \times (-502.897) \times (0.07313)$$

$$= \underline{1.4619 \times 10^7}$$

Problem. Evaluate some of the determinants of Example 23, sec. 13.28. The answers are found in Table 16.

13.28. Solution of Secular Determinants.—In many quantum mechanical problems, it is necessary to find one or more roots of a secular equation (see sec. 10.14):

$$y(\lambda) = \left| a_{ij} - b_{ij}\lambda \right| = 0 \qquad (13\text{-}65)$$

$a_{ij} = a_{ji}$, $b_{ij} = b_{ji}$; $i, j = 1, 2, \cdots, N$. In most cases, $b_{ij} = \delta_{ij}$, but even if this is not true in the original form of the determinant it is usually possible to reduce (65) to this form by suitable addition and subtraction of rows and columns. We shall assume here that λ occurs only in the diagonal elements. The particular method to be used in finding values of λ depends to some extent on the special problem at hand. We present three methods, each of which has certain advantages.

a. *The Polynomial Method.* When (65) is expanded, it obviously gives a polynomial of the N-th degree in λ. Once this polynomial is obtained, either of the methods of sec. 13.25 may be used to find values of λ. Graeffe's method is particularly useful when it is required to find all of them. To convert the determinant into the polynomial, its expansion may be effected by the usual method of reduction of its order (see sec. 10.3) or by a very convenient procedure which has been described by Hicks.[22]

According to the latter method, we substitute $\lambda = 0, 1, 2, \cdots, (N + 1)$ in the given determinant and evaluate each numerically. From these $(N + 2)$ results, $y_0, y_1, y_2, \cdots, y_{N+1}$, a table of differences is formed as described in sec. 13.2. An immediate check on the computation of the determinants is available for the $(N + 1)$-st differences should vanish. The polynomial is then given by

$$y(\lambda) = \sum_{t=0}^{N} p_t \lambda^t \qquad (13\text{-}66)$$

where

$$p_0 = y_0; \quad p_t = \sum_{s=t}^{N} r_{ts}\Delta^s y_0; \quad t \geq 1 \qquad (13\text{-}67)$$

[22] Hicks, B. L., *J. Chem. Phys.* **8**, 569 (1940).

The coefficients r_{ts} are independent of the values of the elements in (65) and may be computed from the following relations:

$$r_{ts} = \frac{c_t(s)}{s!} \qquad (13\text{-}68)$$

$$c_s(s) = 1; \quad c_i(s+1) = c_{i-1}(s) - sc_i(s); \quad c_0(s) = 0$$

where

$$c_1(s+1) = (-1)^s s!, \quad s \geq 1; \quad c_{s-1}(s) = \frac{s(1-s)}{2}$$

The results may be checked by the identities

$$\sum_{i=1}^{s} \left| \frac{c_i(s)}{s!} \right| = 1; \quad \sum_{i=1}^{s} \frac{c_i(s)}{s!} = 0 \qquad (13\text{-}69)$$

Values of the r_{ts} through $t = s = 6$ are given in Table 15.

TABLE 15

t \ s	1	2	3	4	5	6
1	1					
2	$-\frac{1}{2}$	$\frac{1}{2}$				
3	$\frac{1}{3}$	$-\frac{1}{2}$	$\frac{1}{6}$			
4	$-\frac{1}{4}$	$\frac{11}{24}$	$-\frac{1}{4}$	$\frac{1}{24}$		
5	$\frac{1}{5}$	$-\frac{5}{12}$	$\frac{7}{24}$	$-\frac{1}{12}$	$\frac{1}{120}$	
6	$-\frac{1}{6}$	$\frac{137}{360}$	$-\frac{5}{16}$	$\frac{17}{144}$	$-\frac{1}{48}$	$\frac{1}{720}$

Example 23. As an example of the use of this method, we choose the secular determinant whose expanded form served as an example for the Graeffe method (see Example 20). The determinant follows

$$y(\lambda) = \begin{vmatrix} 36 - \lambda & -4.062 & 0 & 0 \\ -4.062 & 16 - \lambda & 8.216 & 0 \\ 0 & 8.216 & 4 - \lambda & 14.49 \\ 0 & 0 & 14.49 & -\lambda \end{vmatrix} = 0$$

Making the substitutions $\lambda = 0, 1, 2, 3, 4, 5$ in turn and evaluating the determinants, we obtain Table 16. The fact that the fifth differences

TABLE 16

λ	y	Δ	Δ^2	Δ^3	Δ^4
0	$-117{,}495$				
1	$-105{,}948$	$+11{,}457$			
2	$-93{,}743$	$+12{,}205$	$+658$		
3	$-81{,}180$	$+12{,}563$	$+358$	-300	
4	$-68{,}535$	$+12{,}645$	$+82$	-276	$+24$
5	$-56{,}060$	$+12{,}475$	-170	-252	$+24$

vanish assures us that the determinants have been computed correctly. From (67), Tables 15 and 16, we find

$$p_0 = -117{,}495$$

$$p_1 = 11{,}547 - \frac{658}{2} - \frac{300}{3} = 11{,}112$$

$$p_2 = \frac{658}{2} + \frac{300}{2} + \frac{11 \times 24}{24} = 490$$

$$p_3 = -\frac{300}{6} - \frac{24}{4} = -56$$

$$p_4 = 1$$

hence the required polynomial is

$$y(\lambda) = \lambda^4 - 56\lambda^3 + 490\lambda^2 + 11{,}112\lambda - 117{,}495$$

in agreement with the result given in Example 20.

b. *Matrix Method.* A matrix method, described by Frazer, Duncan and Collar (loc. cit.) is sometimes useful. It gives the largest value of $|\lambda|$ only, but in quantum mechanical problems this is often all that is required. The method does not converge rapidly unless the largest root is widely separated from the remaining ones. The procedure is as follows. Set $\lambda = 0$ in the secular determinant and multiply the resulting matrix by a matrix of one column. The latter is arbitrary but in its most convenient form it contains unity in one row and zeros in the other rows. Extract a constant scalar quantity from the resulting matrix product and multiply the original matrix with the new one-column matrix. Continue in the same way until the scalar quantity becomes constant. This is the required root of largest amplitude.

Example 24. Find the largest root of the secular determinant of Example 23. The procedure is apparent from the following.

$$
\begin{vmatrix} 36 & 4.062 & 0 & 0 \\ 4.062 & 16 & 8.216 & 0 \\ 0 & 8.216 & 4 & 14.49 \\ 0 & 0 & 14.49 & 0 \end{vmatrix}
\begin{vmatrix} 1 \\ 0 \\ 0 \\ 0 \end{vmatrix}
=
\begin{vmatrix} 36 \\ 4.062 \\ 0 \\ 0 \end{vmatrix}
= 36
\begin{vmatrix} 1 \\ 0.1128 \\ 0 \\ 0 \end{vmatrix}
$$

For the next approximation,

$$
\begin{vmatrix} 36 & 4.062 & 0 & 0 \\ 4.062 & 16 & 8.216 & 0 \\ 0 & 8.216 & 4 & 14.49 \\ 0 & 0 & 14.49 & 0 \end{vmatrix}
\begin{vmatrix} 1 \\ 0.1128 \\ 0 \\ 0 \end{vmatrix}
=
\begin{vmatrix} 36.46 \\ 5.87 \\ 0.93 \\ 0 \end{vmatrix}
= 36.46
\begin{vmatrix} 1 \\ 0.1610 \\ 0.0256 \\ 0 \end{vmatrix}
$$

Continuing in the same way, we obtain the results in Table 17. The sixth, seventh, eighth and ninth approximations give 36.85, 36.87, 36.88, 36.88, hence $\lambda = \underline{36.88}$. Comparison with Example 20 shows that this result is uncertain in the last place. The convergence here is not very rapid since the next largest root is 22.341. More rapid convergence could be obtained by squaring the original matrix several times before commencing the matrix multiplications. The constant value so obtained is then some power of the desired root. Once having found the largest root, the next largest one may be obtained by the same method. Further details are given by Frazer, Duncan and Collar (loc. cit.).

TABLE 17

Successive Column Matrices

Third		Fourth		Fifth	
36.65	1	36.76	1	36.81	1
6.85	0.1869	7.37	0.2005	7.68	0.2086
1.42	0.0387	1.84	0.0500	2.07	0.0562
0.37	0.0101	0.56	0.0152	0.72	0.0196

c. *Iteration Method.* Several iteration methods which do not depend on matrix properties have been described.[23] Crude approximations to the roots of the polynomial are given by the diagonal terms in the secular determinant. Suppose one of these values, say λ_0 is substituted in the determinant for λ in every place except one where the quantity $\lambda_0 - \lambda$ occurs. Now if the determinant is evaluated, the resultant value of λ is the next approximation to the true value. The process may be repeated as often as necessary.

Example 25. Find a root of the determinant of Example 23 by the iteration method. Taking $\lambda_0 = 36$, the determinant becomes

$$\begin{vmatrix} 36 - \lambda & 4.062 & 0 & 0 \\ 4.062 & -20 & 8.216 & 0 \\ 0 & 8.216 & -32 & 14.49 \\ 0 & 0 & 14.49 & -36 \end{vmatrix}$$

When this is evaluated, we obtain $^1\lambda = 36.893$. Substitution of $^1\lambda$ in the original determinant gives $^2\lambda = \underline{36.896}$. The third approximation gives the same result.

Problem. Compute some of the coefficients of Table 15.

[23] See, for example, James and Coolidge, *J. Chem. Phys.* **1**, 825 (1933), Cross and Crawford, *J. Chem. Phys.* **5**, 621 (1937). Another iterative method, which gives both the eigenvalues and the amplitudes for a system of homogeneous linear equations, has been described by W. Kohn, *J. Chem. Phys.* **17**, 670 (1949).

PART 3. ERRORS AND LEAST SQUARES

13.29. Errors.—Measurements are always accompanied by errors. They are of two kinds: *determinate* and *random*. Those of the first type[24] are often constant or systematic, being due to faulty or incorrectly adjusted instruments, mistakes on the part of the observer in reading a scale, recording a number or other similar effects. It is usually possible to discover the causes of such errors and to make corrections for them. Random errors, on the other hand, are indeterminate and due to unknown causes, but they may be treated by statistical methods. As in the previous parts of this chapter, we shall often refer the reader to other sources[25] for proofs of theorems and results to be given here.

Suppose several equally reliable measurements of a physical quantity yield the numbers $X_1, X_2, \cdots, X_n$. The corresponding *errors* are defined by

$$x_1 = X_1 - X, \quad x_2 = X_2 - X, \quad \cdots, \quad x_n = X_n - X \quad (13\text{-}70)$$

where X is the *true* value of the quantity. Actually, we seldom know[26] the true value since any experiment made to determine it will be accompanied by random errors. However, in order to proceed further we must choose *some* quantity which is called the *most probable value*. It will be indicated by $\overline{X}$, the notation anticipating a fact that we prove in sec. 13.30, namely, the most probable value is the average of all the data. Since $\overline{X}$ is not equal to X, the true value, we must distinguish between the error and the *residual* which is defined by

$$d_1 = X_1 - \overline{X}, \quad d_2 = X_2 - \overline{X}, \quad \cdots d_n = X_n - \overline{X} \quad (13\text{-}71)$$

It is assumed that the errors and residuals with which we are concerned are random ones. They are neither systematic nor constant but are equally likely to be positive or negative. Small errors are more frequent than large ones and very large errors do not occur at all. Under these conditions, the errors follow the laws of probability as given by the *normal* " *Gauss* " *distribution* (see sec. 12.3)

$$w(x) = \frac{e^{-x^2/2\sigma^2}}{\sigma\sqrt{2\pi}}$$

---

[24] Errors of this kind are discussed in some detail by Crumpler, T. B. and Yoe, J.H., " Chemical Computations and Errors," John Wiley and Sons, New York, 1940. They may be detected in some cases by methods explained by Birge, R. T., *Phys. Rev.* **40**, 207 (1932).

[25] See references at end of chapter.

[26] An exception is the case where the quantity is exact by definition. For example, the true value of the atomic weight of oxygen is 16.0000 to as many decimal places as may be needed.

It is convenient for our purposes to change the notation, writing $w(x) = N$ and $h^2 = 1/2\sigma^2$. The resulting equation

$$N = \frac{h}{\sqrt{\pi}} e^{-h^2 x^2} \tag{13-72}$$

gives us the relative number of measurements N having an error x. The plot of N vs. x is called the Gauss error curve; it is shown in Fig. 1 for $h = 1$ and $h = 0.6$. From that curve or from eq. (72), we can discover the meaning of the constant h which is called the *precision index*. When it is large, N is large for a given small error x and decreases as x increases.

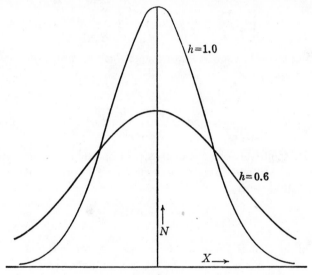

FIG. 13-1

Thus a high precision index means that a large number of the measurements agree closely with the true value of the quantity observed. On the other hand, if h is small, a smaller fraction of the results are close to the true value and more large errors occur than in the previous case.

The probability that the error of a single measurement will lie between the limits $\pm a$ is

$$\frac{h}{\sqrt{\pi}} \int_{-a}^{a} e^{-h^2 x^2} dx = \frac{2h}{\sqrt{\pi}} \int_{0}^{a} e^{-h^2 x^2} dx$$

This integral occurs so often in mathematical physics that it has been given the special name of the *error function*. It is usually denoted by

$$\operatorname{erf}(t) = \frac{2}{\sqrt{\pi}} \int_{0}^{t} e^{-v^2} dy$$

Hence the probability in question is $\mathrm{erf}\,(ha)$. It cannot be evaluated in finite form but must be expanded in a power series and integrated term by term. Values of the integral as a function of t are found in all books on probability.[27]

The special case where the limits of integration are $\pm\infty$ is of considerable interest. The error must lie somewhere within this range, hence probability must be unity. This is readily found to be true when the integration is performed.

The simplest way of evaluating the integral when the limits are $\pm\infty$, is the following. Let

$$I^2 = \left[\int_{-\infty}^{\infty} e^{-u^2}du\right]^2 = \int_{-\infty}^{\infty}\int_{-\infty}^{\infty} e^{-(u^2+v^2)}dudv$$

Transforming to polar coordinates we get: $dudv = rdrd\phi$, $u^2 + v^2 = r^2$,

$$I^2 = \int_0^{2\pi} d\phi \int_0^{\infty} re^{-r^2}dr = \pi$$

Thus we see that the area under the whole curve (72) is unity. This, obviously, is the reason for the constant $2/\sqrt{\pi}$.

13.30. Principle of Least Squares.—Suppose n measurements have been made, the i-th one having the error x_i. The probability that x_i lies between x_i and $x_i + dx_i$ is

$$P_i = \frac{h}{\sqrt{\pi}}\, e^{-h^2 x_i^2}dx_i \qquad (13\text{-}73)$$

The probability that the n errors $x_1, x_2, \cdots, x_n$ occur is the product of n terms like (73), for each measurement is an independent event. Hence we have

$$P = \prod_{i=1}^{n} P_i = \left(\frac{h}{\sqrt{\pi}}\right)^n e^{-h^2(x_1^2+x_2^2+\cdots+x_n^2)}dx_1 dx_2 \cdots dx_n \qquad (13\text{-}74)$$

Clearly the differentials $dx_1, dx_2, \cdots$ are arbitrary, for they may be interpreted as the smallest subdivisions on a scale which is being read. Finally, remembering that h is fixed, we see that the probability P is a maximum when the exponent of e is a minimum; thus we have

$$x_1^2 + x_2^2 + \cdots + x_n^2 = \text{a minimum} \qquad (13\text{-}75)$$

as the criterion for the most probable value obtainable from n equally reliable measurements of a quantity. This result is known as the *Principle of Least Squares.*

[27] See also " Tables of the Probability Functions $P(x)$ and Erf (x)," Works Progress Administration, New York City, 1941.

In accordance with that principle, let us determine the most probable value of the set of measurements $X_1, X_2, \cdots, X_n$. Rewrite (75) in the form

$$(X_1 - X)^2 + (X_2 - X)^2 + \cdots + (X_n - X)^2$$

differentiate with respect to X and equate the derivative to zero in order to obtain a minimum. Since the result is to be the most probable value of X, we replace X by the symbol $\overline{X}$ to indicate that X is chosen to satisfy eq. (75). The answer is

$$\overline{X} = \frac{X_1 + X_2 + \cdots + X_n}{n} \tag{13-76}$$

As might be expected the most probable value is the *arithmetic mean* of all of the experimental results. It is interesting to note that the error law of eq. (72) is, within reasonable limits, the only form of equation which gives the average as the most probable value.[28]

13.31. Errors and Residuals.—If we add n errors, we find, since $x_i = X_i - X$

$$\sum X_i = nX + \sum x_i$$

and from eq. (76)

$$\overline{X} = \frac{1}{n}\sum X_i = X + \frac{1}{n}\sum x_i \tag{13-77}$$

Also, we obtain for the first residual

$$d_1 = X_1 - \overline{X} = X_1 - X - \frac{1}{n}\sum x_i$$

$$= x_1 - \frac{1}{n}\sum x_i = \frac{(n-1)}{n}x_1 - \frac{1}{n}x_2 - \frac{1}{n}x_3 - \cdots \tag{13-78}$$

with similar equations for the others. We thus conclude that as n increases, the second term on the right of (77) becomes smaller and $\overline{X}$ approaches the true value X. In the same way, we conclude from (78) that as n increases, the residuals approach the true errors. Actually, if we square n equations like (78) and add them, we get

$$\sum d_i^2 = \sum x_i^2 - \frac{1}{n}\left(\sum x_i\right)^2$$

so that the sum of the squares of the residuals is slightly less than the sum of the squares of the errors.

Suppose two independent quantities (M_1 and M_2) have been measured and the errors in each case obey the normal law. Then the probability of

---

[28] A proof is given by Plummer, H. C., " Probability and Frequency," Macmillan Co., London, 1940, p. 123.

an error between x_1 and $x_1 + dx_1$ in M_1 is

$$p_1 = \frac{h_1}{\sqrt{\pi}} e^{-h_1^2 x_1^2} dx_1$$

while the probability of an error between x_2 and $x_2 + dx_2$ in M_2 is

$$p_2 = \frac{h_2}{\sqrt{\pi}} e^{-h_2^2 x_2^2} dx_2$$

Since the observed quantities are independent of each other, the probability of the simultaneous occurrence of these errors in M_1 and M_2 is

$$p = p_1 p_2$$

Now suppose that M_1 and M_2 are combined linearly to form a quantity

$$M = \alpha_1 M_1 + \alpha_2 M_2$$

where α_1 and α_2 are constants. The error in M will lie between

$$\alpha_1 x_1 + \alpha_2 x_2 = x \tag{13-79}$$

and

$$\alpha_1 (x_1 + dx_1) + \alpha_2 (x_2 + dx_2) = x + dx$$

We recognize the fact that such an error may be composed of any value of x_1 between $\pm \infty$ together with the corresponding value of x_2 fixed by eq. (79). Thus to compute the probability of an error x in M we integrate $p = p_1 p_2$ with respect to x_1 between the limits $\pm \infty$ and eliminate dx_2 by the relation $dx = \alpha_2 dx_2$ which will be true when the integration has been performed over x_1. Let us first rewrite p in terms of x_1 and x which gives

$$p = C \exp\left[-h_1^2 x_1^2 - h_2^2 \left(\frac{x - \alpha_1 x_1}{\alpha_2} \right)^2 \right] dx_1 dx_2$$

where $C = h_1 h_2 / \pi$. With the further abbreviation

$$\tau = \frac{h_1^2 h_2^2}{\alpha_1^2 h_2^2 + \alpha_2^2 h_1^2}$$

we also have

$$p = C \exp\left[-\tau x^2 - \frac{h_1^2 h_2^2}{\tau \alpha_2^2} \left(x_1 - \frac{\alpha_1 \tau x}{h_1^2} \right)^2 \right] dx_1 dx_2$$

Let $N(x)dx$ be the required probability of an error in M between x and $x + dx$, then

$$N(x)dx = C e^{-\tau x^2} dx_2 \int_{-\infty}^{\infty} \exp\left[-\frac{h_1^2 h_2^2}{\tau \alpha_2^2} \left(x_1 - \frac{\alpha_1 \tau x}{h_1^2} \right)^2 \right] dx_1$$

$$= C e^{-\tau x^2} dx_2 \left[\frac{\alpha_2 \sqrt{\pi \tau}}{h_1 h_2} \right]$$

Since $\alpha_2 dx_2 = dx$ we see that

$$N(x) = \frac{H}{\sqrt{\pi}} e^{-H^2 x^2}$$

where

$$H = \frac{h_1 h_2}{\alpha_1^2 h_2^2 + \alpha_2^2 h_1^2} \tag{13-80}$$

or

$$\frac{1}{H^2} = \frac{\alpha_1^2}{h_1^2} + \frac{\alpha_2^2}{h_2^2}$$

Thus the error law for M is the same as the error law for M_1 and M_2, the only difference being in the precision index. The equation is easily generalized by the same method; in fact it may be shown that if

$$M = \sum \alpha_i M_i$$

the precision index of M is given by

$$\frac{1}{H^2} = \sum \frac{\alpha_i^2}{h_i^2} \tag{13-81}$$

We would like to apply this result to the residuals. From (78), we may write

$$d_i = \frac{(n-1)}{n} x_i - \frac{1}{n} \sum_{j=1}^{n}{}' x_j \tag{13-78a}$$

where the prime on the summation sign means that the term $i = j$ is omitted. The residuals are thus linear combinations of the errors, for d_i corresponds to M in the preceding discussion and

$$\alpha_1 = \frac{(n-1)}{n} ; \quad \alpha_2 = \alpha_3 = \cdots = \alpha_n = -\frac{1}{n}$$

The error law for the residuals is of the form of (72) or (80)

$$\frac{H}{\sqrt{\pi}} e^{-H^2 d^2} \tag{13-80a}$$

and from (81) since h is the precision index for each x_i

$$\frac{1}{H^2} = \frac{1}{h^2} \left[\frac{(n-1)^2}{n^2} + \frac{1}{n^2} + \cdots + \frac{1}{n^2} \right] = \frac{1}{n^2 h^2} [(n-1)^2 + n - 1]$$

or

$$H = h \sqrt{\frac{n}{n-1}} \tag{13-82}$$

From (82), it is seen that the precision index for the residuals depends on both n and h and is always larger than h. Reference to Fig. 1 shows that the curve of (80a) rises higher in the middle and falls off more rapidly than the curve of (72) but as the number of measurements increases the two graphs approach each other more closely.

13.32. Measures of Precision.—Having obtained the most probable value of a series of measurements, we need to find expressions for its reliability. In order to do this we must first consider the case where the true value X of the quantity is known. We may then proceed to the more practical question of expressing the uncertainty of $\overline{X}$ in terms of the residuals. If the precision index were known it would be suitable for our measure of precision for as we have seen in sec. 13.29, erf (hx) is the probability that the error is within the range $\pm x$. However, h has the dimension of a reciprocal error and it proves more convenient to use as a precision measure a quantity which is inversely proportional to h, thus having the same dimension as the error itself. Three such measures are commonly employed; they are the *average error* (a), the *root mean square error* (m) and the *probable error* (r).

The *average error* is the arithmetic mean of all the errors without regard to sign

$$a = \frac{\sum |x_i|}{n} \tag{13-83}$$

From its definition (see sec. 12.3), it follows that

$$a = \int_{-\infty}^{\infty} |x| N dx = \frac{2h}{\sqrt{\pi}} \int_{0}^{\infty} x e^{-h^2 x^2} dx = \frac{1}{h\sqrt{\pi}} \tag{13-84}$$

Let us seek the most probable value of h. We recall that P of eq. (74) is the probability of the simultaneous occurrence of the errors $x_1, x_2, \cdots,$ x_n. Hence we must make P of that equation a maximum. Taking the logarithm of (74) we see that the most probable value of h is that quantity h' which makes

$$\phi = n \log h - h^2 \sum x_i^2$$

a maximum, or

$$\frac{d\phi}{dh} = \frac{n}{h} - 2h \sum x_i^2 = 0$$

hence

$$h' = \sqrt{\frac{n}{2\sum x_i^2}}$$

The quantity m defined by

$$m = \frac{1}{h'\sqrt{2}} = \sqrt{\frac{\sum x_i^2}{n}} \tag{13-85}$$

is called the *root mean square error*. Comparison of (84) with (85) shows that

$$a = m\sqrt{\frac{2}{\pi}} \tag{13-86}$$

The root mean square error is frequently used in mathematical statistics; there it is called the *standard deviation* and indicated by σ (see sec. 12.3, especially problem a).

The *probable error* is defined as that error r such that one half of the errors of n observations are greater than r and one half are less than r. Thus it is given by the integral

$$\text{erf}\,(hr) = \tfrac{1}{2} \tag{13-87}$$

for this says that there is an equal chance that a given error lies within $\pm r$ or outside these limits.   From tables of the integral, we obtain

$$r = \frac{0.4769363 \cdots}{h} \tag{13-88}$$

Combining this result with (85) we get for the probable error

$$r = 0.6745\sqrt{\frac{\Sigma x_i^2}{n}} = 0.6745m \tag{13-89}$$

From eqs. (86) and (89) we can readily obtain all relations between a, m and r.   They are

$$r = 0.4769h^{-1} = 0.6745m = 0.8453a$$

$$m = 0.7071h^{-1} = 1.4826r = 1.2533a$$

$$a = 0.5642h^{-1} = 0.7979m = 1.1829r$$

The geometric significance of the three precision measures is also of interest. The average error a is the abscissa of the center of gravity of the area bounded by the error curve and the axes x and N of eq. (72).   To see this, let x_0 be the center of gravity of that area, then

$$x_0 = \frac{\displaystyle\int xN dx}{\displaystyle\int N dx} = \frac{1}{h\sqrt{\pi}} = a$$

which follows from (84) since $\displaystyle\int N dx = 1$.

The root mean square error is the radius of gyration of the same area about the N axis; it is also the abscissa of the point of inflection of the error curve, as will now be shown.   For the point of inflection, $d^2N/dx^2 = 0$ and from (72)

$$\frac{dN}{dx} = N' = -\frac{2h^3}{\sqrt{\pi}}\, xe^{-h^2x^2}; \quad N'' = N'\left(\frac{1}{x} - 2h^2x\right) = 0$$

Thus,

$$(1 - 2h^2x^2) = 0$$

or

$$x = \pm\frac{1}{h\sqrt{2}} = \pm m$$

From the definition of r, it follows that the abscissa $x = r$ corresponds to the ordinate which bisects the area of the error curve (72) between 0 and ∞.

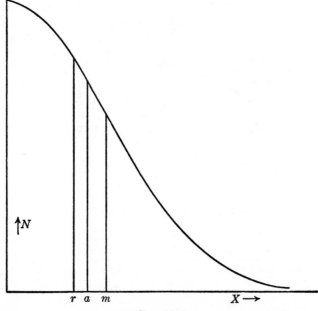

FIG. 13–2

The relative sizes and positions of these three measures are shown in Fig. 2 where we draw only that half of (72) corresponding to positive values of x.   It is perhaps not amiss to comment on the most appropriate measure to use.   The average error recommends itself because of the ease with which it is computed.   The probable error is less easy to calculate[29]

---

[29] Convenient tables of $0.6745/\sqrt{n}$ as a function of n and other quantities useful in the calculation of errors may be found in "Handbook of Chemistry and Physics," Chemical Rubber Publishing Co., Cleveland, Ohio.

but is perhaps more often used than the others in chemical and physical literature.   As may be seen from **Fig. 2** it is the smaller of the three and is thus more flattering than a or m to a set of experimental data.   There is little choice between the three measures on theoretical grounds.

It is often of importance to find some estimate of the probable error of an adopted precision measure itself.   The result has been obtained by Gauss[30] who shows that the relative error of r is

$$\frac{0.4769}{\sqrt{n}}$$

With 10 measurements, it is seen that the probable error is uncertain by about 15 per cent while for even 500 measurements the uncertainty is 2 per cent.   It thus follows that it is seldom if ever of meaning to state the probable error with more than two significant figures, for usually one of these is uncertain.

13.33. Precision Measures and Residuals.—From the equations of the previous section it is a simple matter to express the precision measures in terms of residuals.   Suppose $X_1, X_2, \cdots, X_n$ are n observations.   If they follow the error law, the residual d_i is given by eq. (78a) and the index of precision of the residuals by eq. (82).   Therefore, the average error

$$a = \frac{1}{h\sqrt{\pi}} = \sqrt{\frac{n}{n(-1)}} \frac{\sum |d_i|}{n} = \frac{\sum |d_i|}{\sqrt{n(n-1)}} \qquad (13\text{-}83a)$$

Similarly,

$$m = \frac{1}{h\sqrt{2}} = \sqrt{\frac{n}{(n-1)} \frac{\sum d_i^2}{n}} = \sqrt{\frac{\sum d_i^2}{(n-1)}} \qquad (13\text{-}85a)$$

and

$$r = 0.6745m = 0.6745 \sqrt{\frac{\sum d_i^2}{(n-1)}} \qquad (13\text{-}89a)$$

The differences between eqs. (83), (85), (89) and (83a), (85a), (89a) should be carefully noted.   In many cases, the deviations are used in place of the errors to get a from (83) rather than from the correct eq. (83a).   The difference is negligible, of course, in most cases.

The most probable value or arithmetic mean also follows the error law. Its index of precision is obtained from (81) where $\alpha_i = 1/n$, hence

$$\frac{1}{H^2} = \frac{1}{h^2} \left(\sum \alpha_i^2 \right) = \frac{1}{h^2 n}$$

[30] A derivation of it is given by Plummer, loc. cit.

Thus if a, m and r refer to the individual members of a set of n measurements, the corresponding precision measures relating to the arithmetic mean are

$$A = \pm \frac{a}{\sqrt{n}}, \quad M = \pm \frac{m}{\sqrt{n}}, \quad R = \pm \frac{r}{\sqrt{n}}$$

It will be observed that the precision varies as the square root of n. Therefore comparatively little is gained by increasing n, for in order to change the precision by one decimal point n must be multiplied by 100. This is in accord with common sense which suggests that instead of making 100 measurements it is more economical and reasonable to seek an improvement in the experimental method. A graph of r versus n is shown in Fig. 3. It will be seen from that curve that it is seldom worthwhile to make more than 10 measurements of a given quantity by the same method.

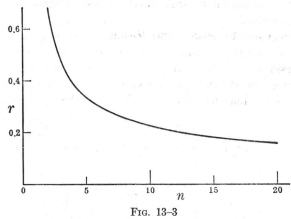

Fɪɢ. 13–3

13.34. Experiments of Unequal Weight.—It often happens that the results of one experimenter are more reliable than those of another. This may be due to superior method or apparatus, to greater experience with the operations involved or to other reasons. Moreover, because of particularly favorable conditions, the *same* investigator may obtain better results at some times than at others. In all such cases, more *weight* is attached to some of the data than to the remainder of them. For example, if one result X_1 has a weight twice that of X_2, then the average $\overline{X} = (2X_1 + X_2)/3$. A result of weight w is thus equivalent to w results of unit weight, or we say that a result of large weight has a high precision index.

If the j-th measurement is of weight w_j, the weighted average or most probable value is

$$\overline{X}_w = \frac{\sum w_i X_i}{\sum w_i}$$

The probable error for the value of weight w_j is

$$p_{w_j} = \pm 0.6745 \sqrt{\frac{\sum w_i d_i^2}{(n-1)w_j}}$$

and for the weighted average

$$P_w = \pm 0.6745 \sqrt{\frac{\sum w_i d_i^2}{(n-1)\sum w_i}}$$

It is possible to determine the relative weights to be attached to the individual measurements since the weight w_j is inversely proportional to $p_{w_j}^2$. The usual custom is to assign weights arbitrarily.

13.35. Probable Error of a Function.—In general the results of several independently measured quantities are combined to give the final value of the physical constant desired. Suppose $X, Y, \cdots$ have been obtained as the average value of certain quantities with probable errors $P_X, P_Y, \cdots$. If they are combined to give Z, where

$$Z = f(X, Y, \cdots)$$

then its probable error is

$$P = \sqrt{(P_X \partial Z/\partial X)^2 + (P_Y \partial Z/\partial Y)^2 + \cdots}$$

We record a few special cases for convenience of reference.

1. $Z = X \pm Y; \quad P = \pm \sqrt{P_X^2 + P_Y^2}$

2. $Z = XY; \quad P = \pm \sqrt{(XP_Y)^2 + (YP_X)^2}$

3. $Z = X/Y; \quad P = \pm \dfrac{1}{Y^2} \sqrt{Y^2 P_X^2 + X^2 P_Y^2}$

4. $Z = a + bX$. Suppose we know the value Z_1 with its probable error p_1 at the point $X = X_1$ and Z_2 with error p_2 at $X = X_2$. We wish to fit the two points to a linear equation. Then

$$P_a = \sqrt{\left(\frac{X_2 p_1}{X_2 - X_1}\right)^2 + \left(\frac{X_1 p_2}{X_1 - X_2}\right)^2}$$

$$P_b = \sqrt{\left(\frac{p_1}{X_1 - X_2}\right)^2 + \left(\frac{p_2}{X_2 - X_1}\right)^2}$$

$$P_Z = \sqrt{\left(\frac{p_1(X_2 - X)}{(X_2 - X_1)}\right)^2 + \left(\frac{p_2(X_1 - X)}{(X_1 - X_2)}\right)^2}$$

where P_a, P_b and P_Z are the probable errors in a, b and Z, respectively.

13.36. Rejection of Observations.—Occasionally a single measurement from a set differs so widely from the others that the experimenter is tempted to discard it. A simple rule in such cases, based on statistical methods is the following: Calculate the average of all the data including the suspected measurement. Find each residual and calculate the probable error of a single determination. If any residual exceeds *five* times the probable error it may be rejected, the supposition being that the error cannot be a random one. The reason for the use of this rule is as follows. Suppose the probability of an error as large as x_i in the quantity measured is 0.001, then the chance that an error as large as x_i will *not* occur is 0.999. Let us then determine the value of hx for which erf $(hx) = 0.999$. From tables of this integral we find [31]

$$hx = 2.326$$

Now from eq. (88) we have

$$hr = 0.4769$$

thus

$$x = 4.9r$$

We conclude that the probability of an error 5 times as great as the probable error of a single measurement is less than 1 in 1000 hence the somewhat dogmatic rule for rejecting such measurements.

13.37. Empirical Formulas.—As mentioned in sec. 13.1, there is considerable advantage in representing experimental data by means of equations, the correct form of them being often suggested by theoretical considerations. In other cases, plots of various functions of the data may indicate a suitable form. When this question is settled, the next step is to determine the constants in the equation. Sometimes a graph may be used for this purpose, for if the equation is linear it is only necessary to determine the slope and intercept of the curve. In more exact work, numerical methods are needed.

a. *The Method of Averages.* Suppose that the quantity y has been observed as a function of another quantity x, the resulting numbers being $y_1, y_2, \cdots, y_n$. It has been decided that a polynomial of the m-th degree, $m < n$ is a suitable equation

$$y = A + Bx + Cx^2 + \cdots \qquad (13\text{-}90)$$

Divide the measurements into groups equal in number to the unknown constants, placing an equal number of results in each group if this is possible. Add the equations in each group thus obtaining a set of simultaneous equations equal in number to the number of unknowns. The equations may be solved by the methods of sec. 13.26.

[31] See, for example, the reference in footnote 29.

It will be found in general that this procedure is quite satisfactory. The resulting constants are different for different groupings of the data, but the simplest such grouping is usually better than any other. If there are a large number of results or if the polynomial is of degree higher than four or five this method is nearly as good as the method of least squares and entails considerably less calculation.

b. *The Method of Least Squares.* Suppose as before that n values are available for y but that the chosen equation is of a more general form than (90),

$$y = f(x,A,B,C,\cdots) \qquad (13\text{–}91)$$

If there are n constants we may obviously fit the data exactly to such an equation but usually there will only be $m < n$ constants. Thus the calculated value of y will not agree with the observed one. Let

$$y_i = y_i \text{ (calc.)} + d_i$$

where y_i is an observed y and y_i (calc.) is the corresponding calculated one using the constants finally adopted. In accordance with the principle of least squares we wish to make

$$\sum d_i^2 = \text{a minimum} \qquad (13\text{–}92)$$

Let us now assume that we have found approximate values of the constants by graphical means or otherwise so that

$$A = A_0 + a; \quad B = B_0 + b; \quad C = C_0 + c; \quad \cdots$$

$a, b, c, \cdots$ being small correction terms. Then the i-th equation of (91)

$$f_i(A,B,C,\cdots) = y_i - d_i$$

may be written as

$$f_i(A_0,B_0,C_0,\cdots) + a\,\frac{\partial f_i}{\partial A_0} + b\,\frac{\partial f_i}{\partial B_0} + c\,\frac{\partial f_i}{\partial C_0} + \cdots = y_i - d_i \qquad (13\text{–}93)$$

where we have discarded derivatives of second and higher order. Using the abbreviations

$$\frac{\partial f_i}{\partial A_0} = u_i; \quad \frac{\partial f_i}{\partial B_0} = v_i; \quad \frac{\partial f_i}{\partial C_0} = w_i; \quad \cdots$$

and

$$y_i - f_i(A_0,B_0,C_0,\cdots) = F_i$$

(93) becomes

$$u_i a + v_i b + w_i c + \cdots - F_i + d_i = 0$$

where u_i, v_i, w_i, F_i are known and a, b, c, d_i are unknown. Since we wish (92) to hold we must require that

$$\sum_{i=1}^{n}(u_ia + v_ib + w_ic + \cdots - F_i)^2 = \phi(a,b,c)$$

be a minimum or that

$$\frac{\partial\phi}{\partial a} = 2\sum(u_ia + v_ib + w_ic + \cdots - F_i)u_i = 0$$

$$\frac{\partial\phi}{\partial b} = 2\sum(u_ia + v_ib + w_ic + \cdots - F_i)v_i = 0 \qquad (13\text{-}94)$$

$$\frac{\partial\phi}{\partial c} = 2\sum(u_ia + v_ib + w_ic + \cdots - F_i)w_i = 0$$

. . . .

These equations (when divided by two) are called the *normal equations*. There will be as many of them as there are unknowns.

In many cases, the chosen relation between x and y is a polynomial, when some simplification in the procedure is possible. The original equations corresponding to (91) will be of the form

$$A + Bx_i + Cx_i^2 + \cdots = y_i \qquad (13\text{-}95)$$

It is still worthwhile to use approximate values of the constants for then the normal equations will be easier to handle. If this is done (95) becomes

$$a + bx_i + cx_i^2 + \cdots = F_i \qquad (13\text{-}95a)$$

In either case, the normal equations may be written down without differentiation. They are found as follows: (1) multiply each equation of (95) or (95a) by the coefficient of the first unknown (unity since we are speaking of A or a) and add the resulting n equations; (2) multiply each equation by the coefficient of the next unknown (x_i) and add these equations; (3) continue in the same way until each equation has been multiplied by the coefficient of each unknown. The resulting normal equations which are identical with those obtained by the procedure leading to eq. (94) may then be solved by the methods of sec. 13.26 to obtain the constants. The final equation should always be checked by using it to compute each known y_i. The sum of the squares of the residuals should be small and the algebraic sum of the residuals themselves should be nearly zero.[32]

Such a procedure will show how closely the curve fits the known points but says nothing about the reliability of the curve at other places. In the

[32] Further details of the method of least squares are given by Brunt, D., " The Combination of Observations," Cambridge Press, 1917. He describes several schemes for checking the calculations and evaluating the constants with their probable errors. See also, Birge, R. T., *Revs. Mod. Phys.* **19**, 298 (1947).

important case of a linear equation, $y = a + bx$ the formulas[33] are comparatively simple.   The probable errors in a and b are

$$P_a = r_e \sqrt{\frac{\sum x_i^2}{D}} \; ; \quad P_b = r_e \sqrt{\frac{n}{D}}$$

$$r_e = 0.6745 \sqrt{\frac{\sum d_i^2}{(n-2)}} \; ; \quad D = n \sum x_i^2 - (\sum x_i)^2$$

The error in y at any point x (x not necessarily a measured value) is

$$P_x = r_e \sqrt{\frac{\sum (x_i - x)^2}{D}}$$

[33] See Birge, loc. cit.

REFERENCES

Numerical calculations of various kinds are discussed in:

Allen, D. N. deG., " Relaxation Methods," McGraw-Hill Book Co., Inc., New York, 1954.

Collatz, L., " Eigenwert Probleme und Ihre Numerische Behandlung," Chelsea Publishing Co., New York, 1945.

Dwyer, P. S., " Linear Computations," John Wiley and Sons, Inc., New York, 1951.

Householder, A. S., " Principles of Numerical Analysis," McGraw-Hill Book Co., Inc., New York, 1953.

Milne, W. E., " Numerical Calculus," Princeton University Press, Princeton, 1949.

Milne, W. E., " Numerical Solution of Differential Equations," John Wiley and Sons, Inc., New York, 1953.

Scarborough, J. B., "Numerical Mathematical Analysis," Second Edition, The Johns Hopkins Press, Baltimore, 1950.

Shaw, F. S., " An Introduction to Relaxation Methods," Dover Publications, Inc., New York, 1953.

Whittaker, E. T., and Robinson, G., " The Calculus of Observations," Second Edition, D. Van Nostrand Co., Inc., New York, 1930.

Willers, F. A., " Practical Analysis.   Graphical and Numerical Methods," translated by R. T. Beyer, Dover Publications, Inc., New York, 1948.

Probability, the theory of errors and related subjects are treated in:

Arley, N. and Buch, K. R., " Introduction to Mathematical Probability," John Wiley and Sons, Inc., New York, 1949.

Beers, Y., " Theory of Errors," Addison-Wesley Publishing Co., Cambridge, 1953.

Deming, W. E., " Statistical Adjustment of Data," John Wiley and Sons, Inc., New York, 1943.

Jeffreys, H., " Theory of Probability," Second Edition, Oxford University Press, New York, 1948.

Kolmogorov, A., " Foundations Of the Theory of Probability," Chelsea Publishing Co., New York, 1950.

Uspensky, J. V., " Introduction to Mathematical Probability," McGraw-Hill Book Co., Inc., New York, 1937.

Wilson, E. B., Jr., " An Introduction to Scientific Research," McGraw-Hill Book Co., New York, 1952.

Youden, W. J., " Statistical Methods for Chemists," John Wiley and Sons, Inc., New York, 1951

CHAPTER 14

LINEAR INTEGRAL EQUATIONS

14.1. Definitions and Terminology.—An *integral equation* is one which contains the unknown function behind the integral sign. Its importance for physical problems lies in the fact that most *differential equations together with their boundary conditions* may be reformulated to give a *single integral equation*. If the latter can be solved, the mathematical difficulties are not appreciably greater even when the number of independent variables is increased, while differential equations, such as Laplace's, are considerably more complex in three dimensions than in two. The theory of integral equations also furnishes a uniform method for the study of the eigenvalue problems of mathematical physics.

A *linear integral equation of the third kind*, the most general type considered, has the form

$$g(x)\phi(x) = f(x) + \lambda \int_a^b K(x,z)\phi(z)dz \qquad (14\text{-}1)$$

The known functions are $g(x)$, $f(x)$ and $K(x,z)$, the latter being called the *kernel* or *nucleus*. The limits of integration a and b are either known functions of x or constants; λ is an absolute constant or a parameter. It is desired to find the unknown ϕ as a function of the independent variable x.

Four special cases of (1) have been most widely studied. In *Fredholm's equation of the first kind*, $g(x) = 0$, and in his equation of the *second kind*, $g(x) = 1$; in both cases a and b are constants. *Volterra's equations* of the first and second kind are like Fredholm's equations except that $a = 0$, and $b = x$. If $f(x) = 0$ in either case, the equation is said to be *homogeneous*. When one or both limits become infinite or when the kernel becomes infinite at one or more points within the range a to b, the equation is called *singular*.

Non-linear integral equations may occur in the form

$$\phi(x) = f(x) + \lambda \int_a^b K(x,z)\phi^n(z)dz$$

or

$$\phi(x) = f(x) + \lambda \int_a^b F[x,z,\phi(z)]dz$$

We limit[1] our discussion here to linear equations in one variable where the unknown ϕ enters only to the first power.   Our plan is to present first the purely formal mathematical methods of solution.   We then show how to convert differential equations into integral equations and apply the theory to certain physical problems.

GENERAL METHODS OF SOLVING INTEGRAL EQUATIONS

14.2. The Liouville-Neumann Series.—a. *Fredholm's Equation of the Second Kind.* Suppose the given integral equation is

$$\phi(x) = f(x) + \lambda \int_a^b K(x,z)\phi(z)dz \tag{14-2}$$

where x and z are real variables with $a \leq x \leq b$, $a \leq z \leq b$; $K(x,z)$ and $f(x)$ are continuous but may be complex.   We attempt to solve (2) by means of a power series in λ:

$$\phi(x) = \sum_{n=0}^{\infty} \lambda^n \phi_n(x) \tag{14-3}$$

Substituting (3) into (2) and equating coefficients of equal powers of λ we obtain

$$\phi_0(x) = f(x)$$

$$\phi_1(x) = \int K(x,z)\phi_0(z)dz$$

$$\phi_2(x) = \int K(x,z)\phi_1(z)dz \tag{14-4}$$

$$\cdots \cdots \cdots \cdots \cdots \cdots \cdots \cdots \cdots$$

$$\phi_n(x) = \int K(x,z)\phi_{n-1}(z)dz$$

Remembering that both x and z are restricted to lie between a and b, we see that the kernel and $f(x)$ must have *maximum* values, for we assumed them to be continuous.   Let these maxima be given by $|K(x,z)| \leq M, |f(x)| \leq N$.   Then it follows that

$$|\phi_0| \leq N, \quad |\phi_1| \leq NM(b-a), \cdots, \quad |\phi_n| \leq N[M(b-a)]^n$$

---

[1] References to more complete accounts of the subject will be found at the end of this chapter.   Integral equations are frequently encountered in current physical and chemical literature, indicating that they are powerful tools for handling a variety of problems. Many examples of such usage are given by Morse, P. M., and Feshbach, H., " Methods of Theoretical Physics," McGraw-Hill Book Co., Inc., New York, 1953.

If

$$|\lambda| < \frac{1}{M(b-a)} \tag{14-5}$$

the series (3) which is called the *Liouville-Neumann series* converges uniformly and is the unique continuous[2] solution of (2) within the range $a \leq x \leq b$.

In order to obtain the solution in more convenient form, we define the *iterated kernels*:[3]

$$K_1(x,z) = K(x,z)$$

$$K_2(x,z) = \int K(x,y)K(y,z)dy$$

$$\dots\dots\dots\dots\dots\dots\dots\dots \tag{14-6}$$

$$K_n(x,z) = \int K(x,y)K_{n-1}(y,z)dy$$

$$= \int\int \cdots \int K(x,y_1)K(y_1,y_2)\cdots K(y_{n-1},z)dy_1 dy_2 \cdots dy_{n-1}$$

Introducing these functions into (4) we may write

$$\phi_1(x) = \int K(x,z)f(z)dz$$

$$\phi_2(x) = \int K_2(x,z)f(z)dz \tag{14-7}$$

$$\dots\dots\dots\dots\dots\dots$$

$$\phi_n(x) = \int K_n(x,z)f(z)dz$$

By the same means as before we see that $|K_n(x,z)| \leq M^n(b-a)^{n-1}$; hence if (5) is fulfilled we can construct a uniformly convergent series called the *resolvent (lösender Kern)*.

$$K(x,z;\lambda) = \sum_{n=0}^{\infty} \lambda^n K_{n+1}(x,z) \tag{14-8}$$

From (3), (6) and (8), it follows that the solution of the integral equation is

$$\phi(x) = f(x) + \lambda \int K(x,z;\lambda)f(z)dz \tag{14-9}$$

[2] Continuous solutions of the equation may exist even if (5) is not true. There may also be discontinuous solutions. For these exceptions, see Lovitt, loc. cit., pp. 13 and 21.

[3] Henceforth, we usually omit limits of integration unless they are different from a and b.

The resolvent and $\phi(x)$ have properties of a reciprocal nature as may be seen by comparing (2) and (9). If $\phi(x)$ is the unknown, (9) is the solution; if $f(x)$ in (9) is the unknown, (2) is the solution. These properties are even more apparent if we rewrite (8) in the form

$$K(x,z;\lambda) - K(x,z) = \lambda \sum_{n=0}^{\infty} \lambda^n K_{n+2}(x,z) = \lambda \sum_{n=0}^{\infty} \lambda^n \int K(x,y)K_{n+1}(y,z)dy$$

or

$$K(x,z;\lambda) - K(x,z) = \lambda \int K(x,y)K(y,z;\lambda)dy \qquad (14\text{-}10)$$

Similarly, we may obtain

$$K(x,z;\lambda) - K(x,z) = \lambda \int K(x,y;\lambda)K(y,z)dy$$

b. *Volterra's Equation of the Second Kind.* Application of the Liouville-Neumann series may also be made in this case. Suppose

$$\phi(x) = f(x) + \lambda \int_0^x H(x,z)\phi(z)dz \qquad (14\text{-}11)$$

is given. Then if

$$K(x,z) \left.\begin{array}{l} = H(x,z); \ 0 \le z \le x \\ = 0; \qquad\quad z \ge x \end{array}\right.$$

we may write an equation similar to (7) for $z \le x$

$$\phi_n(x) = \int K_n(x,z)f(z)dz$$

and also an equation like (6)

$$K_n(x,z) = \int K(x,y)K_{n-1}(y,z)dy$$

$$= \int K(x,y_1)dy_1 \int K(y_1,y_2)K_{n-2}(y_2,z)dy_2$$

The solution of Volterra's equation obtained in this way converges for all values of λ.

c. *Volterra's Equation of the First Kind.* Under certain conditions, Volterra's equation of the first kind may also be solved by the Liouville-Neumann series. With a change of notation, we write this equation as

$$g(x) = \lambda \int_0^x K(x,z)\phi(z)dz \qquad (14\text{-}12)$$

Differentiation with respect to x results in

$$g'(x) = \lambda \int_0^x \frac{\partial K}{\partial x} \phi(z)dz + \lambda K(x,x)\phi(x)$$

which is similar to (11) provided $K(x,x) \neq 0$ and

$$f(x) = \frac{g'(x)}{\lambda K(x,x)} \; ; \quad H(x,z) = \frac{\dfrac{\partial K}{\partial x}}{\lambda K(x,x)}$$

A similar conversion of (12) to an equation of the second kind may be made by partial integration.

When $K(x,x)$ vanishes, the procedure just described gives an equation of the first kind again. Let us consider the situation in more detail, assuming that the kernel is a polynomial of n-th degree in x and that the coefficients of the terms in x are polynomials in z, but not necessarily of n-th degree. It is convenient to express the kernel as a polynomial in $\xi = (x - z)$, so that it may be written

$$K(x,z) = a_0(z) + a_1(z)\xi + \cdots + a_n\xi^n$$

Two special cases are of interest: (1) $a_0(z) = 0$; (2) $a_0(z)$ contains no constant term.

In the first case, $K(x,x)$ vanishes identically; but if the derivative of the kernel does not vanish, which means that $a_1 \neq 0$, two differentiations of eq. (12) will yield an equation of the second kind and a solution is again possible by this method. Further differentiations could be carried out if necessary. Several partial integrations could replace the differentiations, if this were preferred.

In the second case, the kernel vanishes only for $x = z = 0$. However, the integral equation may then be converted into a differential equation. With the same polynomial kernel, differentiate eq. (12) $(n + 1)$-times. The integral on the right will vanish, the differential equation remaining is of order n, and its solution, adjusted to fit the appropriate boundary conditions, is the solution of the integral equation.

An explicit form for the solution can be given, but it is quite awkward in the general case. Note also that the presence or absence of a constant term in $a_0(z)$ is of no consequence. For illustrative purposes, let us take a simpler expression for the kernel. Suppose the polynomial is only of second degree and that $a_0(z) = A_0 + A_1 z + A_2 z^2$; $a_1(z) = B_0 + B_1 z$; $a_2(z) = C_0$, where A_i, B_i, C_i are constants. Three differentiations of (12) will give

$$g'''(x) = A_2 x^2 \phi''(x) + (B_1 + 4A_2)x\phi'(x) + (2A_2 + B_1 + 2C_0)\phi(x)$$

which is a differential equation of the Euler type. **Introduction of a new** variable, $u = \ln x$, will reduce it to a linear, inhomogeneous equation with constant coefficients, as discussed in sec. 2.8. Its form, with proper change in notation, is identical with eq. (2–19), and its solution is eq. (2–22).

An important case arises when the kernel becomes infinite at one or more points within the range of x and z. It is then necessary to transform the equation to remove the singularity. As a typical example, consider the kernel

$$K(x,z) = \frac{1}{(x - z)^\alpha} \; ; \quad 0 < \alpha < 1$$

which is infinite when $x = z$. Substitute this kernel in (12), multiply both sides of the equation by $dx/(u - x)^{1-\alpha}$ and integrate with respect to x from 0 to u. If for simplicity we also take $\lambda = 1$ the result is

$$\int_0^u \frac{g(x)dx}{(u - x)^{1-\alpha}} = \int_0^u \frac{dx}{(u - x)^{1-\alpha}} \int_0^x \frac{\phi(z)dz}{(x - z)^\alpha}$$

$$= \int_0^u \phi(z)dz \int_z^u \frac{dx}{(u - x)^{1-\alpha}(x - z)^\alpha}$$

The justification of the change of limits and order of integration in the last equation is the following. Since x varies from 0 to u, and for every value of x, the variable z goes from 0 to x, the situation is equivalent to the variation of z from 0 to u and the variation of x from z to u for every value of z. The same result is also easily obtained from a figure. If we are integrating $F(x,z)$ over the shaded area of Fig. 1 we see that

$$\int_0^u dx \int_0^x F(x,z)dz = \int_0^u dz \int_z^u F(x,z)dx$$

The definite integral $\int_z^u F(x,z)dx = \int_z^u (u - x)^{\alpha-1}(x - z)^{-\alpha}dx$ may be evaluated as follows. Introduce the new variable $y = (u - x)/(u - z)$ which shifts the limits to 0 and 1, respectively. The result is an Eulerian integral of the first kind[4] or B-function which is simply related to the Γ-function. Explicitly, the result using (3–12) is

$$B(\alpha, 1 - \alpha) = \Gamma(\alpha)\Gamma(1 - \alpha) = \pi/\sin \alpha\pi.$$

The solution of the integral equation is thus

$$\phi(u) = \frac{\sin \alpha\pi}{\pi} \frac{d}{du}\left[\int_0^u g(x)(u - x)^{\alpha-1}dx\right]$$

Equations with singular kernel, especially those where the singularity results from an infinite limit of integration, may usually be solved by integral transforms. In fact, the transforms of Fourier, Laplace, Hankel,

[4] See sec. 3.2.

and Mellin are special cases of integral equations of the first kind. They have been discussed at length by Morse and Feshbach, loc. cit.

Problem. Solve the equation $\phi(x) = x + \int_0^x (z - x)\phi(z)dz$ by the Liouville-Neumann series. *Hint:* substitute $(z - x) = u$; $(y - x) = v$.   *Ans.:* $\phi(x) = \sin x$.

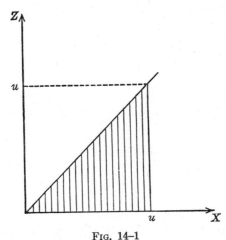

FIG. 14–1

14.3 Fredholm's Method of Solution.—a. *The Inhomogeneous Equation.* Fredholm studied the solution of a system of linear equations in n variables and observed that as n becomes infinite the results are applicable to linear integral equations. Although the reasoning is simple, the derivation of the final formulas requires considerable space. We therefore show only how the method may be used, referring the reader to other sources[5] for the intermediate steps and proofs.

The unique and continuous solution of (2) is of the form (9), where the resolvent is the ratio of two infinite series in λ. In fact

$$K(x,z;\lambda) = \frac{D(x,z;\lambda)}{D(\lambda)} \tag{14–13}$$

where

$$D(x,z;\lambda) = K(x,z) + \sum_{n=1}^{\infty} \frac{(-1)^n}{n!} D_n(x,z)\lambda^n \tag{14–14}$$

$$D(\lambda) = \sum_{n=0}^{\infty} \frac{(-1)^n}{n!} D_n \lambda^n \tag{14–15}$$

The coefficients D_n and the functions $D_n(x,z)$ may be found from the following recurrence relations. Starting with $K(x,z) = D_0(x,z)$ we obtain

[5] See references at end of the chapter.

D_1 from the integral

$$D_m = \int D_{m-1}(x,x)dx \tag{14-16}$$

We then find $D_1(x,z)$ from

$$D_m(x,z) = K(x,z)D_m - m \int K(x,y)D_{m-1}(y,z)dy \tag{14-17}$$

which enables us to determine D_2 from (16). Continuing in this way, all of the coefficients are calculated. In many cases, depending on the explicit form of the kernel, the series (14) and (15) contain only a finite number of terms.

One distinct advantage of the Fredholm method is that (13) is uniformly convergent for all values of λ unless $D(\lambda) = 0$. If that happens, the procedure which we have described is inapplicable since the resolvent vanishes. Actually, there is then no solution unless certain other conditions are met. We omit the necessary extension of the Fredholm theory but return to the problem in sec. 14.4b.

b. *The Homogeneous Equation.* If $f(x) = 0$, so that the given equation is homogeneous,

$$\phi(x) = \lambda \int K(x,z)\phi(z)dz \tag{14-18}$$

Then cursory inspection of the solution (9) leads to the conclusion that $\phi(x) \equiv 0$. This is generally the case but we shall see that when the parameter λ assumes certain special values we are led to a situation similar to the eigenvalue and eigenfunction problem described in Chapter 8. If $D(\lambda) = 0$ and $D(x,z;\lambda) \neq 0$, eq. (13) indicates that $K(x,z;\lambda)$ approaches infinity and we may still find non-vanishing solutions of (18). Equating the right side of (15) to zero, we have a polynomial in λ with n roots, multiple or distinct. They are the *eigenvalues* of the kernel, and the corresponding solutions of (18) are the *eigenfunctions*. Assuming that all eigenvalues are distinct, choose one of them, say λ_i, substitute (13) in (10) and multiply by $D(\lambda_i)$, which gives

$$D(x,z;\lambda_i) = \lambda_i \int K(x,y)D(y,z;\lambda_i)dy \tag{14-19}$$

If we compare this equation with (18), we observe that $D(x,z;\lambda_i)$, for any constant z, is a solution of the homogeneous equation, i.e.,

$$\phi_i(x) = D(x,c;\lambda_i) \tag{14-20}$$

Having found a solution for λ_i, we proceed to find the others for the remaining eigenvalues in the same way. Linear combinations of them form the general solution

$$\phi(x) = \sum_{m=1}^{n} C_m \phi_m(x) \qquad (14\text{-}21)$$

where the C_m are arbitrary constants.

It is true that $D(x,z;\lambda_i)$ may vanish identically in x and z or vanish because of an unfortunate choice of the constant value of z. In the former case, non-trivial solutions may often be found by more complicated methods; in the latter case, we simply choose another $z \neq c$. When the eigenvalues are degenerate further modifications of the method are required.

Problem a. Solve by the Fredholm method:

$$\phi(x) = x + \lambda \int_0^1 (x + z)\phi(z)dz$$

Ans.:
$$\phi(x) = \frac{6x(\lambda - 2) - 4\lambda}{\lambda^2 + 12\lambda - 12}$$

Problem b. Show that $\int D(x,x;\lambda)dx = -\dfrac{dD(\lambda)}{d\lambda}$. *Hint:* use eq. (16).

Problem c. Set $f(x) = 0$ in the equation of Problem a and solve. *Hint:* show that $D(x,c;\lambda) = (2/\epsilon)(2 - \epsilon)(\epsilon x + 1)(\epsilon c + 1)$; $\lambda = 2\epsilon(2 - \epsilon)$; $\epsilon = \pm\sqrt{3}$.
Ans.: $\phi_\pm(x) = C_\pm(1 \pm \sqrt{3}x)$.

14.4. The Schmidt-Hilbert Method of Solution.—In many physical problems, the kernel has the property of being *symmetric*, i.e., $K(x,z) = K(z,x)$. In such cases,[6] the integral equation may be solved by a method which is somewhat different from any of those in the preceding sections. We find it convenient to limit the discussion to kernels which are *real* as well as symmetric.

a. *The Homogeneous Equation. A real symmetric kernel has at least one eigenvalue and it may have an infinite number.* We omit the proof of these facts.

The eigenfunctions of the homogeneous equation (18) *are mutually orthogonal.* Suppose λ_i and λ_j are two different eigenvalues corresponding respectively to eigenfunctions ϕ_i and ϕ_j. Then we may write

$$\phi_i(x) = \lambda_i \int K(x,z)\phi_i(z)dz$$

$$\phi_j(x) = \lambda_j \int K(x,z)\phi_j(z)dz$$

[6] Unsymmetric kernels may often be symmetrized; see sec. 14.7 or Courant-Hilbert, loc. cit.

Multiply the first equation by ϕ_j and the second by ϕ_i, then integrate over x

$$\int \phi_i(x)\phi_j(x)dx = \lambda_i \int K(x,z)\phi_i(z)\phi_j(x)dzdx$$

$$= \lambda_j \int K(x,z)\phi_i(x)\phi_j(z)dzdx \qquad (14\text{-}22)$$

The last integral may be written as $\lambda_j \int K(z,x)\phi_i(z)\phi_j(x)dzdx$ by interchanging x and z. Thus if $K(x,z) = K(z,x)$, the two integrals of (22) are identical and since $\lambda_i \neq \lambda_j$, it follows that

$$\int \phi_i(x)\phi_j(x)dx = 0 \qquad (14\text{-}23)$$

As we know from Chapter 8 such functions may always be normalized. Henceforth, we will assume that this has been done and will indicate the orthonormal solutions of (18) by $\Phi_i(x)$, so that

$$\int \Phi_i(x)\Phi_j(x)dx = \delta_{ij} \qquad (14\text{-}24)$$

The eigenvalues of a real, symmetric kernel are all real. Suppose the solution of the homogeneous equation (18) were of the form $\phi(x) = \phi_1(x) + i\phi_2(x)$ and one of its eigenvalues were also complex, $\lambda = \alpha + i\beta$. We could then take the complex conjugate of (18)

$$\phi^*(x) = \lambda^* \int K(x,z)\phi^*(z)dz$$

But according to (23)

$$(\lambda - \lambda^*) \int \phi(x)\phi^*(x)dx = 0$$

or

$$2i\beta \int (\phi_1^2 + \phi_2^2)dx = 0$$

which means that $\beta = 0$ and the eigenvalues must all be real.

Arbitrary functions of x, including the kernel for fixed z, may be expanded in terms of the eigenfunctions

$$K(x,z) = \sum C_i\Phi_i(x) \qquad (14\text{-}25)$$

The functions $\Phi_i(x)$ form a complete set as explained in Chapter 8. As also shown there, the coefficients of (25) may be found by integrating that equation term by term. Thus, using (24),

$$C_i = \int K(x,z)\Phi_i(x)dx$$

But

$$\Phi_i(z) = \lambda_i \int K(z,x)\Phi_i(x)dx = \lambda_i \int K(x,z)\Phi_i(x)dx$$

since the kernel is symmetric, hence

$$C_i = \frac{\Phi_i(z)}{\lambda_i}$$

and (25) becomes

$$K(x,z) = \sum_i \frac{\Phi_i(x)\Phi_i(z)}{\lambda_i} \tag{14-26}$$

b. *Solution of the Inhomogeneous Equation.* We are now ready to consider the inhomogeneous equation (2); for that purpose we assume that we have found the eigenfunctions of the homogeneous equation by the method of sec. 14.3b. Let them be $\Phi_i(x)$. Then we may write

$$\phi(x) - f(x) = \sum \alpha_i \Phi_i(x)$$
$$\alpha_i = \int [\phi(x) - f(x)]\Phi_i(x)dx \tag{14-27}$$

where $\phi(x)$ and $f(x)$ both come from (2). Now substitute (27) in (2) to give

$$\sum \alpha_i \Phi_i(x) = \lambda \int K(x,z)f(z)dz + \lambda \sum \alpha_i \int K(x,z)\Phi_i(z)dz \tag{14-28}$$

We may also expand $f(x)$:

$$f(x) = \sum \beta_i \Phi_i(x); \quad \beta_i = \int f(x)\Phi_i(x)dx \tag{14-29}$$

and obtain by using (26) and (24)

$$\int K(x,z)\sum \beta_i \Phi_i(z)dz = \int \sum_i \frac{\Phi_i(x)\Phi_i(z)}{\lambda_i}\sum_j \beta_j \Phi_j(z)dz$$
$$= \sum \frac{\beta_i \Phi_i(x)}{\lambda_i}$$

with a similar expression for the last integral of (28). That equation becomes

$$\sum \alpha_i \Phi_i(x) = \lambda \sum \frac{\beta_i}{\lambda_i}\Phi_i(x) + \lambda \sum \frac{\alpha_i}{\lambda_i}\Phi_i(x) \tag{14-30}$$

Because of the independence of the functions Φ_i, the coefficients of each may be equated on both sides of this equation. Hence,

$$\alpha_i = \left[\frac{\beta_i}{\lambda_i} + \frac{\alpha_i}{\lambda_i}\right]\lambda$$

or if $\lambda \neq \lambda_i$,

$$\alpha_i = \beta_i \frac{\lambda}{\lambda_i - \lambda} \tag{14-31}$$

This method, which was devised by Schmidt and Hilbert, thus gives a solution for $\lambda \neq \lambda_i$, for we may substitute (29) and (31) into (27) and obtain

$$\phi(x) = f(x) + \lambda \sum_i \frac{\beta_i}{\lambda_i - \lambda} \Phi_i(x)$$

$$= f(x) + \lambda \sum_i \left[\frac{\Phi_i(x)}{\lambda_i - \lambda} \int f(z) \Phi_i(z) dz \right] \tag{14-32}$$

As we have noted before, the homogeneous equation for $\lambda \neq \lambda_i$ has the solution $\phi(x) \equiv 0$ since $f(x) = 0$.

We must still consider the exceptional case when λ is one of the eigenvalues of the kernel. Suppose, for example, that $\lambda = \lambda_0$ is an m-fold degenerate eigenvalue, i.e., $\lambda_0 = \lambda_1, \lambda_2, \cdots, \lambda_m$. Then (2) reads

$$\phi(x) = f(x) + \lambda_0 \int K(x,z) \phi(z) dz$$

and by the preceding method we obtain

$$\alpha_i = \frac{\beta_i \lambda_0}{\lambda_i - \lambda_0}$$

where i is not one of the numbers $1, 2, \cdots, m$. When i equals one of these integers we have, if α_i is to remain finite,

$$\beta_1 = \beta_2 = \cdots = \beta_m = 0$$

which in turn requires that

$$\beta_j = \int f(x) \Phi_j(x) dx = 0; \quad j = 1, 2, \cdots, m \tag{14-33}$$

Thus if λ_0 is an m-fold degenerate eigenvalue, the inhomogeneous equation has solutions only if $f(x)$ is orthogonal to the corresponding eigenfunctions $\Phi_j(x)$. The general solution of the equation is then

$$\phi(x) = f(x) + \lambda_0 \sum' \left[\frac{\Phi_i(x)}{\lambda_i - \lambda_0} \int f(z) \Phi_i(z) dz \right] + C_1 \Phi_1(x) + \cdots$$
$$+ C_m \Phi_m(x) \tag{14-34}$$

where the prime on the summation sign means that the terms $i = 1, 2, \cdots, m$ are to be omitted from the sum.

Problem. Find the solution of the equation of Problem a, sec. 14.3, by the Hilbert-Schmidt method for λ not equal to an eigenvalue. Show that there are no solutions when λ is an eigenvalue.

14.5. Summary of Methods of Solution.—a. *The Homogeneous Equation.*

1. $D(\lambda) \neq 0$. No solution except $\phi(x) \equiv 0$.
2. $D(\lambda) = 0$; $D(x,z;\lambda) \neq 0$. Solution is given by (20) and (21). The resulting eigenfunctions are orthogonal and may be normalized. To each solution belongs an eigenvalue.

b. *The Non-homogeneous Equation.*

$$\lambda \neq \lambda_i$$

1. Solution given by (9) provided (5) holds.
2. For all values of $\lambda \neq \lambda_i$ solution is given by (9) and (13).
3. If $K(x,z) = K(z,x)$, solution is (32).

$$\lambda = \lambda_i$$

4. $K(x,z) = K(z,x)$; solution is (34). Special methods have been given for Volterra's equations of the first and second kinds.

USE OF INTEGRAL EQUATIONS

14.6. Relation between Differential and Integral Equations.—We have shown in the previous sections how integral equations of the more common types may be solved. We now propose to study the relation between differential and integral equations so that we may state physical problems in either form at will. For this purpose consider as a simple example the second order differential equation

$$y'' = f(x,y) \tag{14-35}$$

Integration results in

$$y'(x) = \int_0^x f\{z,y(z)\}\,dz + C_1$$

$$y(x) = \int_0^x \left[\int_0^x f\{z,y(z)\}\,dz\right] dx + C_1 x + C_2 \tag{14-36}$$

An alternative form of the last expression[7] is

$$y(x) = \int_0^x (x - z)f\{z,y(z)\}\,dz + g(x) \tag{14-37}$$

$$g(x) = C_1 x + C_2$$

which is recognized as a non-linear Volterra equation of the second kind with $y(x)$ as the unknown.

[7] To show that the two equations for $y(x)$ are identical, differentiate the last one with respect to x; the result is (36).

The boundary conditions which are needed to determine the two integration constants, C_1 and C_2, may be either of two types: (a) y and y' are fixed at one point within the range of integration, say at $x = 0$; (b) y is fixed at two points. The first case is simple, for if $y(0) = a$, $y'(0) = b$, (37) becomes

$$y(x) = \int_0^x (x - z)f\{z,y(z)\}dz + bx + a$$

The second case leads to greater difficulties. Suppose $y(0) = a, y(1) = b$; then $C_2 = a$, as before. For $x = 1$, we have

$$b = y(1) = \int_0^1 (1 - z)fdz + C_1 + a$$

or

$$C_1 = (b - a) - \int_0^1 (1 - z)fdz$$

where we abbreviate $f\{z,y(z)\}$ by the single symbol f. Substituting the values of C_1 and C_2 into (37) we obtain

$$y(x) = h(x) + \int_0^x (x - z)fdz + x\int_0^1 (z - 1)fdz$$

$$= h(x) + \int_0^x (x - z)fdz + x\int_0^x (z - 1)fdz + x\int_x^1 (z - 1)fdz$$

$$= h(x) + \int_0^x z(x - 1)fdz + \int_x^1 x(z - 1)fdz \qquad (14\text{-}38)$$

where $h(x) = a + (b - a)x$. We thus see that in this case, if we are willing to divide the range of x into two parts with a different kernel for each part,

$$K(x,z) \begin{cases} = z(x - 1) & x \geq z \\ = x(z - 1) & x \leq z \end{cases}$$

eq. (38) becomes an integral equation of the Fredholm type

$$y(x) = h(x) + \int_0^1 K(x,z)f\{z,y(z)\}dz$$

Problem. Convert the following differential equation and its boundary conditions to an integral equation.

$$y'' + y = 0; \quad y(0) = y''(0) = 0; \quad y'(0) = 1$$

Ans.: $y(x) = x + \int_0^x (z - x)y(z)dz$

14.7. Green's Function.—Our problem now is to find a general method of constructing such kernels. For this purpose we consider the *inhomogeneous Sturm-Liouville equation*

$$L(u) = (pu')' - qu = -\phi(x) \qquad (14\text{--}39)$$

the homogeneous form of which has been discussed in Chapter 8. We will later prove that a certain function $G(x,z)$ called *Green's function* is the kernel of a homogeneous integral equation which is equivalent to (39) and its boundary conditions. At the moment we study the means of finding Green's function. For reasons which will presently be clear, it is defined to have the following properties:

a. For fixed z, it is a continuous function of x and satisfies all of the boundary conditions to be imposed on u.

b. Both G' and G'' are continuous at every point within the range of x except at $x = z$, where it is discontinuous[8] so that

$$G'(z + 0) - G'(z - 0) = -1/p(z) \qquad (14\text{--}40)$$

c. Except at $x = z$, $G(x,z)$ satisfies the differential equation $L(G) = 0$. We now proceed to find such a function G. Suppose two linearly independent solutions of

$$L(u) = 0 \qquad (14\text{--}41)$$

are known. If these are $u_1(x)$ and $u_2(x)$ their independence may be recognized by the fact that the Wronskian, $u_1 u_2' - u_1' u_2 \neq 0$ (see sec. 3.13), and the general solution of (41) is

$$u(x) = C_1 u_1 + C_2 u_2$$

Let us divide the range of x into two portions; $a \leq x \leq z$, $z \leq x \leq b$, and write

$$u = \begin{cases} u_I = (A - \alpha)u_1(x) + (B - \beta)u_2(x); & x \leq z \\ u_{II} = (A + \alpha)u_1(x) + (B + \beta)u_2(x); & x \geq z \end{cases} \qquad (14\text{--}42)$$

where A, α, B, β are constants to be so chosen that u, which will later be taken as our Green function, satisfies conditions a, b and c. If we impose on this function the requirements a and b, we must have

$$\begin{aligned} u_I(z) &= u_{II}(z) \\ u_{II}'(z) - u_I'(z) &= -1/p(z) \end{aligned} \qquad (14\text{--}43)$$

---

[8] The notation $G'(z + 0)$ means that G' is evaluated at the discontinuity when it is approached from values of $x > z$ while $G'(z - 0)$ is evaluated when the discontinuity is approached in the opposite direction. It is necessary to make this distinction in order that the magnitude of the discontinuity will be determined with respect to sign.

or, because of (42)

$$\alpha u_1(z) + \beta u_2(z) = 0$$

$$\alpha u_1'(z) + \beta u_2'(z) = -1/2p(z)$$

Solving these equations for α and β, we obtain

$$\alpha = -\frac{1}{2p}\frac{u_2}{u_1'u_2 - u_1u_2'} ; \quad \beta = \frac{1}{2p}\frac{u_1}{u_1'u_2 - u_1u_2'}$$

and hence

$$u(x) = f(x,z) + Au_1(x) + Bu_2(x) \tag{14-44}$$

where

$$f(x,z) = \pm\frac{1}{2p(z)}\left[\frac{u_1(z)u_2(x) - u_2(z)u_1(x)}{u_2'(z)u_1(z) - u_1'(z)u_2(z)}\right]$$

Here and in the remainder of this chapter, when two equations are given or when there is a choice of sign, the first always refers to $x \leq z$ and the second to $x \geq z$. The two constants A and B of (44) are determined so that $u(x)$ satisfies the boundary conditions of the problem. The resulting function, which we henceforth indicate by $G(x,z)$, is *Green's function*.

We now prove that if $\phi(x)$ is a continuous function of x, then the function which will satisfy the differential equation (39) is given by

$$u(x) = \int_a^b G(x,z)\phi(z)dz \tag{14-45}$$

Differentiation of (45) with respect to x gives

$$u'(x) = \int_a^b G'(x,z)\phi(z)dz$$

$$u''(x) = \int_a^x G''(x,z)\phi(z)dz + \int_x^b G''(x,z)\phi(z)dz$$

$$+ G'(x,x-0)\phi(x) - G'(x,x+0)\phi(x)$$

$$= \int_a^b G''(x,z)\phi(z)dz + [G'(x+0,x) - G'(x-0,x)]\phi(x)$$

$$= \int_a^b G''(x,z)\phi(z)dz - \frac{\phi(x)}{p(x)}$$

Therefore

$$pu'' + p'u' - qu = L(u) = \int_a^b (pG'' + p'G' - qG)\phi(z)dz - \phi(x)$$

Requirement c causes the first term on the right to vanish. Hence, we have established (39) and completed the proof that $G(x,z)$, calculated as described, is the kernel of (45), and that the latter is equivalent to (39) and its boundary conditions.

An important consequence of the properties of Green's function is that it is symmetric. The proof proceeds as follows: Let us integrate the identity

$$vL(u) - uL(v) = \frac{d}{dx}[p(vu' - uv')]$$

This results in a relation known as *Green's formula*:

$$I_a^b = \int_a^b [vL(u) - uL(v)]dx = [p(vu' - uv')]_a^b = S_a^b \qquad (14\text{–}46)$$

Now let $G(x,z_1) = v$; $G(x,z_2) = u$; and consider the three ranges $a \le x \le z_1$; $z_1 \le x \le z_2$; $z_2 \le x \le b$. Evaluate the integral, dividing it into three parts $a, z_1 - \delta$; $z_1 + \delta, z_2 - \delta$; $z_2 + \delta, b$, where δ is a small increment which will approach zero in the limit.

We thus may write

$$\begin{aligned} I_a^b &= S_a^{z_1-\delta} + S_{z_1+\delta}^{z_2-\delta} + S_{z_2+\delta}^b \\ &= S_a^b - S_{z_1-\delta}^{z_1+\delta} - S_{z_2-\delta}^{z_2+\delta} \end{aligned} \qquad (14\text{–}47)$$

According to (46) $I_a^b = S_a^b$ and both must be zero, because from c, $L(u) = L(v) = 0$. This in turn requires that $I_a^b = 0$ and $S_a^b = 0$ since otherwise Green's function will not satisfy the boundary conditions. If in (47) we let $\delta \to 0$ and use (46) we obtain

$$0 = -p(z_1)\{[v(z_1)u'(z_1) - v'(z_1+0)u(z_1)] - [v(z_1)u'(z_1) - v'(z_1-0)u(z_1)]\}$$
$$-p(z_2)\{[v(z_2)u'(z_2+0) - v'(z_2)u(z_2)] - [v(z_2)u'(z_2-0) - v'(z_2)u(z_2)]\}$$

In writing these equations it must be remembered that u and v are continuous for the whole range while u' is discontinuous only at z_2 and v' only at z_1, so that for example $u'(z_1 + 0) = u'(z_1)$. Finally from (40) we obtain

$$u(z_1) = v(z_2)$$

or

$$G(z_1,z_2) = G(z_2,z_1)$$

Since the points z_1 and z_2 are arbitrary we write in general

$$G(x,z) = G(z,x)$$

The symmetry of Green's function is of considerable importance, since it permits application of the Hilbert-Schmidt theory.

It frequently happens that the two constants A and B of (44) cannot be adjusted to satisfy the given boundary conditions. In this case, a *modified Green function*[9] can be found in the following way. Suppose $u_0(x)$ is a solution of (41) that satisfies both boundary conditions. Then $cu_0(x)$ will also satisfy the conditions. No loss of generality occurs if we determine the constant so that $u_0(x)$ is normalized,

$$\int u_0^2(x)dx = 1$$

and we shall suppose that this is done. We now set

$$L(u) = u_0(x)u_0(z)$$

and determine a function $G(x,z)$ that has the same properties as we required of the simple Green function, except that it satisfies the equation $L(G) = u_0(x)u_0(z)$ instead of $L(G) = 0$. We finally require that

$$\int G(x,z)u_0(x)dx = 0 \qquad (14\text{--}48)$$

The resulting modified Green function, which is symmetric, satisfies the inhomogeneous differential equation (39) including its boundary conditions. The proof of these facts is similar to that used in the case of the simple Green function.

Problem. Find Green's function for $L(u) = u''$ with $u(0) = u(1) = 0$. *Hint:* let $u_1(x) = x$; $u_2(x) = 1$.
Ans.: See Table 1, sec. 14.9.

Example. Suppose $L(u) = u'' = 0$; $u(1) = u(-1)$; $u'(1) = u'(-1)$. If we substitute the two linearly independent solutions of the preceding problem in (44) we see that $\partial G(x,z)/\partial x = \pm\frac{1}{2} + A$, hence the second boundary condition cannot be satisfied. A solution of the differential equation which does satisfy the boundary conditions is $u_0 = $ constant or when normalized $u_0(x) = 1/\sqrt{2}$. Hence we seek a solution of the equation $L(u) = u'' = u_0(x)u_0(z) = \frac{1}{2}$. This is $u = x^2/4$. Using (44) and the results of the last problem we see that

$$G(x,z) = \pm\frac{(x-z)}{2} + Ax + B + \frac{x^2}{4}$$

which gives $A = -z/2$ when the further condition $G(x,z) = G(-x,z)$ is imposed. Omitting the constant factor $u_0(x) = 1/\sqrt{2}$ we now determine B so that (48) is satisfied. This requires that

$$\int_{-1}^{z} \frac{(x-z)}{2}\,dx - \int_{z}^{1}\frac{(x-z)}{2}\,dx + \int_{-1}^{1}\left\{-\frac{xz}{2} + B + \frac{x^2}{4}\right\}dx = 0$$

[9] A different procedure is possible in some cases; see Lovitt, loc. cit.

The result is $B = \frac{1}{6} + z^2/4$, so that finally

$$G(x,z) = \pm \frac{(x-z)}{2} + \frac{(x-z)^2}{4} + \frac{1}{6}$$

This will satisfy all of the boundary conditions.

14.8. The Inhomogeneous Sturm-Liouville Equation.—Having proved that we can convert (39) to an integral equation, we wish to give explicit forms of the latter for different $\phi(x)$. Suppose

$$\phi(x) = \lambda wu - \chi(x)$$

so that (39) becomes

$$L(u) + \lambda wu = \chi(x) \tag{14-49}$$

The resulting integral equation is

$$u(x) = \lambda \int G(x,z)w(z)u(z)dz + g(x)$$

$$g(x) = -\int G(x,z)\chi(z)dz \tag{14-49a}$$

which is equivalent to (49) and its boundary conditions. Finally if $\chi(x) = 0$, the homogeneous differential equation

$$L(u) + \lambda wu = 0 \tag{14-49b}$$

and its boundary conditions become equivalent to

$$u(x) = \lambda \int G(x,z)w(z)u(z)dz \tag{14-50}$$

but the kernel in this case is not symmetric unless $w(x) = 1$. If that is true (50) is a homogeneous integral equation and can be solved by the methods of sec. 14.3b. If $w(x) \neq 1$, we may introduce a new unknown function

$$y(x) = u(x)\sqrt{w(x)}$$

multiply the integral equation by $\sqrt{w(x)}$ and obtain

$$y(x) = \lambda \int H(x,z)y(z)dz$$

where we now have a symmetric kernel $H(x,z) = G(x,z)\sqrt{w(x)w(z)}$. Eq. (49b) forms the basis of the Sturm-Liouville theory which was discussed in sec. 8.5.

Let us consider (41) and (49b) further. We write

$$L(v) + \lambda v = 0; \quad L(u) = 0 \tag{14-51}$$

and suppose that their Green functions are known so that

$$u = G(x,\eta); \quad v = \Gamma(x,\xi) \qquad (14\text{--}52)$$

Substitute these relations in Green's formula (46), use (40) and arguments similar to those which proved that Green's function is symmetric. The result is

$$\Gamma(\eta,\xi) = G(\xi,\eta) + \lambda \int G(x,\eta)\Gamma(x,\xi)dx$$

For fixed ξ, this is recognized as identical with (2) where Γ is the unknown, $G(\xi,\eta) = f(\eta)$ and $G(x,\eta)$ is the kernel. If we now change x, η, ξ to z_1, x, z and remember that the kernel is symmetric we obtain

$$\Gamma(x,z;\lambda) - G(x,z) = \lambda \int G(x,z_1)\Gamma(z_1z;\lambda)dz_1$$

which shows by comparison with (10) that $\Gamma(x,z;\lambda)$ is the resolvent of the kernel $G(x,z_1)$. We may thus use equations of the form of (2) or (10) to find the solution of either form of (51) when the appropriate Green function (52) is known. Finally, referring to (17) and the result of Problem b, sec. 14.3, we see that

$$\frac{D'(\lambda)}{D(\lambda)} = \frac{d \ln D(\lambda)}{d\lambda} = -\int \Gamma(x,x;\lambda)dx$$

which will give $D(\lambda)$ by integration over λ and hence the eigenvalues from the relation $D(\lambda) = 0$.

Problem. Find Green's function for $L(u) = u'' + k^2 u$ with the boundary conditions of the previous problem. *Hint:* take $u_1(x) = \cos kx$; $u_2(x) = \sin kx$.
Ans.: See Table 1.

14.9. Some Examples of Green's Function.—For convenience of reference, we list in Table 1 Green's function for some important differential equations. The following boundary conditions include those most often encountered:

a. $u(0) = u(1) = 0$

b. $u(-1) = u(1); \quad u'(-1) = u'(1)$

c. $u(0) = u'(1) = 0$

d. $u(-1) = u(1) = 0$

e. $u(0) = -u(1); \quad u'(0) = -u'(1)$

f. $u(0) = u(1) = u'(0) = u'(1)$

g. $u(x)$ finite; $-\infty < x < \infty$

When the limits are a and b, the appropriate Green function $G(X,Z)$ may be found from our results by the transformations

$$x = \frac{X - a}{b - a} ; \quad z = \frac{Z - a}{b - a} \tag{14-53}$$

for if $G(x,z)$ is bounded by $(0,1)$ then $G(X,Z)$ is bounded by (a,b). The method of calculating Green's function in each case is identical with that described in the preceding sections. When only one equation is given for $G(x,z)$ it refers to $x \leq z$; for $x \geq z$, interchange x and z.

In addition to the results found in Table 1, Green's function for several other differential equations will be given (see also Table 1 in sec. 8.5).

For the *Legendre differential equation*

$$L(u) = [(1 - x^2)u']'; \quad -1 \leq x \leq 1$$

The boundary conditions are that the solutions remain finite at $x = \pm 1$. Green's function is

$$G(x,z) = -\tfrac{1}{2} \ln [(1 - x)(1 + z)] + \ln 2 - \tfrac{1}{2} \tag{14-54}$$

The *associated Legendre differential equation* is

$$[(1 - x^2)u']' - \frac{m^2 u}{1 - x^2} = 0$$

and

$$G(x,z) = \frac{1}{2m} \left\{ \frac{(1 + x)(1 - z)}{(1 - x)(1 + z)} \right\}^{m/2} ; \quad m \neq 0 \tag{14-55}$$

For $m = 0$, the proper Green function is (54).

The *zero-th order Bessel equation* is $L(u) = (xu')' = 0$. With the boundary conditions $u(1) = 0$; $u(0)$ finite

$$G(x,z) = - \ln z \tag{14-56}$$

The *n-th order equation* is

$$(xu')' - \frac{n^2}{x} u = 0$$

and

$$G(x,z) = \frac{1}{n} \left[\left(\frac{x}{z} \right)^n - (xz)^n \right]$$

with the same boundary conditions as for the zero-th order equation.

TABLE 1

$L(u)$	Boundary Condition	$G(x,z)$
1. u''	(a)	$(1 - z)x$
2. u''	(b)	$\frac{1}{4}(x - z)^2 + \frac{1}{8} - \frac{1}{2}\lvert x - z\rvert$
3. u''	(c)	x
4. u''	(d)	$-\frac{1}{2}\{\lvert x - z\rvert + xz - 1\}$
5. u''	(e)	$-\frac{1}{2}\lvert x - z\rvert + \frac{1}{4}$
6. u''	(g)	none exists
7. $u'' + \lambda u$	(a)	$\dfrac{\sin kx \sin k(1 - z)}{k \sin k}$; $k^2 = \lambda > 0$
8. $u'' + \lambda u$	(b)	$\dfrac{-1}{2k \sin k}\cos k(x - z + 1)$
9. $u'' - \lambda u$	(a)	$\dfrac{\sinh kx \sinh k(1 - z)}{k \sinh k}$
10. $u'' - \lambda u$	(b)	$\dfrac{1}{2k \sinh k}\cosh k(x - z + 1)$
11. $u'' - \lambda u$	(c)	$\dfrac{\cosh kx \cosh k(1 - z)}{k \sinh k}$
12. $u'' - u$	(g)	$\frac{1}{2}e^{-\lvert z - s\rvert}$
13. u^{IV}	(f)	$\dfrac{x^2(z - 1)^2}{6}(2xz + x - 3z)$

APPLICATION TO PHYSICAL PROBLEMS

14.10. Abel's Integral Equation.—One of the earliest applications of integral equations to a physical problem was made by Abel (1823). Consider a particle which falls along a smooth curve in a vertical plane. Let its original position above a given horizontal plane be z_0, its position at time t be z and at the end of its fall be $z = 0$. Let ds be the distance travelled in time dt. Then if the particle moves under no force but mg, the force of gravity, its velocity

$$v = \frac{ds}{dt} = \sqrt{2g(z_0 - z)} \qquad (14\text{-}57)$$

The whole time of descent is

$$T(z_0) = \int_0^T dt = \int \frac{ds}{\sqrt{2g(z_0 - z)}} = -\frac{1}{\sqrt{2g}}\int_0^{z_0} \frac{s'(z)dz}{\sqrt{z_0 - z}}$$

If the shape of the curve is given in terms of z,

$$s = s(z)$$

then the time of descent may be calculated. The reverse problem studied by Abel is to find a curve for which the time T is a given function of x, $T(z_0) = f(z_0)$ (compare the brachistochrone problem, sec. 6.1b). We thus wish to find

$$\phi(z) = -\frac{s'(z)}{\sqrt{2g}} > 0$$

or

$$f(z_0) = \int_0^{z_0} \frac{\phi(z)dz}{\sqrt{z_0 - z}} \tag{14-58}$$

which is a Volterra integral equation of the first kind. The presence of the singularity at $z = z_0$ makes it necessary to solve the equation in the manner of sec. 14.2c. The details may be left to the reader.

14.11. Vibration Problems.—a. *The homogeneous string* treated in Chapter 7 was reduced to the eigenvalue problem (cf. eq. 7–33),

$$S''(x) + k^2 S(x) = 0$$

If we make the proper change of variable so that the boundary conditions are $S(0) = S(1) = 0$ we see that the differential equation is similar to (49b), the boundary conditions lead to Green's function (1) from Table 1 and the resulting homogeneous integral equation is of the form of eq. (50) when $\lambda = k^2$ and $w = 1$.

b. *Forced Vibrations.* Suppose the string is subjected to a periodic force $f(x) \cos (\beta t + \delta)$. Then if we set $v = 1$ in eq. (1) of Chapter 8 we have

$$\frac{\partial^2 U}{\partial t^2} = U'' + f(x) \cos (\beta t + \delta) \tag{14-59}$$

with boundary conditions $U(0,t) = U(1,t) = 0$. We seek a solution of the form

$$U = S(x) \cos (\beta t + \delta)$$

which reduces (59) to

$$S''(x) + \beta^2 S(x) = -f(x) \tag{14-60}$$

if we remember that $S(0) = S(1) = 0$. This differential equation is like (49) and the integral equation like (49a) with kernel identical with that of the homogeneous string. The integral equation may be solved provided β^2 is an eigenvalue and $f(x)$ is orthogonal to the eigenfunctions of the homogeneous equation. We know from Chapter 8 that the latter are $\sin n\pi x$, hence the required condition is

$$\int f(x) \sin n\pi x \, dx = 0$$

If β^2 is not an eigenvalue, solutions are still possible. Following the procedure of sec. 14.8, we look for Green's function of eq. (60) which is given as item (7) in Table 1. This is the resolvent of our integral equation, hence from eq. (10) the unique solution of (60) is

$$S(x) = g(x) + \beta^2 \int_0^1 \Gamma(x,z)g(z)dz$$

$$g(x) = -\int G(x,z)f(z)dz$$

c. *The Suspended Rope.* Let a rope of unit length hang in its equilibrium position from the point $x = 1$. If it executes small vibrations in a vertical plane, its equation of motion is

$$\frac{\partial^2 U}{\partial t^2} = \frac{\partial}{\partial x}\frac{(x\partial U)}{\partial x}$$

with U as its displacement. The horizontal component of its tension at x is $x(\partial U/\partial x)$, so the boundary conditions are $U(1) = 0$, $U(0)$ finite. Writing $U = u(x)\phi(t)$ we obtain

$$[xu'(x)]' + k^2u(x) = 0$$

$$\phi''(t) + k^2v(t) = 0$$

The proper Green function for the homogeneous differential equation in x is eq. (56).

REFERENCES ON INTEGRAL EQUATIONS

Chapters on the subject may be found in:

Courant, R., and Hilbert, D., " Methods of Mathematical Physics," Vol. I, First English Edition, Revised, Interscience Publishers, Inc., New York, 1953.

Horn, J., " Partielle Differentialgleichungen," de Gruyter, Berlin, 1929.

Kowalewski, G., " Determinantentheorie einschliesslich der Fredholmschen Determinanten," Third Edition, Chelsea Publishing Co., New York, 1942.

Morse, P. M., and Feshbach, H., " Methods of Theoretical Physics," 2 vols., McGraw-Hill Book Co., Inc., New York, 1953.

Murnaghan, F. D., " Introduction to Applied Mathematics," John Wiley and Sons, Inc., New York, 1948.

Whittaker, E. T., and Watson, G. N., " Modern Analysis," Fourth Edition, Cambridge University Press, 1927.

More extended treatments are:

Bôcher, M., " Introduction to Integral Equations," Cambridge Mathematical Tracts, No. 10, 1909.

Hamel, G., " Integralgleichungen," J. Springer, Berlin, 1937; Edwards Brothers, Ann Arbor.

Hellinger, E., and Toeplitz, O., " Integralgleichungen und Gleichungen mit Unendlichen Unbekannten," Chelsea Publishing Co., New York, 1928.

Kneser, A., "Integralgleichungen und ihre Anwendung in der Mathematischen Physik," Second Edition, Vieweg, Brunswick, 1922.

Kowalewski, G., " Integralgleichungen," deGruyter, Berlin, 1930.

Lovitt, W. V., " Linear Integral Equations," McGraw-Hill Book Co., Inc., New York, 1924; reprinted by Dover Publishing Co., New York, 1950.

Muskhelishvili, N. I., translated by J. R. M. Radok, " Singular Integral Equations," Groningen, 1953.

Vivanti-Schwank, " Lineare Integralgleichungen," Helwingsche Verlagsbuchhandlung, Hannover, 1929.

CHAPTER 15

GROUP THEORY

PROPERTIES OF A GROUP

Group theory has become so vital a part of modern physical and chemical analysis that the inclusion of its basic structure seemed inevitable to the authors of this book. Because of the great volume of available material arbitrary selection had to be made, and many proofs had to be omitted or given only in outline. Care has been taken, however, to insure that the attentive reader of the present chapter will be able to familiarize himself with all the tools needed for handling the simpler problems of group theory, such as those arising in quantum mechanics and in the field of molecular structure. A certain amount of material, easily obtained by the methods discussed in this chapter, but of somewhat lengthy derivation, has been collected at the end in Table 7.

15.1. Definitions.—A *group*[1] is a set of abstract *elements A, B, C,* $\cdots$ finite or infinite in number, with a law of combination for any two elements A and B to form a *product*[2] AB such that:

a. Every product of the two elements and the square of every element is a member of the set.

b. The set contains a *unit element E* for which $EA = AE = A$ for every member of the set.

c. The associative law holds: $A(BC) = (AB)C$.

d. Every element has an *inverse*, $X = A^{-1}$, so that $AX = AA^{-1} = A^{-1}A = E$.

The set of all integers, positive, negative and zero, forms a group if the law of combination is addition. The unit element is zero and the negative of every element is its inverse. These numbers do not form a group if the law of combination is multiplication. In this case, $E = 1$, but the element zero has no inverse hence (d) cannot be satisfied. For any law of combination, we always speak of a product and write the two elements as if they were multiplied together.

---

[1] For general treatises on group theory, see references at end of this chapter.

[2] Following the convention of sec 10.10, it is to be understood throughout this chapter that the elements of a product are to be taken in the order from *right* to *left*.

A *finite group* of *order g* contains a finite number of elements, g. A simple example of such a group (of order four) is furnished by the numbers ± 1, $\pm i$. If n is the smallest integer for which $X^n = E$, n is called the *order of the element* X. The n elements X, X^2, X^3, $\cdots$, X^{n-1}, $X^n = E$ form the *period* of X, indicated by $\{X\}$. The period of a single element is thus a finite group; it is called a *cyclic group*.

All of the groups so far mentioned have the property that $AB = BA$ for every element. When this condition is fulfilled, the group is said to be *Abelian*. Two or more cyclic groups (they are also Abelian) may be combined to form a single group which is non-Abelian. Suppose

$$A^3 = E; \quad C^2 = E; \quad CA = A^{-1}C \qquad (15\text{-}1)$$

then the group, which we designate by $\mathbf{D}_3$ (for reasons which appear later) is of order six with elements E, A, A^2, C, AC, A^2C. The products of these elements may be arranged in a multiplication table; CA, for example, is found at the intersection of row C and column A. If we let $A^2 = B$, $AC = D$, $A^2C = F$ and use (1) we obtain for the group $\mathbf{D}_3$

	E	A	B	C	D	F
E	E	A	B	C	D	F
A	A	B	E	D	F	C
B	B	E	A	F	C	D
C	C	F	D	E	B	A
D	D	C	F	A	E	B
F	F	D	C	B	A	E

$$(15\text{-}2)$$

It should be noticed that each element occurs once and only once in each row or column.

Problem a. Use (15-1) to derive the multiplication table of (15-2).

Problem b. Show that if any element occurs more than once in a row or column of a multiplication table for a group then the group postulates (a)–(d) could not be fulfilled.

15.2. Subgroups.—A group whose elements are contained in another group is called a *subgroup*. Thus we may always find subgroups in any group by forming the period of each of its elements. For example, in $\mathbf{D}_3$ a subgroup of order three is obtained from $\{A\} = \{B\} = E, A, B$. Similarly, three different subgroups, each of order two, may be found: $\{C\} = E$, C; $\{D\} = E, D$; $\{F\} = E, F$. In addition to these subgroups, the single element E is a subgroup of order one while the group itself is a subgroup of order six. In this case, each subgroup, except the group itself, is cyclic. It does not follow, however, that all subgroups are cyclic.

Suppose a given group is of order g and a subgroup of it is of order h

with elements $A_1, A_2, \cdots, A_h$. Now take B, an element of the group which is not contained in the subgroup, and form the products $BA_1, BA_2, \cdots BA_h$. These must all be in the group but none can be in the subgroup, for if $BA_i = A_j$ were one of the members of the subgroup then $B = A_jA_i^{-1}$ would also be in the subgroup which is contrary to our assumption concerning the selection of B. We have now found $2h$ members of the group. If $2h < g$, it will be possible to find a new element C contained neither among the elements $A_1, \cdots, A_h$ nor among the elements $BA_1, \cdots, BA_h$. Repeating the operations of multiplication and using the same arguments as before we obtain h new elements $CA_1, \cdots, CA_h$. Since the group is of finite order, the procedure must end when we have found $kh = g$ elements (k an integer). It thus follows that the order of the subgroup must be a divisor of the order of the whole group. In the example of the preceding paragraph, we see that we have found all possible subgroups since the only divisors of 6 (the order of the group) are 1, 2, 3, and 6.

15.3. Classes.—Let A, B and X be any three elements of a group; then if $B = X^{-1}AX$, B is said to be the *transform* of A by the element X; A and B are *conjugate* to each other. The following properties of conjugate elements may be proved; it is easy to verify them for $\mathbf{D}_3$ by the use of the group table (2).

a. Every element is conjugate with itself.

b. If A is conjugate with B, then B is conjugate with A.

c. If A is conjugate with both B and C, then B and C are conjugate with each other.

The complete set of elements $\mathcal{C} = A_1, A_2, \cdots, A_r$, which are conjugate with each other, is called a *class* of the group. If the group contains the elements $A_1 (= E), A_2, \cdots, A_g$ the class of A may be found by calculating

$$E^{-1}AE = A, \quad A_2^{-1}AA_2, \quad \cdots, A_g^{-1}AA_g$$

although not all of these elements will be distinct as may be seen from the following example. Clearly $\mathcal{C}_1 = E$ always forms a class by itself. In (2), $\mathcal{C}_2 = A, B$, for

$$E^{-1}AE = A; \quad B^{-1}AB = A; \quad D^{-1}AD = B$$
$$A^{-1}AA = A; \quad C^{-1}AC = B; \quad F^{-1}AF = B$$

Similarly $\mathcal{C}_3 = C, D, F$. By arguments similar to those used in discussing subgroups it follows that the whole group may be separated into a number of different classes none of which contain any elements in common. Moreover, if there are h elements of a group which transform a given element into another element of the same class, then the number of elements in that class $r = g/h$ where g is the order of the group.

15.4. Complexes.—A set of elements from a group, considered as a whole, is called a *complex*. If the complex $\mathcal{Q}$ contains A, B, C then $C\mathcal{Q}$ contains CA, CB, C^2. By the product of two complexes $\mathcal{Q}\mathcal{B}$ we mean the product of every element in $\mathcal{Q}$ with every element in $\mathcal{B}$, but products occurring more than once are only taken once. By the complex $\mathcal{G}$ we mean the whole group. If $\mathcal{H}$ is a subgroup, then

$$\mathcal{H}\mathcal{H} = \mathcal{H}^2 = \mathcal{H} \tag{15-3}$$

If X is an element of $\mathcal{G}$ not contained in $\mathcal{H}$ then the complex $\mathcal{H}X$ is called a *right coset* (*Nebengruppe*) and $X\mathcal{H}$ is a *left coset*. Cosets are not groups since $\mathcal{H}X$ does not contain E. It is easy to see that if another element Y is neither in $\mathcal{H}$ nor in $\mathcal{H}X$, then the coset $\mathcal{H}Y$ will contain no element common with $\mathcal{H}$ or $\mathcal{H}X$, so that the whole group may be written as a sum of a finite number of cosets

$$\mathcal{G} = \mathcal{H} + \mathcal{H}X + \mathcal{H}Y + \mathcal{H}Z + \cdots$$

The group may also be divided in this way by means of left cosets. In $\mathbf{D}_3$, we may write

$$\mathcal{G} = \mathcal{H} + \mathcal{H}C = \mathcal{H} + \mathcal{H}D = \mathcal{H} + \mathcal{H}F = \mathcal{H} + C\mathcal{H} = \mathcal{H} + D\mathcal{H}$$
$$= \mathcal{H} + F\mathcal{H}$$

where $\mathcal{H} = E$, A, B. The *index* of a subgroup equals the order of the group divided by the order of the subgroup. It also equals the number of complexes obtained by splitting a group into that particular subgroup and its cosets; two, in the example just given.

15.5. Conjugate Subgroups.—If a subgroup $\mathcal{H}$ contains the elements $H_1 (= E)$, H_2, $\cdots$, H_h then it also contains $EH_j = H_j$, H_2H_j, $\cdots$, H_hH_j for every H_j in $\mathcal{H}$, and it contains $H_j^{-1}E = H_j^{-1}$, $H_j^{-1}H_2$, $\cdots$, $H_j^{-1}H_h$. In fact these arrangements of the h elements of $\mathcal{H}$ are identical except for the sequence in which the members are written. Still another arrangement is $H_j^{-1}EH_j = E, H_j^{-1}H_2H_j, \cdots, H_j^{-1}H_hH_j$. To see this, sort out the arrangement $EH_j = H_j$, $\cdots$, H_hH_j so that the natural order H_1, H_2, $\cdots$, H_h is regained and multiply each element by H_j^{-1}. A similar argument will show that for X, any member of the group (not necessarily contained in $\mathcal{H}$) $X^{-1}\mathcal{H}X$ is also a subgroup, but $X^{-1}\mathcal{H}X$ and $\mathcal{H}$, called *conjugate subgroups*, may be different if X is not in $\mathcal{H}$. When $\mathcal{H}$ and $X^{-1}\mathcal{H}X$ are identical for every X in the group, $\mathcal{H}$ is called an *invariant subgroup* or a *normal divisor*. To illustrate these statements choose $\mathcal{H} = E$, C and $\mathcal{H} = E$, A, B from $\mathbf{D}_3$. It is easily verified that the only invariant subgroup of $\mathbf{D}_3$ is $\mathcal{H} = E$,

A, B. The invariant subgroup and its cosets form a group[3] called the *quotient* (or *factor*) *group* with the invariant subgroup as unit element. In $\mathbf{D}_3$, if $\mathcal{F} = \mathcal{H}C$, then the multiplication table of the quotient group $\mathcal{G}/\mathcal{H}$ is

$$
\begin{array}{c|cc}
 & \mathcal{H} & \mathcal{F} \\
\hline
\mathcal{H} & \mathcal{H} & \mathcal{F} \\
\mathcal{F} & \mathcal{F} & \mathcal{H}
\end{array}
\qquad (15\text{-}4)
$$

15.6. Isomorphism.—Two groups $\mathcal{G}$ and $\mathcal{G}'$ are said to be simply *isomorphic* if to each element A, B, C, $\cdots$ of $\mathcal{G}$ there corresponds an element A', B', C', $\cdots$ of $\mathcal{G}'$ so that if $AB = C$, then $A'B' = C'$ for every product. In the general case, two or more elements of one group may be isomorphous with a single element of another group. Thus the quotient group (4) is multiply isomorphous with $\mathbf{D}_3$, for $\mathcal{H}$ corresponds to E, A, B and $\mathcal{F}$ to C, D, F.

In order to find a group which is simply isomorphous with $\mathbf{D}_3$, we consider the $n!$ permutations of n symbols. By $(acbed)$ we shall mean a replaced by c, c replaced by b, b by e, e by d and d by a. This may also be written as $(bedac)$ or $(dacbe)$ as long as we do not change the cyclic order of the symbols. When a single letter occurs in a parenthesis, that letter is unaffected by the permutation, hence we will write $(bce)(a)(d)$ as (bce). By the product of two permutations, we mean the permutation directed in the right parenthesis followed by the permutation in the left parenthesis. For example, in the product $(acbed)$ (bce), b is replaced by c and then c by b, the net result for b being that it returns to its original position. Continuing in this way, we obtain $(acbed)$ $(bce) = (cda)$. If we use only three letters and write

$$
\begin{array}{lll}
E = (a)(b)(c) & B = (abc) & D = (ac)(b) \\
A = (acb) & C = (a)(bc) & F = (ab)(c)
\end{array}
\qquad (15\text{-}5)
$$

the resulting operations form a group which is simply isomorphic with $\mathbf{D}_3$, for

$$
AB = (acb)(abc) = (a)(b)(c) = E
$$
$$
BC = (abc)(bc) = (ab) \qquad = F; \quad \text{etc.}
$$

Problem. Derive the complete multiplication table for the group of permutations on three letters.

---

[3] Note that the elements of the quotient group are complexes, i.e., collections of the original elements of $\mathcal{G}$.

15.7. Representation of Groups.—If to every member of a group $A_1, A_2, A_3, \cdots$, we can associate a square matrix $D(A_1), D(A_2), D(A_3), \cdots$ in such a way that if $A_i A_j = A_k$ and $D(A_i)D(A_j) = D(A_k)$, then the matrices themselves form a group isomorphous with $\mathcal{G}$. Such matrices are a *representation* of the group; their order is the *degree* or *dimension* of the representation. One trivial example of a representation is the unit matrix E associated with every element of the group. A representation for $\mathbf{D}_3$ may be obtained from its quotient group if we associate $\mathcal{H}$ with the matrix $[1]$ and $\mathcal{F}$ with the matrix $[-1]$.

To find another representation of $\mathbf{D}_3$, let us think of the symbols a, b, c as the components of a vector $\mathbf{x}$ and the elements of the group as operations which change $\mathbf{x}$ into a new vector $\mathbf{x}'$ with the same components but in a different order. Hence the required representation D will be a matrix such that $\mathbf{x}' = D\mathbf{x}$ where the rows and columns are labelled with the components a, b, c. Now E is the operation which replaces each component by itself so $D(E)$ is the unit matrix. On the other hand, A replaces a by c, but a itself becomes b, etc., so unity will appear in $D(A)$ at the intersection of the a-th row and the b-th column, etc. Continuing in this way, we find.

$$D(E) = \begin{bmatrix} 1 & 0 & 0 \\ 0 & 1 & 0 \\ 0 & 0 & 1 \end{bmatrix}; \quad D(A) = \begin{bmatrix} 0 & 1 & 0 \\ 0 & 0 & 1 \\ 1 & 0 & 0 \end{bmatrix}; \quad D(B) = \begin{bmatrix} 0 & 0 & 1 \\ 1 & 0 & 0 \\ 0 & 1 & 0 \end{bmatrix};$$

$$D(C) = \begin{bmatrix} 1 & 0 & 0 \\ 0 & 0 & 1 \\ 0 & 1 & 0 \end{bmatrix}; \quad D(D) = \begin{bmatrix} 0 & 0 & 1 \\ 0 & 1 & 0 \\ 1 & 0 & 0 \end{bmatrix}; \quad D(F) = \begin{bmatrix} 0 & 1 & 0 \\ 1 & 0 & 0 \\ 0 & 0 & 1 \end{bmatrix} \quad (15\text{-}5a)$$

By multiplying the matrices together, it will be seen that the multiplication table (2) is reproduced. For example, $D(A)D(B) = D(E)$ and $D(A)D(C) = D(D)$. Thus (5a) is a representation of $\mathbf{D}_3$.

Suppose a representation of a group has been found, consisting of matrices $D = D(A_1), D(A_2), \cdots, D(A_g)$, each matrix being of dimension n. Then it is often possible to find a new coordinate system, i.e., a transformation of the type $Q^{-1}DQ$, such that every matrix D is changed to the form

$$\begin{bmatrix} D_1 & 0 \\ 0 & D_2 \end{bmatrix} \quad (15\text{-}6)$$

where D_1 is of order m, $m < n$ and D_2 is of order $(n - m)$. Under these conditions, the representation D is said to be *reducible*[4] into D_1 and D_2.

---

[4] In the more general case, the matrices are converted to the triangular form of (10–40). If the form obtained is that of (6), the representation is said to be *completely reducible*.

We now examine D_1 and D_2 to see if they are reducible, continuing until D is completely reduced. When this has been accomplished, we will have a relation between the original and final coordinate systems such as $z = Qx$ and

$$Q^{-1}DQ = \text{diag}\,[\boldsymbol{\Gamma}^{(1)},\ \boldsymbol{\Gamma}^{(2)},\ \cdots,\ \boldsymbol{\Gamma}^{(s)}] = \boldsymbol{\Gamma} \qquad (15\text{-}7)$$

where the $\boldsymbol{\Gamma}^{(i)}$ are themselves matrices.

It should be understood that if there are g elements in the group, there will be g equations like (7), one for each element; D means the set of g matrices in the original coordinate system and $\boldsymbol{\Gamma}$ means the same matrices in the new coordinate system. Suppose there are s irreducible representations in (7), $\boldsymbol{\Gamma}^{(1)},\boldsymbol{\Gamma}^{(2)},\cdots,\boldsymbol{\Gamma}^{(s)}$; each one of these is a set of g matrices, one for each element of the group, $\boldsymbol{\Gamma}^{(j)} = \boldsymbol{\Gamma}^{(j)}(A_1),\boldsymbol{\Gamma}^{(j)}(A_2),\cdots,\boldsymbol{\Gamma}^{(j)}(A_g)$. Each $\boldsymbol{\Gamma}^{(j)}$ is isomorphous with the corresponding D in the original coordinate system since the two sets of matrices are related to each other by a collineatory transformation (cf. sec. 10.11).

It may happen that some $\boldsymbol{\Gamma}^{(j)}$ may appear more than once or not at all in the reduction of a given representation. To indicate this, we rewrite (7) as

$$\boldsymbol{\Gamma} = c_1\boldsymbol{\Gamma}^{(1)} + c_2\boldsymbol{\Gamma}^{(2)} + \cdots + c_s\boldsymbol{\Gamma}^{(s)} \qquad (15\text{-}8)$$

where the c's are positive integers or zero. Such an expression, called the *direct sum*, is not meant to imply that the $\boldsymbol{\Gamma}^{(j)}$ are to be added. It is simply a shorthand method of showing that the matrices D have been reduced to the form (7).

It is of considerable advantage to choose unitary or orthogonal matrices as the representations of groups and we shall suppose that this is always done. Under these conditions the following statements may be proved.[5] Two irreducible representations will be orthogonal, and if d_j is the dimension of $\boldsymbol{\Gamma}^{(j)}$, then

$$\sum_A \Gamma^{(i)}_{\alpha\beta}\Gamma^{(j)*}_{\mu\nu} = \frac{g}{(d_id_j)^{1/2}}\,\delta_{ij}\delta_{\alpha\mu}\delta_{\beta\nu} \qquad (15\text{-}9)$$

the summation to be made over the g elements of the group, $A_1, A_2, \cdots, A_g$. Moreover if there are s classes of elements in a group, there will be exactly s different irreducible representations and

$$d_1^2 + d_2^2 + \cdots + d_s^2 = g \qquad (15\text{-}10)$$

It is not always possible to obtain all s of the irreducible representations from a single set of reducible matrices D since some of the c_j in (8) may be zero. If this is the case, another set of matrices D' must be found and these must be reduced in the same way until the complete set is obtained.

[5] See texts on group theory cited at end of this chapter.

15.8. Reduction of a Representation.—We now wish to show how it is possible to find all of the irreducible representations for $\mathbf{D}_3$. Since $g = 6$ and $s = 3$, it follows from (10) that they are of dimension 1, 1 and 2. To find the two representations[6] of degree one, we consider the quotient group (4) with two classes, $\mathcal{C}^+$ containing $\mathcal{H}$ and $\mathcal{C}^-$ containing $\mathcal{F}$. Its two representations are $\Gamma^{(1)}(\mathcal{C}^+) = \Gamma^{(1)}(\mathcal{C}^-) = 1$; $\Gamma^{(2)}(\mathcal{C}^+) = 1$; $\Gamma^{(2)}(\mathcal{C}^-) = -1$. While these are almost trivial, it is seen that they satisfy all of the requirements for a representation of $\mathbf{D}_3$. They are therefore taken as its two representations of degree one.

In order to obtain the representation of dimension two, we attempt to reduce the matrices of (5a). We expect to get, as a result, matrices of the form of (6) where D_1 is either $\Gamma^{(1)}$ or $\Gamma^{(2)}$, and D_2 is a set of two-dimensional matrices. We note that each of the matrices of (5a) is orthogonal and from the discussion of sec. 10.17 we see that another real orthogonal matrix will reduce any of them to the desired form. The columns of the reducing matrix will be composed of the eigenvectors of one of the matrices to be reduced. If we choose $D(A)$ we find that its eigenvalues are 1, $e^{\pm i\phi}$, where $\phi = 2\pi/3$. Taking linear combinations of the complex eigenvectors and normalizing them the result is $3^{-1/2}[1, 1, 1]$; $6^{-1/2}[1, -2, 1]$; $2^{-1/2}[-1, 0, 1]$. They form the columns of a matrix Q which will reduce each matrix of (5a) by the transformation indicated in (7). The diagonal elements will be $\Gamma^{(1)}$ and two-dimensional matrices which were sought. A typical result is

$$\tilde{Q}D(A)Q = \begin{bmatrix} 1 & 0 & 0 \\ 0 & -1/2 & -\sqrt{3}/2 \\ 0 & \sqrt{3}/2 & -1/2 \end{bmatrix} \tag{15-11}$$

Other methods[7] of reducing a given representation may be found. Consider, for example, the effect of the matrices (5a) in changing a vector $\mathbf{x}$ into another vector $\mathbf{x}'$ by the relation $\mathbf{x}' = D\mathbf{x}$. In such an operation two components of the vector are simply interchanged in their original plane or else both of them are transferred to a plane perpendicular to the one in which they originally lay. These relations could be examined in a new coordinate system in which x_1 is along the normal to the plane

---

[6] The reader will recall that the complex $\mathcal{H}$ must be regarded as a *single* element of the quotient group $\mathcal{G}/\mathcal{H}$. It is true that $\mathcal{H}$ is made up of the elements E, A, and B of the original group $\mathbf{D}_3$, but it acts as the unit element of $\mathcal{G}/\mathcal{H}$. The other element of the quotient group, $\mathcal{F}$, contains the elements C, D, F, of $\mathbf{D}_3$.

[7] A formal method, based on hypercomplex numbers in a general type of algebra, called Frobenius algebra, is described by Speiser, Littlewood, and other references cited at the end of this chapter.

determined by x_2 and x_3. Calling the new system $\mathbf{y}$, not necessarily a rectangular Cartesian one, x_2 could be taken in the plane of y_1, y_2 and x_3 in the plane of y_2, y_3. The relation between the new and old coordinates is then $\mathbf{x} = S\mathbf{y}$, where $x_1 = y_1 + y_2 + y_3$; $x_2 = y_1 - y_2$; $x_3 = y_2 - y_3$. In this new system, $\mathbf{x}' = S\mathbf{y}'$ and, since S is non-singular, $\mathbf{y}' = S^{-1}DS\mathbf{y}$, according to sec. 10.13. When the reciprocal matrix is found (note that S is not orthogonal)

$$S^{-1}D(A)S = \begin{bmatrix} 1 & 0 & 0 \\ 0 & -1 & 1 \\ 0 & -1 & 0 \end{bmatrix}$$

with similar results for the remaining matrices of (5a). We note again that we have found $\Gamma^{(1)}$ and two-dimensional representations.

Although these two reductions do not give identical results, their matrices are related by a similarity transformation and their traces are identical as can be seen by comparing $S^{-1}D(A)S$ with eq. (11). The importance of this property is explained in the next section.

In the usual case, it is easier to find another representation of the required dimension than to reduce one already known. Consider a plane, equilateral triangle with apexes labeled a, b, c and located in a Cartesian coordinate system so that the coordinates of its apexes are $a = (1,0)$; $b = \frac{1}{2}(-1, \sqrt{3})$; $c = -\frac{1}{2}(1, \sqrt{3})$. The elements of the permutation group (5) will then be seen to correspond with the following operations on this triangle: (E) identity; (A) rotation of the triangle about the origin of the coordinate system through the angle $2\pi/3$ in the counter-clockwise direction; (B) rotation by $4\pi/3$ in the same direction, or by $2\pi/3$ in the clockwise direction; (C) rotation through the angle π about an axis lying in the plane of the triangle and passing through $y = 0$; (D) a similar rotation about an axis through $y = -\sqrt{3}x$, which passes through the apex b of the triangle; (F) rotation about an axis passing through the apex c, or $y = \sqrt{3}x$.

Since we are considering a space-fixed coordinate system and we are moving the triangle rather than the coordinate system, the appropriate two-dimensional matrices for A and B are the transforms of eq. (52) given later in this chapter, with $\phi(A) = 2\pi/3$ and $\phi(B) = 4\pi/3$. Operation C merely changes the sign of the y-coordinates and its matrix is that of eq. (61). The two remaining matrices for D and F are easily obtained from the multiplication table for the group, for example, $D = CB$; $F = CA$. They could also be found from geometric considerations.

We now have the three irreducible representations for D_3. The one-dimensional representations have been given in the first paragraph of this section. Using the matrices obtained from operations with the triangle, the abbreviated notations E, A, B, $\ldots$, instead of the more explicit forms $\Gamma^{(3)}$ (E), etc., and writing $c = \cos \phi = -1/2$; $s = \sin \phi = \sqrt{3}/2$, $\phi = 2\pi/3$, the two-dimensional representations are given in Table 1.

TABLE 1

$$E = \begin{bmatrix} 1 & 0 \\ 0 & 1 \end{bmatrix}; \qquad A = \begin{bmatrix} c & -s \\ s & c \end{bmatrix}; \qquad B = \begin{bmatrix} c & s \\ -s & c \end{bmatrix};$$

$$C = \begin{bmatrix} 1 & 0 \\ 0 & -1 \end{bmatrix}; \qquad D = \begin{bmatrix} c & s \\ s & -c \end{bmatrix}; \qquad F = \begin{bmatrix} c & -s \\ -s & -c \end{bmatrix}.$$

15.9. The Character.—The task of finding all the irreducible representations of a given group is usually very laborious. However, for most physical applications, it is sufficient to know only their trace, a quantity called the *character*[8] in group theory. We shall indicate the trace of $\Gamma^{(i)}$ by $\chi^{(i)} = \chi^{(i)}(A_1)$, $\chi^{(i)}(A_2)$, etc. A further simplification is afforded by the fact that elements in the same class are obtained from each other by a similarity transformation, hence the character of every element in a single class is identical. This follows from the fact that elements in the same class are related to each other by a similarity transformation and, as we have shown in sec. 10.14, the trace of two quantities so related is identical. Therefore, if we know all the characters of one element from every class of the group, we have all of the information concerning the group which is usually needed. We shall indicate the particular class to which we refer by a subscript, so that the s characters $\chi_1^{(i)}$, $\chi_2^{(i)}$, $\cdots$, $\chi_s^{(i)}$ refer to the i-th irreducible representation.

The following properties of the characters may be derived[9] or verified using tables of characters given in later sections.

a. The class $\mathcal{C}_1 = E$ is always represented by the unit matrix, thus $\chi_1^{(i)}$ equals the dimension of the representation and hence must be a divisor of the order of the group. We also see from (10) that

$$\sum_{i=1}^{s} [\chi_1^{(i)}]^2 = g \qquad (15\text{--}12)$$

---

[8] The character (especially of permutation groups) is treated in detail by Littlewood, D. E., " The Theory of Group Characters," Oxford University Press, 1940.

[9] Cf. Speiser, loc. cit., Chapter 12.

If g and s are known, it will usually be found that there is but a single way in which this equation can be satisfied.

b. From (9) it follows that the s characters also form an orthogonal system. Summing over the classes we obtain

$$\sum_{q=1}^{s} r_q \chi_q^{(i)} \chi_q^{(j)*} = g\delta_{ij} \qquad (15\text{--}13)$$

where r_q is the number of elements in the q-th class.

c. If Ξ is the character of a *reducible* representation, then from (8), we have

$$\Xi = c_1 \chi^{(1)} + c_2 \chi^{(2)} + \cdots + c_s \chi^{(s)} \qquad (15\text{--}14)$$

On multiplying this by $\chi_q^{(j)*}$ and summing over q, we obtain, using (13),

$$c_j = \frac{1}{g} \sum_{q=1}^{s} r_q \Xi_q \chi_q^{(j)*} \qquad (15\text{--}15)$$

When the complete multiplication table for a group is known, the following procedure[10] may be used to obtain the characters. First calculate the product of all elements in the class $\mathcal{C}_i$ by all elements in $\mathcal{C}_k$. It will be found that the resulting set of elements may be uniquely arranged in classes and that the same results are obtained irrespective of whether we multiply $\mathcal{C}_i$ by $\mathcal{C}_k$ or the reverse. Now a given class may occur in the products several times or not at all. Let us use $h_{ik,j}$ to indicate the number of times the j-th class appears. Then if we abandon our earlier rule for the multiplication of complexes (cf. sec. 15.4) and take each element of the product as many times as it occurs, we may write

$$\mathcal{C}_i \mathcal{C}_k = \mathcal{C}_k \mathcal{C}_i = \sum_{j=1}^{s} h_{ik,j} \mathcal{C}_j$$

where we sum over the total number of classes, s. Having found the numbers $h_{ik,j}$ it is then possible to find the characters from the relations

$$r_i r_k \chi_i \chi_k = \chi_1 \sum_{j=1}^{s} h_{ik,j} r_j \chi_j \qquad (15\text{--}16)$$

where r_i is the number of elements in $\mathcal{C}_i$.

As an example of the use of this equation, we find for $\mathbf{D_3}$

$$\mathcal{C}_2^2 = A^2, \quad B^2, \quad AB, \quad BA = 2\mathcal{C}_1 + \mathcal{C}_2$$

$$\mathcal{C}_3^2 = 3\mathcal{C}_1 + 3\mathcal{C}_2; \quad \mathcal{C}_2 \mathcal{C}_3 = 2\mathcal{C}_3$$

[10] Proof of the statements in this paragraph may be found in Murnaghan, p. 83 or Speiser, p. 170, loc. cit. They may be verified by using the multiplication table for $\mathbf{D_3}$.

(The other products are not needed.)   Since $r_1 = 1$; $r_2 = 2$; $r_3 = 3$, we have

$$4\chi_2^2 = \chi_1(2\chi_1 + 2\chi_2)$$
$$9\chi_3^2 = \chi_1(3\chi_1 + 6\chi_2) \qquad (15\text{--}17)$$
$$6\chi_2\chi_3 = 6\chi_1\chi_3$$

From (12), we know that χ_1 has the values 1, 1 and 2.   Solving (17) with each of these quantities in turn we obtain the entries in Table 2.   They are identical with the trace of the matrices of the last section.

TABLE 2

	$\mathcal{C}_1$	$\mathcal{C}_2$	$\mathcal{C}_3$
$\Gamma^{(1)}$	1	1	1
$\Gamma^{(2)}$	1	1	-1
$\Gamma^{(3)}$	2	-1	0

Let us apply eq. (15) to the matrices (5a) and confirm a fact that we already know, namely, that these reducible representations contain $\Gamma^{(1)}$ and $\Gamma^{(3)}$ once each but not $\Gamma^{(2)}$.   From (5a), we see that $\Xi_1 = 3$; $\Xi_2 = 0$; $\Xi_3 = 1$, hence, using eq. (15),

$$c_1 = (1.3.1 + 2.0.1 + 3.1.1)/6 = 1$$
$$c_2 = (1.3.1 + 2.0.1 - 3.1.1)/6 = 0$$
$$c_3 = (1.3.2 - 2.0.1 + 3.1.0)/6 = 1$$

We have shown how a reducible representation of the group $\mathbf{D}_3$ may be found (cf. sec. 15.7).   Now elements occur on the diagonal of the matrices of eq. (5a) only when the symbols are unchanged by the permutations with which $\mathbf{D}_3$ is isomorphous.   Since these diagonal elements are all unity, the reducible character of an element of a permutation group is equal to the number of symbols unchanged by the permutation.   This result is very useful, for every group is isomorphous with some permutation group; hence when the latter is known it is a simple matter to find Ξ_q.

Problem.   Derive Table 2 by the method described in the text.

15.10. The Direct Product.—Two cyclic groups were combined in (1) to form a single larger group.   We now describe another method of augmenting the order of a group.   Suppose $\mathcal{G}'$ is of order m with elements A_1, A_2, $\cdots$, A_m and $\mathcal{G}''$ is of order n with elements B_1, B_2, $\cdots$, B_n and that every A commutes with every B.   Then the mn elements A_iB_j form a group of order mn called the *direct product* of $\mathcal{G}'$ and $\mathcal{G}''$, $\mathcal{G} = \mathcal{G}' \times \mathcal{G}''$.   If the matrices $\Gamma(A)$ and $\Gamma(B)$ are irreducible representations of $\mathcal{G}'$ and $\mathcal{G}''$, then their direct product

$$\Gamma(A) \times \Gamma(B) = \Gamma(AB)$$

is a representation of $\mathcal{G}$. Moreover, if $\chi_s^{(i)}$ is a character of $_s$ in $\mathcal{G}'$ and $\chi_t^{(j)}$ belongs to $\mathcal{C}_t$ in $\mathcal{G}''$, then the st characters of $\mathcal{C}_{st}$ in $\mathcal{G}$ are given by

$$\chi_{st}^{(ij)} = \chi_s^{(i)} \chi_t^{(j)}$$

If one or both of the representations $\Gamma(A)$ and $\Gamma(B)$ are of the first degree, the direct product $\Gamma(AB)$ is irreducible. If both are of degree higher than one, $\Gamma(AB)$ is reducible. The reduction is very simple provided the table of characters for both groups is known, for multiplication of one set of characters by another will give a sum of characters already contained in the table. This can always be uniquely resolved into its component parts. An illustration of such reduction will be given in sec. 15.18.

SOME SPECIAL GROUPS

15.11. The Cyclic Group.—If a cyclic group is formed from $\{A\}$, $A^n = E$ and X is any element of the group defined by $X = A^m$, $m = 1, 2, \cdots, n$, then $X^{-1}AX = A$. It thus follows that every element of a cyclic group or any other Abelian group is in a class by itself. Moreover, we see from (10) that the n irreducible representations will each be of degree one so that each representation is also a character. Now if $\epsilon = \exp(2\pi i/n)$, then ϵ will be a representation and a character for A and ϵ^m will be a character for A^m ($m = 1, 2, \cdots, n$), since these n numbers will satisfy the multiplication properties of the group elements. Moreover, ϵ^{2m} will also serve as a set of characters for the same reason. In fact the n distinct powers of ϵ^m ($m = 1, 2, \cdots, n$) will give the n characters for each of the n elements. They are shown in Table 3. We can simplify such a table by using de Moivre's theorem: $\epsilon^p = \cos 2\pi p/n + i \sin 2\pi p/n$. For example, if $n = 4$, the only numbers that will occur are ± 1 and $\pm i$.

TABLE 3

	$\mathcal{C}_1 = A^n = E$	$\mathcal{C}_2 = A$	$\mathcal{C}_3 = A^2$	$\cdots$	$\mathcal{C}_n = A^{n-1}$
$\Gamma^{(1)}$	1	1	1	$\cdots$	1
$\Gamma^{(2)}$	1	ϵ	ϵ^2	$\cdots$	ϵ^{n-1}
$\cdots$	$\cdots$	$\cdots$	$\cdots$	$\cdots$	$\cdots$
$\Gamma^{(m)}$	1	ϵ^{m-1}	$\epsilon^{2(m-1)}$	$\cdots$	$\epsilon^{(n-1)(m-1)}$
$\cdots$	$\cdots$	$\cdots$	$\cdots$	$\cdots$	$\cdots$
$\Gamma^{(n)}$	1	ϵ^{n-1}	$\epsilon^{2(n-1)}$	$\cdots$	$\epsilon^{(n-1)^2}$

15.12. The Symmetric Group.—Consider a particular permutation of five letters which we write as

$$P_e = \begin{pmatrix} a & b & c & d & e \\ c & e & b & a & d \end{pmatrix}$$

This is to be interpreted as meaning: a is replaced by c, b by e, c by b, d by a, and e by d. A more convenient and equivalent form for such a permutation is

$$P_e = (acbed)$$

which we have already used in sec. 15.6. The one-line form is called a *cycle*; its *degree* equals the number of letters in the parenthesis. It will be found that any permutation may be written as a *single* cycle with no letter repeated, or as a *product* of two or more cycles, none of which has a letter in common. Provided their proper sequence is retained, the letters in a cycle may be rearranged, but the number and degree of all cycles corresponding to a given permutation is unique. For example,

$$P_0 = \begin{pmatrix} a & b & c & d & e \\ c & e & a & b & d \end{pmatrix} = (ac)(bed) = (bed)(ac) = (dbe)(ca), \quad \text{etc.}$$

A cycle of degree two is called a *transposition*. A cycle of higher degree may be rewritten as a product of two or more transpositions in several different ways, but then the product will contain the same letter or letters in two or more parentheses. However, if the original cycle contained an *even* number of letters, the product of transpositions will be composed of an *odd* number of transpositions and if the original cycle contained an *odd* number of letters, the product will have an *even* number of transpositions. Since any permutation may be decomposed into a product of cycles, and each of the latter may be written as a product of transpositions, it follows that any permutation may be factored into a product of transpositions. Moreover, all the different products corresponding to a given permutation contain either an even number or all contain an odd number of transpositions. This property of a permutation is unique, and permits us to speak of *even* and *odd* permutations, P_e and P_0. As examples, we see that

$$P_e = (ac)(ab)(ae)(ad) = (ac)(cb)(be)(ed), \quad \text{etc.}$$

$$P_0 = (ac)(be)(ed) = (ca)(be)(bd), \quad \text{etc.}$$

The *symmetric group* of order $n!$ is defined as the group of all permutations, both even and odd, of n letters. The set of $n!/2$ even permutations of n letters forms a subgroup of the symmetric group, of order $n!/2$; it is called the *alternating group*. A simple consideration shows it to be an invariant subgroup. The odd permutations contained in the symmetric group do not alone form a group, since the product of two odd permutations is even.   However, the complex of odd permutations is one of the elements of the quotient group of order two which is isomorphous with the symmetric group, the other element being the complex of even permutations.

Problem a.   Construct elements and group table for the symmetric group on four letters.   Decompose all elements into transpositions.

Suppose a permutation has been factored into α cycles of degree one, β cycles of degree two, etc.   We describe this arrangement by the symbol $(1^\alpha 2^\beta 3^\gamma \cdots)$ which is called a *partition*.   It is easy to see that any permutation P and its inverse P^{-1} will belong to the same partition, for P^{-1} is formed from P by reversing the order of the letters in the cycles of P. Thus

$$P_0 = (ac)(bed); \quad P_0^{-1} = (ca)(deb)$$

It is also true that elements in the same class belong to the same partition and that there are as many classes as partitions (cf. Problem a).   Now if the total number of letters in a permutation is n, we must have

$$\alpha + 2\beta + 3\gamma + \cdots = n$$

hence the number of possible partitions or the number of classes equals the number of distinct solutions of this equation in positive integers or zero.

In order to find the number of elements in a class we must find the number of permutations having the same cycle structure.   Suppose there are n letters and that the particular class under consideration belongs to the partition $(1^\alpha 2^\beta 3^\gamma \cdots)$.   There are $n!$ ways of arranging the n letters but not all of the arrangements will lead to a different permutation.   For instance, we may start a given cycle with any letter in it; i.e., (abc), (bca) and (cab) are identical.   This fact means that $1^\alpha 2^\beta 3^\gamma \cdots$ arrangements will differ only by cyclic permutation within the various cycles.   There is still another possibility of duplication.   It does not matter whether we write $(ab)(cd)$ or $(cd)(ab)$, hence there are $\alpha!\beta!\gamma!\cdots$ interchanges of this kind, each corresponding to the same permutation.   We thus conclude that the number of different arrangements or the number of elements in a class symbolized by the partition $(1^\alpha 2^\beta 3^\gamma \cdots)$ equals

$$r = \frac{n!}{1^\alpha \alpha! 2^\beta \beta! 3^\gamma \gamma! \cdots} \tag{15–18}$$

Application of the methods just described will show that for $n = 4$, there are 5 classes corresponding to the partitions (1^4), $(1^2, 2)$, $(1, 3)$, (2^2), (4).   Typical elements of each class are $E = (a)(b)(c)(d)$; (ab); (abc);

$(ab)(cd)$; $(abcd)$. The number of elements in each class is 1, 6, 8, 3 and 6, respectively. The complete class of (2^2) is $\mathcal{C}_4 = (ab)(cd)$; $(ac)(bd)$; $(ad)(bc)$.

Problem b. Verify the statements of the preceding paragraph.

Two irreducible representations of the symmetric group are found immediately from the quotient group, for if the even and odd classes are indicated by $\mathcal{C}^+$ and $\mathcal{C}^-$, we have[11]

$$
\begin{aligned}
\mathbf{\Gamma}^{(1)}(\mathcal{C}^+) &= \mathbf{\Gamma}^{(1)}(\mathcal{C}^-) = 1 \\
\overline{\mathbf{\Gamma}}^{(1)}(\mathcal{C}^+) &= 1; \quad \overline{\mathbf{\Gamma}}^{(1)}(\mathcal{C}^-) = -1
\end{aligned}
\tag{15–19}
$$

All other irreducible representations are of higher degree. From each one of these (and also from $\mathbf{\Gamma}^{(1)}$ as shown in (20)), a new representation called the *associated representation* can be obtained by forming the direct product

$$
\mathbf{\Gamma}^{(j)} \times \overline{\mathbf{\Gamma}}^{(1)} = \overline{\mathbf{\Gamma}}^{(j)}
\tag{15–20}
$$

Both $\mathbf{\Gamma}^{(j)}$ and $\overline{\mathbf{\Gamma}}^{(j)}$ have the same dimensions and $\overline{(\overline{\mathbf{\Gamma}}^{(j)})} = \mathbf{\Gamma}^{(j)}$. If $\mathbf{\Gamma}^{(j)} = \overline{\mathbf{\Gamma}}^{(j)}$, the two representations are *self-associated*. Since $\overline{\mathbf{\Gamma}}^{(1)} = +1$ for even classes and -1 for odd classes, it follows that

$$
\begin{aligned}
\overline{\chi}^{(j)}(\mathcal{C}^+) &= \chi^{(j)}(\mathcal{C}^+) \\
\overline{\chi}^{(j)}(\mathcal{C}^-) &= -\chi^{(j)}(\mathcal{C}^-)
\end{aligned}
\tag{15–21}
$$

In order to satisfy (21), the character of $\mathcal{C}^-$ for a self-associated representation must be equal to zero.

Provided $n < 5$, a simple method may be used to obtain the complete table of characters for the symmetric group. When $n \geq 5$, this procedure will not give the characters for all the classes but actually it still gives the characters which are of interest for physical problems.[12] The restriction on n is not a defect of the theory, since eq. (22) which follows is a simplified form of the general polynomial which applies for any value of n.

Suppose there are in a given class of the group ρ cycles of degrees $\lambda_1, \lambda_2, \cdots, \lambda_\rho$ with $\lambda_1 + \lambda_2 + \cdots \lambda_\rho = n$. Then $\chi^{(k)}$ is the coefficient of x^k, $k \leq n/2$, in the polynomial

$$
(1 - x)(1 + x^{\lambda_1})(1 + x^{\lambda_2}) \cdots (1 + x^{\lambda_\rho}) = \sum_k \chi^{(k)} x^k
\tag{15–22}
$$

The coefficients of the highest power of x^k, that is, $k = 1$ for $n = 3$ and $k = 2$ for $n = 4$, are the characters of the self-associated representation.

---

[11] We originally denoted $\overline{\mathbf{\Gamma}}^{(1)}$ by the symbol $\mathbf{\Gamma}^{(2)}$. It is convenient here to use a different notation in order to show the relation between $\mathbf{\Gamma}^{(1)}$ and $\mathbf{\Gamma}^{(2)}$.

[12] For proof of this statement and a derivation of the method with $n < 5$, see Wigner, E., " Gruppentheorie und ihre Anwendung auf Quantenmechanik der Atomspektren," Braunschweig, 1931, Chapter XIII.

We thus obtain from (22) the characters of k representations of the group while those of the remaining $(s - k)$ representations are the associated ones which may be obtained by using (21). We illustrate the procedure for the symmetric group of order 4!.

For the partition (1^4), $\lambda_1 = \lambda_2 = \lambda_3 = \lambda_4 = 1$ and the polynomial is $(1 - x)(1 + x)^4$. Since $k \leq n/2$, we take the coefficients of x^0, x and x^2 which are 1, 3 and 2. The last value, 2, is the character of a self-associated representation as previously pointed out. The class under consideration is even, hence the associated characters are 3 and 1, completing the first column of the character table. For the next class $(1^2,2)$, we have $\lambda_1 = \lambda_2 = 1$, $\lambda_3 = 2$; the polynomial is $(1 - x)(1 + x)^2(1 + x^2)$ and the coefficients of x^0, x and x^2 are 1, 1, 0. The class is odd and the associated characters are -1, -1 The remaining polynomials are $(1 - x)(1 + x)$ $(1 + x^3)$; $(1 - x)(1 + x^2)^2$ and $(1 - x)(1 + x^4)$. All of the characters are given in Table 4. We have added the number of elements in each class and indicated by signs the even and odd classes.

TABLE 4

Class	$(1^4)^+$	$(1^2,2)^-$	$(1,3)^+$	$(2^2)^+$	$(4)^-$
No. of Elements	1	6	8	3	6
$\Gamma^{(1)}$	1	1	1	1	1
$\Gamma^{(2)}$	3	1	0	-1	-1
$\Gamma^{(3)}$	2	0	-1	2	0
$\Gamma^{(4)} = \bar{\Gamma}^{(2)}$	3	-1	0	-1	1
$\Gamma^{(5)} = \bar{\Gamma}^{(1)}$	1	-1	1	1	-1

15.13. The Alternating Group.—If two elements A_i and A_j of the symmetric group are in the same class, it does not follow that they will belong to the same class of the *alternating* group. Any even class of the symmetric group which contains none or one cycle of odd order or no cycles of even order will split into two classes in the alternating group, each of the new classes containing half as many elements as it contained in the symmetric group. For example A and B of (5) belong to the same class of the symmetric group with $n = 3$, but to different classes of the alternating group, as may be verified from (2).

The characters of the symmetric group which are not self-associated are also characters of the alternating group. Every character of a self-associated representation is the sum of two equal characters for the alternating group except for the two classes which have been obtained by splitting a class of the symmetric group. Thus if $n = 3$ or 4 and the character table is known for the symmetric group, we can fill the character table for the alternating group except for four blank spaces. Suppose the two classes whose entry in the table is blank are obtained from the partition $(\lambda_1, \lambda_2, \lambda_3, \cdots)$. Then if $\mu = \lambda_1 \lambda_2 \lambda_3 \cdots$, the character $(-1)^{(\mu-1)/2}$ will occur in the symmetric group at the intersection of the row corresponding to the self-associated representation and the column of the class in question while in the alternating group we will have

$$\frac{(-1)^{(\mu-1)/2} \pm \sqrt{\mu} \cdot i^{(\mu-1)/2}}{2} \tag{15-23}$$

The two remaining vacant places in the table are filled by interchanging the two characters given by (23).

For $n = 4$, there are 4 classes since $(1,3)^+$ splits into $(1,3)'$ and $(1,3)''$. The self-associated representation is $\Gamma^{(3)}$. Its characters become $(1,x,y,1)$ and $(1,y,x,1)$ where x and y obtained from (23) are $(-1 \pm i\sqrt{3})/2$ since $\lambda_1 = 1$, $\lambda_2 = 3$, $\mu = 3$. Writing $\epsilon = \exp(2\pi i/3)$, we thus have $x = \epsilon$, $y = \epsilon^2$. This completes the calculation as shown in Table 5.

TABLE 5

Class	(1^4)	$(1,3)'$	$(1,3)''$	(2^2)
No. of Elements	1	4	4	3
$\Gamma^{(1)}$	1	1	1	1
$\Gamma^{(2)}$	3	0	0	-1
$\Gamma^{(3)}$	1	ϵ	ϵ^2	1
$\Gamma^{(4)}$	1	ϵ^2	ϵ	1

15.14. The Unitary Group.—The collection of all *non-singular* matrices of order n, with matrix multiplication as the law of combination, is the representation of a group called the *full linear group* (**FLG**). The order of the group is infinite, for its elements are the infinite number of linear transformations that change a vector **x** into a new vector. This group has many subgroups obtained by imposing certain restrictions on the matrices of its transformations. Thus, we might exclude all matrices except those with determinant equal to ± 1 or we might require that the matrices be orthogonal. Such groups are *discrete*, if the elements are infinitely denumerable (an example of a discrete group of this type is given in sec. 15.1); *continuous*, if the elements are non-denumerable. An example is the group of rotations about an axis. One may also have *mixed-continuous* groups such as $\mathbf{R}^{\pm}(2)$ discussed in sec. 15.16. Infinite groups have many of the properties of finite groups, although naturally some modifications[13] in their treatment are necessary.

[13] See, for example, Wigner, loc. cit., Chapter X.

We first consider a subgroup of **FLG**, which is called the *two-dimensional unimodular unitary group* (**SUG**, *special unitary group*). Its elements are square unitary matrices of order two with determinant of $+1$. Let us take a matrix

$$\begin{bmatrix} a & b \\ c & d \end{bmatrix}$$

and modify it so that these conditions are met. Referring to eq. (10–50), we see that we must have $c = -b^*$ and $d = a^*$. Thus a typical element of **SUG** is

$$U = \begin{bmatrix} a & b \\ -b^* & a^* \end{bmatrix}; \quad |\, U \,| = aa^* + bb^* = 1 \qquad (15\text{-}24)$$

When this matrix is applied to a column vector $\mathbf{x} = \{x_1, x_2\}$ so that $U\mathbf{x} = \mathbf{x}'$, we have

$$\begin{aligned} x_1' &= ax_1 + bx_2 \\ x_2' &= -b^*x_1 + a^*x_2 \end{aligned} \qquad (15\text{-}25)$$

It will also transform any function of $\mathbf{x}$ into a linear combination of x_1, x_2; for example,

$$Uf(\mathbf{x}) = f(\mathbf{x}') = f(ax_1 + bx_2, -b^*x_1 + a^*x_2) \qquad (15\text{-}26)$$

Thus if U operates on a set of $(n + 1)$ homogeneous products

$$f_p^{(n)} = x_1^p x_2^{n-p}; \quad (p = 0, 1, 2, \cdots, n) \qquad (15\text{-}27)$$

the result is a homogeneous polynomial of the same degree

$$\begin{aligned} Uf_p^{(n)} &= (ax_1 + bx_2)^p(-b^*x_1 + a^*x_2)^{n-p} \\ &= \sum_{k=0}^{n} U_{pk}^{(n)} x_1^k x_2^{n-k} \end{aligned} \qquad (15\text{-}28)$$

Clearly, the two-dimensional matrices U are themselves representations of **SUG**. But the matrices with elements $U_{pk}^{(n)}$, being isomorphous with U because of eq. (28), must also be representations, provided we can show that they are unitary. As a matter of fact, they are not unitary, but if each element is multiplied by $[p!(n - p)!]^{-1/2}$, they become so. Multiplication of the elements by this constant factor is, of course, equivalent to multiplying $f_p^{(n)}$ by the same quantity. When we do this, we find it convenient to set $n = 2j$; $p = j + m$. The purpose of the latter substitution is to enable us to prove in sec. 15.15 that **SUG** is isomorphous with the three-dimensional rotation group.

When these changes have been made, $f_p^{(n)}$ becomes

$$f_m^{(j)} = \frac{x_1^{j+m} x_2^{j-m}}{\sqrt{(j + m)!(j - m)!}} \qquad (15\text{-}29)$$

where $j = 0, \frac{1}{2}, 1, \frac{3}{2}, \cdots;$ $m = -j, -j + 1, \cdots, j - 1, j$. At the same time, eq. (28) becomes

$$Uf_m^{(j)} = \frac{(ax_1 + bx_2)^{j+m}(-b^*x_1 + a^*x_2)^{j-m}}{\sqrt{(j + m)!(j - m)!}}$$

$$= \sum_{q=-j}^{j} U_{mq}^{(j)}f_q^{(j)} \tag{15-30}$$

The resulting matrices whose elements are $U_{mq}^{(j)}$ will be indicated by $U^{(j)}$. They are unitary and irreducible; furthermore, there are no other irreducible representations[14] of **SUG**.

In order to obtain the elements of $U^{(j)}$, we develop (30) by the binomial theorem and pick out the coefficient of $f_q^{(j)}$. It is found to be

$$U_{mq}^{(j)} = \sum_{t} \frac{(-1)^t\sqrt{(j + m)!(j - m)!(j + q)!(j - q)!}}{(j - m - t)!(j + q - t)!(t - q + m)!t!}$$
$$\times \, a^{j+q-t}a^{*j-m-t}b^{t-q+m}b^{*t} \tag{15-31}$$

In this expression, t takes the values $0, 1, 2, \cdots$ and the summation breaks off automatically when negative powers of the a's and b's appear because the denominator will then contain the factor $(-)!$ which is ∞.

Since m and q have $(2j + 1)$ possible values, it follows that the matrices of the representations have dimensions of $(2j + 1)$. If $j = 0$, $U^{(0)} = 1$. If $j = \frac{1}{2}$, m and q can take the values $\pm\frac{1}{2}$, hence if the elements of the matrix are characterized by $+\frac{1}{2}$ and $-\frac{1}{2}$, in that order, we have $U^{(1/2)}$ identical with U of (24).

In order to determine the characters of **SUG** let us select a typical matrix of the group and transform it to diagonal form. A unitary transformation is required and it is certain that among the infinite number of unitary matrices in the group, one may be found, say V, that will effect the diagonalization

$$V^{-1}UV = U_1 = \begin{bmatrix} a' & 0 \\ 0 & a'^* \end{bmatrix} \tag{15-32}$$

Finally, since we require $|U_1| = 1$, the coefficients of U_1 may be determined[15]

$$U_1 = \begin{bmatrix} e^{i\phi/2} & 0 \\ 0 & e^{-i\phi/2} \end{bmatrix} \tag{15-33}$$

All other matrices of the group belong to the same class as U and U_1, for the class is composed of elements which are obtained from each other by

---

[14] The proof of these facts will be found in Murnaghan, loc. cit., Chapter 3 or Wigner, loc. cit., Chapter XV.

[15] The reason for choosing $e^{i\phi/2}$ instead of $e^{i\phi}$ will become apparent in sec. 15.15.

similarity transformations or, in this particular case, by unitary transformations. Since each matrix is unitary it remains unitary when it undergoes such a transformation. We also know that it is only necessary to calculate the character of one element from a class; thus using U_1

$$\chi^{(1/2)} = e^{i\phi/2} + e^{-i\phi/2}$$

is the character for the representation of degree $(2j + 1) = 2$.

Now the matrices $U^{(j)}$, which are identical with U_1 when $j = \frac{1}{2}$, must be transformable in such a way that the characters will be identical for $j = \frac{1}{2}$, and when this is done the characters should apply to **SUG** for any value of j. If we substitute $a = e^{i\phi/2}$, $b = 0$ in (31), the result is of diagonal form since all elements disappear unless $t = 0$ and $m = q$,

$$U^{(j)}_{mq} = e^{im\phi}\delta_{mq} \tag{15-34}$$

The required characters for **SUG**, infinite in number, are thus

$$\chi^{(j)} = \sum_{m=-j}^{j} e^{im\phi} \tag{15-35}$$

A simpler form of the last expression may be obtained as follows. Let $\rho = e^{i\phi}$ so that

$$\chi^{(j)} = e^{-ij\phi}(1 + \rho + \rho^2 + \cdots + \rho^{2j}) = e^{-ij\phi}\frac{(1 - \rho^{2j+1})}{(1 - \rho)}$$

Multiply numerator and denominator by $e^{-i\phi/2}$ and use the relation $\sin x = i(e^{-ix} - e^{ix})/2$; then

$$\chi^{(j)} = \frac{\sin (2j + 1)\phi/2}{\sin \phi/2} \tag{15-36}$$

The irreducible representations and characters satisfy certain orthogonality and normalization conditions[16] as in the case of finite groups, but the summations in (9) and (13) are replaced by integrals.

15.15. The Three-Dimensional Rotation Groups.—Another important subgroup of **FLG** (as well as of the n dimensional unitary group) is the n dimensional full, real orthogonal group which consists of all unitary matrices with real elements. If we further restrict this subgroup, choosing all real unitary matrices with determinant equal to $+1$, we have the n-dimensional proper, real orthogonal group or the *rotation group*. It should be remembered that an orthogonal matrix need not be unitary but a real orthogonal matrix and a real unitary matrix are synonymous terms. For the moment, we consider the *three-dimensional rotation group* $R^+(3)$ whose elements are real orthogonal matrices of order three.

---

[16] See Wigner, loc. cit., Chapter XV or Eckart, Carl, *Rev. Mod. Phys.* **2**, 344 (1930).

Assume that we have a sphere of unit radius, the center of which coincides with the origin of a coordinate system $OXYZ$ fixed in space. Now let the coordinate of some point on the surface of the sphere be (x,y,z) and rotate the sphere in any manner whatsoever leaving its center fixed. The new coordinates of the point (x',y',z') will be related to (x,y,z) by some matrix $R(\alpha,\beta,\gamma)$ which is an element of $\mathbf{R}^+(3)$. As we have shown in sec. 9.5, such a rotation may be factored into a product of three plane rotations described by the Eulerian angles (α,β,γ); i.e., we may write

$$R(\alpha,\beta,\gamma) = R_z(\gamma)R_x(\beta)R_z(\alpha) \tag{15-37}$$

where $\mathbf{R}_z$ and $\mathbf{R}_x$ are rotations about the Z- and X-axes respectively.

In order to find the representations of $\mathbf{R}^+(3)$ we could use a method similar to that of sec. 15.14 and study the effect of transforming a function of (x,y,z) by the elements of the group. A simpler method[17] is available for we will show that $\mathbf{R}^+(3)$ is isomorphous with $\mathbf{SUG}$. Since we know the representations of the latter, we may use the same results for $\mathbf{R}^+(3)$. We recall, however, that the elements of $\mathbf{SUG}$ are two-dimensional matrices while the elements of $\mathbf{R}^+(3)$ are three-dimensional, hence the proof of the isomorphism depends upon finding some relation between these two kinds of matrices. The problem is an old one which occurred in classical mechanics; it was solved by Klein and by Cayley, who made use of a special kind of transformation in the complex plane.[18] We prefer to proceed in another way.

We first observe that any two-dimensional matrix may be written as a linear combination of the four matrices[19]

$$P_1 = \begin{bmatrix} 0 & 1 \\ 1 & 0 \end{bmatrix}; \quad P_2 = \begin{bmatrix} 0 & -i \\ i & 0 \end{bmatrix}; \quad P_3 = \begin{bmatrix} 1 & 0 \\ 0 & -1 \end{bmatrix}; \quad P_4 = \begin{bmatrix} 1 & 0 \\ 0 & 1 \end{bmatrix}$$
$$\tag{15-38}$$

For example, if

$$H = \begin{bmatrix} H_{11} & H_{12} \\ H_{21} & H_{22} \end{bmatrix}$$

we may write

$$H = c_1 P_1 + c_2 P_2 + c_3 P_3 + c_4 P_4$$

where

$$c_1 = (H_{12} + H_{21})/2; \quad c_2 = i(H_{12} - H_{21})/2$$
$$c_3 = (H_{11} - H_{22})/2; \quad c_4 = (H_{11} + H_{22})/2$$

[17] Both methods are discussed by Wigner, loc. cit., Chapter XV.

[18] The details are given by Whittaker, E. T., " Analytical Dynamics," Third Edition, Cambridge University Press, 1927, p. 12. The quantities a and b which appear in our eq. (24) are identical with the Cayley-Klein parameters. Eckart, loc. cit., and Bauer, loc. cit., have used a similar method in the group theory problem.

[19] The first three of these are the Pauli spin matrices, discussed in sec. 11.29.

Let us take $c_1 = x$, $c_2 = y$, $c_3 = z$, $c_4 = 0$. Then we have,

$$H(x,y,z) = xP_1 + yP_2 + zP_3$$

$$= \begin{bmatrix} z & x + iy \\ x - iy & -z \end{bmatrix} \tag{15-39}$$

Clearly if x, y, z are real, H is Hermitian. Moreover, its trace is zero; in fact, any two dimensional matrix with trace of zero may be put into this form, P_4 not being needed. If H is now subjected to a unitary transformation by the matrix U of (24) its trace is unchanged and we obtain

$$H'(x',y',z') = U^\dagger HU = x'P_1 + y'P_2 + z'P_3 \tag{15-40}$$

If we can prove that the relation between x, y, z and x', y', z' is a rotation, we may conclude that the matrices U of **SUG** perform the same transformations as the matrices of the group $\mathbf{R}^+(3)$ and that the two groups are isomorphous. To do this, we note (see the problem in sec. 10.14) that $\mid H \mid = \mid H' \mid$, hence

$$x^2 + y^2 + z^2 = x'^2 + y'^2 + z'^2 \tag{15-41}$$

which means that the length of a vector is unchanged by the transformation of eq. (40) and the latter must be a rotation.

Let us study some special forms of the matrix U whose general form is given by eq. (24). We first put $a = e^{i\alpha/2}$, $b = 0$, that is, we use U_1, the diagonal matrix of eq. (33). We easily find

$$U_1^\dagger P_1 U_1 = \cos \alpha P_1 + \sin \alpha P_2$$

$$U_1^\dagger P_2 U_1 = -\sin \alpha P_1 + \cos \alpha P_2 \tag{15-42}$$

$$U_1^\dagger P_3 U_1 = P_3$$

With these results, (40) becomes

$$x' = x \cos \alpha + y \sin \alpha$$

$$y' = -x \sin \alpha + y \cos \alpha$$

$$z' = z$$

This clearly represents a rotation through an angle α about Z; it may be suitably represented by

$$\mathbf{r}' = R_z(\alpha)\mathbf{r}$$

where $\mathbf{r}'$ and $\mathbf{r}$ are the vectors having components (x',y',z') and (x,y,z), respectively and

$$R_z(\alpha) = \begin{bmatrix} \cos \alpha & \sin \alpha & 0 \\ -\sin \alpha & \cos \alpha & 0 \\ 0 & 0 & 1 \end{bmatrix} \tag{15-43}$$

We have thus identified the element of **SUG** which corresponds to the last factor on the right of (37); it is U_1. Obviously, $R_z(\gamma)$ corresponds to a matrix like (43) but with α replaced by γ.

In order to find $R_x(\beta)$ we take

$$U_2 = \begin{bmatrix} \cos \beta/2 & i \sin \beta/2 \\ i \sin \beta/2 & \cos \beta/2 \end{bmatrix} \qquad (15\text{--}44)$$

It is obtained by putting $a = \cos \beta/2$, $b = i \sin \beta/2$ in (24). We now find

$$U_2^\dagger P_1 U_2 = P_1$$

$$U_2^\dagger P_2 U_2 = \cos \beta P_2 + \sin \beta P_3$$

$$U_2^\dagger P_3 U_2 = -\sin \beta P_2 + \cos \beta P_3$$

and (40) may be written $\mathbf{r}' = R_x(\beta)\mathbf{r}$, where

$$R_x(\beta) = \begin{bmatrix} 1 & 0 & 0 \\ 0 & \cos \beta & \sin \beta \\ 0 & -\sin \beta & \cos \beta \end{bmatrix} \qquad (15\text{--}45)$$

Our notation, $R_x(\beta)$, is meant to exhibit the fact that (45) represents a rotation through β about X. Thus we have shown that by a proper choice of the elements of U, **SUG** and $\mathbf{R}^+(3)$ are isomorphous since $U = U_1(\gamma)U_2(\beta)U_1(\alpha)$ corresponds to $R(\alpha,\beta,\gamma)$.

Let us write $U(\alpha,\beta,\gamma) = U_1(\gamma)\ U_2(\beta)U_1(\alpha)$

$$= \begin{bmatrix} e^{i\gamma/2} & 0 \\ 0 & e^{-i\gamma/2} \end{bmatrix}\begin{bmatrix} \cos \beta/2 & i \sin \beta/2 \\ i \sin \beta/2 & \cos \beta/2 \end{bmatrix}\begin{bmatrix} e^{i\alpha/2} & 0 \\ 0 & e^{-i\alpha/2} \end{bmatrix}$$

$$= \begin{bmatrix} e^{i(\alpha+\gamma)/2} \cos \beta/2 & ie^{-i(\alpha-\gamma)/2} \sin \beta/2 \\ ie^{i(\alpha-\gamma)/2} \sin \beta/2 & e^{-i(\alpha+\gamma)/2} \cos \beta/2 \end{bmatrix} \qquad (15\text{--}46)$$

On comparing this with (24), we see that we have $a = e^{i(\alpha+\gamma)/2} \cos \beta/2$ and $b = ie^{-i(\alpha-\gamma)/2} \sin \beta/2$, so that (31) becomes

$$U_{mq}^{(j)}(\alpha,\beta,\gamma) = \sum_t \frac{(-i)^{t-q+m}\sqrt{(j+m)!(j-m)!(j+q)!(j-q)!}}{(j-m-t)!(j+q-t)!(t-q+m)!t!}$$

$$\times e^{iq\alpha} \cos^{2j-m+q-2t}\beta/2 \cdot \sin^{m-q+2t}\beta/2 \cdot e^{im\gamma} \qquad (15\text{--}47)$$

As before, $j = 0, \frac{1}{2}, 1, \frac{3}{2}, \cdots$. For $j = 0$, we get $U^{(0)}(\alpha,\beta,\gamma) = 1$; for $j = \frac{1}{2}$, we obtain (46). It may be shown[20] that the matrices whose elements are given by (47) are irreducible representations and that there are no further ones. The characters of the representations are found from

[20] See footnote 17.

(35).   Remembering that $e^{\pm ix} = \cos x \pm i \sin x$, they may be written as

$$\chi^{(j)}(\alpha) = 1 + 2 \cos \alpha + \cdots + 2 \cos j\alpha; \quad \text{if } j = 0, 1, 2, \cdots$$
$$\chi^{(j)}(\alpha) = 2 \cos \alpha/2 + 2 \cos 3\alpha/2 + \cdots + 2 \cos j\alpha;$$
$$\text{if } j = \tfrac{1}{2}, \tfrac{3}{2}, \cdots \qquad (15\text{–}48)$$

Although $\mathbf{R}^+(3)$ is isomorphous with $\mathbf{SUG}$, the isomorphism is not simple.   If $0 \le \alpha \le 4\pi$, $0 \le \beta \le \pi$, $0 \le \gamma \le 2\pi$, then as α, β and γ take all values between these limits, a and b of (24) will take all pairs of values satisfying the requirement $aa^* + bb^* = 1$ once only.   On the other hand, if α, β and γ are Eulerian angles their limits are $0 \le \alpha \le 2\pi$, $0 \le \beta \le \pi$, $0 \le \gamma \le 2\pi$.   But the angles occur in (46) divided by 2, hence the trigonometric functions are undetermined with regard to sign.   In other words, every matrix $R(\alpha,\beta,\gamma)$ is isomorphous with two matrices $U(\alpha,\beta,\gamma)$.   We must thus discard half of the representations of $\mathbf{SUG}$ in order to find the ones appropriate to $\mathbf{R}^+(3)$.   It is easy to see which ones we want.   From (47) it follows that

$$U_{mq}^{(j)}(\alpha + 2\pi,\beta,\gamma) = e^{2\pi iq} U_{mq}^{(j)}(\alpha,\beta,\gamma)$$

Now when j is integral, q is also integral, for $-j \le q \le j$ and then $e^{2\pi iq} = 1$. If j were half integral, the identical rotations α and $\alpha + 2\pi$ would have representations differing in sign.   However, $R(\alpha,\beta,\gamma)U(\alpha,\beta,\gamma)$ for both integral and half-integral j values is a group which is isomorphous with $U(\alpha,\beta,\gamma)$ and all matrices $U^{(j)}(\alpha,\beta,\gamma)$ are representations.   This group is of importance in the Pauli spin theory.[21]

If we take as elements of an infinite group, all real unitary matrices of order three with determinant equal to $+1$ as well as -1, we have the *three-dimensional full real orthogonal group* $\mathbf{R}^{\pm}(3)$.   The quotient group isomorphous with it has two elements.   The unit element, which is also an invariant subgroup of $\mathbf{R}^{\pm}(3)$ contains E and all proper rotations R such as (43) or (45) with $|R| = +1$.   The other element of the quotient group is an infinite number of improper rotations T with $|T| = -1$, a typical one (cf. sec. 10.17) being

$$T_z(\phi) = \begin{bmatrix} \cos\phi & \sin\phi & 0 \\ -\sin\phi & \cos\phi & 0 \\ 0 & 0 & -1 \end{bmatrix} \qquad (15\text{–}49)$$

The simplest member of the class of T is an improper rotation by the angle π, the operation called *inversion*

$$T(\pi) = I = \begin{bmatrix} -1 & 0 & 0 \\ 0 & -1 & 0 \\ 0 & 0 & -1 \end{bmatrix} \qquad (15\text{–}50)$$

[21] Cf. Chapter 11.

It is always possible to find some improper rotation T which will convert any other improper rotation T' into an inversion, $T^{-1}T'T = I$, just as it is always possible to find an inverse to a proper rotation, $R^{-1}R'R = E$. The group $\mathbf{R}^{\pm}(3)$ may thus be considered as the direct product of $\mathbf{R}^{+}(3)$ and the group $\mathbf{I}$, the latter having elements E and I. It will have two irreducible representations for every value of j, each being of dimension $(2j + 1)$. The element R has two representations both equal to $U^{(j)}$ while T has representations, $\pm U^{(j)}$.

15.16. The Two-Dimensional Rotation Groups.—The *two-dimensional pure rotation group* $\mathbf{R}^{+}(2)$ is a subgroup of $\mathbf{R}^{+}(3)$. Its elements are the proper rotations in a plane perpendicular to a fixed axis. Let $R(\phi)$ be one of the elements where $0 \leq \phi \leq 2\pi$, then if $\mathbf{x}$ is a vector with components x_1 and x_2, the element $R(\phi)$ may be represented by the matrix $\mathbf{C}(\phi)$

$$\mathbf{x}' = \mathbf{C}(\phi)\mathbf{x}; \quad 0 \leq \phi \leq 2\pi \tag{15-51}$$

with

$$\mathbf{C}(\phi) = \begin{bmatrix} \cos \phi & \sin \phi \\ -\sin \phi & \cos \phi \end{bmatrix} \tag{15-52}$$

If $R(\phi')$ is another element of the group, which is represented by $\mathbf{C}(\phi')$, then

$$\mathbf{C}(\phi)\mathbf{C}(\phi') = \mathbf{C}(\phi + \phi') = \mathbf{C}(\phi')\mathbf{C}(\phi) \tag{15-53}$$

and the group is Abelian. Referring to sec. 15.11, we see that for such groups, each element is in a class by itself and the irreducible representations are one-dimensional. Thus (52) is reducible, a unitary matrix of eigenvectors of $\mathbf{C}$ being required for that purpose since $\mathbf{C}$ itself is an orthogonal matrix. The normalized eigenvectors of $\mathbf{C}$ are found to be

$$\mathbf{u}_1 = \frac{1}{\sqrt{2}}\{1,i\}; \quad \mathbf{u}_2 = \frac{1}{\sqrt{2}}\{1,-i\} \tag{15-54}$$

and the eigenvalues are $e^{\pm i\phi}$. These, then, are characters of an irreducible representation. However, there are an infinite number of classes, so there must be an infinite number of representations for each class. The corresponding characters may be taken as

$$\chi^{(m)} = e^{im\phi}; \quad m = 0, \pm 1, \pm 2, \cdots \tag{15-55}$$

for each will satisfy the multiplication requirement of the group, as indicated by eq. (53).

The *two-dimensional rotary reflection group* $\mathbf{R}^{\pm}(2)$ is composed of both *proper* and *improper* rotations. A typical element of it is represented by the matrix

$$\mathbf{A}(\phi,d) = \begin{bmatrix} \cos \phi & \sin \phi \\ -d\sin \phi & d\cos \phi \end{bmatrix} \tag{15-56}$$

where d equals the determinant of $A(\phi,d)$ and may be either $+1$ or -1. If $d = +1$, we have a proper rotation with matrix $C(\phi)$; if $d = -1$, an improper rotation, the matrix of which will be indicated by $S(\phi)$

$$S(\phi) = \begin{bmatrix} \cos \phi & \sin \phi \\ \sin \phi & -\cos \phi \end{bmatrix} \tag{15-57}$$

Clearly,

$$S^2(\phi) = S(\phi)S(\phi) = E \tag{15-58}$$

but the group is not Abelian for $C(\phi)$ and $S(\phi)$ do not commute. In fact,

$$A(\phi,d)A(\phi',d') = A(d'\phi + \phi',dd') \tag{15-59}$$

Let us reduce $S(\phi)$ to diagonal form (cf. sec. 10.17). Its eigenvectors are found to be

$$\mathbf{v}_1 = \{\cos \phi/2, -\sin \phi/2\}; \quad \mathbf{v}_2 = \{\sin \phi/2, \cos \phi/2\} \tag{15-60}$$

and the eigenvalues are ± 1. The resulting diagonal matrix,

$$\sigma = \begin{bmatrix} 1 & 0 \\ 0 & -1 \end{bmatrix} \tag{15-61}$$

which corresponds to a reflection through the axis of rotation, is that obtained from $S(\phi)$ when ϕ equals 0 or 2π. It is interesting to observe that the matrix of eigenvectors, eq. (60), is actually a proper rotation by the angle $\phi/2$. Moreover,

$$S(\phi) = C^{-1}(\phi/2)\sigma C(\phi/2) \tag{15-62}$$

hence, an improper rotation is equivalent to a proper rotation by the angle $\phi/2$, followed by a reflection and finally by a proper rotation of $\phi/2$ in the opposite direction.

It will be remembered that every element of the group $\mathbf{R}^+(2)$ is in a class by itself. It does not follow, however, that the proper rotations of $\mathbf{R}^\pm(2)$ are each in a separate class. Thus the element represented by $C(\phi)$ is in the same class with the element represented by $C'(\phi)$, since

$$S^{-1}(\phi')C(\phi)S(\phi') = C'(\phi)$$

where

$$C'(\phi) = \begin{bmatrix} \cos \phi & -\sin \phi \\ \sin \phi & \cos \phi \end{bmatrix}$$

and

$$C'(\phi) = C^{-1}(\phi) = C(-\phi)$$

There are an infinite number of classes as before but each class contains the proper rotation by ϕ and the proper rotation by $-\phi$.

On the other hand, all improper rotations are in the same class. If

$S(\phi)$ and $S(\phi')$ are representations of two improper rotations, we find that

$$C^{-1}(\phi'')S(\phi)C(\phi'') = S(\phi') \qquad (15\text{-}62a)$$

where $\phi' = \phi + 2\phi''$. This could have been inferred from eq. (62), for it is a special case of (62a) when we set $\phi'' = \phi/2$, $\phi = 0$ or 2π and $\phi' = \phi$.

If the representations of eq. (56) are transformed by the matrix of eigenvectors (54), the result is

$$\Gamma^{(m)}(\phi,1) = C^{(m)}(\phi) = \begin{bmatrix} e^{im\phi} & 0 \\ 0 & e^{-im\phi} \end{bmatrix} \qquad (15\text{-}63)$$

$$\Gamma^{(m)}(\phi,-1) = S^{(m)}(\phi) = \begin{bmatrix} 0 & e^{-im\phi} \\ e^{im\phi} & 0 \end{bmatrix} \qquad (15\text{-}64)$$

with $m = 1$. However, when $m = 0, 1, 2, \cdots$ the same matrices also satisfy the multiplication requirements of the group. They are irreducible except when $m = 0$. There, we obtain (see problem at the end of this section).

$$\begin{aligned} C^{(0)}(\phi) &= 1; \quad S^{(0)}(\phi) = 1 \\ C^{(0')}(\phi) &= 1; \quad S^{(0')}(\phi) = -1 \end{aligned} \qquad (15\text{-}65)$$

A slightly different procedure is sometimes desirable. We see from eq. (62) that any improper rotation may always be written as a combination of a proper rotation and a reflection. The elements of the group could thus be considered as an infinite number of proper rotations and the single improper rotation which is represented by σ. When the latter is transformed by means of (54) we obtain

$$\Gamma(0,-1) = \begin{bmatrix} 0 & 1 \\ 1 & 0 \end{bmatrix} \qquad (15\text{-}64a)$$

Thus the irreducible representations are those of (63) and the single one of (64a). When $m = 0$, we again get (65).

Problem. Show that both (63) and (64) may be reduced to diagonal form with the matrix $\begin{bmatrix} 1 & 1 \\ 1 & -1 \end{bmatrix}$.

15.17. The Dihedral Groups.—An important subgroup of $\mathbf{R}^{\pm}(2)$ is obtained by restricting the values of ϕ. Consider a regular polygon in the XY-plane with coordinates of the n corners

$$x_k = r\cos 2\pi k/n; \quad y_k = r\sin 2\pi k/n; \quad (k = 0, 1, 2, \cdots, n-1)$$

where r is the radius vector from the origin to the corner. Now if ϕ in (56) takes the value $2\pi/n$, the matrix $A(\phi,d)$ will transform the polygon into itself by either a proper or an improper rotation. The elements of the group will be indicated by C in the former case and by S in the latter. The corresponding matrices are $A(2\pi/n,d)$ with the appropriate choice of d, but we will find it convenient again to use C and S for the matrices, distinguishing between the abstract element and its representation by means of different type. The whole group is finite and of order $2n$; it is called the *dihedral group* $\mathbf{D}_n$. It may be generated by the relations

$$C^n = E; \quad S^2 = E; \quad SC = C^{-1}S \tag{15-66}$$

We now see why the group of sec. 15.1 was called $\mathbf{D}_3$. If we let $n = 3$ in eq. (66), we will have the generating relation of eq. (1), provided we reletter the elements C and S of (66) so that they read A and C, respectively.

Suppose q is an integer; then we may write $n = 2q + 1$ if n is odd or $n = 2q$ if n is even. There will be $(q + 1)$ classes among the proper rotations for both n even and n odd. These will correspond to C, C^2, C^3, $\cdots$, C^q, $C^n = E$. For n odd, there will be one additional class, that of S. For n even, there will be two classes involving an improper rotation. The separation into classes for both cases is illustrated in the problem in this section.

If n is odd, the classes for proper rotations will be represented by $\mathbf{C}^{(0)}$ and $\mathbf{C}^{(0')}$ of eq. (65) and q matrices of (63) with $m = 1, 2, \cdots, q$. The

TABLE 6

n odd; $q = (n - 1)/2$; $\phi = 2\pi/n$

	$\mathcal{C}(E)$	$\mathcal{C}(C)$	$\cdots$	$\cdots$	(C^q)	$\mathcal{C}(S)$
$\Gamma^{(0)}$	1	1	$\cdots$	$\cdots$	1	1
$\Gamma^{(0')}$	1	1	$\cdots$	$\cdots$	1	-1
$\Gamma^{(1)}$	2	$2\cos\phi$	$\cdots$	$\cdots$	$2\cos q\phi$	0
$\Gamma^{(2)}$	2	$2\cos 2\phi$	$\cdots$	$\cdots$	$2\cos 2q\phi$	0
$\cdots$	$\cdots$	$\cdots\cdots$	$\cdots$	$\cdots$	$\cdots\cdots$	$\cdots$
$\Gamma^{(q)}$	2	$2\cos q\phi$	$\cdots$	$\cdots$	$2\cos q^2\phi$	0

n even; $q = n/2$

	$\mathcal{C}(E)$	$\mathcal{C}(C)$	$\cdots$	$\mathcal{C}(C^{q-1})$	$\mathcal{C}(C^q)$	$\mathcal{C}(S)$	$\mathcal{C}(S')$
$\Gamma^{(0)}$	1	1	$\cdots$	1	1	1	1
$\Gamma^{(0)}$	1	1	$\cdots$	1	1	-1	-1
$\Gamma^{(q)}$	1	-1	$\cdots$	$(-1)^{q-1}$	$(-1)^q$	1	-1
$\Gamma^{(q')}$	1	-1	$\cdots$	$(-1)^{q-1}$	$(-1)^q$	-1	1
$\Gamma^{(1)}$	2	$2\cos\phi$	$\cdots$	$2\cos(q-1)\phi$	$2\cos q\phi$	0	0
$\Gamma^{(2)}$	2	$2\cos 2\phi$	$\cdots$	$2\cos 2(q-1)\phi$	$2\cos 2q\phi$	0	0
$\cdots$	$\cdots$	$\cdots$	$\cdots$	$\cdots$	$\cdots$	$\cdots$	$\cdots$
$\Gamma^{(q-1)}$	2	$2\cos(q-1)\phi$	$\cdots$	$2\cos(q-1)^2\phi$	$2\cos q(q-1)\phi$	0	0

class of S is represented by the remaining one-dimensional matrices of (65) and q matrices like (64) or (64a). If n is even, the situation is similar, except for the case $m = n/2$ when (63) and (64) become

$$C^{(n/2)}(\pi) = \begin{bmatrix} -1 & 0 \\ 0 & -1 \end{bmatrix}; \quad S^{(n/2)}(\pi) = \begin{bmatrix} 0 & 1 \\ 1 & 0 \end{bmatrix} \qquad (15\text{-}67)$$

This representation may be reduced to give $C^{(q)} = -1$; $S^{(q)} = 1$; $C^{(q')} = -1$; $S^{(q')} = -1$. Hence, when n is even there are four representations of degree one and $(q - 1)$ of degree two. The characters of dihedral groups are shown in Table 6.

Problem. Show that if $n = 6$, the classes of the group are $\mathcal{C}(E) = E$; $\mathcal{C}(C) = C$, C^5; $\mathcal{C}(C^2) = C^2$, C^4; $\mathcal{C}(C^3) = C^3$; $\mathcal{C}(S) = S$, C^2S, C^4S; $\mathcal{C}(S') = CS$, C^3S, C^5S. If $n = 5$, show that the classes are $\mathcal{C}(E) = E$; $\mathcal{C}(C) = C$, C^4; $\mathcal{C}(C^2) = C^2$, C^3; $\mathcal{C}(S) = S$, CS, C^2S, C^3S, C^4S.

15.18. The Crystallographic Point Groups.[22]—By considering all operations which transform certain solid geometric figures into themselves, we obtain a number of finite subgroups of $R^{\pm}(3)$, called the *crystallographic point groups*. They are of considerable importance in the study of crystal and molecular structure. We assume that one point of the figure is fixed in space so that if we know the position of two more points which are not collinear with the fixed point, the position of the figure is completely determined. Under these conditions, the only possible types of motion are rotations around an axis passing through the fixed point and reflections in a plane containing that point. All other motions may be reduced to a combination of these two, for as we have seen in sec. 15.16 any improper rotation may always be written as a product of two proper rotations and a reflection. When the improper rotation is an inversion (i.e., improper rotation by the angle π) a point will be collinear with its original position and some fixed point on the axis of rotation, hence an inversion is uniquely determined by the position of this fixed point and is independent of the position of the axis. The fixed point is called a center of inversion.

We thus have four fundamental operations: (a) a proper rotation C_n by an angle $\phi = 2\pi/n$ (n is an integer) about an n-fold axis of rotation; (b) reflection in a plane, indicated by σ_h, σ_d, σ_v (subscripts h, d and v refer to horizontal, diagonal and vertical planes); (c) an improper rotation, S_n; (d) inversion, indicated by I.

Selected sets of such operations, together with a unit element which leaves every point of a figure unchanged, are the elements of the crystallo-

---

[22] More details about the crystallographic groups are given by Schoenflies, A., " Theorie der Kristallstruktur," Gebrüder Borntraeger, Berlin, 1932, and in other references cited later in this chapter. The geometric arguments given here, which we do not prove, are discussed in detail by Schoenflies; see also, Rosenthal, J. and Murphy, G. M., *Revs. Mod. Phys.* **8**, 317 (1936).

graphic groups. The number of these groups which is of interest is limited by the fact that we need to consider only those types of symmetry which occur in crystals or molecules. It may be shown, from geometric arguments, that crystals in nature may have axes of rotation only for $n = 1, 2, 3, 4, 6$, and this fact restricts the total number of crystallographic point groups to 32. Some gaseous molecules may exist with axes $n = 5, 7, 8$ and the appropriate character tables are readily found. For the linear molecule without a center of symmetry, like CO or HCl, the symmetry group is $\mathbf{C}_{\infty v}$, isomorphous with $\mathbf{D}_{\infty}$ and $\mathbf{R}^{+}(2)$. If a linear molecule has a center of symmetry, like H_2 or C_2H_2, the group is $\mathbf{D}_{\infty h}$.

The crystallographic groups may be generated in an elegant way from group theory considerations but we present the results without proof. Consider first the cyclic groups, designated by $\mathbf{C}_n$ ($n = 1, 2, 3, 4, 6$) and of order n. A new group of order $2n$ may be obtained by adding n two-fold axes of symmetry to $\mathbf{C}_n$, in a plane perpendicular to the principal n-fold axis of the cyclic group. These are the dihedral groups, $\mathbf{D}_n$ ($n = 2, 3, 4, 6$), but only four in number since $\mathbf{D}_1$ duplicates $\mathbf{C}_2$. Two more, containing proper elements of symmetry only, are the cubic groups, $\mathbf{T}$ of order 12 and $\mathbf{O}$ of order 24, having the symmetry of the tetrahedron and the octahedron, respectively.

Of the required 32 groups, 11 have now been found. They contain nothing but proper elements of symmetry, so the remaining 21 groups must contain both proper and improper elements, which could be planes (improper rotations by the angle zero), the inversion (improper rotation by π), or rotary reflections (rotation by the angle $2\pi/n$, followed by a reflection in a plane perpendicular to the axis of rotation). Let us first add horizontal planes of symmetry, σ_h to the groups $\mathbf{C}_n$ and require that they be perpendicular to the principal axis of the proper rotation. The results are $\mathbf{C}_{nh}$ ($n = 1, 2, 3, 4, 6$) of order $2n$, but when n is even a center of symmetry also exists and the group could also be written as the direct product, $\mathbf{C}_{nh} = \mathbf{C}_n \times \mathbf{I}$. When σ_h is added to $\mathbf{D}_n$, we get $\mathbf{D}_{nh}$ ($n = 2, 3, 4, 6$) and again for n even, $\mathbf{D}_{nh} = \mathbf{D}_n \times \mathbf{I}$, but $n = 1$ duplicates $\mathbf{C}_{2h}$. The two cubic groups $\mathbf{T}_h$ and $\mathbf{O}_h$ also have centers of symmetry.

Now add vertical planes of symmetry, σ_v to $\mathbf{C}_n$, through the n-fold axes to get $\mathbf{C}_{nv}$ ($n = 2, 3, 4, 6$) of order $2n$. When $n = 6$, the group is isomorphous with $\mathbf{D}_{3h}$. When $n = 1$, duplication occurs since $\mathbf{C}_{1v}$ and $\mathbf{C}_{1h}$ are identical in configuration, differing only in orientation. Addition of vertical planes of symmetry to $\mathbf{D}_n$ adds nothing new, for the planes would coincide with the existing two-fold axes. However, if the added planes are diagonal, σ_d and if they bisect the angle of the two-fold axes, we get $\mathbf{D}_{nd}$ ($n = 2, 3$) and $\mathbf{T}_d$. When $n = 2$, the group is isomorphic with $\mathbf{C}_{4v}$; when $n = 3$, $\mathbf{D}_{3d} = \mathbf{D}_3 \times \mathbf{I}$; $n = 4$ or 6 would require 8-fold and 12-fold axes of symmetry, hence they are impossible for crystals; $n = 1$ duplicates C_{2v}.

Three more groups will complete the list of 32. Let us try improper rotations, S_n but with n even, for an n-fold improper axis implies a proper axis, C_p $(p = n/2)$. We are thus limited to $n = 2, 4, 6$. When $n = 2$, $S_2 = C_1 \times I$, usually designated C_i; similarly, $S_6 = C_3 \times I = C_{3i}$. There is no center of symmetry, however, for S_4.

For convenient reference, these results are collected in Table 7. The first column contains the proper groups $(\overline{G})$. The cyclic groups are of order n; the dihedral groups of order $2n$; T is of order 12 and O of order 24. Improper groups $(\overline{G})$ on the same line in the table with a proper group (G) have the same order as (G) and the groups are isomorphous with each other. An improper group $(G) \times I$, which is the direct products of (G) and I, has an order twice that of (G). Other relations between the various groups would have been apparent if they had been derived in other possible ways. These relations may also be found by study of appropriate solid models or plane diagrams. Stereographic projections of the solid figures are suitable for such a study.[23]

The symbols given in Table 7 are those generally used in molecular problems and devised by Schoenflies. Some alternative, but lesser used symbols, are shown in parenthesis. The dihedral group, D_2, for example is often called V for "Vierergruppe." It is that of the Cartesian coordinate system with three mutually perpendicular two-fold axes. The meaning of the other alternative forms will be obvious. Crystallographers, unfortunately, have used a bewildering variety of systems[24] for designating the groups.

In the tables at the end of this section we present the characters for all of these groups. It is convenient to indicate a class by means of symbols like C_n, S_n or σ, a typical element of it. If a number precedes the symbol it is the number of elements in that class; otherwise the class in question contains but one element. Representations of degree one are indicated[25] by A or B; of degree two by E (except for certain cases, where two one-dimensional representations occur in pairs); of degree three by T.

---

[23] They are given, for example, by Eyring, H., Walter, J., and Kimball, G. E., "Quantum Chemistry," John Wiley and Sons, Inc., New York, 1944. Easily understood perspective drawings of the group symmetries may be found in Davey, W. P., "A Study of Crystal Structure and Its Applications," McGraw-Hill Book Co., Inc., New York, 1934.

[24] See Davey, loc. cit. for a discussion of these nomenclatures.

[25] Further description of the designation of representations, especially the usage of molecular spectroscopists, is given by Herzberg, G., "Molecular Spectra and Molecular Structure, Vol. II, Infrared and Raman Spectra of Polyatomic Molecules," D. Van Nostrand Co., Inc., New York, 1945.

When two one-dimensional representations A and B occur in the same group, it will be found that the character of A is $+1$ for the class representing rotation by $2\pi/n$ around the principal n-fold axis and -1 for B. The principal axis is always taken in the direction of Z. Different representations of similar symmetry to reflection in a plane perpendicular to the principal axis are indicated by $'$ and $''$ while subscripts g and u refer to positive and negative characters for the class of I.

TABLE **7**

THE CRYSTALLOGRAPHIC POINT GROUPS

Proper Groups (G)	Improper Groups	
	$(\tilde{G})$	$(G) \times I$
Cyclic Groups		
C_1	—	$C_i(S_2)$
C_2	$C_{1h}(C_s)$	C_{2h}
C_3	—	$C_{3i}(S_6)$
C_4	S_4	C_{4h}
C_6	C_{3h}	C_{6h}
Dihedral Groups		
$D_2(V)$	C_{2v}	$D_{2h}(V_h)$
D_3	C_{3v}	D_{3d}
D_4	$C_{4v}; D_{2d}(V_d)$	D_{4h}
D_6	$C_{6v}; D_{3h}$	D_{6h}
Cubic Groups		
T	—	T_h
O	T_d	O_h

Methods of finding the characters[26] for cyclic and dihedral groups have already been described in detail. The cubic groups **O** and **T** are the symmetric and alternating groups on four letters; they have been discussed as examples of permutation groups. The remaining groups, which are indicated as a direct product will have twice as many classes and representations as appear in our tables. Each representation given there will occur once with the subscript g and once with u (except for $\mathbf{C}_{3h}$ where the representations are A', A'', E', E''). For example, $\mathbf{C}_{3i} = \mathbf{C}_3 \times \mathbf{I}$ will have classes E, C_3, C_3^2, I, IC_3, IC_3^2. The classes which are found in $\mathbf{C}_{3i}$ will have the same characters as $\mathbf{C}_3$, once as g and once as u while the new classes will have the same characters as $\mathbf{C}_3$ for g-representations and the negative of those for u-representations. Groups having the same character table are isomorphous.

For convenience of reference, we also include the infinite group $\mathbf{D}_\infty$ which is isomorphous with both $\mathbf{R}^+(2)$ and $\mathbf{C}_{\infty v}$ and the group $\mathbf{D}_{\infty h} = \mathbf{D}_\infty \times \mathbf{I}$ which is isomorphous with $\mathbf{R}^\pm(2)$.

One further question of interest here concerns the transformation properties of a vector when subjected to the operations of a crystallographic group. We have shown, in sec. 15.15, how a vector is transformed by the elements of the group $\mathbf{R}^\pm(3)$. The representations from which this effect is immediately seen are given by (43) and (49), the characters of which are

$$\Xi_R = 1 + 2 \cos \phi; \quad \Xi_T = -1 + 2 \cos \phi$$

The same characters must also apply to the crystallographic groups since they are subgroups of $\mathbf{R}^\pm(3)$, but it does not follow that the characters remain irreducible. As an example, consider the group $\mathbf{C}_4$ where all of the classes involve proper rotations. The angles for the classes of E, C_2, C_4, C_4^3 are 0, π, $\pi/2$, $3\pi/2$, hence $\Xi_R = 3, -1, 1, 1$. Comparison with the character table for $\mathbf{C}_4$ shows that these numbers are the sums of the characters for the representations A and E. The reader should draw a figure of the appropriate symmetry which in this case is a square. Let the Z-axis be perpendicular to the plane of the paper, then it is immediately obvious that z transforms like A for z is unchanged by the operations of the group. When the operation C_2 is applied to the figure (i.e., rotation by π) x is transformed into $-x$ and y into $-y$, hence $x + iy$ becomes $-(x \pm iy)$. Proceeding in this way with the other elements of the group, it will be seen that $x + iy$ transforms like the first set of characters for E in Table 7 and $x - iy$ like the second set. For $\mathbf{S}_4$, the last two classes are improper rotations with $\Xi_T = -1, -1$, hence the reducible characters are $3, -1$, $-1, -1$ or $B + E$; z transforms like B and $x + iy$ like E. We have indi-

[26] The reducible representations of all of the crystallographic groups are given by Seitz, *Z. Kristallographie*, **A88**, 433 (1934).

cated all of these transformation properties in our tables. When two or more groups are isomorphous and the representations are the same (examples, $\mathbf{D_4}$, $\mathbf{C_{4v}}$ and $\mathbf{D_{2d}}$ or $\mathbf{C_4}$ and $\mathbf{S_4}$), the characters for the coordinates refer to the first group of that table. To obtain them for the other groups, one must change the sign of the characters for the improper rotations, for example, z transforms like A_2 for $\mathbf{D_4}$, like A_1 for $\mathbf{C_{4v}}$ and like B_2 for $\mathbf{D_{2d}}$.

TABLE 8

CYCLIC GROUPS

$\mathbf{C_1}$	E
A; x,y,z	1

$\mathbf{C_i}$	$\mathbf{C_2}$	$\mathbf{C_{1h}}$	E E E	I C_2 σ_h
A_g	A; z	A'; x,y	1	1
A_u; x,y,z	B; x,y	A''; z	1	-1

$\mathbf{C_3}$	E	C_3	C_3^2
A; z	1	1	1
E; $x \pm iy$	$\begin{cases} 1 \\ 1 \end{cases}$	ϵ^* ϵ	ϵ ϵ^*

$$\epsilon = e^{2\pi i/3}$$

$$\mathbf{C_{3h}} = \mathbf{C_3} \times \sigma_h; \qquad \mathbf{C_{3i}} = \mathbf{C_3} \times \mathbf{I}$$

$\mathbf{C_4}$ $\mathbf{S_4}$	E E	C_2 C_2	C_4 S_4	C_4^3 S_4^3
A; z	1	1	1	1
B	1	1	-1	-1
E; $x \pm iy$	$\begin{cases} 1 \\ 1 \end{cases}$	-1 -1	$-i$ i	i $-i$

$$\mathbf{C_{4h}} = \mathbf{C_4} \times \mathbf{I}$$

$\mathbf{C_6}$	E	C_6	C_3	C_2	C_3^2	C_6^5
A; z	1	1	1	1	1	1
B	1	-1	1	-1	1	-1
E_2	$\begin{cases} 1 \\ 1 \end{cases}$	$-\epsilon^*$ $-\epsilon$	$-\epsilon$ $-\epsilon^*$	1 1	$-\epsilon^*$ $-\epsilon$	$-\epsilon$ $-\epsilon^*$
E_1; $x \pm iy$	$\begin{cases} 1 \\ 1 \end{cases}$	ϵ^* ϵ	$-\epsilon$ $-\epsilon^*$	-1 -1	$-\epsilon^*$ $-\epsilon$	ϵ ϵ^*

$$\epsilon = e^{2\pi i/6} \qquad \mathbf{C_{6h}} = \mathbf{C_6} \times \mathbf{1}$$

TABLE 8 (*Continued*)

DIHEDRAL GROUPS

$D_2 = V$	C_{2v}	C_{2h}	E E E	C_2^z C_2 C_2	C_2^y σ_v σ_h	C_2^x σ_v' I
A_1	A_1; z	A_g	1	1	1	1
B_3; x	B_2; y	B_g	1	-1	-1	1
B_1; z	A_2	A_u; z	1	1	-1	-1
B_2; y	B_1; x	B_u; $x \pm iy$	1	-1	1	-1

$$D_{2h} = D_2 \times I$$

D_3 C_{3v}	E E	$2C_3$ $2C_3$	$3C_2'$ $3\sigma_v$
A_1	1	1	1
A_2; z	1	1	-1
E; $x \pm iy$	2	-1	0

$$D_{3d} = D_3 \times I$$

D_4 C_{4v} D_{2d}	E E E	C_2 C_2 C_2	$2C_4$ $2C_4$ $2S_4$	$2C_2'$ $2\sigma_v$ $2C_2'$	$2C_2''$ $2\sigma_d$ $2\sigma_d$
A_1	1	1	1	1	1
A_2; z	1	1	1	-1	-1
B_1	1	1	-1	1	-1
B_2	1	1	-1	-1	1
E; $x \pm iy$	2	-2	0	0	0

$$D_{4h} = D_4 \times I$$

D_6	C_{6v}	D_{3h}	E E E	C_2 C_2 σ_h	$2C_3$ $2C_3$ $2C_3$	$2C_6$ $2C_6$ $2S_3$	$3C_2'$ $3\sigma_d$ $3C_2'$	$3C_2''$ $3\sigma_v$ $3\sigma_v$
A_1	A_1; z	A_1'	1	1	1	1	1	1
A_2; z	A_2	A_2'	1	1	1	1	-1	-1
B_1	B_2	A_1''	1	-1	1	-1	1	-1
B_2	B_1	A_2''; z	1	-1	1	-1	-1	1
E_2	E_2	E'; $x \pm iy$	2	2	-1	-1	0	0
E_1; $x \pm iy$	E_1; $x \pm iy$	E''	2	-2	-1	1	0	0

$$D_{6h} = D_6 \times I$$

TABLE 8 (*Continued*)

DIHEDRAL GROUPS (*Continued*)

$\mathbf{D}_\infty$	E	$2C(\phi)$	C_2
$\mathbf{C}_{\infty v}$	E	$2C(\phi)$	σ_v
A_1; z	1	1	1
A_2	1	1	-1
E_1; $x \pm iy$	2	$2\cos\phi$	0
E_2	2	$2\cos 2\phi$	0
...	...	...	...
E_k	2	$2\cos k\phi$	0
...	...	...	...

$$\mathbf{D}_{\infty h} = \mathbf{D}_\infty \times \mathbf{I}$$

CUBIC GROUPS

$\mathbf{T}$	E	$3C_2$	$4C_3$	$4C_3'$
A	1	1	1	1
E	$\begin{cases}1 \\ 1\end{cases}$	$\begin{matrix}1 \\ 1\end{matrix}$	$\begin{matrix}\epsilon \\ \epsilon^*\end{matrix}$	$\begin{matrix}\epsilon^* \\ \epsilon\end{matrix}$
T; x,y,z	3	-1	0	0

$$\epsilon = e^{2\pi i/3} \qquad \mathbf{T}_h = \mathbf{T} \times \mathbf{I}$$

$\mathbf{O}$	E	$8C_3$	$3C_2$	$6C_2$	$6C_4$
$\mathbf{T}_d$	E	$8C_3$	$3C_2$	$6\sigma_d$	$6S_4$
A_1	1	1	1	1	1
A_2	1	1	1	-1	-1
E	2	-1	2	0	0
T_2	3	0	-1	1	-1
T_1; x,y,z	3	0	-1	-1	1

$$\mathbf{O}_h = \mathbf{O} \times \mathbf{I}$$

15.19. Applications of Group Theory.—Since group theory is concerned with symmetry properties, its mathematical methods should be useful in many physical problems. Its most obvious application consists in the classification of crystals and polyatomic molecules according to a group of the appropriate symmetry. It is natural to inquire whether group theory may also be used in quantum mechanics. For systems containing a number of particles, calculations by the usual methods are difficult; hence it is fortunate that the symmetry of such problems can be utilized to some extent in their study.[27]

The Schrödinger equation for a system of n identical particles (electron, protons, etc.) may be written as follows:

$$H(1,2,\cdots n)\psi(1,2,\cdots n) = E\psi(1,2,\cdots n) \qquad (15\text{-}68)$$

where the numbers $1, 2, \cdots n$ appearing in both the Hamiltonian operator H and the state function ψ indicate that these quantities depend on the coordinates of particles $1, 2, \cdots n$. Now it is clear that, if the coordinates of particles i and j are interchanged everywhere in eq. (68), the latter remains valid, for such an exchange amounts to no more than a relabelling of the particles. This fact might be indicated formally by applying to (68) the operator P_{ij} defined as effecting an interchange of particles i and j:

$$P_{ij}H(1,2,\cdots n)P_{ij}\psi(1,2,\cdots n) = EP_{ij}\psi(1,2,\cdots n)$$

But the functional form of H is unaltered when P_{ij} is applied, regardless of its specific form, provided the particles are identical, hence this equation reads

$$HP_{ij}\psi = EP_{ij}\psi$$

In other words, $P_{ij}\psi$ is also an eigenfunction of H, and one belonging to the same eigenvalue E.

Now P_{ij} is an element of the symmetric group on n particles. Therefore the state of affairs described above is usually expressed by saying that the Schrödinger equation is invariant under the symmetric group. Permutations are not the only operations under which the wave equation is invariant. Suppose the nucleus of an atom is considered as a fixed field of force; then rotations and reflections at this point leave the energy of the system unchanged (i.e., the operator H is invariant with respect to them). The groups in question are those of sec. 15.15. If the atom is in a homogeneous electric or magnetic field, the appropriate group is the subgroup of rotations about a fixed axis (see sec. 15.16). For a diatomic molecule, the two nuclei are the centers of force (as a first approximation) and the groups are those of rotation about, and reflection in a plane through, the line joining the nuclei. If the nuclei are identical (as in hydrogen, oxygen or nitrogen) reflections in a plane perpendicular to the internuclear line (i.e., exchange of the nuclei) must also be included. For a polyatomic molecule, the potential energy has the same symmetry as the molecule itself, hence the wave equation is invariant to some one of the crystallo-

---

[27] Such usage has been discussed in detail, especially by Wigner and Weyl in references cited at the end of this chapter. Many of the books listed in Chapter 11 on quantum mechanics, particularly that of Dirac, avoid the formalism of group theory but obtain equivalent results by a more physical procedure.

graphic groups. These examples should be sufficient to indicate the kinds of groups which are of importance in quantum mechanical problems. Each case must be studied individually and all groups under which the particular Schrödinger equation to be studied is invariant must be taken together to form the complete group of the Schrödinger equation.

As a simple example of the method, consider the one-dimensional wave equation[28]

$$\left\{-\frac{\hbar^2}{2m}\frac{d^2}{dx^2} + V(x)\right\}\psi(x) = W\psi(x) \tag{15-69}$$

where the potential energy is of such a form that

$$V(x) = V(-x)$$

and where the energy state W is non-degenerate, as is nearly always true in such one-dimensional problems. Suppose P_I is an operation which replaces x by $-x$ wherever it occurs in (69). Then

$$P_I\psi(x) = \psi'(x) = \psi(-x)$$

the result being a new ψ-function, ψ', which has the same value at x as the old one, ψ, had at $-x$. The new ψ-function will, however, satisfy the wave equation as well as the old one, with the same value of W. Hence it must be a constant multiple of $\psi(x)$, i.e., $\psi' = c\psi$. But if both ψ and ψ' are to be normalized, c can only be $+1$ or -1. This result recalls the well-known fact that all eigenfunctions of eq. (69) are either *even* or *odd* functions of x.

To exhibit the connection with group theory we note here the following facts which will be illuminated subsequently. Let P_E be an operator that replaces x by itself. Then

$$P_E\psi(x) = \psi(x)$$

Combining P_E with P_I we obtain a group,

$$P_I P_E = P_E P_I = P_I; \quad P_I^2 = P_E$$

which is isomorphous with $\mathbf{C}_i$ (sec. 15.18), and others mentioned in preceding sections. It has two irreducible representations both of degree one (see Table 8). The representation A_g corresponds to even eigenfunctions and A_u to odd ones.

Next, let us suppose that the Hamiltonian operator is invariant to a group of linear substitutions, such as R, S, etc., and that the ψ-function depends on n coordinates $x_1 \cdots x_n$. These may be combined to form a vector $\mathbf{x}$. If, then,

$$\mathbf{x}' = R\mathbf{x}$$

[28] We now use W for the eigenvalue in this section, reserving E for the unit element of a group.

we may define an operator P_R which changes $\psi(\mathbf{x})$ into $\psi(\mathbf{x}')$:

$$P_R \psi(\mathbf{x}) = \psi(\mathbf{x}')$$

Now consider two cases:

a. $\psi(\mathbf{x})$ is non-degenerate. Since, from the invariance of H, $P_R \psi$ must also satisfy the Schrödinger equation with the same W, it must be identical with ψ (except for a constant multipler).

b. $\psi(\mathbf{x})$ belongs to an eigenvalue W which possesses an α-fold degeneracy. We may then label the α linearly independent functions

$$\psi_1, \psi_2, \cdots, \psi_\alpha$$

The effect of P_R on ψ will then be to convert it into a linear combination of ψ_i, for such a combination is the most general function belonging to W. Hence

$$P_R \psi_i = \sum_{k=1}^{\alpha} \psi_k D(R)_{ki}$$

$D(R)$ being a certain matrix associated with the operator P_R. Similarly,

$$P_S \psi_k = \sum_{j=1}^{\alpha} \psi_j D(S)_{jk}$$

and

$$P_S P_R \psi_i = \sum_{j,\,k} \psi_j D(S)_{jk} D(R)_{ki}$$
$$= \sum_j \psi_j [D(S) D(R)]_{ji} \qquad (15\text{--}70)$$

From sec. 15.7, it should be clear that the matrices whose elements appear on the right of eq. (70) are representations of the group of the operators P_R, P_S. The dimension of each representation equals the number of linearly independent ψ-functions, hence it is also equal to the degeneracy of the eigenvalue. If the original set of ψ-functions were not linearly independent the resulting representations would be reducible. When the complete set of irreducible representations is obtained we see that each one would correspond to an eigenvalue of the quantum mechanical problem. The value of group theory in quantum mechanics is thus evident. From the symmetry of the system and without solving the wave equation we may obtain the possible number of eigenvalues and the degeneracy of each. Moreover the eigenvalues may be classified with regard to the particular representation to which it belongs. This fact is of considerable interest to the spectroscopist in studying the possible number and the symmetry of the energy levels to be expected in a given case. For example, as indicated in an earlier paragraph of this section, the group for the diatomic molecule is $R^{\pm}(2)$ of sec. 15.16. This has an infinite number of representations $m = 1, 2, 3, \cdots$ and two representations for $m = 0$. These correspond to the electronic energy levels[29] Π, Δ, Φ, $\cdots$ for $m = 1, 2, 3, \cdots$ and Σ^+, Σ^- for $m = 0$.

The selection rules for allowed transitions in atomic and molecular spectroscopy may be determined readily from the symmetry alone. As shown in sec. 11.28 these depend on the matrix elements of the electric moment. The latter is itself a vector and its components will transform under the operations of the group like x, y, z or some combination of these components as shown in Table 8 for the various symmetry groups. The ψ-function of a given state will also belong to some irreducible representation of the group. The product of a component of the electric moment and the ψ-function will transform like the direct product of the representations for each. This direct product will often be reducible and when reduction is effected, the result will be a sum of representations of the symmetry group. Transitions are allowed only to these states. Actually it is not necessary to know the representations themselves as a knowledge of the characters alone is sufficient. The reader should refer to other sources[30] for the details of the theory. A simple example will show how the method is used.

Suppose a given energy level is known to have a ψ-function which transforms like E_2 in the group D_6. Then for an electric moment along z, the direct product of the characters of A_2 and E_2 is 2, 2, -1, -1, 0, 0, hence the only allowed transition from E_2 is to another state of the same symmetry. If the component of electric moment $(x \pm iy)$ is of interest, the characters are those of E_1 times E_2 or 4, -4, 1, -1, 0, 0 which is a sum of characters for B_1, B_2 and E_1. Transitions are allowed from E_2 to either B_1, B_2 or E_1 but to no others for the $(x \pm iy)$ component of electric moment.

Selection rules for the Raman effect depend in a similar way on the transformation properties of the polarizability tensor. Its characters are $2 \pm 2 \cos \phi + 2 \cos 2\phi$, the upper sign referring to a proper rotation and the lower sign to an improper one.

As shown in sec. 9.10, the instantaneous position in space of a polyatomic molecule containing n atoms is specified by $3n$ coordinates. Three of them locate the center of gravity of the molecule and are thus associated with translational motion. Three more (or two, if the molecule is a linear one) describe orientation relative to principal axes of inertia and the motion

---

[29] These are the customary symbols in molecular spectroscopy; see, for example, Herzberg, G., "Molecular Spectra and Molecular Structure; Diatomic Molecules," D. Van Nostrand Co., Inc., New York, 1950.

[30] See, for example, Eyring, Walter, and Kimball, loc. cit. or Meister, A. G., Cleveland, F. F., and Murray, M. J., *Am. J. Phys.* **11**, 239 (1943).

is rotation. The remaining $(3n - 6)$ or $(3n - 5)$ coordinates are descriptive of internal motions or vibrations. Now the latter, as well as the three translations, transform like vectors and the activity of the vibration in the infrared or the Raman effect may be determined as we have indicated. The transformation properties of rotation are like that of angular momentum and from eq. (9–19) it may be shown that the reducible characters are $1 \pm 2 \cos \phi$, the upper sign again referring to proper rotations and the lower sign to improper ones. Use of these transformation properties makes it possible to predict in considerable detail the spectroscopic behavior of the polyatomic molecule, provided its symmetry is known or a reasonable one assumed.

REFERENCES

General treatments of group theory:

Burnside, W., " The Theory of Groups," Cambridge University Press, 1927.

Kowalewski, G., " Einführung in die Theorie der Kontinuierlichen Groppen," Chelsea Publishing Co., New York.

Kurosh, A., " Group Theory," Second Edition, Chelsea Publishing Co., New York, 1954.

Ledermann, W., " Introduction to the Theory of Finite Groups," Second Edition, Interscience Publishers, Inc., New York, 1953.

Littlewood, D. E., " The Theory of Group Characters," Oxford University Press, New York, 1940.

Murnaghan, F. D., " The Theory of Group Representations," The Johns Hopkins Press, Baltimore, 1938.

Weyl, H., " The Classical Groups," Princeton University Press, 1939.

Zassenhaus, H., " The Theory of Groups," Chelsea Publishing Co., New York, 1949.

Treatises on quantum mechanics, which use the methods of group theory:

Bauer, H., " Introduction a la Theorie des Groupes," Les Presses Universitaires de France, Paris, 1933.

Casimir, H. B. G., " Rotation of a Rigid Body in Quantum Mechanics," J. B. Wolters, Groningen, 1941.

Corson, E. M., " Perturbation Methods in the Quantum Mechanics of n-Electron Systems," Second Edition, Hafner Publishing Co., New York, 1953.

Corson, E. M., " Introduction to Tensors, Spinors, and Relativistic Wave-Equations," Hafner Publishing Co., New York, 1953.

van der Waerden, B. L., " Die Gruppen Theoretische Methode in der Quantenmechanik," J. Springer, Berlin, 1932; Edwards Brothers, Inc., Ann Arbor.

Weyl, H., " Theory of Groups and Quantum Mechanics," Methuen and Co., Ltd., London, 1931.

Wigner, E. P., " Gruppen Theorie und Ihre Anwendung auf die Quantenmechanik der Atomspektren," Vieweg, Braunschiverg, 1931; Edwards Brothers, Inc., Ann Arbor.

Application of group theory to crystal structure:

Burckhardt, J. J., " Die Bewegungsgruppen der Kristallographie," Birkhäuser, Basel, 1947.

Phillips, F. C., " Crystallography," Longmans, Green and Co., Inc., New York, 1947.

Schoenflies, A., " Theorie der Kristallstruktur," Gebrüder Borntraeger, Berlin, 1932.

Zachariasen, W. H., " Theory of X-ray Diffraction in Crystals," John Wiley and Sons, Inc., New York, 1945.

INDEX

Abelian group, 546
Abel's integral equation, 541
Absolute temperature, 30
  velocity, 289
Accessory conditions, 209
Adams, E. P., 177
Adams, N. I., Jr., 140
Addition of matrices, 306
  of vectors, 140
  theorem for Legendre polynomials, 109
Adiabatic expansion, 36
Adjacent path, 198
Adjoint matrix, 309
Aggregate, probability, 435
Aitken, 316, 322, 332
Algebraic calculations, 491
Allen, 519
Alternating group, 558, 561
Amplitudes, probability, 347
Analogies between thermodynamic and statistical quantities, 459
Analogues statistical, of thermodynamic quantities, 452
Analysis, indirect chemical, 314
Anchor rings, 190
Angles, Eulerian, 282, 286
Angular momentum, 285
  eigenfunctions of, 360
  eigenvalues of, 360
  in quantum mechanics, 338
  internal, 293
  velocity, 145, 285
Antisymmetric eigenfunctions, 416, 455
Approximate quadrature, 474
Approximation in the mean of functions, 276, 279
Approximation method, for algebraic equations, 493
  for differential equations, 484
  for secular determinants, 503
Arbitrary constants, 33
Areas, vector, 144

Arley, 519
Arrangements, 431, 432
Arrays of numbers, 301
Assemblage of identical particles, 417
Assignment of statistical weights, 456
Associate matrix; 309
Associated Laguerre function, 80
  Laguerre polynomial, 78, 106, 132
  Legendre functions, 68
    differential equation for, 68
  representation, 560
  spherical harmonics, 68, 69
    differential equation for, 68, 69, 223
  tensor, 167
Associative law, 545
Atmospheric pressure, 36
Atom in a magnetic field, 408
Auxiliary equation, 49
Average error, 510
  weighted, 514
Axes, coordinate, 172
Axial vector, 165
Axiomatic foundation of quantum mechanics, 335
Axis of rotation, 285
  of symmetry, n-fold, 575
Azimuth, 177

Bacteria, 33
Baggott, 483
Balls in boxes, 438
Barrier problems, 353
Base vectors, 193
Bateman, 245
Bauer, 566, 586
Beattie, 477
Beers, 519
Bent, 17
Bernoulli's equation, 44
  numbers, 474
Bessel coefficients, 113